The
British Overseas
Airways Corporation

A History

Graham M Simons

AIR WORLD

British Overseas Airways Corporation
A History

First published in Great Britain in 2019 by Air World Books,
an imprint of Pen & Sword Books Ltd,
Yorkshire - Philadelphia

Copyright © Graham M Simons
ISBN: 9781473883574.

The right of Graham M Simons to be identified as Author of this work has been asserted by him in accordance with the Copyright, Designs and Patents Act 1988. A CIP catalogue record for this book is available from the British Library All rights reserved.

No part of this book may be reproduced or transmitted in any form or by any means, electronic or mechanical including photocopying, recording or by any information storage and retrieval system, without permission from the Publisher in writing.

Typeset in 9pt Times by GMS Enterprises
Printed and bound by Replika Press Pvt. Ltd., India.

Pen & Sword Books Ltd incorporates the imprints of Air World Books, Pen & Sword Archaeology, Atlas, Aviation, Battleground, Discovery, Family History, History, Maritime, Military, Naval, Politics, Social History, Transport, True Crime, Claymore Press, Frontline Books, Praetorian Press, Seaforth Publishing and White Owl.

For a complete list of Pen & Sword titles please contact:

PEN & SWORD BOOKS LTD
47 Church Street, Barnsley, South Yorkshire, S70 2AS, UK.
E-mail: enquiries@pen-and-sword.co.uk
Website: www.pen-and-sword.co.uk

Or

PEN AND SWORD BOOKS,
1950 Lawrence Roadd, Havertown, PA 19083, USA
E-mail: Uspen-and-sword@casematepublishers.com
Website: www.penandswordbooks.com

Contents

Acknowledgements		4
Introduction		5
Chapter One	Predecessors and Origins	7
Chapter Two	The War Years	15
Chapter Three	Planning For Peace	45
Chapter Four	Becoming Competitive	67
Chapter Five	Entering a New Age	95
Chapter Six	Associates, Mail, Cargo and the Independents	165
Chapter Seven	Enter the 707	191
Chapter Eight	Try a little VC Tenderness	221
Chapter Nine	Management Crisis As The Debts Mount Up	239
Chapter Ten	A New Chairman - And The 747	261
Chapter Eleven	The Edwards Report - And Concorde	283
Chapter Twelve	From BOAC to BA	303
Appendices		306
Index		314

Acknowledgements

A book of this nature would not have been possible without the help of many people and organisations. I am indebted to many people and organisations for providing photographs for this story, but in some cases it has not been possible to identify the original photographer and so credits are given in the appropriate places to the immediate supplier. If any of the pictures have not been correctly credited, please accept my apologies.

Thanks are also offered to the usual 'guilty suspects' for their endless, unfailing support over the years: John Stride, David Lee, John Hamlin, Martin Bowman, Vince Hemmings, Brian Cocks, Mick Oakey, Ian Frimston, Warrant Officer Paddy Porter BEM, Peter Green and many others to numerous to mention, but not forgotten! Finally, thanks also go to Laura Hirst, Martin Mace, Charles Hewitt and everyone else at Pen & Sword!

I would also like to thank those long-departed members of BOAC's Press Office - I well remember as a teenager sending postcards to the Press Office of many of the major airlines and aircraft manufacturers, asking for photographs of this or that latest edition. Without fail a parcel - and indeed sometimes a box - of pictures, advertising material and timetables arrived on my parent's doorstep; sometimes I think the Press Office was happy just to have a reason for a clear-out!

Whilst mentioning photographs, it may seem strange to see something that was applied to the reverse of photographs used in this book - a simple publication clearance stamp. Unfortunately nowadays this is especially important to show due to the growth of a plethora of websites who seem to think that it is perfectly acceptable to 'harvest' any image they come across and apply their own watermark to it and then demand exorbitant 'fees' from authors. Well guys, I have news for you - this book was compiled with prints obtained before the internet and you opportunists were around - and to prove it, here are some of the rubber stamps from the back of the prints. Some state *'De Havilland Photograph. This photo may be published without payment of any fee'*, and is just a representative of the many hundreds used from this, and other manufacturers. Many others are stamped *B.O.A.C. Photograph Free Reproduction*. However the one that really amuses me is the delightful - if somewhat hopeful - secondary stamp applied to some alongside the standard BOAC stamp - *'A Credit to BOAC Would Be Appreciated'*.

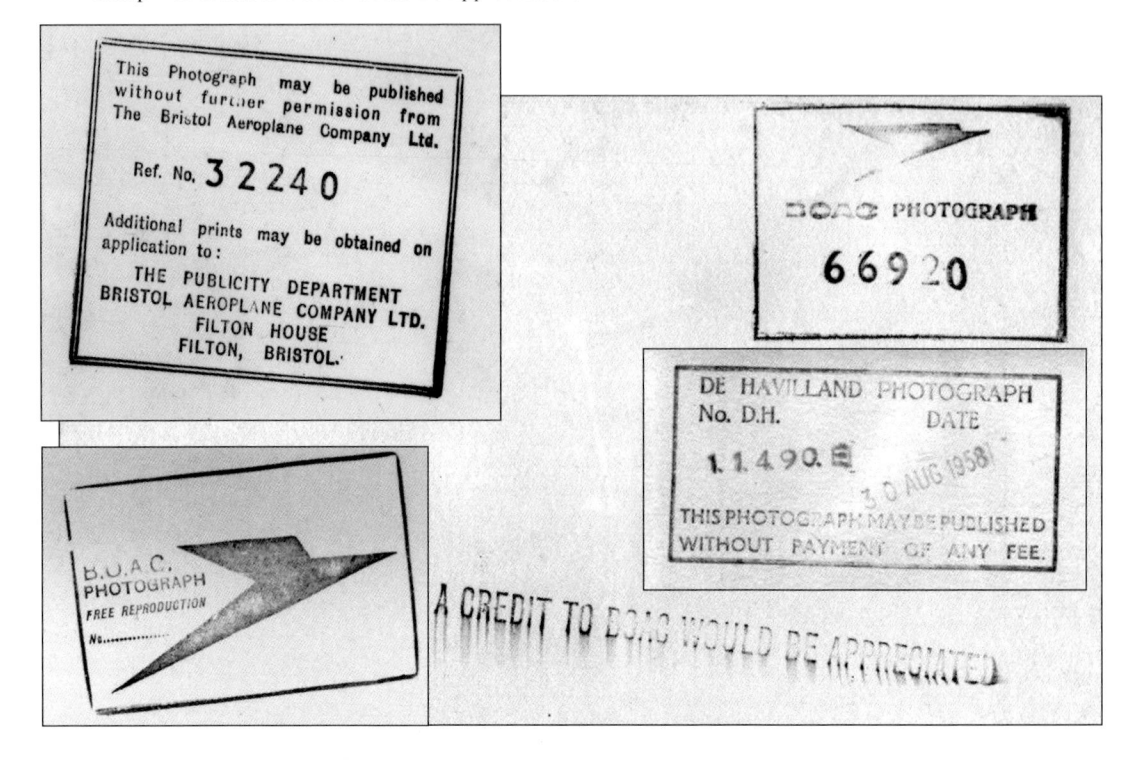

Introduction

British Overseas Airways Corporation came into existence as that rarest of creatures - an organisation nationalised by the Conservative Party. However, unlike most other state-owned industries it had to operate in a competitive international market - a fact that public opinion rarely recognised. Government interference and pressure were present throughout a procession of Ministers, some fifteen over BOAC's last twenty years. In the same period, BOAC had seven Chairmen of whom four were also Chief Executives; only the previous two had been brought up in the airline business. Continuity of policy and purpose invariably suffered and made it easier for Ministers and their officials to exert undue influence upon BOAC.

BOAC began with World War Two and ended as cheap aviation fuel vanished. In between lay a firestorm of political, technical, and commercial change. Almost all the British Colonies achieved independence; aircraft range and capacity grew from four hundred to four thousand miles or more, and passenger numbers rose from under forty to over four hundred per aircraft. As the economic power of countries defeated in the war revived, world tourism developed, and as previously emerging nations in the Middle East became oil-rich. For the six war years, BOAC was inevitably the servant of Government, achieving with distinction its role of maintaining Commonwealth communications and in particular directly supporting the Allied war effort throughout the Middle East.

For the next ten years, BOAC concentrated on building up a worldwide commercial organisation and fought to obtain and bring into use competitive aircraft, particularly on the fast developing Atlantic routes. Sales organisations were created at home and overseas. The QANTAS and SAA partnerships enhanced competitive strength. Newly independent countries and those about to become independent turned to BOAC for help in creating and building up national carriers; no effort was spared and staff both commercial and technical were seconded until the new airlines could operate in their own right as partners along the routes to London.

Throughout, the main bone of contention remained the purchase of aircraft. British manufacturers had built no genuinely civil transport aircraft during the war years. BOAC had, therefore, to make do with such flying boats as had survived the War but were now reaching the end of their useful lives and were expensive to operate. The use of uncompetitive aircraft remained all that BOAC could look forward to until a new generation of British airliners planned by the Brabazon Committee was ready.

Foreign-built aircraft had to be obtained, so the Government gave reluctant permission for the purchase of Constellations, Stratocruisers, and Argonauts. The turning point of BOAC's post-war fortunes was the introduction into service of the Comets in 1952. This first jet established utterly new standards and quickly became the most popular aircraft in world operation by any airline. Its triumph was short. Two years later it was struck down by the unsuspected menace of metal fatigue, and BOAC was back on its heels.

Ministers had a responsibility to support the British manufacturing industry; BOAC's interest, on the other hand, lay in obtaining aircraft best suited to its route plans and competitive with foreign carriers. The main result was that BOAC found itself the first to order a particular British aircraft to be built for long-haul operations - and had the task of introducing it and making it fit for service on long world routes. The theoretical advantage of being the pioneer proved very costly and not unreasonably the Corporation felt it should have been repaid for initial costs either by the Government or by the manufacturers. Involved, too, was the risk of being pressed by Government to order more aircraft of a new type that was needed to enable the manufacturers to set up a proper production line. This, in fact, happened with the VC10s, an outcome highly expensive to BOAC. True that Chairmen and Boards should at times have acted with more resolution against such ministerial pressures, but it was not easy to resist appeals to the national interest.

The aircraft ordering issue was not resolved until Julian Amery gave Sir Giles Guthrie in January 1964 what became a famous letter establishing once and for all that the Corporation was responsible for looking after its interests and the Government for any overriding matters of national importance. After that BOAC was able to order the aircraft it wanted and at last, had a fleet with only two main types and their variants; it was this and the ability to order aircraft as required in small numbers off the production line which made the last years so profitable. It must, however, be noted that the Amery provision was sidetracked when the Concorde order was placed by BOAC, which was concerned not only about the economics of the aircraft but also about the absence of any permissions to fly the aircraft along world routes.

BOAC remained effectively the British chosen instrument for longhaul scheduled services until the Edwards committee worked out a future for the Independents. The history of BOAC is the history of British longhaul scheduled services until the establishment of a second force airline and of the Civil Aviation Authority in 1971 and the free transfer thereafter of BOAC routes to British Caledonian Airways.

No immediate help from British aircraft was to be found. The Britannias planned for introduction in 1955 were not available until 1957. This loss of capacity was partially solved by the purchase of DC-7Cs in March 1955, but deliveries could not begin before October 1956. In London, BOAC had the most extensive traffic generating point in the world for long-haul traffic, with North America, Africa, and Australia as the other mainstays. The USA and

Canada had been energetically developed from the war's end, but in the last ten years or so of BOAC's life new traffic generating areas arose, and BOAC grabbed the opporunities with both hands.

The Far East - with Japan in the lead - began to provide much new traffic, and BOAC spread its services into Japan by three different routes; Hong Kong and Singapore increasingly came to the forefront. The Middle East for BOAC's first twenty-five years had been served primarily on transit services, but with the increasing oil markets, Iran and the Gulf became new and essential destinations for terminating services to tap the growing oil traffic.

As these new services came into being, so loss-making routes were dropped: the North Pacific route was withdrawn, and the South Pacific route lasted only for a short while after BOAC's separate existence ceased. In the last few years, domestic competition significantly increased, fostered by the CAA from the very start of its existence, following the policy of the American CAB. The justification was the demand from the public for cheaper travel, a demand which sprang from the radical change taking place in the type of passengers seeking air travel. It came from the cheap end of the market, extending first to the shorter distances within Europe and then spreading out to Africa, the Caribbean, and North America, and even to Australasia and the Far East.

BOAC was well aware of this trend - it was one of the leaders trying to persuade other members of IATA to accept promotional fares geared to the new markets. It pioneered the concept of advance purchase of tickets, particular airfare components of inclusive tour prices, and low fares for inclusive tours through wholesale travel agents. In later years it preached the gospel of making each category of passenger traffic stand on its own feet regarding pricing and the facilities provided for it and pushed through these ideas within the industry by applying them with UK Government support to cabotage and Commonwealth routes thus forcing other international carriers to follow suit. Not surprisingly all did not go smoothly with the way in which BOAC sought to build up its revenue. The threat posed by the aspirations of British independent operators at times brought overreaction: the BOAC-Cunard marriage, for instance, was a defensive move which brought little advantage. Rightly BOAC had invested early in airlines overseas to act as focal points for the collection of traffic for the trunk routes.

Over its life, BOAC more than paid its way. In its last year, it made its best-ever operating surplus - £36.4 million; while over its last ten years achieving a total operating surplus of £200 million and a profit of £131million after taxation and interest on borrowings. Moreover, in those years it paid dividends on its equity capital of some £62 million.

BOAC handed over to British Airways not only an operating fleet tailored to meet long-haul needs for at least the next five years but also reserves of £83m amounting to over 80% of all British Airways' initial reserves. This achievement was that of a management team built up and remaining mostly the same individuals over the last twenty years of its life. Though Trade Union power grew, on the whole it was used responsibly, and stoppages due to strikes were comparatively few. The perennial problem was staff numbers, which compared poorly with other airlines - a problem seemingly built into British industry since the war ended.

The perception of the Corporation was complex; domestically public opinion tended to be unfavourable, naturally with a Press, predominantly Tory orientated and thus inimical to any nationalised undertaking. Critics often stated that the company had 'that Imperial Airways mentality that went back to the days of Lord Reith'. Overseas it was different. BOAC held to the end the respect of the international civil aviation world - respect earned by its operation and technical skills, by its determination to ensure its due place in the world, and by its reputation for fair dealing.

The story was not easy to write, or to put together, resulting in a book that is akin to squeezing a quart into a pint pot. I could have - and indeed did - write more than double the finished result, and even after severe editing the only way space could be saved was by bringing forth many abbreviations; changing words such as 'Managing Director' and 'Ministry of Civil Aviation' for MD and MCA to save a lot of space!

The reader will also discover a lot of advertising posters and promotional material used by the Corporation over the years; beautiful uncaptioned artworks with the message in the image sum up the corporate perception of itself.

BOAC was from the era of dressing up to travel; when one flew from London Airport, not LHR, cabin announcements always started with *'this is your captain speaking...'* seats were still well spaced apart, and the cabin staff were never flight attendants, but always stewards or stewardesses. The image may have been conservative, and slightly staid, but it is from a time of air travel glamour that is never to return.

Graham M Simons
Peterborough
18 October 2018

Chapter One

Predecessors and Origins

The Parliamentary Act setting up the British Overseas Airways Corporation (BOAC) received the Royal Assent on 4 August 1939, the Corporation formally being established by the Secretary of State for Air on 24 November 1939, although it did not take over Imperial Airways and British Airways until 1 April 1940. To understand why this development came about and to see what both companies brought to the table, it is necessary to examine their background.

Imperial Airways
In 1923 there were four private British air companies receiving financial help in the form of subsidies from the Government: Daimler Airways - which included Aircraft Transport and Travel - the Instone Air Line, Handley Page Transport and British Marine Air Navigation Company. All were finding survival difficult in spite of the subsidies. The then Secretary of State for Air, Sir Samuel Hoare set up an inquiry into the future of air transport with reference to how any subsidy money allocated by Government could best be used. Accordingly, the Civil Air Transport Subsidies Committee was appointed under the chairmanship of Sir Herbert Hambling, then the Deputy Chairman of Barclays Bank. It reported on 20 February 1923, recommending the merger of the four companies into one commercial organisation with a privileged position as to subsidy.

The government accepted the report and asked for a proposal from the existing companies. Later that year, all four had agreed to form a joint company known as the Imperial Air Transport Company, and later it was announced that the new company would receive a guaranteed total subsidy of £1m over a period of ten years from its formation on 1 April 1924.

The new company, now registered as Imperial Airways Limited (IAL), confined its services to European routes without much success in the face of competition from continental operators such as Germany's Deutsche Lufthansa and France's Air Union. This bedeviled IAL's European network to the beginning of the Second World War, a situation which opened the way for other British operators to step in, so allowing IAL to concentrate on the development of Empire routes, and the North Atlantic.

In an attempt to make up for lost ground, IAL placed an order in 1935 for fourteen four-engined Armstrong Whitworth AW.27 Ensigns, a type that was introduced into service in 1938 on the London-Paris route. The aircraft was underpowered and was withdrawn only to reappear shortly before the outbreak of war. Its place on the London-Paris service was taken by the new De Havilland DH.91 Albatross. Five of these four-engined machines were ordered: they were comfortable and fast enough to be more than a match for the competition.

If IAL demonstrated a pedestrian approach toward European operations, it had a positive approach to the development of Empire routes. By 1929 a route to India had been pioneered and established with flights terminating

An historic day at Croydon as a ceremony in one of the hangars is broadcast to the nation. This was the inauguration of the weekly mail flight to Australia on 8 December 1934. HP.42 *Hengist* is about to carry the mail on the first leg of the journey – Croydon to Karachi. Here the directors of Imperial Airways and senior representatives of the General Post Office including the Postmaster General sit on a platform. Sir Kingsley Wood (Postmaster General) hands Lord Londonderry, Secretary of State for Air, a token mail package in front of the BBC microphone. Imperial Airways' chairman Sir Eric Geddes is seated to the left.
(author's collection)

Handley Page HP.42 G-AAXO comes in to land at Croydon Airport. *(author's collection)*

in Karachi. It was not long before the route was extended to Delhi, this sector under charter to the Indian Government. Calcutta was reached by July 1933, and by the end of the year, a further step forward brought the route into Singapore in sight of the final goal - Australia. This last hurdle was overcome through the formation of a British/Australian company jointly owned by IAL and the Australian domestic airline, Queensland and Northern Territories Aerial Services, and hence the birth of QANTAS as an international operator. In December 1934 the first regular airmail service was opened between England and Australia in which the new DH.86 Express four-engined aircraft - suitable for the crossing of the Timor Sea - played a significant part. But perhaps it was on the link between Southeast Asia and Hong Kong where the DH.86 best demonstrated its flexibility and reliability.

After a series of survey flights between Penang and Hong Kong via Saigon and Tourane, DH.86s began regular mail services in March 1936 connecting with the England-Australia flights at Penang. By the close of 1937, the Hong Kong link was improved when Imperial Airways were permitted to connect with the Australian route at Bangkok by the Siamese Government.

IAL was also involved in opening up the second Empire route down Africa to Capetown. It was a massive task involving the surveying and preparation of aerodromes, landing grounds and rest houses where almost nothing existed before the first regular operation could commence in February 1931. This extended as far south as Mwanza in Tanganyika but Capetown was reached by the end of the year and in 1932 scheduled mail services were operating into Capetown, and was opened up to passengers by April. However, when the South African Government took over Union Airways and formed South African Airways, it insisted on its mails being carried by its airline and by 1936 had required Imperial Airways to omit Capetown and terminate its service at Johannesburg and, when the flying boats were introduced, at Durban.

West Africa needed to be covered if the Empire links were to be complete: IAL decided to operate a link from Khartoum on the South African route across Sudan and Central Africa to Lagos. February 1936 saw the start of

DH91 Albatross G-AEVV - one of the mailplane versions of this elegant machine *(DH Hatfield)*

DH.86 Express VH-USC *Canberra*, the fourth DH.86A to be built was first registered in September 1934 for QANTAS Empire Airways; the aircraft crashed near Darwin in October 1944. *(DH Hatfield)*

scheduled services from Khartoum to Kano via El Obeid, El Fasher, El Geneina, Ft. Lamy and Maiduguri. In September services were extended to Lagos. IAL worked in conjunction with Elders Colonial Airways formed by the Elder Dempster Shipping Line whose boats ran between the UK and Takoradi. It was providential that this trans-Africa route was already developed and operating when war broke out as in due course it became a vital strategic link with the Middle East when the entry of Italy into the war closed the Mediterranean.

One innovation that had far-reaching effects on IAL's aircraft requirements, routes and finances were the Empire Airmail Scheme under which all first-class mail to Empire countries would eventually be carried by air. The scheme, conceived by IALs own staff in cooperation with the Post Office, was announced to Parliament by the Under Secretary of State for Air on 20 December 1934.

It was an understandable decision to opt for flying boats as the mainstay of its future fleet able to carry the volume of mails along the Empire routes expected from the introduction of the Empire Airmail scheme. At the time 'hard' runways were few and far between, but surveys had shown that operations on the water were feasible all along the route to South Africa and the Middle East, India and Australia and eventually it was hoped across the Atlantic.

In 1935, with Government support, IAL placed an order for twenty-eight flying boats straight off the drawing board from Short Brothers of Rochester. Deliveries began in 1936, and the original order was completed in 1938. These aircraft began to come into service early in 1937, and so made possible the introduction of the EAMS when on 23 February 1938 one operated through to Singapore with the Empire mails; by June passengers were able to travel right through to Sydney without change of aircraft. At the same time, all-flying boat operations began to Durban and so eliminated all the time-wasting changes en route.

IAL had plans in other directions as well. As far back as 1930, they had begun to consider how a North Atlantic route could be established with a New York-Bermuda operation forming part of the through route. Suitable aircraft with adequate range were not available, but when the new C-Class boats were ordered, IAL had made provision for extra tankage to be fitted to two of them, *Caledonia* and *Cambria*. Flights were begun in July 1937, and further flights followed in 1938 with Empire boats to test out refuelling in the air to increase range and payload. The composite *Mercury/Maia* also took part in these experimental flights. Unfortunately as the result of aircraft problems IAL was not able to start regular services via Foynes, Botwood and Montreal to New York until August 1939, a few days before war broke out, and the operation survived only a few weeks.

The New York-Bermuda connection enjoyed a slightly longer run. Started by *Cavalier,* one of the S.23 boats, on a regular basis in June 1937, the service operated successfully through 1938, Baltimore being used instead of New York in the winter months. This was not to continue, for, in January 1939, *Cavalier* ditched on the way south to Bermuda due to icing problems and the service never restarted before war came.

British Airways
The second parent of BOAC was a much smaller company than IAL, but like IAL was itself created out of several regional airlines. Its existence was short, being formed in October 1935 and swallowed up into BOAC only four years later. The company came into being in the merger of Spartan Airlines and United Airways under the name of Allied British Airways, but a few weeks later the name was changed simply to British Airways (BA). Almost immediately the new company's size was increased when Hillman Airways was acquired in December 1935. Hillmans was backed by the private banking firm of d'Erlangers, so the new British Airways had the support of two powerful financial groups, d'Erlangers and the Pearson Group through Whitehall Securities. In 1936 the company was further strengthened by the incorporation of British Continental Airways. It was clear from the start to the management of BA that the company's immediate future lay in building a network of services within Europe,

particularly as IAL had not been over-successful in its European operations and had left a gap for a new operator to exploit. There was also Government encouragement for the venture which enabled BA to receive a subsidy and mail contracts. IAL, who were heavily committed to the development of Empire routes, were agreeable to BA taking over European operations north of a line London-Berlin and BA moved quickly to get a Government assisted service underway to Malmo via Amsterdam, Hamburg and Copenhagen by February 1936. By July the route was extended to Stockholm and a night mail service was opened up to Cologne, Hanover and eventually Berlin.

BA faced competition from Continental airlines and to meet this, they had to acquire faster aircraft than the Fokker F.12s then in use. Government permission was granted to obtain American Lockheed 10A Electras for passenger services and three German Ju.52s for the night mail flights to Germany. The new Lockheed 10As were placed on the service to Sweden in 1937 and also on the London-Paris service in competition with IAL.

Two more developments to its route pattern were made before war came, a non-stop operation London-Berlin with an extension to Warsaw and a London-Frankfurt-Budapest service; also the service to Sweden was re-routed to omit

Below: Albert Brenet's 1935 poster advertising the Imperial service to Australia, flown in part by QANTAS DH.86s.

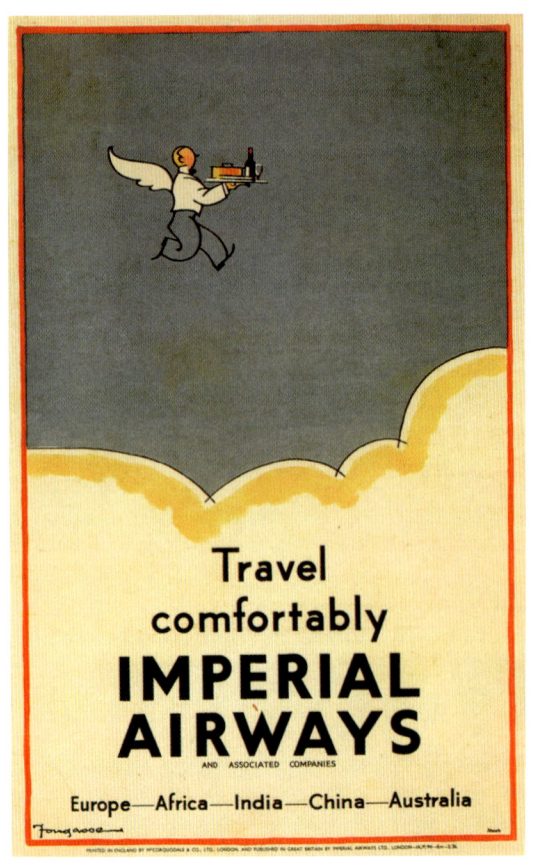

'Fougasse' was the nom de plume for Kenneth Bird, who later became editor of *Punch*. whose 'travel comfortably' poster was used to adverise Imperials services.

Amsterdam and Stockholm became the terminus instead of Malmo. It had been hoped to start a service to Lisbon following the Government's decision that BA was to be the chosen instrument for opening up operations to South America via Lisbon and Africa. There were two major problems, aircraft and over-flying rights. With the delivery in the autumn of 1938 of Lockheed 14s, BA had the capability of flying direct London-Lisbon, and in December of that year, a survey flight reached Bathurst over Lisbon, with calls at Casablanca, Agadir, Port Etienne and Dakar. That was as far as the proposed South American service progressed until after the war, partly because of the imminent merger with IAL but also there were political difficulties to overcome in securing the necessary traffic rights from South American countries.

By the summer of 1939 BA was well set for the future. It had modern American aircraft that were fast and competitive, it had Government support in the way of subsidy and mail contracts, and it had an established route pattern in Europe. Moreover, it had the exciting prospect of pioneering the new route to South America over West Africa and for this was in the process of preparing to order an entirely new British type from Fairey, the FC1 four-engined pressurised machine with tricycle undercarriage. That was the situation when war intervened.

Below: Mounting G-ADAJ *Mercury* on top of G-ADHK *Maia*.

Left: Major Mayo, designer of the Mayo composite design congratulates John Lancester Parker and Harold Piper after the first flight of the aircraft pair from the River Medway on 20 January 1938. *(both author's collection)*

Politics and Repercussions

By 1937 the outlook for IAL was looking good with the C Class flying boats coming along for Empire routes, the Armstrong Whitworth Ensigns about to appear on the European routes and the prospect of the DH.91 about to enter services after its maiden flight in May, but all was not well on the political front.

There was growing criticism of its European operations which had been virtually stagnant for years: there were difficulties with pilots over-representation and pay: and there was a feeling that the subsidies it was receiving were being used to bolster up dividend payments to shareholders. The Management was reluctant to recognise the newly formed British Airline Pilots Association (BALPA) or to discuss with the Association some outstanding problems. This was unfortunate since one of the founders of BALPA, Robert Perkins, was a Member of Parliament where in October he not only raised the question of BALPA representation but also seized the opportunity to bring into the open some pilot grievances. The Government claimed these matters were for the management of IAL and that HMG was not prepared to have an inquiry. This attitude did little to quell the criticism, and it was not long before the House of Commons was listening to an even more pungent attack from Perkins. On 17 November he introduced a motion on civil aviation calling for a public inquiry. His speech was critical both of the Air Ministry and of IAL, for which he received considerable support.

The outcome was that the Conservative Government made a statement in the House on 24 November announcing that a committee of inquiry into civil aviation was to be set up under the chairmanship of Lord Cadman of the Anglo Iranian Oil Co. No formal terms of reference were given to the Committee other than that they were to examine any matters raised in the debate of 17 November whether affecting the Air Ministry or IAL.

The Committee submitted their report to Viscount Swinton, Secretary of State for Air, in February 1938 and this, along with the Government's views was published a month later. It made important recommendations on the organisation of the Air Ministry so that greater emphasis and attention could be given to civil aviation. On international air services, the Committee pressed for the establishment of services between London and all the

Pre-war airline personalities.
Right: Mr and Mrs Whitney Straight are received at a Royal Aeronautical Society Garden party by Lord Brabazon of Tara.

Below left: John Charles Walsham Reith, later Baron Reith of Stonehaven KT, GCVO, GBE, CB, (*b*. 20 July 1889, *d*. 16 June 1971. Centre: Walter Leslie Runciman, 2nd Viscount Runciman of Doxford (*b*. 26 August 1900, *d* 1 September 1989. Right: The Right Honourable Sir Eric Campbell Geddes PC, GCB, GBE, KCB. (*b*. 26 September 1875, *d*. 22 July 1937. *(all author's collection)*

principal capitals of Europe financially assisted by the State if necessary. They also urged the inauguration of services to South America as expeditiously as possible, and that other routes, including those to the West Indies and the Pacific, should be developed.

As to subsidies, the Report referred to the annual limit of £1.5m permitted under the Air Navigation Act of 1936 in any one year and strongly urged that if British civil aviation was to compete with the subsidized foreign competition, Parliament must sanction larger sums urgently. The Report then turned to the subject of how external air services should be organised and recommended the concentration into a small number of '...*well-founded and substantial organisations rather than dissipated among a large number of competing companies of indifferent stability*', and pointed out that foreign competition would do all that was necessary to give a spur to improve service and equipment.

As a logical outcome, there should be only one British company on each overseas route and uncontrolled competition should thus be avoided - a significant policy which remained the centre of argument ever since. The Committee confirmed the selection of Imperial Airways to operate the routes to Africa, India and Australia and to develop a North Atlantic service - a task so extensive that routes elsewhere should be dealt with by another company. For this, the Committee nominated British Airways to run services within Europe and to pioneer a British operation over West Africa to South America. The Report recommended one exception to this policy, the London-Paris route where it felt traffic to be adequate to support both IAL and British Airways services, but in practice no exception since it went on to suggest the formation of a single company set up by the two airlines. This proposal may well have had considerable influence on subsequent developments more especially when taken in conjunction with the Committee's insistence that there should be close co-operation between the two chosen instruments over the whole field of their activities.

Given the policy laid down by the Hambling Committee in 1923 that IAL should be run on commercial lines without any direct control by Government, the Report reminded both the Air Ministry and IAL of their obligations to maintain close liaison and to be open and frank as to information and plans.

The Committee concluded that in future the Company should have a full-time Chairman given its increasing responsibilities and they recommended the same for BA with its widened scope proposed in the Report. The wording made it clear that the Chairman would also act as chief executive. Lord Cadman and his Committee were asked by the Minister to look into the matter of staff representation to the Management of IAL with particular reference to flying staff. They concluded the time had come due to the increase of flying personnel for formalised collective representation and bargaining, thus reinforcing the arguments put forward by Robert Perkins in the 17 November debate that BALPA should be the negotiating body for the pilots. When the Report was finally issued in March 1938, the Government saw fit to publish with the Report a series of observations making it clear that they accepted the bulk of the findings and those relating to the operating companies. As to subsidies the Government were prepared to recommend to Parliament an increase in the statutory limit on the annual amount payable from £1.5m to £3m. On the 16 March the Prime Minister, Neville Chamberlain, formally submitted the Report to the House for approval.

Not unsurprisingly, the Committee`s Report was received with rather different feelings in the two companies. IAL was distressed at the criticisms in the Report of its Managing Director, Woods Humphery, and unhappy at the restrictions placed on its operations, whilst British Airways was delighted at the prospect of being the main operator in Europe and with the scope offered with its choice as the pioneer of the new route to West Africa and South America. Both companies agreed to accept full-time Chairmen. In the case of British Airways, there was no difficulty as Hon. Clive Pearson was already acting as chairman and would continue to do so.

Major Hon. Bernard Clive Pearson was born on 12 August 1887. He was the son of Weetman Dickinson Pearson, 1st Viscount Cowdray and Annie Cass. He married Hon. Alicia Mary Dorothea Knatchbull-Hugessen, daughter of Edward Hugessen Knatchbull-Hugessen, 1st Baron Brabourne of Brabourne and Ethel Mary Walker, on 14 October 1915. The Hon. Bernard Clive Pearson, who usually went by his middle name of Clive, was educated at Rugby School, Warwickshire. He graduated from Trinity College, Cambridge in 1908 with a Bachelor of Arts. He gained the rank of Major in the service of the Sussex Yeomanry. He died on 22 July 1965.

However, at IAL a different situation existed. Sir Eric Campbell Geddes, Chairman since its start, had passed away on 22 June 1937 after some years of declining health and Sir George Beharrell was now acting Chairman.

The Government finally decided that Sir John Reith from the BBC was the man to tackle the task of reviving staff morale and of building up the Empire routes. Exactly how this choice was decided on remains unclear but from Reith's autobiography *'Into the Wind'* it appears that he made known to Sir Kingsley Wood, then the Postmaster-General, that he was dissatisfied because his job at the BBC did not keep him at full stretch. He had in any case already offered his resignation to the Board of the BBC in November 1937. By May 1938 Kingsley Wood become Air Minister, and the Cadman Report had been issued.

Kingsley Wood asked to see Reith, where again Reith said he was not at full stretch. It appears that Reith was

On 30 September 1938, Prime Minister Neville Chamberlain returned to Heston Airport from Munich abaoard a British Airways Lockheed 14 after signing the 'peace in our time' agreement with Adolf Hitler. *(PHT Green collection)*

hoping to go to the War Office but on 3 June Sir Horace Wilson, Labour Adviser to the Prime Minister, told Reith both the Prime Minister and the Air Minister required him to take over the chairmanship of IAL from Sir George Beharrell and that the Managing Director, Woods Humphery, must go. That same day Reith saw Beharrell and made it clear that he would be taking over both as Chairman and Chief Executive.

On 4 July Reith became, according to his own words, '...*the unwilling Chairman of IAL*'. That may well be true, but there can be no doubt that he intended from the start to work for an amalgamation of IAL and British Airways into one worldwide company and that this had been a condition of his acceptance of the post. It was also his declared intention to make IAL a Commonwealth airline with representatives on the Board from the Commonwealth countries particularly Australia, New Zealand, Canada, Central and South Africa. The new company was to be free of any private shareholding and ran as a public corporation, with service to the public rather than profits and dividends as the driving force.

Reith pressed on with this objective and had early meetings with Clive Pearson and his Managing Director, Ronald McCrindle. By August 1938 an agreed scheme for amalgamation had been worked out, an outcome far removed from the recommendations of the Cadman Report. Kingsley Wood was advised of the plans early in October so that incongruously it fell to a Conservative Government to put to Parliament on 4 November the proposals for a nationalised air corporation. The Secretary of State for Air made a case for merger and the setting up of a public corporation.

Before Parliament's approval of the new Corporation, Reith as Chairman designate had already started planning the new organisation. He decided not to combine the positions of Chairman and Chief Executive, so the search began for a suitable candidate. Reith's choice fell on Hon. Walter Leslie Runciman. Apart from the family's shipping interests, Runciman was a director of the London and North Eastern Railway and Lloyds Bank. He accepted the offer, subject of course to the eventual approval of the BOAC Board when appointed. He immediately joined Reith, Pearson and McCrindle in the preparatory work for setting up the Corporation.

Reith continued discussions with the Air Minister and with representatives of some of the Dominions to attempt to achieve his objective of a Commonwealth airline with a representative Board. This imaginative project never progressed very far, and the war finally killed it but not before it had received the Air Minister's blessing and according to Reith's autobiography, some encouragement from the main Commonwealth countries.

Reith set about explaining to the shareholders of IAL and British Airways the Government's intention to merge the two companies and so that by June 1939, both groups of shareholders had agreed to sell, at 32/9d per £1 share with interest at 4% up to the actual date of payment, for IAL, and at 15/9d per £1 share with interest from 30 September 1938 to the date of sale, for British Airways,

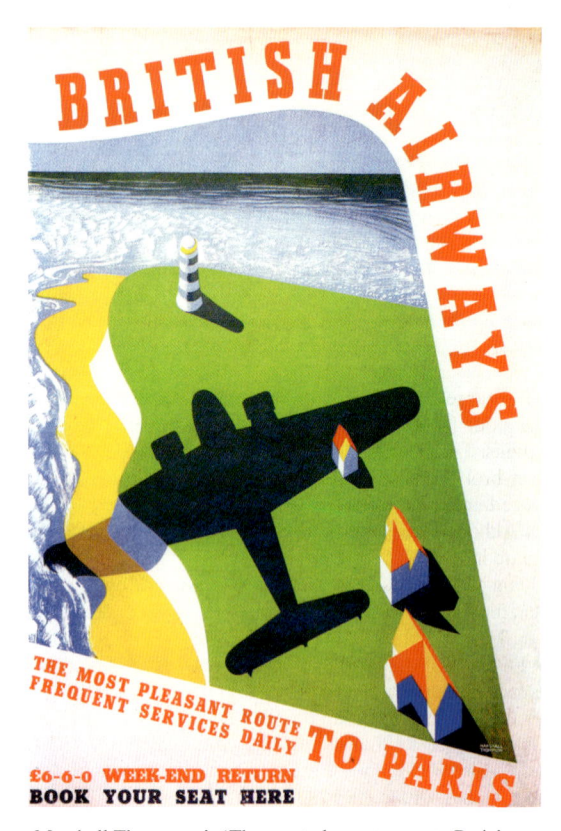

Marshall Thompson's 'The most pleasant route to Paris' was a 1936 British Airways poster.

thus clearing the way for the Minister to introduce a BOAC Bill to Parliament on 12 June. This Bill received the Royal Assent in August although BOAC was not established until 24 November 1939.

The Act provided that BOAC was to be a publicly owned corporation with an issue of Airways stock to the public bearing interest at a fixed rate - but no stock was ever issued to the public because of the ban on stock issues during the war years.

The function of the new Corporation was set out in Section 2(i) of the Act, namely '...*that it shall be the duty of the Corporation subject as herein provided, to secure the fullest development, consistent with economy, of efficient overseas air transport services to be operated by the Corporation and to secure that such services are operated at reasonable charges*'.

The Act stipulated a great number of things, including that no aircraft designed or manufactured '*outside His Majesty's Dominions should be used on the Corporation's services, unless with the approval of the Secretary of State*'.

Before the Corporation could be established under the Act, war broke out on 3 September 1939, and as from that date the two existing companies were operated largely as one concern and both were immediately placed under the control of the National Air Communications, an organisation set up under the Air Ministry for all civil air operations.

Chapter Two

The War Years

The outbreak of World War Two came at a difficult time for both Imperial Airways and British Airways. On 4 August the BOAC Bill became law, just four weeks before war broke out when both companies were in a state of uncertainty as to effects of the approaching merger.

Although new aircraft had just come or were about to come into service - the Ensigns, the full fleet of C-Class flying boats, including the three longer range G Class boats and the F-Class machines for IAL, and the Lockheed 10As and 14s for British Airways - both companies were looking for new types to undertake the tasks set out in the Cadman Report. BA was planning to order more Lockheed 14s and by the development of an entirely new British type suitable for the projected South American service, the Fairey FC.1, while Sir John Reith was seeking Air Ministry approval to the purchase of Douglas DC-5s and additional flying boats. All were cancelled as soon as the war started.

The threat of war led to tripartite discussions between the two airlines and the Air Ministry in early 1939 to draw up a set of detailed policies and plans to go into a 'War Book'. A set of priorities was established, first being the provision of transport for the RAF, the second the carriage of critical loads of both passenger and freight and third the operation of a surcharged Empire Mail Scheme in place of the all-up scheme in force.

A National Air Communications (NAC) section of the Air Ministry was set up to control operations of overseas air services by the civil airlines. It was foreseen at these discussions that enemy action could involve the need to have alternative routes, in particular, if the Mediterranean were to be closed and if it became necessary to reach Australia via a route over East Africa and the Indian Ocean.

In place of Croydon and Heston airports, Whitchurch, near Bristol, and Exeter would become the wartime bases for landplanes and Pembroke Dock, Falmouth or Poole for the flying boats in place of Southampton. The headquarters of the new NAC, IAL and BA would be Bristol.

In the expectation that any UK base for flying boats might be too vulnerable or that the link between the UK and the Middle East might be broken, the war emergency plans provided for the setting up of an alternative overseas flying boat base at Durban. Selected ground staff in both IAL and BA were to be positioned overseas as required by the war plans and were to be alerted by the respective managements when instructed to do so by the Ministry.

Both airlines were greatly concerned during discussions with the Air Ministry for there appeared to be a complete lack of any clear-cut indications as to the task for civil aircraft in the event of war; it seemed that the RAF was to be given the first call on the airlines' resources regardless of any commercial considerations. In the months to come to this hard-line approach by the Air Ministry was to create severe difficulties for the new BOAC.

As 1939 progressed, it became clear that war was inevitable. On 31 August both IAL and BA were instructed to put the War Plan into effect, and by 2 September both companies began moving their headquarters from London to Bristol to set up in a joint headquarters in the Grand Spa Hotel, Clifton, and action was taken to ensure that aircraft were not immobilized in enemy territory. All commercial

IAL acquired a number of Armstrong Whitworth AW27 Ensigns - termed the E-Class, with G-ADSU *Euterpe*, being the fourth delivered in December 1938. The type suffered a number of operational problems, resulting in withdrawal from use for modification - the eventual return to service clearance came too late for them to resume flying before the outbreak of war. The aircraft suffered a number of landing accidents, but did not have its airworthiness certificate withdrawn until 20 February 1946.
(author's collection)

Lockheed Lodestar G-AGIL was one of thirty-eight of the type taken over from RAF stocks by IAL for supplying the troops in North Africa and the Middle East. *(author's collection)*

operations were abandoned, and the entire civil fleet was put at the disposal of the Air Ministry. The flying boats moved from Southampton to Poole and landplanes from Croydon and Heston to Whitchurch.

On 3 September staff in the Grand Spa Hotel were summoned to the ballroom to hear Chamberlain's broadcast at 11am announcing a state of war between Great Britain and Germany. Reith had been holidaying in Canada and did not land in England until 6 September. Runciman told Reith that there was little to do in Bristol, so they arranged for weekly progress meetings. The Government Minister concerned with civil aviation in the Air Ministry was Captain Harold H Balfour, Under Secretary of State for Air, and so airline contacts were with him, with the Permanent Under Secretary for Air, Sir Arthur Street, and with Sir Francis Shelmerdine, the Director-General of Civil Aviation. Although Balfour had been on the Board of BA, he was unsympathetic to Reith's earlier approaches for help from the RAF, and this attitude was made even more evident at a meeting on 14 September 1939 with Reith, Pearson and Runciman, the last named acting as Chief Executive designate of BOAC. Reith summed up the meeting in his autobiography with the comment that it was a '*...wholly unsatisfactory discussion. We warned them there would be no civil aviation left'*. Balfour made it clear that civil aviation in war was to be entirely subservient to the military needs, that no new types could be proceeded with and that current types would have to die when cannabilisation yielded no more spares. He did, however, agree that no civil aircraft should be requisitioned without the Air Minister's consent.

The Corporation, whose full title Reith claimed to have conceived, was established by the Secretary of State on 24 November 1939. A week earlier directives appointing Sir John Reith to be a Member of the Board and first Chairman, Hon. Clive Pearson to be a member and Deputy Chairman and the Hon. Walter Leslie Runciman to be a member had been issued. On 28 November Harold G. Brown was appointed a board member and the Board confirmed Runciman's appointment as Chief Executive.

The Chief Executive had under him six executives at Headquarters. In each of the overseas regions, there was a regional director responsible directly to the Chief Executive. The regional director had under him functional officials who would follow procedures laid down by their functional chiefs at headquarters but who would be responsible in all other respects to their regional director.

From the outset, the new Corporation's task was to carry passengers, cargo and mail for the Government along Empire and international routes. All the available aircraft space on all flights was at the disposal of Government, and the Air Ministry instituted a system of priorities for loads

Armstrong Whitworth Atalantas were built for the pre-war Middle East and African routes. The type played an important wartime role with BOAC and later the Indian Air Force. *(author's collection)*

Whitchurch Airfield, Bristol shortly after the declaration of war. These, and many other aircraft were hastily evacuated there from the London aerodromes. (author's collection)

which was administered by the staff of the Director-General of Civil Aviation who had also been evacuated to Bristol and were housed close to the airline headquarters.

This system applied throughout the war and ensured that best use was made of the limited space available. Only after all priority loads had been provided for could any space be made available for commercial traffic paying its way. One of the first casualties was the Empire Mail Scheme with its heavy demands on aircraft capacity. It was substituted by surcharged mail and then when enemy action made things far more difficult, by miniaturised letters - airgraphs - and flimsy air letter cards.

Accounting arrangements were complex. The net costs - the cost of providing the services less payments for the carriage of sponsored loads received from Government departments, and less the revenue from any commercial load BOAC was able to fit into spare capacity - of the Corporation's activities during the war were charged to the Secretary of State. Payment for the carriage of mails was paid by the Post Office direct to the Air Ministry. However, the actual financial position was obscured by various benefits received by BOAC from Government without

charge, such as spares, certain accommodation and fuel etc.; and on the other hand by some disbenefits such as the carriage of RAF personnel, Air Ministry officials or stores for which the Air Ministry made no payment.

The new Corporation had hardly settled down under the new situation - and even before the assets of IAL and BA had been formally taken over on 1 April 1940 and the shareholders paid off - when Sir John Reith resigned the chairmanship in January 1940 and was appointed as Minister of Information. His time in civil aviation was a mere eighteen months, but his influence on the shape of the industry was considerable.

Following Reith's resignation, the Hon. Clive Pearson was appointed Chairman and Eric C. Geddes Deputy-Chairman as from 5 March, and Gerard d'Erlanger became a Member of the Board from the same date. On 1 April 1940 Imperial Airways and British Airways disappeared, and BOAC became the sole instrument for international overseas routes, but completely at the Secretary of State's disposal. It inherited a fleet of sixty-nine aircraft of thirteen different types with twelve different kinds of engines.

Leslie Runciman, the Chief Executive, who held the

C-Class flying boat G-AFBL Cooee in seen undergoing maintenance at Rod El Faray, Cairo. BOAC received this machine from QANTAS in August 1940. (author's collection)

The DH.91 Albatross G-AFDK *Fortuna* of BOAC in full wartime civil aviation camouflage and markings. *(author's collection)*

title of Director-General (DG), had under him two Assistant Director-Generals (ADG), Colonel Burchall being responsible for services in the west and Major McCrindle for services in the east, a division which corresponded roughly with non-Empire and Empire services. A Management Committee, consisting of the Director-General, the two Assistant Director-Generals and the six Departmental Directors came into being. Overseas there were eleven Regional Directors. A huge problem for the Director-General was to keep a hold on enough staff, both flying and ground, to enable BOAC to carry out its tasks. This was a constant worry, especially when the war entered a more acute phase and demands grew for release of staff to the Services, although the Air Ministry in consultation with the Ministry of Labour ensured as far as possible that the operation of the National Services (Armed Forces) Act 1939 should not reduce staff below a level which would impair the effective operation of the Corporation.

A further strain was put on the Corporation when the Air Transport Auxiliary (ATA) was set up under the aegis of the Ministry of Aircraft Production (MAP) to ferry service aircraft from factory to squadron and so save the use of combat pilots. Gerard d'Erlanger, a BOAC Board Member, was put in command and some flying, engineering and other staff were seconded from BOAC. In June 1941 the MAP placed the administrative responsibility for the ATA in BOAC's hands. It continued its work throughout the war with conspicuous success but at the cost of a considerable number of casualties among flying personnel.

Even more work came BOAC's way when in May 1940 the MAP asked the Corporation's technical staff to undertake the modification and repair of RAF aircraft and the overhaul of engines and propellers. As a measure of the size of the task, by as early as March 1941 more than half the time and effort of Corporation staff was engaged in this work. Initially, the workshops at Croydon and Hythe were used, but by the end of June 1940, a group of buildings had been taken over in South Wales at the Treforest Trading Estate and equipment transferred there from Croydon. By the end of July, engine testbeds had been built and the production of overhauled RAF engines put in hand.

BOAC also undertook for MAP the task of assembling American aircraft for the RAF arriving by sea at Liverpool, carrying out this work for a time at Speke airfield. BOAC staff also did similar work at Colerne, near Bristol, until the RAF was ready to undertake the work. There was much work to be done on the repair of propellers, and two factories were set up at Bath. To be able to cope with the volume and range of work for the MAP, BOAC formed a separate organisation on 1 September 1942 known as the Propeller & Engine Repair Auxiliary (PERA) to take over the factories and staff. BOAC's Administration Director, A. J. Quin Harkin, was seconded to take charge. A year later the arrangements with Government were terminated, and BOAC began to rationalise its repair work into one large factory unit in the Treforest Trading Estate, and Treforest remained the Corporation's main engine base after that.

Runciman was pressing the Ministry for greater capacity from aircraft with sufficient range to give flexibility under wartime conditions. He made a case for American civil aircraft being the only adequate types and asked for five flying boats and nine Lockheed 18s or six DC-3s.

The saga of 'The Balfour Boeings'
The Under Secretary of State, Captain Harold Balfour, crossed to USA in *Clare* in the summer of 1940, as he recalled in his biography *'Wings over Westminster'*: 'Early

AW.27 Ensigns were called the E-Class. G-AFZU *Everest* carries the wartime civil registration style of outlined letters, underlined with red, white and blue stripes.*(author's collection)*

Harold Balfour, Lord Balfour of Inchrye, PC, MC.
(author's collection)

in August I left the field of the Battle of Britain for two weeks to do the first of a good many flying visits to Canada and the U.S.A. I was to report on the progress of the Empire Training Scheme then go on to Washington to see how quickly we could get on with harnessing the US training facilities to our needs. This was the first of about twenty-six air crossings I did during the war. It may not have been the riskiest but it was certainly the most uncomfortable of all including a couple of trips to the Middle East and South Africa. When I left this country, I had no authority or intention of buying aircraft. It had been agreed that if I heard of suitable available communications types, I should investigate and bring back particulars. Consideration would then be given to possible purchase by Beaverbrook's MAP.

Pan American Airways continued operating pre-war trans-Atlantic Services to Lisbon using six giant Boeing Clipper flying boats, at that time far the most advanced commercial aircraft in design, performance and comfort. Each boat weighed nearly 40 tons fully loaded and was nearly twice the size of Britain's biggest civil flying boat.

In New York, I heard a whisper of the possibility of getting hold of three new, improved Clippers then under construction for Pan American in Boeing's Seattle works on the Pacific coast. Pan American had intended to duplicate the service and ordered six boats at the cost of approximately six million dollars.

Then came the war. France had fallen, traffic had fallen, and now their Lisbon stage was threatened. If Germany occupied Spain, or Spain joined the Axis, Portugal as a base would either be overrun or made unavailable. Traffic on the Pacific route alone would not justify the use of a further six giants. Pan American directors were of a mind to 'hedge' on their commitment to the extent of selling three of the six under construction. Here was something tremendously worthwhile for Britain. These boats would fulfil a vital war need.'

Balfour knew that if he referred the purchase to the Secretary of State he would have to go to the Chancellor and dollars were scarce for, as yet, there was no Lease-Lend. Even if the Chancellor had agreed, the purchase would still have to be sanctioned by the Minister of Aircraft Production, and he felt certain Beaverbrook would oppose the project. Balfour was not willing to take the risk.

'I lunched with Juan Tripp, President, and members of the Board of Pan American Airways in a private room at the top of a skyscraper. With New York spread around us, we discussed the sale. I put my cards on the table; why we wanted the boats as help to our war effort. I asked for the agreement to sell reasonably and at once, not on commercial but National grounds. Tripp and his colleagues responded. By the end of lunch, we had settled outlines of the bargain. Britain should have numbers one and two and number four of the production line of six. The British Government would pay Pan American net cost to the Company of the boats, engines and other equipment, plus 5% commission as consideration for selling.

For 5% P.A.A. would continue inspection and supervision during manufacture and carry out flying training of our first air crew sent over to take delivery.

Straight from lunch, I took two executive directors to the offices of the British Purchasing Commission. I told Sir Henry Self, Civil Servant, then in charge, of the deal, I had done. Policy was my responsibility, and I asked only for his help in putting agreed terms in fair and proper contract form. The complete boats worked out at something over one million dollars each. That evening I told Lord Lothian, our then UK Ambassador, what I had done. He cleared it at once with the State Department as a private commercial transaction in which they held no interest.

One of the 'Balfour Boeings', G-AGBZ *Bristol* seen touching down in a flurry of spray. Three of these machines made up the BOAC B-Class fleet. *(author's collection)*

I knew I was wrong in assuming the authority to buy without reference to my Government. I did not take this sort of risk lightly nor did I derive pleasure in entering into obligations beyond those proper to my Office. But it is the man on the spot who gets the chances'.

On his return to the UK, Balfour was handed a letter by special messenger from Sir Archibald Sinclair. It seems his action over the Boeings had indeed shocked both the Chancellor and the Minister of Aircraft Production. Questions were asked, the Prime Minister became involved, and papers flew in furious rounds of discussions, especially between Lord Beaverbrook and Harold Balfour.

The first of the three Boeings was delivered to BOAC on 22 May 1941 and only four days later left on its first flight to West Africa, By July all three Boeings were in service, being christened *Berwick, Bristol,* and *Bangor*. The aircraft were based in the USA to have ready access to spares, and so Baltimore became their maintenance base. In the summer of 1941, the Boeings flew on a round route from Baltimore to Botwood in Newfoundland and then to Foynes and their loads were ferried between the UK and Eire by a Poole - Foynes and Whitchurch - Collinstown shuttle service, the landplane airfield near Dublin. From Foynes, the Boeings joined the C-Class and Catalinas on the West African route via Lisbon and Bathurst to Lagos and back to Foynes: the round trip was completed by another crossing of the Atlantic via Botwood to Baltimore for maintenance. As from the onset of winter when weather made Botwood unusable, the boats switched to southerly routeing, Baltimore - Bermuda - Azores - Lisbon - Foynes, before returning to Baltimore via Lisbon - Bathurst - Lagos and across the South Atlantic to Belem in Brazil, thence to Trinidad and Bermuda and so to Baltimore. They carried many VIPs, thousands of passengers, diplomatic bags and mails without accident. Their contribution to the war effort was great. They finished their careers with BOAC in the Summer of 1946, operating a USA - Bermuda service and were sold still as flying concerns.

But in summer 1942 all was forgiven when Churchill and Beaverbrook returned from Bermuda in Berwick saving days as against the originally planned return by warship, as Balfour was later to discover. *'In January 1942 the Prime Minister carried out his first war mission to the United States. Our new battleship, Duke of York, was to pick up and bring back his party from Bermuda. They were to fly from Baltimore in Berwick under conditions of great secrecy. At Bermuda, there was a favourable weather report across the whole Atlantic, so the party decided to continue in Berwick all the way home. On January 16th the Prime Minister with Portal, Admiral Sir Dudley Pound, and Beaverbrook took off piloted by Captain Kelly Rogers. On January 17th Churchill completed his first Atlantic air crossing of 3365 statue miles in 17 hours 55 minutes from Bermuda to Plymouth. On 27 January Churchill spoke in*

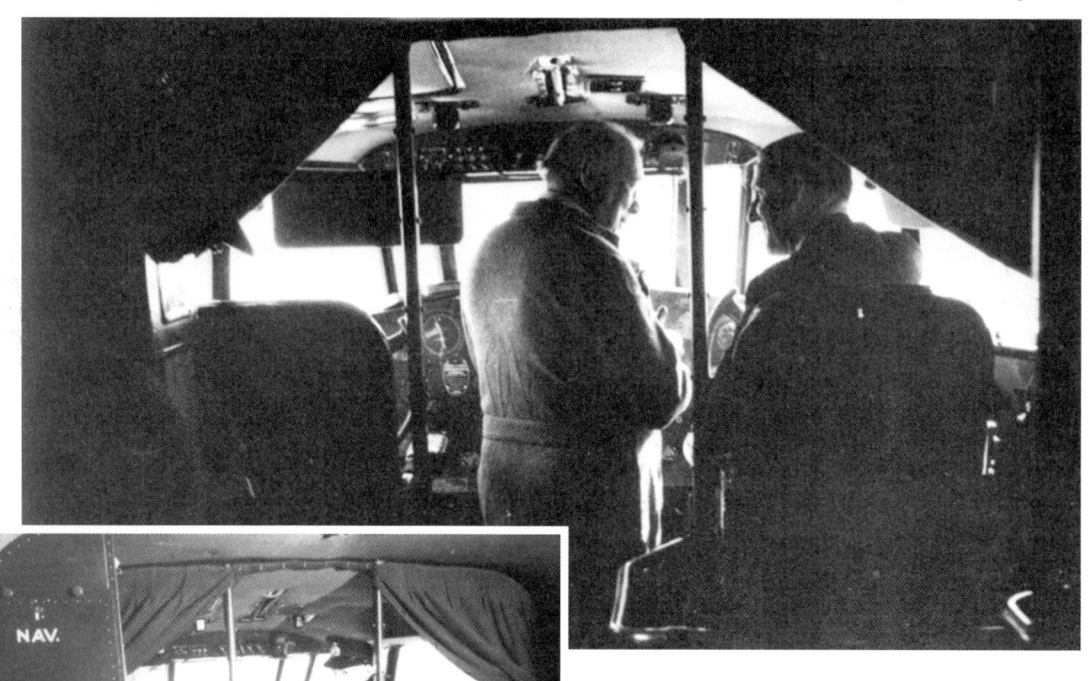

Above: Prime Minister Winston Churchill is seen on the flightdeck of BOAC's Boeing 314 G-AGCA *Berwick* talking to Captain Shakespere.

Left: The flightdeck was reached via a staircase from Cabin B on the starboard side. Aft of the Captains chair was the radio operator's table. *(both author's collection)*

BOAC's Boeing 314 G-AGCA *Berwick* landing at Lagos, Nigeria. Note the crewman riding in the nose mooring hatch - this must have been an exhilarating ride! *(author's collection)*

the House. At dinner that night he took his usual place in the Members' Dining Room, at the head of the Ministers' Table. The Chancellor of the Exchequer and some other Ministers, including myself, had already commenced. Churchill turned to me saying that he had just returned in one of the 'Balfour Boeings' and that he withdrew his censure entirely on myself for having bought them. They were, he said, 'fine ships'. 'You apologised to me,' he added. 'It is rather I who should apologise to you.' Though I had not 'obeyed the rules', he was glad I had done the transaction. I was grateful for this generous absolution. A few months later, on his submission to the King, I was sworn of the Privy Council.

Ferrying and Frustration
The summer of 1940 saw the birth of another venture with lasting repercussions for BOAC. There was a desperate need for American bomber aircraft, and in spite of Air Ministry opposition, the MAP under Beaverbrook pressed ahead with a scheme for ferrying American bombers across the Atlantic. The task of organising the Canadian end was given to Sir Edward Beatty of the Canadian Pacific Railway assisted by George Woods Humphery, a familiar name to former IAL staff. BOAC was asked to help with experienced staff both technical and flying personnel. Four senior Captains, Arthur S. Wilcockson, Donald C. T. Bennett, R. H. Page and I. C. Ross, were released for this purpose. The General Manager was Col. H. Burchall, who

had resigned as Deputy Director-General of BOAC in June. BOAC made a further contribution by supplying twelve complete crews in September. The first delivery flight took place on 10 November from Montreal via Gander with a group of seven Hudson bombers, led by Bennett, all arriving safely at Aldergrove in Northern Ireland.

In March 1941 the government decided this should be directly controlled by MAP and given the title of the Atlantic Ferry Organisation; by July a further change was made when control passed to the RAF. On 19 December 1940, the Chairman, Clive Pearson, wrote a letter to the Secretary of State for Air, appealing for clearer authority for BOAC's wartime operations and for greater resources to carry them out. The letter reveals a sense of frustration in BOAC at being held back by lack of aircraft and personnel. Pearson asked for directions to operate flights between the UK and North America and between the UK and Sweden, to maintain existing Empire routes and to open up a new landplane service between South Africa and Egypt, Turkey and other countries in the Balkans.

Pearson asked for five B.24s, eight Lodestars, ten or more DC-3s plus the full fleet of eight DH.95 Flamingos already ordered before the war, and the three Boeing 314A flying boats - all these to be in addition to BOAC's existing fleet of C-Class boats. Sir Archibald Sinclair replied that BOAC needed no directions to maintain existing services and that the War Cabinet had now laid down a priority for air routes and agreed that certain aircraft should be made

When Germany invaded Denmark in April 1940, DDL's Fw 200B four-engined airliner OY-DAM *Dania* was at Shoreham on a scheduled flight. It was siezed and the aircraft and pilot, Capt H. Hansen, were flown to Whitchurch and seconded to BOAC as G-AGAY and given the BOAC name of *Wulf*. It was planned that the Condor would be flown out to the Middle East, but permission was refused and the Condor, with a fighter escort, flew over to White Waltham on 9 January 1941, for use by the ATA. After a test flight on 12 July, Capt Hansen landed at White Waltham, and on applying the Condor's brakes the wheels locked and it skidded across the wet grass, causing the undercarriage to collapse. Damage was severe, and the Condor was written-off. *(author's collection)*

A BOAC B-24 Liberator takes off from Dorval, Montreal bound for Prestwick in Scotland. *(author's collection)*

available to undertake them. Apart from the North Atlantic and Scandinavian routes for which five Liberators and possibly two other American civil types were to be handed over, the order of route priority should be first UK-West Africa, second the Horseshoe route South Africa-Egypt-Australia followed by new routes between Capetown and Cairo, and between Egypt and Turkey and other countries in the Balkan peninsula. The Cabinet had decided BOAC should receive eight Lodestars, eight DH.95 Flamingos and ten or more DC-2s, DC-3s or Lockheeds, all provided by the MAP. Not all these arrangements worked out in practice, but BOAC's position was eased somewhat both as to crews and ground staff by the return to the Corporation of some of those hitherto engaged in the ATA and in the Return Ferry Service.

BOAC was struggling to make a maximum contribution to the war effort, and for eight months the Empire routes were operated over virtually the same route pattern as applied in peacetime. The pattern was extended in April 1940 when the newly formed Tasman Empire Airways opened up the Sydney-Auckland route with the S.30 boat ZK-AMA *Aotearoa,* joined by ZK-AMC *Awarua,* and they kept the route going throughout the war. Even the mail service across the North Atlantic inaugurated in August 1939 with G-ADCU *Cabot,* and G-ADCV *Caribou* continued to 30 September although both were later taken over by Coastal Command.

The landplane operations in Europe presented a different picture. All services were halted, but on 13 October the Paris service was restarted with Ensigns from Heston to Le Bourget and continuing until 11 June 1940 when they were withdrawn due to the German advance on Paris. The Scandinavian route saw only a brief break in operations as the base was moved from Heston to Perth in Scotland, services being resumed via Stavanger and Oslo to Stockholm on 4 September. No other European services in those early days of the war could be operated apart from some ad hoc flights to Lisbon via Bordeaux and, in addition to regular flights, some valuable ad hoc work was carried out by Ensigns transporting personnel and equipment to the forces in France.

Although the flying boats flew routes between England and the Middle East through the Mediterranean, the military in Cairo was pressing for an increase in capacity. As a result, a landplane service by DH.91s was begun in September 1939 from Shoreham via Malta to Alexandria; from April 1940 flights were transferred from Shoreham to Heston and the Lockheed 14s took over. This service continued until the entry of Italy into the war when for a short time an attempt was made to divert the route over North African territory via Oran, Gao, Maiduguri and then over the Central African route to Khartoum and up to Cairo. The French banned overflying their Central African territories, and the service ceased at the end of June. This was the pattern of operations up to the invasion of France and the entry of Italy into the war in June 1940.

The Bristol headquarters and the overseas headquarters at Durban had both been successfully established, and although life at Bristol had been disturbed when it became a target for enemy bombing raids, work carried on in spite of damage to buildings and staff billets. With increased tasks, Headquarters had to spread itself to seven nearby houses in Clifton. The new bases at Whitchurch and Poole and the engineering base at Treforest were all functioning adequately, new West African bases had been opened and steps were in hand to improve hotel and resthouse accommodation for the increased flows of priority passengers, particularly on the trans-African routes.

The return to normal operations except within Europe did not last long. Just before the storm broke, BOAC gained permission from the Portuguese to start a service to Lisbon via Bordeaux with DH.91s; but Bordeaux became

K-Class DH.95 Flamingo G-AFYF *King Alfred* is seen at Bramcote in August 1940. (author's collection)

Douglas C-47 Dakota I G-AGGB seen on approach to Whitchurch. It was one of six such machines diverted to BOAC from RAF stocks. *(author's collection)*

unusable, so the service was maintained with C-Class flying boats linking at Lisbon with Pan American's Clippers; thereafter a regular landplane operation was run throughout the war by Dutch KLM crews using DC-3s. The only other service in Europe, to Scandinavia, came to a halt after the invasion of Norway on 9 April but it was essential to maintain some link with neutral Sweden, so nine flights with Lockheed 14s took place from Leuchars direct to Stockholm up to December, flying over enemy-held territory. Once France had fallen, and Italy entered the war, the main Empire lifeline through the Mediterranean was closed to the flying boats, and it became BOAC's priority to re-establish the links between the UK and the Middle East. This had been foreseen in the War Plan and to overcome the lack of a maintenance base in the UK for the flying boats operating from Cairo to Durban, and from Cairo to Australia - detailed plans were ready to establish an overhaul base at Durban. As early as mid-June 1940 preparations at Durban were in hand and by the end of the month engineering staff, stores and equipment arrived by sea to set up the base. With overhaul facilities available at Durban and Sydney, the flying boats could continue to operate although unable to return to their original home base at Hythe, and so the Horseshoe route came into being. This did not solve the problem of how to fill the gap between the UK and the Middle East. BOAC's long-range capability had been made more difficult by the handing over to the RAF of *Cabot* and *Caribou* at the beginning of November 1939, both being destroyed in the Norwegian campaign in May 1940. However, at the beginning of August the North Atlantic service was reopened with two other S30 boats, *Clyde* and *Clare,* thus proving to the Americans that even at the height of German air attacks, unarmed civil aircraft could fly in and out of England.

G-AFCX *Clyde* was used to test a direct route to West Africa to link up with the landplane trans-African route to the Middle East. Under the command of Captain Anthony C. P. Loraine with Captain W. S. May as second pilot 'CX

A tender makes it's way out to BOAC S.25 Sunderland III G-AGER which has one of it's engines running.
(author's collection)

BOAC Sunderland IIIs wore RAF Transport Command radio codes. OQZC (G-AGHW) and OQZH (G-AGKZ) are photographed moored in Poole Harbour after their flights from Calcutta. The tender is bringing passengers ashore. *(author's collection)*

was at Poole on 5 August ready to depart the following day to America but that same evening Captain Loraine received instructions to fly to Lagos with important passengers.

Although no previous crossing from Lisbon to Bathurst had been undertaken by flying boat, *Clare* landed at Bathurst after a flight of 15 hours from Lisbon, and flew on via Freetown to Lagos, demonstrating the feasibility of flying boat operations from Poole to Lagos.

Captain Loraine with *Clyde* then continued to the Belgian Congo, alighting on a stretch of the Congo river. His passengers were Free French officers who then made contact with certain French officials across the river in French Brazzaville. A meeting between the two groups took place aboard *Clyde,* which led to the territory of French Equatorial Africa coming over to De Gaulle and the Free French. Apart from its political significance, the decision meant that the air route across French territory to the Middle East was now open for Allied flights. This trans-African route between Khartoum and Lagos - Accra - Takoradi had been maintained in conjunction with Elders Colonial Airways Ltd by DH.86s, but after 16 June 1940, this had to cease as the French refused permission for operations through French Equatorial Africa. However, from August this restriction no longer applied as the territory had declared for De Gaulle; services were restarted with Lockheed 14 aircraft, and shortly afterwards the DH.86s resumed flights. By the end of the year operations were stepped up using Lockheed 14s, DH.86s and Ju.52s and by the use of Sabena machines under charter to BOAC but with a routeing from Cairo over Khartoum, through the Belgian Congo to Lagos and Takoradi. So the link between the UK and the Middle East could be reforged although it was September before *Clare* and *Clyde* were released from the North Atlantic operations to start up a regular service from Poole to Lagos to make the connection with the trans-African landplanes.

In the Far East, the connection between Bangkok and Hong Kong with DH.86 aircraft was maintained right through to October 1940 when the French authorities stopped British aircraft flying through Indo-China, so the service was suspended on 15 October.

The fleet of aircraft taken over by BOAC in April 1940 was largely intact. Three A.W. Ensigns had been lost by enemy action, but ten were still in England. The DH.91s had suffered losses with only three aircraft surviving *Fingal* being badly damaged after a crash landing near Bristol,

The former IAL DH.86A G-ADUI *Denebola* was impressed in Egypt in 1941. *(author's collection)*

Imperial's DH86B G-AEAP *Demeter* served with No. 1 Air Ambulance Unit RAAF as HK843 in the Western Desert in 1943. *(author's collection)*

Shorts S.30 G-AFCX *Clyde* in full wartime markings seen over Long Island Sound. The underlining of the registration on the upper surface of the wing is noticable. *(author's collection)*

Frobisher destroyed, and *Faraday* damaged on the ground by arson at Whitchurch. Of the Atalantas based on Karachi, three had been requisitioned by the RAF, and they were no longer an effective fleet. Of the HP.42s. three were requisitioned, and two had been wrecked in a gale at Whitchurch. The flying boat fleet which remained on 1 April 1940 was seventeen S.23s, four S.30s and two S.33s. Of the Lockheed aircraft, one of the fleet had crashed early in 1940 in Scotland, but the two Lockheed 10As were still operational. Also the Poles had flown in two Lockheed 14s to Whitchurch at the start of the war, and these were on hire to BOAC. Of the three British Airways Ju. 52s, one had been captured by the Germans in Oslo. However, by the end of 1940, it was clear there had to be some increase in capacity, particularly of the longer range types.

It is an interesting to note that early in the year Director-General, Leslie Runciman, issued an instruction that the name of the Corporation for use in signals, on aircraft, vehicles and launches should be shortened, and that the Corporation was to be known simply as 'British Airways'. Operations in 1941 were considerably stepped up.

The Near East Region, based on Cairo, under the Regional Director Robert Maxwell, had a particularly daunting task with the increasing demands made on it by new services in the area. The enlarged Repair Unit was set up in August, and by the end of the year, over one thousand men were employed. On top of this Maxwell was given the overall responsibility for organising the plan proposed by Air-Marshal Tedder to build up a network of communications in the Near East Area. No adequate base with maintenance facilities existed to service a heavily increased fleet of civil aircraft, and so a new base at Asmara in Eritrea had to be created.

At the Bristol Headquarters, BOAC's day-to-day management was coordinated and directed by the Director-General's Committee with Heads of Departments represented. With all this new work in hand, the staff problem became increasingly serious. There was a shortage of crews on the Return Ferry Service and to man the G Class boats; to make matters worse the Essential Works Order was not working effectively to protect existing staff.

1941 saw the war with Germany and Italy in full swing with the ground forces engaged in a struggle for the control of the Middle East. Early in the year *Clare* and *Clyde* maintained the only link with West Africa, but on 15 February *Clyde* was sunk in a gale at Lisbon. A replacement was available in the shape of a Consolidated PBY flying boat bought from America and flown across the Atlantic. This aircraft (with the old name of *Guba*) joined *Clare* on the Poole-Lisbon-West Africa route. The payload situation eased somewhat in February when the Irish Government allowed the use of Foynes on the Shannon.

By June the situation had improved with the arrival of a second PBY boat on loan and two more C-Class available. But even more important to BOAC's war effort was the introduction into service of the three Boeing 314A flying boats. They began to operate in West Africa in May 1941.

At the same time as the Boeing 314As transferred to their winter routeings in October 1941, the C-Class boats, *Cathay, Champion* and *Clare*, together with *Guba* and the second Catalina were re-routed through the Mediterranean, an indication of how the war situation there had improved. Furthermore, the revised routeing, Poole, Lisbon, Gibraltar, Malta, Cairo, enabled badly needed supplies to be flown into Malta, already under air attack.

By June 1941 the RAF had put nine Lockheed Lodestars at BOAC's disposal, and by December this number had built up to twenty-one machines, so the availability of aircraft had eased, although BOAC was still without the use of the Ensign and Flamingo fleets which were still undergoing modification and tests in Britain.

By October the Asmara base had become the primary overhaul centre for all BOAC's landplanes in Africa. From the autumn a network of communication flights was established under the Tedder Plan with aircraft based on Asmara, all of which were operated by Lodestars. Regular routes operated were Cairo - Wadi Halfa - Port Sudan - Asmara, Cairo - Habbaniya - Teheran, Cairo - Lydda - Adana and Asmara - Khartoum. The Teheran route became a supply route to the Russian war effort.

At the end of 1941, the Horseshoe Route came under attack as the Japanese moved into Siam and then into Malaya. Early in 1942, the route had to be terminated at Calcutta, and so the link with Australia and New Zealand was broken. The possibility of this happening had been

Right: a pair of Model 28 Consolodated Catalinas during their service with QANTAS. The rearmost aircraft is G-ACIE *Antares Star*. The tail closest to the camera belongs to G-AGFM *Altair Star*. As can be seen, the aircraft carried both military and civil registrations, as well as a fleet number.

Below: Consolodated 28 Catalina G-AGFM *Altair Star* was photographed on the Nile. Its military service number FP244 is visible just in front of the fleet number '2' on the fin. *(both author's collection)*

foreseen, and the emergency routeings worked out from Rangoon via the Andaman Islands and Sumatra over Penang to Singapore, or if Malaya became unusable, then via the Andaman Islands and Sumatra to Batavia and on to Australia. QANTAS crews were flying the whole sector of the Horseshoe Route from Karachi to Sydney because of BOAC's crew shortages, and it was the QANTAS crews who had to cope with the Japanese threat from 7 December onwards. At first, it was still possible to route into Singapore via the Andamans, but shortly afterwards Malaya itself had to be abandoned from the trunk route, although for a time a shuttle service under hazardous conditions was flown between Batavia and Singapore.

In the spring of 1942, Maxwell reported to London that the Tedder Plan of June 1941 had not been fully implemented; the aircraft expected from the USA by the end of 1941 had not materialised; the Asmara base was proving disappointing and the flying hours operated were only about one-third of those then proposed. The shortfall was caused by lack of spares, equipment, aircraft and engineering management. Early in April, a senior RAF officer had commented upon the lack of resourcefulness shown by BOAC at Asmara to get over its problems and an unwillingness to depart from peacetime methods.

Later that month BOAC was confronted with an explosive signal from Tedder to the Minister complaining about BOAC's shortcomings and stating that if it could not improve he would withdraw aircraft, crews and facilities made available to it in the Middle East. A month later Tedder signalled again that he was upset that BOAC had felt obliged to refer to the Air Ministry for approval before any new services could be operated. He found this odd given the arrangements under the Tedder plan; after all, he added, *'...there is, in fact, no purely commercial service now operated by your company in the Middle East'*.

In parallel to these bolts from the Middle East, BOAC's Director-General had written in February to the Ministry stressing the fact that BOAC's operations were being restricted by a shortage of pilots, spares, equipment and administrative and technical engineering staff; this needed correction to enable even existing services to be continued.

The issues between BOAC with the Air Ministry were considered at a meeting late in March between BOAC's Deputy Director-General and Balfour. As a result, Balfour wrote to the Director-General stressing the value of the splendid work done by the Corporation to the war effort and asking that the Corporation extend considerably its operations based on the UK; though he appreciated this could mean lowering the standards normally applicable to good airline operation in peacetime. He accepted that this was a matter of policy for the Air Ministry.

That year the Corporation had been pressing for a centralised landplane base in the UK; by the end of 1942, the Air Ministry decided on making Hurn available.

On the Horseshoe route, the emergency routeings put into force at the close of 1941 did not last long. Singapore was under massive Japanese attack by the end of January, and the last QANTAS flight into the city landed on 3 February 1942. Singapore surrendered on 15 February, and with only one day to spare the Airways ground staff with some army officers and civilians managed to get away in the only remaining airways launch.

The DH.91 Albatross was termed the F-Class with BOAC. G-AEVW *Franklin* was handed over to the RAF as AX904 and appeared in full RAF markings. *(author's collection)*

An attempt was made to continue flying via Sumatra and Java, but Japanese fighters made this impossible, and on 23 February the last flying boat service reached Sydney. After that, there were no operations on the Horseshoe route south of Calcutta. Heroic efforts had been made by the QANTAS crews to keep the route open and to ferry passengers out of beleaguered Singapore, but suffered heavy losses. Of the C-Class boats, six BOAC machines had been caught east of Rangoon and two QANTAS aircraft west of Rangoon. The latter was transferred to BOAC, but of the former *Circe*, *Corinna* and *Corio* were all destroyed by Japanese action. Two more were lost by QANTAS from flying accidents, *Cassiopeia* at Sabang and *Corinthian* at Darwin. So QANTAS suffered a shattering blow, and their flying boat fleet was decimated. For the rest of the year, BOAC continued to operate Durban-Calcutta.

In North Africa, Allied forces had had to retreat to the Nile Delta, and the Mediterranean had again become very dangerous, especially Malta landings. Consequently the S.30 boats *Cathay, Champion* and *Clare* were moved back to the UK - West African route in February to rejoin the three Boeing 314As. As a desperate measure, fifteen Armstrong Whitworth Whitley bombers were transferred to BOAC from the RAF. These twin-engined aircraft had limited load carrying ability and a nasty habit of overheating. Only one flight was operated to West Africa which proved the aircraft's unsuitability for tropical conditions and they were withdrawn. It was not until mid-summer that more capacity was provided by the use of the two S.25 boats *Golden Hind* and *Golden Horn* which had been returned by the RAF the previous December and by the return of the two PBY Consolidated boats from the Mediterranean. This improvement was short-lived as *Clare* was lost off Bathurst in September and it was decided to transfer the remaining two S30 boats, *Cathay* and *Champion,* to Durban for use on the Horseshoe route.

The position became even more acute in October when the three Boeings had to revert to their long winter routeings reducing their frequency between Foynes and Lagos. Shuttle services were run by the G class boats between Poole and Lisbon with loads for the Boeings, so cutting out the Lisbon-Foynes-Lisbon leg of the Boeing route. At the end of the year, more capacity was produced when the RAF made available two more PBY boats and one Liberator specifically for the West African route.

Somehow loads had to be got into the Middle East from England, and the RAF were making direct flights from their base at Lyneham to Cairo in Liberators; to supplement these BOAC undertook similar flights with Liberators loaned from the North Atlantic ferry service, but on the return from Cairo on 15 February 1942 the first flight under the command of Captain Page was mistakenly shot down off the south coast of England with the loss of all on board. It was not until July that the operation was resumed

G-Class G-AFCK *Golden Horn* suspended from a crane at Cabo Ruivo, Lisbon in April 1942. *(author's collection)*

with BOAC crews but with RAF markings that continued until mid- December normally with a call at Gibraltar.

During this year Malta was threatened by enemy attack from the air and various methods were tried to keep communications open. Loads were flown into Gibraltar by the Liberators en route to Cairo and by Whitleys and Hudson IIIs from Whitchurch and were then shuttled from Gibraltar to Malta by Catalina. Eventually, the situation in Malta became too severe for civil aircraft to attempt to fly in any longer and military machines took over, but BOAC continued to fly loads into Gibraltar for onward carriage by the RAF; also from October the Dutch helped by extending the KLM service to Lisbon on to Gibraltar.

On the trans-African route, it had been intended to cease BOAC Operations with the advent of PAN AM (Africa) operating a high frequency, but with the United States now in the active fighting there would have been inadequate capacity for British loads, so the British services were reintroduced between Takoradi and Khartoum with Lockheed Lodestars. During the year a variety of services were operated on this route, some terminating at Khartoum and some at Cairo with a number of different aircraft types - Lodestars, Flamingos (eight being delivered and flown out to the Asmara base by September although one crashed at Adana in September and a second at Addis Ababa in November). Lockheed 14s, Wellingtons and Ensigns. The last named had slowly been re-engined and by June five had been flown to the Middle East and were based at Asmara, and by the end of the year, this number had increased to eight, leaving only one aircraft still under modification at the Bramcote base in England.

Two other sources of capacity continued to be available. Sabena continued its flights under charter to BOAC, linking at Juba with a once-weekly S.23 boat Juba-Cairo; when this link ceased in February Sabena did the whole trip from Cairo to Takoradi via the Belgian Congo as before, using Ju.52s and Lockheed 14s. From November Sabena included a call at Entebbe in Uganda to make a connection with the Horseshoe services calling at nearby Port Bell on Lake Victoria. The second source was provided by the S.23 boats which continued to fly across Africa from Cairo to Lagos via Lake Victoria and the Congo River, with intermediate calls to get the optimum payload. This service suffered in frequency at the close of the year with the loss of *Ceres* at Durban in December and to the need for a flying-boat survey of a proposed new route between East Africa and Diego Suarez in Madagascar.

The 1941 revolt in Iraq affected the Horseshoe route; there were even fears that the link between Cairo and India via the Persian Gulf might be broken. As an emergency measure, the Government decided to explore alternative routeings avoiding the Persian Gulf, passing along southern Saudi Arabia, the Hadramaut coast, crossing the Persian Gulf just south of Muscat, thence to Jiwani and with a proposed terminus at Karachi. A flight was accordingly made from Karachi on 29 May arriving at Aden on the following day. A BOAC party was already in Aden to test out the prospects of the route for flying boats but this came to nothing, and it was eventually agreed that steps should be taken to set up the ground organisation for a landplane route. The route selected was from Aden via Riyan, Salala, Masira Island, Ras el Hadd in Oman, thence to Jiwani in Baluchistan and on to Karachi. Work on the necessary ground arrangements was split between the RAF and the Indian Government with BOAC assisting the latter with the organisation and provision of ground staff. BOAC's Clive Adams was directly involved in the work at Salala and Masira, both far off the beaten track with virtually no contact with the outside world, so it took time and ingenuity to get equipment and supplies into position and to provide some quarters for passengers. By May 1942 the route was ready, and the first flight through the new stations was made by a Lockheed Lodestar on 11 May with a once-monthly service after that. Lockheed 14s and Ensigns were also used on this route during the year, and all services commenced from the overhaul base at Asmara.

More capacity was needed between the Middle East and India for the war effort against Japan and flights additional to the regular Horseshoe flying boats were provided for a few weeks by extra C-Class operations but mainly by landplane from Cairo to Lydda, Habbaniya, Basra, Bahrein, Sharjah and Jiwani to Karachi and by the end of the year a once-weekly service was being flown with Wellington aircraft. The communications flights in the Middle East started in 1941 continued on an even more significant scale throughout 1942 with the various types based at Asmara - Lockheed 14s, Lockheed Lodestars, Flamingos, Ensigns, and to a limited extent Wellingtons. Kamaran, Aden and Addis Ababa and also East Africa were all added to these operations which formed an involved network with frequent re-routeings and changes of frequency, stretching BOAC's resources to the limit. BOAC was also required to operate as many flights as possible between Cairo and the Western Desert during November in support of the advancing Eighth Army; to do this something had to give and for a time BOAC's trans-African operations were suspended.

On the North Atlantic, the Boeing 314As continued on the same pattern of operations as in the previous year. They did the eastbound crossings via Bermuda and Lisbon to Foynes returning via West Africa to Baltimore on the southern routeing over Brazil up to May and then reverted to direct crossings. Baltimore-Botwood-Foynes-Botwood-Baltimore. When the autumn came, they returned to the southern routeings. As before, loads were shuttled between England and Eire to connect not only with BOAC Boeings but also later in the year with the Pan American and American Export Lines flying boats. Initially, the Frobisher class, used on this shuttle, flew from Whitchurch to Collinstown, near Dublin, but from the end of February the landplane base at Rineanna, Shannon, was available and made a more convenient transfer point for loads.

On the other North Atlantic operation, the Return Ferry Service between Prestwick and Montreal, BOAC made 136 flights all with Liberators without any serious mishap, using Gander for refueling. Some services were routed through Reykjavik in Iceland and Goose Bay in Labrador, and as experience was gained a variety of routeings through Iceland and alternates in Canada were used.

During the year there was no let up on the dangerous

run to Sweden mainly by Lockheed Hudson IIIs and Lodestars on irregular frequencies. The four Whitleys allocated to the service in August had only a brief run, proving as unsuitable as they had on the route to West Africa; it was necessary to fly overloaded on these vital supply routes and with no single engine performance, the Whitleys failed to meet even wartime operational standards. To satisfy the need for capacity even the one and only Curtiss Wright CW.20 was called into service for a few trips. In all over 170 flights in each direction were completed during the year with the loss of only one aircraft, a Hudson on 23 June off Sweden although all on board were saved and some of the cargo salvaged.

BOAC was involved in the opening up of another new route towards the end of the year along a track it had never previously flown and moreover across the enemy held territory. Although there was air communication with the USSR from Britain on indirect routeing through Cairo to Teheran, where a connection was made with the Russian flights to Kuibyshev, the benefits of a direct flight were obvious, and early in the year, consideration was given as to how to achieve this. It was decided to detach a Liberator from the Return Ferry Service for the purpose, and a BOAC crew under Captain Percy were nominated to manage the flight. It was clearly impossible to avoid enemy territory as the route selected was from Prestwick north to the Arctic circle and then east crossing the northern coast of Norway, over Sweden then south across Finland to Riga, continuing over German-occupied Russia and actually over the German-Russo front line to Yaroslavl and then south to the airfield of Ramenskoye, outside Moscow. The problem was to fly over German-held territory as far as possible in darkness and preferably in cloudy conditions. The first experimental flight left Prestwick on 21 October with Captain Percy in command, a crew of four and eight passengers plus mail and freight.

Although the temperatures experienced were extreme, the aircraft arrived at Ramenskoye at the time planned after a flight of 13 hours 9 minutes. Only one further round trip was operated by the end of the year, and altogether eight round trips had been completed successfully in spite of the hazards when the service on this routeing ceased in April 1943 with the onset of the long daylight hours in Northern latitudes.

It is worth noting that at 31 December 1942 BOAC had 113 aircraft in the fleet, twenty-two boats and ninety-one landplanes; of these seven were based at Hythe, three at Baltimore, twelve at Durban, nineteen at Whitchurch, forty-eight at Asmara, four at Cairo, six at Leuchars, twelve at Montreal plus one Ensign still at Bramcote in England under modification and one Lodestar at Vaalbank, near Johannesburg, for training - a sum of sixteen different types at ten widely separated bases. This gives some measure of the immense task BOAC faced in terms of personnel, maintenance of aircraft and day-to-day administration under the frustrating and often dangerous wartime conditions.

1943 - takeover fears
The year opened with BOAC's Management again concerned about the lack of suitable aircraft and the shortages of flying and technical staff. Although unofficial indications from the DGCA held out hopes of more and better aircraft - particularly Yorks DC3s - this could not fill immediate needs, particularly for crews where the requirement was estimated at twenty new crews per month. The sense of frustration felt by Management was expressed in a letter from the DDG to Air Marshal Tedder on 29 January. McCrindle pointed out that BOAC had to cope with many different types of aircraft, to use some bases, mainly inadequate, to take in untrained crews rather than use, as would be preferred, operationally tired type crews

First flown at St Louis on 26 March 1940 as NX19436/41-21041, this prototype 36-seater airliner was bought by the British Purchasing Commission in America. Registered as G-AGDI in September 1941, it was ferried to Prestwick by Capt A. C. P. Johnstone in 9.40 flying hours on 12 November. It entered service with BOAC as a 24-seater with long-range tanks for the long haul routes. Named *St Louis,* it received a C of A on 22 November 1941, and is known to have been used on the Leuchars to Stockholm service, and on the UK-Gibraltar-Malta run between May and September 1942. *(author's collection)*

from the RAF. Moreover, too many of BOAC's technical and administrative staff were on loan to MAP and elsewhere. It was also felt by the Board that they were not adequately informed about the intentions and requirements of the Government nor were they being given responsibility for carrying out specific orders, and that the existing procedure where all matters had to be channelled through the Department of Civil Aviation caused delays and waste of effort. Accordingly, Clive Pearson wrote to the Secretary of State for Air, Sir Archibald Sinclair, on 22 February setting out the views of the Board and making suggestions for specific changes in procedure. The letter also asked for the Board to be informed whether it was the intention that the Corporation should remain the sole British instrument for overseas air transport.

The Secretary of State did not reply, but a meeting was arranged on 1 March between him and Clive Pearson at which the latter was told of the intention to set up an RAF Transport Command, but no answers were received to the Chairman's points in his letter. On 10 March Clive Pearson wrote to Harold Balfour, the Under Secretary of State, attaching a memorandum drawn up by the DG, setting out the principles which in the view of BOAC should govern the relationship between Transport Command and BOAC, and made the point that BOAC should be directly responsible to the Secretary of State in such a position that it would not become subservient to Transport Command. The Board was seeking an assurance that BOAC would be given the specific task of operating all regular trunk services although willing that such a task would be '...*subject to political and military requirements*'.

On 11 March the Minister informed the House of Commons of the decision to establish an RAF Transport Command and had a statement drawn up setting out the relationship between this and BOAC. This did nothing to allay the fears of the BOAC Board that they would be subservient to the C-in-C Transport Command and on the 19 March Clive Pearson wrote to the Secretary of State regretting that no definite role had been allotted to BOAC thus in no way alleviating '...*the difficult conditions in which the Corporation has operated throughout its existence*'. The entire Board - except d'Erlanger - tendered their resignations, which the Minister accepted on 20 March. The DG offered to carry on as Chief Executive until a new Board could be appointed and this the Minister accepted. d'Erlanger was absent when these decisions were being made but on his return wrote to Clive Pearson that in his view the arrangements with Transport Command should be given a trial and it was not his intention to resign. Suspicions that a takeover by the new Command were rife in BOAC.

On 24 March the Minister announced to the House the resignation of four out of five of the Board Members and tabled a White Paper setting out the points of difference. He also announced the appointment of a new Board with Sir Harold Howitt as temporary Chairman and Simon Marks and John Marchbank in place of the members who had resigned. On that date, the old Board as their final word set out their views emphasising especially the lack of provision by the authorities for long-term post-war matters.

Above: Sir Harold Brewer Hartley KVO, CBE. Hartley had been chairman of Railway Air Services and later moved over to BOAC and BEA.

Below: Major John Ronald McCrindle started in the airline industry with Hillman Airways, but worked for many years with BOAC. *(both author's collection)*

The senior management of BOAC was alarmed at the way things were going and through Dennis Handover, the Traffic Director, a signal was sent to the Minister requesting the return as Chairman of Sir John Reith. The signal was later withdrawn, and Handover left BOAC to join SAS.

The new Board accepted the Minister's conditions in respect of cooperation with Transport Command. Alan Campbell-Orde was nominated as BOAC's liaison with the Command. On 3 May the C-in-C Transport Command issued an order of the day on relationships with BOAC, while Sir Harold Howitt expressed his conviction of the Command's goodwill towards BOAC.

On 26 May the Secretary of State announced that Viscount Knollys would be the new Chairman of BOAC in place of Sir Harold Howitt who would now become Deputy Chairman. He also announced the appointment of Brigadier-General A. C. Critchley and Pauline Gower, head of the Women's Section of the ATA as Board Members.

Critchley was to be the DG in place of Runciman.

Edward George William Tyrwhitt Knollys, 2nd Viscount Knollys was born on 16 January 1895. He was the son of Francis Knollys, 1st Viscount Knollys and Hon. Arden Mary Tyrwhitt. Educated at Harrow School, London, he fought in the First World War gaining the rank of Captain in the 16th Battalion, London Regiment and later the Royal Air Force. He was director of Barclays Bank and succeeded as the 2nd Viscount Knollys of Caversham on 15 August 1924. He held the office of Governor and Commander-in-Chief of Bermuda between 1941 and 1943 until being appointed chairman of BOAC.

On 28 May, Critchley took the chair at the Executive Committee meeting for the first time although he was present at the previous meeting where it was reported that BOAC was to be given the same priority as Transport Command for all spares. Liaison with Transport Command continued to improve as the earlier fears were allayed. On 11 August, Critchley told the Executive Committee that he had met with the C-in-C Transport Command who confirmed willingness for BOAC to continue to operate all main trunk routes. However, to do so it would be necessary for certain flying and ground staff to be in RAF uniform. It was therefore agreed that on the main military trunk route from the UK through the Mediterranean to the Middle East and on to India BOAC crews on Sunderlands should be in uniform and the aircraft have RAF markings; in due course, the Command issued a directive as to the control and operation of the UK-India route.

In June Critchley wrote to the DGCA suggesting the transfer of BOAC's landplane base at Asmara to Almaza, Cairo, because with the opening up of the Mediterranean and North African air routes, Cairo had become the hub of these routes and the RAF was a main source of supply for landplane spares and equipment; furthermore the division of effort between Asmara and Cairo hindered efficiency because of the very limited number of administrators and supervisors available to BOAC. These arguments were accepted by the Air Ministry, and Almaza became the chief overseas landplane base.

The Mystery of Leslie Howard
On 1 June 1943, A BOAC DC-3 G-AGBB, the former KLM PH-ALI operating under the designation Speedbird 777 was shot down by eight German fighters over the Bay of Biscay. There were no survivors among the seventeen passengers and crew on board.

Amongst those was British actor and celebrity film star Leslie Howard, most famous for his role as Ashley Wilkes in the 1939 blockbuster movie *Gone With The Wind*, though he also played opposite Ingrid Bergman in the 1939 romantic film *Intermezzo*. BOAC Flight 777 was unarmed, and the attack by the Luftwaffe was regarded as a violation of the laws of warfare.

The flight was going from Lisbon, Portugal to the BOAC's wartime terminal at Whitchurch, near Bristol. Leslie Howard was extremely prominent in the British film industry, making and promoting pro-Allied films and propaganda in Portugal, having been invited there by the British government to promote a film called *The Lamp Still Burns*. He was on his way back to England from this tour. Other passengers on board the aircraft included Wilfrid B. Israel, son of a German department store magnate, who had spent most of the war trying to get Jews out of Germany. Legend has it that Howard was not supposed to be flying that day, but the day after. Impatient to get home to see his family, he had pulled strings to get on the flight at the last minute, and a seven-year boy and his companion were asked to vacate their seats. One of those was a young boy, Derek Partridge, who in his adult life went on to appear on the original *Star Trek* show and narrated a documentary

Above: Before the war, Pauline Gower was a barnstorming pilot. She was appointed head of the Womens Section of the ATA, and a BOAC board member.

Below: Edward George William Tyrwhitt Knollys, 2nd Viscount Knollys, was the chairman of BOAC between 1943 and 1947. *(both author's collection)*

about Leslie Howard's life. Partridge was later to recall: *'We were on the plane waiting for take off. Some crew members came to the door and asked us if we would please vacate our seats because two VIPs needed to travel urgently. We were escorted back to the terminal. From there I remember watching the plane taking off.'* Partridge could not believe how lucky he was that Leslie Howard took his place. *'He saved my life that day.'*

Flight 777 took off from Lisbon at 7:35 AM, five minutes late due to the tardy delivery of a package to Leslie Howard. A bit more than three hours later, over the Bay of Biscay, eight German Messerschmitt fighters appeared and began firing at the DC-3. The Dutch pilot radioed the ground that he was being followed by *'strange aircraft'* and then that cannon tracers and shells were ripping through the fuselage. His last words were, *'Wave-hopping and doing my best'*. One of the airliner's engines was severely damaged in the first salvo. One of the German pilots years later reported seeing three people try to bail out of the crippled aircraft, but they were on fire and had no chance of survival. The aircraft crashed into the water and sank. Its remains have never been located.

There were numerous theories as to why the Germans shot down the DC-3. Possibly the most famous one was that German intelligence believed Winston Churchill was on board. Churchill spent the end of May 1943 in North Africa conferring with General Eisenhower. Rumours of a portly man in a hat with a cigar boarding Flight 777 might have been misinterpreted as a suspicion that Churchill was on board. Indeed, rumours soon began to circulate that the Luftwaffe had targeted the airliner because they believed that Churchill was on it. The British prime minister was indeed supposed to be returning to the UK from Lisbon that day, on a later flight. But it doesn't seem likely that agents from the Abwehr, German military intelligence, could have confused the two flights – not least because Churchill would not have taken an unescorted scheduled flight at that stage in the war.

There was a suggestion that one of the Germans thought his Howard's manager, who was accompanying him, was Churchill – because he looked quite like him, a bald, pudgy man who smoked big cigars. Adding to the mystery is the speculation that Bletchley Park - the central site for British codebreakers during World War Two - may have intercepted Luftwaffe plans to attack the civilian plane, but not warned the airline, to avoid arousing German suspicions that their Enigma coding machines had been cracked. Another theory is that Leslie Howard was a spy and that he was the Germans' intended target. It is well-known that Howard was deeply involved with pro-Allied propaganda, and he did insult Nazi propaganda minister Joseph Goebbels. The question of whether Howard was a spy, though, is less clear. He may have passed intelligence information to the governments of Spain and Portugal. Several books written about him state that he was the target of the attack and that the Germans thought he was a spy; not all endorse the conclusion that he was so that the Germans might have been mistaken.

These theories are exotic, but the truth is probably more prosaic. In the 1980s and 1990s, several historians tried to track down the surviving German pilots of the fighter group that downed the aircraft. At that time several were still alive, and to a man, all insisted that they didn't know about the BOAC flights and assumed when they saw the aircraft, which it was a British warplane. Only after they'd already fired did they realise it was a civilian flight, and they were shocked that their superiors hadn't warned them about the

Left: Leslie Howard Steiner (*b*. 3 April 1893, *d*.1 June 1943) English stage and film actor, director, and producer.

Below: C-47 G-AGKN in full wartime markings.
(both author's collection)

G-AGDD *Loch Losna* was a Lockheed 18-10 Lodestar which formed part of the L-Class series. In BOAC service the tyype was usually flown by Norwegian crews from Leuchars. *(author's collection)*

possibility of this happening.

The Brabazon Committee.
Mention must be made of the Committees set up by the Secretary of State under Lord Brabazon of Tara to recommend what new types of civil aircraft were required for the post-war period.

The first Brabazon Committee was small, consisting solely of Lord Brabazon and five officials from the Air Ministry and the MAP. There were no representatives from the manufacturers, or BOAC, although Alan Campbell-Orde, BOAC's Operations Director, was invited to sit in as from the second meeting. The Committee produced its report on 9 February 1943. In its report, the Committee recommended that immediate action should be taken on the design of five post-war types:-
1. A multi-engined pressurised landplane for the North Atlantic.
2. A medium-sized unpressurized twin-engined landplane for European and feeder services.
3. A four-engined landplane for Empire trunk routes.
4. A jet-propelled mailplane for the North Atlantic.
5. A small twin-engined landplane for internal services in the UK, the Dominions and Colonies.

The Committee also recommended there should be conversions of military types, specifically Yorks, Halifaxes and the Shetland flying boat and that these types should be used during the interval of three to six years between design and production of the new models.

BOAC was in general agreement with the findings of the first Brabazon Committee as to new types which had been listed in such a general way that *'they are likely to cover any specification we may desire to submit'*.

To progress the conclusions of the first Committee, a second Committee was set up under Lord Brabazon on 25 May 1943 and first met on 2 June. Its membership was an improvement on the first and included representatives of the manufacturing and operating sides of the air industry as well as of the shipping world. McCrindle and Campbell-Orde represented BOAC and Captain Geoffrey de Havilland the manufacturing interests, with Major Thornton from shipping. Apart from the Chairman, Sir William Hildred CB, OBE was the only member from the first Committee.

It was not until this second Committee met that the Corporation learnt officially of the first Committee's recommendation that converted military types should be used for civil air transport for up to six years until production of the new types came on stream. This so concerned the Corporation that at the first meeting of the second Committee the BOAC representatives protested that it would be at a serious disadvantage compared with its competitors who would be using American civil aircraft. They told the Committee BOAC should be able to buy American types to cover the gap until the new British

BOAC had a total of fifteen Armstrong A.W.38 Whitley Vs on their books; G-AGUY shown here. Some of these were used on Scandanavian services for a short time, but due to their poor performance and weight-lifting ability they were soon withdrawn. *(author's collection)*

aircraft were available. Lord Brabazon commented that a policy of purchasing American aircraft would not contribute to the problem of finding a use for British productive capacity after the war and BOAC would have to make any representations to the Air Ministry.

BOAC's attitude had been supported by one of the conclusions reached by the first Committee which stated *'this country is far behind America in the design, production and operation of transport types. Converted bombers will never compete successfully'*.

The possibility of the purchase of American stop-gap aircraft was being considered as BOAC was asked by Lord Brabazon to say if it would purchase any of a list of American aircraft put before the Committee - subject of course to Government approval. The American types preferred by BOAC if the purchase was allowed the Lockheed Constellation and the Douglas DC4.

By the autumn of 1943, BOAC felt the time had come to summarise its post-war aircraft needs both to the Air Ministry and to the MAP. Accordingly, a memorandum was submitted by the Chairman attached to his letter of 21 October to Balfour listing the following types:

1. North Atlantic non-stop:	all-up weight approx 150/175,000 lbs Type A
2. North Atlantic with stop: South Atlantic Australia and South Africa Express	all-up weight approx 100/110,000 lbs Type B
3. Other Empire Services: Long European Services	all-up weight approx 70,000 lbs Type C
4. Medium and short: European services:	all-up weight approx 40,000 lbs Type D

In setting out a summary of its needs, BOAC stated that it aimed to obtain aircraft suitable for the above purposes in three to four years' time but at the same time, experimental designs ought to proceed with the aim of producing aircraft in perhaps seven years ahead of anything else available. BOAC summarised the situation at that date in practical terms for each of these types. Type A, if it were to be the massive Bristol project as per Brabazon Type 1, must be regarded as an experimental aircraft ready for service within five to seven years. Type B, perhaps the one most urgently needed by BOAC, could not be in service before 1947 or 1948. However, if there were to be a developed York with pressurisation, this might cover the need. The Committee had not yet approved type C, but it could be that the existing type of York might temporarily fill the gap; while Type D had not yet passed through the Committee.

BOAC drew both Ministers attention to the critical post-war period, the gap between 1945 and 1950, warning that unless it could count on a long-range pressurised York for type B and if no manufacturer was available to build type A on conventional lines, it would have to seek permission to procure American aircraft. BOAC regarded the vast Bristol project as a valuable experimental type for later use: indeed, BOAC had severe doubts about its practicability and made this clear in correspondence and at meetings with the Air Ministry. In writing to the Under Secretary of State on 28 October 1943, Knollys, while promising full support for the project, again hedged by commenting *'we should always feel it our duty at any time to ask for... The type of aircraft which we as users consider the best suited for operations on the North Atlantic route in keen competition'*.

The Air Ministry's response to BOAC's list of aircraft needs was to repeat the Committee's recommendation that

A BOAC stewardess stands at the entrance to De Havilland DH.91 Albatross, G-AFDK *Fortuna*, to greet boarding passengers, at Whitchurch, Hampshire. Clearly taken during wartime - note the aircraft windows have been 'whited out' to prevent passengers seeing any secret installations they were flying over - the picture is thought to have been posed for the camera. *(author's collection)*

converted military types, the Yorks, the civil Halifax - the Halton - the Shetland and the Sunderland must be used to fill the time gap, and that for the North Atlantic studies were in hand to see if a pressurized Lancaster IV conversion - subsequently named the Lincoln - could be produced to bridge this gap, which would give birth of the Tudor.

Increasing the service.
Some steps were taken on the internal organisation by the new DG, within BOAC. He brought into the Corporation several senior personnel who had been in the RAF with him, causing concern among existing senior staff. On the other hand, BOAC was short of administrators, and some infusion of new blood was inevitable. An Air Safety Committee was set up, a Medical Service (in cooperation with Transport Command) was put in place, and in December the Chairman established the Chairman's Planning Committee with a range of sub-committees on post-war planning as this became more urgent. During the year there was a gradual move back to London headquarters by the staff at Bristol and, as the possibilities of more normal operation began to appear, a Commercial Director, R. D. Stewart, and a Traffic Director, John Brancker were appointed.

BOAC's operations continued to be influenced by the changing fortunes of the war fronts. The year brought further easing of the dangers through the Mediterranean as the Eighth Army cleared the North African coast of enemy forces and there was a greater need for communications with the USSR and for a build-up of capacity from Britain and the Middle East to India to help in the struggle against Japan. There was also the need, by some means or other, to re-establish the link with Australia and New Zealand, broken when the Japanese took Singapore and the Dutch East Indies. The Allied successes in North Africa opened up a new area where fast communications with Britain were required, and BOAC was called on to introduce new services to French Morocco and Algeria.

There were important additions to the fleet during the year in the shape of seven DH Mosquitos, three more Liberators, five more Hudson VIs, four more Lodestars, twelve Sunderlands and above all twenty-five Dakotas. Thus once again the number of different types of BOAC's fleet increased, so aggravating even more the airline's problems of crewing and maintenance.

On the UK-West Africa route the main effort early in the year was again provided by the flying boats, the three Boeing 314As, the G Class *Golden Hind* and the Catalinas, but one Catalina crashed on training at Poole, and as Sunderlands became available, they replaced the G Class and the remaining two Catalinas. It had been intended to step up capacity by the use of a Liberator, but only two round trips were completed before the aircraft was diverted to the UK-Cairo route. As usual, the Boeing 314As were re-routed from October during the winter and provided only southbound outlets to Lagos; at the same time the Sunderlands were needed to move the Christmas mail to the troops in the Middle East and were placed on the UK-Cairo route through the Mediterranean. This left the West African course with no northbound outlet from Lagos to the UK. To maintain the West African outlet in the meantime, RAF Transport Command provided a service between Lagos and Rabat to make the connection at Rabat with the new BOAC service with Dakotas between Rabat and the UK.

The main burden of the trans-African service was borne by the Ensigns now that all nine aircraft were available in the Middle East; a wide variety of routeing was followed according to priority demands, but a typical service would be Cairo-Wadi Halfa- Khartoum-El Fasher- El Geneina - Maiduguri - Kano - Lagos - Accra - Takoradi. Sabena continued to fly on charter to BOAC from Cairo through the Belgian Congo to West Africa with increased frequency, supplemented by flying boats at a reduced rate from September onwards to enable the Cairo-India frequency to be doubled. As with other services, the Middle East communications flights were influenced by the war situation. The advance of Eighth Army to Tunis had been so rapid that the need for trips to the Western desert practically ceased in June. Cairo was becoming important in BOAC's plans, with operational staff moving there from Asmara by January and before long the overhaul and maintenance base at Asmara would move to Cairo. With the fall in importance of Eritrea as an alternative war base, Port Sudan ceased to be a major gateway for supplies coming in by sea, as ships could now use the Mediterranean. Consequently, except the Asmara - Addis Ababa service with Lodestars, all the remaining flights originated in Cairo and greater emphasis was now put on the connections with Turkey and especially with Teheran as a supply route to the USSR. The Lodestars dominated these routes as the Ensigns did the trans-African route; as for the disappointing DH.95s, a third aircraft crashed at Asmara in February, and the remaining aircraft continued flying between Cairo and Asmara and Cairo and Adana until August but were then taken out of service.

On the direct UK-Cairo route, Liberators from the West African route were brought into service, being transferred there in February and placed on a routeing Lyneham- Lisbon - Gibraltar - Algiers - Cairo, but in June with Libya cleared of the enemy Castel Benito (Tripoli) took the place of Algiers and this service continued throughout the remainder of the year, occasionally on a faster routeing omitting Lisbon and Tripoli. This landplane operation was augmented from October to the end of the year by Sunderlands, also transferred from the UK-West African route, to take out Christmas mail for the troops in the Middle East and India. When this task was completed, the Sunderlands operated a through service from Poole on a Mediterranean routeing to India continuing from Cairo via Habbaniya, Bahrain and Jiwani to Karachi. One Sunderland was lost near Sollum, the second to be lost in the year, one having crashed in Eire in July.

The Horseshoe service remained unchanged throughout the year, operating at a twice-weekly frequency with the C-Class boats from Durban, but still being unable to fly east of Calcutta.

The capacity between Cairo and India was increased by the end of the year. In addition to the Horseshoe flights, there were extra C-Class trips between Cairo and India from the summer, and also the Sunderlands from the UK to India from late December. The landplane service from Cairo via the Persian Gulf was carried on from the previous year with Wellingtons and later with Lodestars, as also was the Hadramaut Coast operation although this now commenced in Cairo through Port Sudan to Asmara and then on the same routeing as previously. The Wellingtons found to be unsuitable for this route, were returned to the RAF and replaced by the well-tried Lodestars.

Following the successful invasion of Madagascar, the C-Class boats were tasked to run a regular service as a spur from the African section of the Horseshoe route between Kisumu and Diego Suarez, which added to the strain on the crews and the C-Class fleet.

Ever since the Horseshoe route had been broken in February 1942 and Australia and New Zealand cut off from air communication, there had been an outstanding priority requirement to fill the gap and provide an outlet for urgent Government despatches for the Australian and New Zealand authorities. The only aircraft type capable of flying the immensely long stage between Ceylon and Perth in Western Australia, over 3,500 miles, was the Catalina and then only with extra tankage fitted in part of the cabin space. Initially, the RAF flew seven experimental flights, but when QANTAS crews had been trained, the operation was handed over to BOAC with QANTAS as its agent doing the flying. The first civil flight left the flying-boat base at Kogalla in Ceylon on 10/11 July under the command of Captain Russell B. Tapp and arrived safely at Perth after a marathon non-stop flight of 27 hours 50 minutes. There were no alternates the aircraft could use in an emergency - the only possibility being the Cocos Islands but these were vulnerable to Japanese attack. Initially only one passenger could be carried after allowing for the weight of the higher priority diplomatic mail, but eventually, the passenger load was increased to three. Four Catalinas were made available by September, and a thrice-fortnightly service then became feasible. To avoid delay in getting urgent despatches across India and Ceylon to connect there with these flights, the Catalinas were extended from 30 October up to Karachi to make the connection with the main Empire services.

After a long period of uninterrupted operations between Whitchurch and Lisbon, KLM suffered a blow when on 1 June one of its aircraft was shot down by enemy aircraft in the Bay of Biscay. Frequency suffered as a result, and BOAC introduced its own Dakota service to make up for the loss of capacity. KLM had been extending two of its services per week to Gibraltar, but with BOAC's service now operating, this extension was withdrawn. At the request of the RAF, BOAC extended the Gibraltar service to Fez in French Morocco during the year, substituting Rabat for Fez as the wet season approached. A second operation commenced in November with a new service from Whitchurch to Gibraltar and thence to Algiers. Dakotas were used on both these North African operations.

It had been hoped to start a regular service to Moscow during the year, but Government talks with the USSR made little headway towards an air agreement. Consequently, British flights continued on an ad hoc basis; up to April six of these were operated on the direct northern route from Prestwick across Norway to Moscow; after that Lyneham became the UK departure point and a Mediterranean routeing via Gibraltar, Algiers or Tripoli, Cairo, Habbaniya, Teheran to Moscow was adopted. Several special flights carrying VIPs were carried out successfully.

G-AGFV was the first civil example of the DH.98 Mosquito and is seen here at Hatfield waiting delivery to BOAC. This aircraft made the first flight on the Leuchars - Stockholm run on 4 February 1943, and was the machine in which Captain Gilbert Rae (left) forced landed at Barkaby, Sweden on the night of 22/23 April 1943 after being attacked by German fighters. Captain Rae was killed when another BOAC Mosquito, G-AGKP crashed into the sea off Leuchars on 19 August 1944. His radio Operator D. T. Roberts and passenger Capt B W B Orton were also killed. *(DH Hatfield)*

Above: Mosquito HJ720 - now marked as G-AGGF taxies out at Leuchars ready for a flight to Stockholm early in 1943. The two lines of dark writing at the base of the fin states the ownership 'British Overseas Airways Corporation'.

Right: How the lucky travelled by BOAC - a priority passenger in the converted bomb bay of a Mosquito, with his own oxygen supply and an intercom to the crew. *(both author's collection)*

The main weight of the North Atlantic operations fell on the Liberator services between Prestwick and Montreal, and nearly two hundred flights were completed in each direction. Normal routeing was Prestwick to Gander and thence to Montreal, but Goose Bay was used as an alternative to Gander when weather conditions required and as a refuelling point when routeings westbound were through Iceland. A few westbound flights during the late winter were routed through the Azores thence to Gander and on to Montreal. With the formation of RAF Transport Command in March 1943, the operational control of these Return Ferry Flights passed to them from the RAF Ferry Command. It was during the summer of this year that Trans Canada Airlines, who had been so helpful to BOAC over the servicing of the BOAC Liberators at Dorval, came on to the route with a converted Lancaster bomber with the object of carrying mail from the homeland to the Canadian troops in the European war theatre and so laid claim to a post-war Atlantic operation.

The remaining Atlantic capacity was provided by the three Boeing 314As flying boats between June and October when they were returning to their Baltimore base via Botwood. One notable change in their operations involved their flying into and out of Poole, made acceptable by the improved war situation. From October the winter route again had to be used via the South Atlantic.

Throughout the year the shuttle services from the UK continued to carry loads for transfer to the PAA and American Export Lines flying boats at Shannon. These shuttles were flown out of Poole by *Golden Hind* and from mid-August on by *Golden Hind* and Sunderlands, and out of Whitchurch by Frobishers, Whitleys and Dakotas to the landplane aerodrome at Rineanna. With the Boeing 314As no longer needing shuttles, the requirement was reduced and the landplane shuttles were terminated in the summer. The Whitleys, again being found unsuitable, were returned to the RAF, while the structure of the DH.91s had deteriorated so much that they were no longer airworthy. In fact the flight of *Fortuna* from Whitchurch to Rineanna on 16 July was more than eventful. On approaching the Irish airfield the aircraft lost a flap and crash-landed, fortunately with no serious injury to crew and passengers who were members of BOAC's new Board only recently appointed. That was the final chapter in the life of the DH.91 which had done sterling service since the outbreak of war. The remaining two aircraft, *Falcon* and *Fiona*, were scrapped in September.

As in previous war years, the whole of the undertaking of the Corporation remained at the disposal of the Secretary of State for Air during 1944, but increasing revenue use was being made of the capacity not taken up by priority loads. As an end to hostilities seemed to be drawing nearer some steps were taken by the Board to prepare for post-war operations. On the important subject of planning for post-war aircraft and routes, a great deal of preliminary work was done internally especially for use at the Brabazon Committee. It was becoming clearer to the Board during the year as a result of planning studies that in the years immediately following the end of the European war available landplanes were likely to fall into four main periods, the first from end 1944 to 1945 where BOAC's requirements would have to be met by Lancasters and Yorks, the second from end 1945 to 1949 when Avro Tudor Is and IIs, still only developments from the military Lancaster Mark IV, should be available, the third and fourth covering the years 1949 to 1954 would be met by suitable genuinely new types based on civil needs coming out of the Brabazon Committee's recommendations, which

Dak at The Rock! A dramatic night-time shot of a BOAC Dakota staging through the airfield at Gibraltar. *(author's collection)*

included a proposed gas turbine or jet propelled aircraft.

Steps were taken during the year to improve the effectiveness of management. A Board Development Policy Committee was set up to receive recommendations and information from the Chairman's Planning Committee, to oversee future developments in respect of aircraft and routes; a Board Staff Committee held its first meeting on 27 January. In April a Test and Development Flight was created with Government encouragement, and a Lancaster aircraft was allocated to test out the potential of Rolls-Royce Merlin engines for eventual civil use. In March a lease was signed on accommodation in Berkeley Square for Head Office, to be known as Airways House, the former Airways House in Buckingham Palace Road being renamed Airways Terminal. There was a welcome return to the use of the Speedbird as the official emblem and the use of the full name British Overseas Airways Corporation.

As an indication of changes to come, Critchley announced the appointment of Air Marshal Sir William Welch as Regional Director for European and Home Stations; this was well-timed as June saw the formation of 110 Wing to RAF Transport Command especially for operating services into liberated Europe; the Wing would be completely militarized, although some BOAC staff were to be seconded to it. Eventually, it was the intention to transfer operations to BOAC. By the autumn 110 Wing had already commenced operating out of Croydon to Paris, Lyons and Brussels. Co-operation with Transport Command was now good and where important joint load control centres had been set up with the two staffs working side by side.

In March BOAC was relieved of the responsibility for running the aircraft repair and overhaul base at Heliopolis. By agreement, it was handed over to the RAF. The provision of an adequate landplane base in England was still exercising the minds of the Chairman and the DG, and serious concern was expressed about conditions at Whitchurch and Lyneham. Although Hurn would be available towards the end of the year and thus make feasible some co-ordination with the flying boat base at Poole, BOAC had its eyes firmly on the airport to be developed for the RAF at Heathrow, close to London; in April it was agreed that the DG should submit the Corporation's proposals to the Government. In June the Chairman had to tell the Board that construction of a military aerodrome at Heathrow had actually begun, although BOAC had not been consulted on whether it would be suitable for adaptation to post-war civil needs and it soon became clear that the Air Ministry runway plans cut across those submitted by BOAC's experts. There were two important items in October affecting policy. On the 6th, the Minister issued a new Directive to BOAC under which BOAC would be notified of the services and frequencies it was to operate by the Secretary of State for Air and the aircraft it was to use. In the Middle East BOAC was to follow instructions issued by the Middle East Air Transport Board, thus eliminating some of the time wasting caused by referring back to London for authority. The second item was a fundamental change in the Government's control of BOAC and domestic operators in that on 12 October the Government announced the appointment of a Minister for Civil Aviation, Viscount Swinton. Henceforward the new Minister would oversee the activities of BOAC in place of the Secretary of State for Air.

By now Africa had been cleared of the enemy and the emphasis switched to the three fronts: Europe, the East and Russia. The Mediterranean was at last free from enemy intervention and was becoming once more BOAC's main highway to the East, taking over from the indirect route over West Africa. Supplies of men and material were needed in the Middle East, mainly to link with the Russians over Teheran, and in India, for the build-up to throw the Japanese back from Burma and Malaya. This required BOAC to concentrate its provisioning of capacity on direct routeings between the UK and the Middle East and India and to reduce its effort to West and North Africa and on the trans-African routes. Direct landplane operations between the UK and Cairo with Liberators continued from the previous year as the proposed service through to Moscow for which these aircraft were earmarked did not materialise. By June these Liberator flights ceased to be ad hoc and were scheduled on a regular thrice-weekly basis through Lisbon, Gibraltar or Rabat thence via Castel Benito, Tripoli. Flights commenced at Lyneham but from 1 November BOAC's new landplane base at Hurn, near Christchurch, replaced Lyneham. In April it became possible to increase the capacity for Cairo with the introduction into BOAC's fleet of the much-awaited Yorks, five of which were delivered to BOAC between January and September. The first flights left the UK for Cairo on 23 April with ad hoc flights to follow, but by November a regular service was mounted on the same routeing as the Liberators; by December the Yorks had replaced the Liberators which were withdrawn and returned to the RAF. A further increase in capacity in Cairo came in June when the Dakota service between the UK and Algiers was extended to Cairo; this too was transferred to Hurn from 1 November, so that at long last there were clear signs of real consolidation of home bases. Over and above all these services to Cairo, capacity was also being provided on Sunderland flying boats which had commenced a through operation between the UK and India via Cairo at the close of 1943. This was a militarised service with RAF roundels, and it became possible in May to extend some of the frequencies on from Karachi to Calcutta, thus restoring a direct UK-Calcutta link, broken as far back as 1940. At the close of the year, another milestone in recovery was passed when the Sunderlands were re-routed on the old direct track across liberated France, so eliminating the time-wasting detour via Gibraltar and releasing enough spare hours to enable the frequency to Karachi to be increased and to add extra capacity between Cairo and Karachi. On top of these operations between Cairo and India, the reliable C-Class boats maintained the Horseshoe service between Durban and Calcutta as did the weekly Lodestar flight from Cairo along the Hadramaut coast route to Karachi. There remained one more effort BOAC could make to the build-up in India, namely a new thrice-weekly service by Ensigns between Cairo and Karachi or Calcutta which started up in March and continued for the rest of the year.

Although the C-Class boats were unable to extend beyond Calcutta on the Horseshoe route, the link with Australia and New Zealand was assured by a continuation of the service flown by Catalinas between Karachi and Perth via Colombo. Capacity was minimal and inadequate to cope with the priority demands for space; the answer came in the use of Liberators which were based at Colombo to fly Colombo-Perth starting in June. All these Australian operations were manned by QANTAS crews.

Also by the end of the year at the request of the military authorities in East Africa and in Ceylon a link was forged between East Africa and Ceylon by *Golden Hind* flying from Kisumu via Mombasa and Seychelles and so opening up a connection between East Africa and Australia. After nearly four years of the highest priority, the route to the Middle East from the UK over West Africa and across Africa to Khartoum and thence Cairo had now been overtaken by events. Already at the close of 1943 only the Boeing flying boats provided the UK-Lagos link and that only in a south-bound direction. They continued to the end of April when they were concentrated on the North Atlantic and so ceased serving West Africa. The connection between UK and Lagos was provided by a BOAC Dakota operation which routed via Rabat so that by October the existing terminal service at Rabat was discontinued. As compared with the two-stop schedule of the Boeings, the Dakotas from Whitchurch made eight and sometimes nine intermediate calls on a very long schedule through St Mawgan, Lisbon, Rabat, Port Etienne, Bathurst, Freetown, Abidjan, Takoradi, Accra to Lagos.

At the start of the year the trans-African route was being covered by Dakota and Ensign services between Cairo and Lagos or Accra, together with the C-Class service via the Congo and Lagos or Accra, and via the Congo plus the SABENA charters. This provided more capacity than was needed and in February the Ensigns were

G-AGJA *Mildenhall* was one of five Avro Yorks diverted to BOAC from RAF stocks. It was delivered to BOAC in January 1944 and made the first York service flight from Lyneham to Cairo on 22 April.

Getting airborne - a 'bare metal' Shorts Sunderland up on the step for another departure. *(author's collection)*

transferred to the higher priority Cairo-India route.

There were few changes from 1943 in the programme of services radiating from Cairo. The highest frequency was given to the Cairo-Teheran connection which was re-routed via Damascus and Baghdad instead of via Lydda and Habbaniyah at the request of the Syrian authorities. The only other changes of significance were the substitution of Ankara for Adana in November on the Cairo-Turkey service and also the call at Nicosia in Cyprus on the same service and the extension of the Cairo-East African flights in mid-summer to Gwelo to accommodate pilots under an RAF training scheme set up in Southern Rhodesia. All these services were operated with Lockheed Lodestars.

Operations to both Lisbon and Gibraltar were continued both by BOAC and KLM, the main purpose of the Gibraltar call being to transfer loads there on to the Sunderlands to the Middle East, thus getting round the payload bottleneck on the Sunderlands over the Poole-Gibraltar sector; with a routeing over France available to the Sunderlands in the winter the need for the Gibraltar extension by the landplanes was reduced. From 24 October BOAC began a new service with Dakotas through Madrid to Lisbon. This service commenced operations from Croydon although like other landplane services were transferred to Hurn from 1 November.

On the North Atlantic, it became possible to mount a considerable increase in capacity provided by the three Boeing 314A flying boats now that they were no longer required to serve West Africa and the frequency was stepped up to thrice weekly. Shannon was still the terminus for most of the year, but Poole was used once the area was free of the restrictions placed on it while the invasion build-up was proceeding. As usual, the Boeings reverted at the end of October to their winter routeing via South America to Baltimore and via Bermuda and Lisbon to Shannon or Poole. The American operators, PAA and the American Export Lines, continued to serve Shannon and loads were shuttled from and to the UK initially by Sunderland and G Class flying boats and by the Hudsons and Dakotas based on Croydon although for nine weeks during the summer there was a return to Whitchurch as Croydon was thought to be too vulnerable to flying bomb attack. On the Canadian route, frequencies operated were double those in the previous year with the majority of flights using the route through Gander or Goose Bay. Where conditions made it necessary, an extra stop would be made at Reykjavik in Iceland, or the southern route through the Azores and Bermuda would be followed. All flights were operated with Liberators.

The only other European service operated in addition to Lisbon and Gibraltar was the priority service to Sweden. This was increased considerably, and nearly 500 flights in each direction were completed, mainly by Dakotas, but the much faster Mosquito took over during the summer months of long daylight hours when the risk of enemy interception would be at its height. Operations were marred by the loss of two Mosquitos during that period, one in the sea off Leuchars and the other failed to arrive at destination. Both the Swedes and Norwegians were flying their services on the route now, although the former's operations were intermittent for reasons of security until in October they acquired American converted B-17 aircraft. This year saw the final closure of Whitchurch as departure airport for BOAC's flights overseas. However, with the opening of Hurn to BOAC in November 1944, all BOAC's landplane services (other than to Scandinavia) and KLM's were

BOAC had a Development Flight that used a number of military aircraft often converted to freighters. One such machine was Avro Lancaster I G-AGJI which was used by the flight from January 1944 until December 1947. *(author's collection)*

BOAC also briefly had a fleet of fourteen Vickers Warwick transports, including G-AGFK as seen here. Unfortunately conversion from bomber to transport was intricate and lengthy and the aircraft were not delivered until 1943. They were used to carry frieght, mail and passengers - in that order of priority - at Bathhurst across Africa to Cairo to avoid the fighting in the Mediterranean and Western Desert. The did not remain in BOAC service long, being transferred to the RAF Transport Command. *(author's collection)*

transferred there, Whitchurch then remaining as a maintenance base.

By late 1944, Whitchurch operation had solidified into what was known as No.1 Line BOAC, using a large number of what was termed C-47A or B aircraft, which the RAF called Dakotas. Mixed in were a few pre-war civian DC-3s . These aircraft were based at Whitchurch, near Bristol, but picked up passengers and freight from Hurn. No 1 Line BOAC operated mainly military flights to Cairo, with occasional extensions to Karachi and Lagos. It also ran a few civil services to Rineanna (Shannon, Eire), Madrid (Spain), Lisbon (Portugal) and Stockholm (Sweden), as those European countries had remained neutral through the war.

By the end of 1944, a number of RAF pilots had been seconded to No 1 Line BOAC to increase its size and services. By this time, the aircraft used could appear in one of two forms. For military service, they were in RAF camouflage with roundels and serials. For the civil flights, they appeared with British civil registration letters. Likewise, the pilots would wear uniforms to match. One day it was possible for a pilot to don his squadron leader's uniform, and the next wear the smart new BOAC dark blue with which he had been supplied. This uniform brought many unwarranted salutes from Navy-minded personnel in a period when civil airline uniforms were relatively rare!

The main No.1 Line conversion pilot was Captain Maurice Haddon, the Line Manager was Ernest Hessey, and the operations were in the safe hands of Captain Jimmy James, an elegant senior pre-war captain. Before any newly seconded pilots were let loose with passengers most of the RAF pilots was detailed to take an empty aircraft 'down the line'. Once such pilot was Tony Spooner, who was tasked with a Dakota to Rabat-Sale, Morocco, and back via Lisbon. On the return leg, he was astonished to see the very senior wireless operator calmly unscrew the emergency hydraulic panel and there conceal a couple of large bottles of Portuguese brandy. As Tony Spooner later recalled: *'I was not having this; I had no wish to restart my civil aviation career with a Customs and Excise charge hanging over me. It was, sadly, indicative of the change of spirit which was creeping over the country now that Germany was no longer a threat. The spivs and black marketeers were beginning to cash in, and the glorious mood of the country so evident during 1940-41 was fast becoming a memory.'*

On one of my first flights to Rineanna, the landplane base alongside the Shannon flying-boat terminal of the Pan American Airways and American Export Airlines (AEA) Services, one of our stewardesses underwent a strange physical transformation. No 1 Line could boast just two of BOAC's total of six stewardesses. They were, if memory serves me, Viva Barker and Helen Wigmore. Helen was a normal sized blonde, but during the Rineanna stop, there was time for a quick trip to Limerick, where the shops were full of goods that could only be obtained, if at all, with clothing coupons in the UK. I do not know how many layers Helen had underneath her very severe BOAC uniform, but it must have been plenty.'

Early in 1945 Croydon reopened and again became London's premier airport. It also became a sub-base of No 1 Line. All civil flights operated from there, but the military ones continued to pick up the passengers and depart from Hurn, the small grass field at Whitchurch being retained as the base airfield. Our military passengers were either a few civilian VIPs or relatively senior Service officers heading for the war in the Far East. On the return flights, sadly, there were liable to be severely wounded persons. The full route to Cairo was Hurn – Istres – Malta - El Adem – Cairo and then, if the flight were to continue to India, the stops would be at Baghdad West, Sharjah and Karachi.

Tony Spooner again: *'The crew consisted of two pilots, often both captains, and a wireless operator. Non-directional beacons using Bendix needles were the sole navigation aid other than map-reading. I cannot recall that we carried oxygen, but I do remember that we once took as a VIP the 20-stone-plus Aga Khan, and I had been advised that he did not like to go above 10,000 feet. As was quite usual in the Mediterranean, a storm lay across our path, and we had to find a way around or above it. At 18,000 feet I sent the Wireless Operator back to see how His Highness was taking it, and to assure him that we were still below 10,000 feet. He was quite happy, and advised that if he went above 10,000 feet, his heart would stop!*

1945 saw not only the end of the war in Europe but also unexpectedly of the hostilities against Japan. Even so, BOAC remained throughout the year under complete Government control, by the Secretary of State for Air up to the passing of the Ministry of Civil Aviation Act on 25 April and after that by the Minister of Civil Aviation

Avro Lancastrian G-AGLF in flight. It was a conversion of Lancaster VB873, and was handed over to the BOAC Development Unit at Hurn on 7 February 1945. *(author's collection)*

(MCA). The Priorities Board continued to have claim over all capacity produced with one exception; on the new landplane service to South Africa with York aircraft which commenced in November in partnership with South African Airways (SAA), BOAC was allowed partial control so as to be able to sell space to the public, this then becoming the first-ever commercial operation by the Corporation since its formation.

It was a difficult year for top management. Apart from keeping the airline functioning from day to day with a variety of uneconomic aircraft types from a number of scattered bases including the new landplane base at Hurn, much of the time of the Board and Management had to be spent on the pressing problems of post-war planning, on urging the Corporation's case wherever and whenever possible with officialdom steeped not surprisingly in the atmosphere of wartime conditions. It did not make it any easier that the first Minister to lead the much-welcomed Ministry of Civil Aviation, Viscount Swinton, only lasted a few months, being displaced by Lord Winster in July when the Labour Government came to power; so the Corporation's views had to be put across afresh. Nor did it help the airline to formulate plans when two Government White Papers on the organisation of British air transport appeared within nine months of each other, each putting forward radically different policies for BOAC's future!

Although there had been changes at Head Office when Brigadier-General Critchley came in as DG, including the innovation of three new posts of Assistant Director-General (ADG), namely Technical, Administration and Commercial - the basic Corporation remained much as it had been throughout the war with regional controls overseas. These had been streamlined with the disappearance of the Iraq and Persian Gulf Region into the Indian and the East African merged with the South African while in due course the West African Region was linked to Home Stations and ceased to be a Region itself. The regional set-up may well have been that suited to wartime conditions where communications with headquarters were often difficult and where decisions had to be made locally. With the approach of commercial operations and the need to examine technical possibilities of aircraft and equipment, it was decided to establish a Project Branch which would act as a channel between the Corporation and the aircraft industry regarding prototype aircraft, engines and equipment. Later in the year on the force of a case made by Alan Campbell-Orde, based on the need for strengthened resources to deal with the manufacturing industry, with the MCA and the MAP, and in addition to act as advisers for Commonwealth operators, the Board approved the formation of a Technical Development Department to include not only the Project Branch but also the Development Flight, all reporting to Campbell-Orde. This was a wise step since none of the Government Departments understood the needs of commercial operations after six years of war.

At the beginning of 1945 staff shortages was still a problem and the DG informed the Board in February that BOAC was still short of 42 crews and there was a serious risk of future services if this shortage could not be eliminated. Associated with this problem was the question of adequate aircrew training facilities, which still had not been acquired. Some improvement over the previous year had been achieved with the use of the RAF's 6 Lancaster Finishing School at Ossington, near Newark, but the rate of failures was extremely high. The situation was made far worse by the demands of the Air Ministry for the return of all seconded aircrew already in BOAC when their secondment dates ran out. A Director of Training was appointed in November, and with representations made to the Minister, and this problem was largely overcome early in the following year with the opening of a central training school at Aldermaston, in Berkshire.

The question of the use of stewardesses on BOAC aircraft was given consideration by the Board, but it was decided to defer for the time being their employment on long-haul routes with the possible exception of the North Atlantic but that on European flights when possible the policy would be to carry one stewardess for every three stewards and that all stewardesses should be given first-aid and nursing training.

With the start of the through UK-Australia Lancastrian service in May, there were exchanges between the Chairman and Sir William Hildred at the Department of Civil Aviation about BOAC's shareholding in QANTAS following the known desire of the Australian Government to make QANTAS an entirely Australian company: it was agreed on both sides that there would be no objection to the disposal of BOAC's holding as being a logical development if the formal request were made by Australia.

BOAC's objective - that of one main landplane base - remained unchanged. The plan was to use Hurn until Heathrow became available as a central base for Training, Development and Maintenance and Overhaul besides being the departure point for BOAC's landplane operations from London. The need for such a centralised point was becoming urgent as commercial operations drew nearer, particularly as costs would then be a major consideration. As to flying boats the objective was to return to Southampton as the departure terminal when this became possible, so in August the Board authorized the DG to place an order for two additional flying boat docks, although it was already recognized by the Corporation that flying boats would have to give way before long to the faster landplanes when overseas facilities for the latter were available. During the war years the boats built up a fine record on the Horseshoe route, and as the Deputy Chairman found on his tour to South Africa, there would be considerable resistance from local people to their eventual disappearance. Developments in the Government's policy towards civil aviation, the emergence of three airline corporations, the re-forming of international bodies concerned with aviation and BOAC's forward plans for aircraft linked with the Brabazon committee all gave indications as to the thinking of 'the powers that be'.

The emphasis in the provision of capacity was still on the routes into India and to Australia, and this tendency became even more marked after the defeat of Germany as the war effort concentrated on the Far Eastern front. The sudden ending of the war with Japan in August required a re-adjustment of plans to cater for returning forces personnel to the UK and caused a slackening in the eastbound traffic flows to India and the Far East although to a limited extent this fall off was offset by business traffic from the UK seeking to re-establish commercial links.

As to aircraft, apart from a few extra Dakotas, the main weight of increase lay in Yorks, a further eleven being handed over to BOAC making the fleet fifteen in all by the end of December, and in Lancastrians, a conversion from the Lancaster bomber, twenty-one of which becoming available over the year. The amount of extra passenger capacity provided by the Lancastrians was restricted because they carried only nine seats with six bunks above, the seats being arranged down one side of the aircraft facing the starboard windows.

Even though the war was over by the mid-summer, the whole of BOAC's capacity except services to South Africa continued at the disposal of the Government and only after all Government priority demands had been met was BOAC able to offer space to genuinely commercial load.

On the UK-Cairo route the Yorks continued to operate but by February had been extended from Cairo on to Karachi and by the end of the year to Calcutta, the reduction in the capacity for Cairo being then supplied by stepping up the existing Dakota frequency. With the more direct route across France and the use of Augusta, in Sicily, the Sunderland's capacity to Karachi and Calcutta could be increased, and the route was extended in October by a return to Rangoon, not served since February 1942, but now open again after the defeat of Japan. With more Dakotas available, a new service was begun in June from Hurn to Karachi via Marseilles, Malta, El Adem, Cairo, Baghdad and Sharjah, and this like the Sunderland service was a 'militarised' operation. In addition to these through operations from the UK, the Ensigns continued to fly between Cairo and' India supported by an extra C-Class flying boat operation from Cairo over the normal routeing followed by that section of the Horseshoe route; Karachi was the initial terminal, but the route was extended to Calcutta in October, and until they extended to Rangoon the Sunderlands were also able to provide extra flights over the Cairo-Karachi Sector. There was no change in the Horseshoe services which continued their fine record of dependability between Durban and Calcutta without any routeing changes and with a doubling up once a week over the Kisumu-Cairo sector.

Although priority demand between the UK and West Africa remained high, this was not so on the trans-African section, and the needs were met by the build-up of the

Curtiss C.46 *St Louis*, G-AGDI regularly transitted through Gibraltar on the UK-Gibraltar-Malta run between May and September 1942. *(author's collection)*

A line-up of BOAC Dakotas just at the end of the war. (author's collection)

Dakota services between UK and Lagos with a reduction across Africa and Cairo. Dakotas operating these flights were now based in the UK for the entire routeing through to Cairo. To help out over the heavier sector UK-Lagos, some converted Halifaxes were flown by BOAC for a limited period. Sabena, still on charter to BOAC, carried on with their trans-African services through the Belgian Congo, maintaining their unbroken record of operation since the difficult days of l940.

The most significant route developments in this year were the re-start of through flights between UK and Australia and between the UK and South Africa. The Australian services were resumed on 31 May with a Lancastrian from Hurn routed through Lydda, Karachi, Ratmalana (Ceylon), Learmonth (Western Australia) to Sydney on a flight time of about 66 hours, a great improvement in time on the ten days by C-Class flying boat before the war. The Lancastrian was extremely limited in carrying capacity, but at least it made the long sector over the Indian Ocean a possibility. The operation was a joint one, BOAC crews took the aircraft as far as Karachi, and the QANTAS crews then took over the Karachi-Sydney section. Although this new Lancastrian service had started in May, there was still a need for capacity which required continuation of the Catalina service but with more Liberators becoming available with a far higher payload, the Catalina operation ceased in July, and by the end of the year the Liberators had extended to Sydney.

The second important route development concerned South Africa. Following discussions in London in the summer between negotiating teams from BOAC and South African Airways, it was agreed to set up a partnership operation between UK and Johannesburg using York aircraft. This was inaugurated on 10 November being the first-ever landplane service between the two countries. The route followed was from Hurn via Castel Benito (Tripoli), Cairo, Khartoum, Nairobi, Johannesburg, Malta taking the place of Tripoli later on. In the Middle East BOAC with its fleet of Lodestars was still required to provide the range of communication flights based on Cairo. The only important change from the previous year was the cessation of the service to Teheran in October, now that Allied Forces had withdrawn from Persia; also by the end of September there was no longer a need to extend the Cairo-Nairobi service to Gwelo with the RAF flying training school winding up there; and finally the C-Class flying boat connection between Kisumu and Diego Suarez could be withdrawn.

On the North Atlantic, the Boeing 314As kept to the usual routeings but were now able to fly direct from Bermuda to Foynes omitting Lisbon. An innovation that winter was the start of a Baltimore-Bermuda shuttle to show the flag in competition with PAA. The Return Ferry Service between Prestwick and Montreal with Liberators was again increased in frequency, and a daily service became possible. Following the decision by PAA and the American Export Lines to fly into the UK instead of terminating at Foynes, there was no longer a requirement to shuttle loads over to connect, and the end of October saw the last of the shuttle operations which had been a regular feature since July 1941. In the autumn BOAC took on a new task to assist the East African Governments by operating a range of local services within the East African Territories pending the formation of East African Airways. Six DH.89s from RAF stocks were used for this job for later transfer to East African Airways.

In spite of the wartime operational problems, BOAC succeeded in keeping open the vital trunk routes linking the home country with the overseas Empire and in creating new links across the Atlantic Ocean with Canada and the United States and had pioneered new routes where ground facilities were non-existent. Moreover, the airline had not hesitated to fly unarmed into war zones. A comparison between its report to the Secretary of State for Air for its financial year 1940/41 and 1944/45 demonstrates the magnitude of the task completed. At the close of 1940/41, BOAC had 61 aircraft of 14 different types, by 31 March 1945 the fleet had grown to 169 aircraft of 19 different types. Capacity provided had increased from 8,641,926 ton-miles to 55,458,344 ton-miles over the same period.

Chapter Three

Planning for Peace

Sir John Reith, as Chairman of BOAC, called for the formation of an all-embracing Commonwealth airline in 1939 – a concept not proceeded with because of the outbreak of war, but one that received support from the then Air Minister and some Commonwealth sources. The idea remained, for the Under-Secretary of State of Air, Harold Balfour, in a speech at the Overseas Club in January 1942 called for the formation post-war of a British Commonwealth Air Council to plan overseas services; by September the Secretary of State for the Colonies was writing to the overseas territories for their views.

Little planning was done in 1942 by the Government although pressure to make some statement on post-war civil types built up both from questions in the House of Commons and from the Press. BOAC had set up a small post-war planning team under the Chairman, then Pearson, assisted by R. H. Mayo, the flying-boat designer, and J. B. Scott, and were expressing concern about the likely absence of competitive British civil aircraft when the war ended, mainly as the Americans were already operating the Douglas DC-4 by 1942 and were about to fly the Lockheed Constellation. Of course, the shipping companies had an eye to the main chance and sent a statement to the Minister of War Transport in August 1942 saying that sea and air should work together if subsidised air services were not to kill passenger traffic by sea. This led the Government to set up a committee under Lord Brabazon on 23 December 1942 to prepare specifications for post-war types.

During 1943 thoughts on post-war civil aviation began to occur in Government and BOAC, with a series of discussions, statements and debates taking place as to what should be the future shape of UK civil aviation in the world. This heightened interest added its strength to the fear of being left well behind the Americans who had set up in February 1943 a special committee under Adolf A Berle, the Assistant Secretary of State with the Commerce Department, and the Civil Aeronautics Board (CAB) to concentrate on post-war issues. BOAC were well aware of this and the DG, Leslie Runciman, made a reference during a talk at the Institute of Transport on 8 March to the strides being made by the United States and urged that as in America there should be no international barriers in Europe after the war. Already it was clear that shipping interests would make a determined bid to play a part and in reply to a question in the House of Commons in February as to whether the Minister would grant powers to assist shipping companies to operate air services.

During the Commons Debate on 14 April 1943 when the announcement was made of the Board changes in BOAC, the concept of a Commonwealth airline was revived and considered an interesting idea by the Secretary of State for Air, Sir Archibald Sinclair. In BOAC the post-war planning team had been strengthened when in May the DDG, Major McCrindle, was given the responsibility for thinking out post-war problems with the assistance of Mayo and Scott.

In addition the new Board member, Simon (later Sir Simon) Marks took an interest in the development work. Numerous statements were put forward pointing out the dangers of leaving the field to the Americans and appealing to Parliament to formulate a civil aviation policy. In October further s urging that there should be no monopoly

The fifteen seat Avro Lancastrian G-AGMG *Nicosia* surrounded by maintenance ladders. It was scrapped in 1951. *(author's collection)*

of civil aviation and that the shipping companies should share in the development appeared, with proposals for setting up an Empire Air Authority to coordinate and develop Empire overseas air routes.

Prior to a Commons debate on civil aviation on 1 June 1943, the annual Conservative Conference called on the Government to expedite the preparation and planning of civil flying and at the debate itself a motion signed by more than 140 members asked for an Empire Conference on civil aviation and its removal from Air Ministry control.

With Lord Knollys' appointment as Chairman at the end of May 1943, the tempo of post-war thinking in BOAC increased. He tabled a directive on post-war planning at the Board meeting on 10 August. He said its purpose was '...*to enable the Corporation to be in a position to answer certain questions on future policy which it was to be expected might be put to them by the Government at any time*'. It gave a clear lead to the planners as to what they should assume about the post-war scenario in compiling route and traffic studies and aircraft requirements. Some of these seem to reveal a somewhat idealistic view of the aviation world after the war; for example, the first was that there would be international freedom of commercial air transit through and commercial outlets in foreign territory and that there would be no cabotage. This seems strange in that British Colonial territory would provide immensely valuable exclusive rights for the UK and so for BOAC. The second assumption had the policy of the single chosen instrument in mind, a policy BOAC would be expected to favour, since it recognized the right of countries, other than ex-enemy, to operate international services '...*subject to the limitation on commercial outlets to one company nominated by each country*'.

The planners were instructed to assume that a kind of international civil aviation board would be established, having the power to define spheres of preferential operating rights depending on public need, and that a reformed International Air Transport Association (IATA) would control fares and rates and make inter-country agreements along the recognized lines of the shipping conferences. It was assumed there would be a close association with the Dominions and some sectionalization of routes regulated by an Empire Council - a reference to Balfour's earlier thinking. Surprisingly, in view of pre-war experience, an all-up mail scheme was envisaged, possibly with separate aircraft confined to mail carriage. Finally, the Chairman recognized that BOAC would have to use converted military aircraft until new civil types coming out of the

Below: a BOAC Lancastrian is prepared for a flight at Eastleigh Airport, Nairobi. The luaggage is being loaded into the nose cargo compartment. *(author's collection)*

Right: *Flight* magazine described the Lancastrian as '...*one of the very few satisfactory bomber conversions*'. The Lancastrian's slim fuselage provided just enough room for two rows of armchair type seats. *author's collection)*

By day and by night - sideways-facing bench seats in a row with their backs to the fuselage wall provided accomodation for nine passengers. Bunks folded down from above from the Lancastrian's roof for use on overnight sleeper services. (author's collection)

Brabazon Committee recommendations were available, although he included a rider that the possible purchase of American aircraft should be considered.

The Board accepted these assumptions, but agreed to the Deputy Chairman's suggestion that they should be amended to provide, as he put it, *'that crazy competition would not be permitted'*.

On 3 July 1943 Sir William Hildred wrote from the Air Ministry to McCrindle asking for a list of BOAC's suggested post-war routes, adding a comment *'I do not know, but you must assume for the present that you will be the sole chosen instrument for overseas operations and plan accordingly'*. In the Lords Debate on 13 July, Lord Londonderry complained that there was no Government statement of policy. How would civil aviation be controlled after the war, by a Ministry of Civil Aviation, by the Board of Trade or what?: he received the stock answer that the Government were still considering the matter. *The Times* came out with an article advocating a policy of having both a chosen instrument and private Companies if civil aviation were to flourish.

In October the Government acceded to pressure following the June Debate in the Commons on Civil Aviation and held, in camera, an Empire Conference under the chairmanship of Lord Beaverbrook, Minister of Aircraft Production. It was known that whereas Australia and New Zealand were in favour of some form of Commonwealth airline, Canada was opposed, preferring to manage its own affairs and to run its own services. Even in the Lords Debate on 20 October 1943 Beaverbrook gave no details of what took place at the Conference and in response to questions as to Government policy, he was only prepared to say that the policy of a chosen instrument was tied up with the whole question of private enterprise and the shipping companies.

Two good stories went around Whitchurch. As the war neared its end, BOAC began to explore future possible airliners. A BOAC senior Captain, Charles W.A. Scott, was detailed to try some possibilities, one design being what he clearly regarded as an absurd civil version of the Lancaster. This four-engined aircraft carried only nine seats, with passengers sitting sideways. Scotty, a man of few words, dismissed this with the comment, *'...With a few modifications, it would seem to make a good bomber!'*

The other story concerns the Irish Shuttle, as the linking service with the Pan-American Clippers was called. A distinguished metallurgist was invited to fly to Washington, DC, to give the annual Wilbur Wright Memorial Lecture at the prestigious Smithsonian Institution; a great honour. However, when he duly arrived at Shannon he was informed that his Pan-American seat had been allocated instead to an army officer with a higher wartime priority. To get aboard the Clipper, then the one and only commercial transatlantic service, one had to be very special. As the service departed only about two or three times a week, the metallurgist was unable to get to Washington on time. He was so incensed that he investigated to find out who the army officer was who had been given such a high priority. He discovered that he was being sent over by the War Office to attend the very lecture

The departure of BOAC's first Lancastrian service from a very wet London to Sydney, Australia on 28 May 1945.

Right: A menu card from the days of the Lancastrian - in this case BOAC's G-AGLS. The entire luncheon menu comprised of food that could either be served cold or warm from Thermos flasks. (both BOAC via author's collection)

he was supposed to give!

BOAC gave more importance to post-war planning; the Chairman announced to the Board on 25 November 1943 that he was setting up a central planning committee to be known as the Chairman's Planning Committee, and that *'in addition a Board Development Policy Committee would be constituted to receive the recommendations of this Planning Committee'*. This was completed by the establishment of an Inter-departmental Planning and Progress Committee to ensure that departmental views were co-ordinated before planning material was considered by the Chairman's Planning Committee.

Some of the routes flown were less than luxurious and somewhat lacking on hygiene. Baghdad could be almost impossibly hot, and with no air conditioning the crews usually took their bedrolls on to the flat roof of the hotel. It was hot enough to fry an egg on the pavement as one wag was to prove – he bought some and deliberately dropped one. It fried in seconds.

Legend has it that two pilots - after being roasted in Dakota's glasshouse up front at Baghdad, had to climb immediately to about 15,000 feet to get out of the uncomfortable thermal bumpiness. There they were frozen their sweat almost turned to icicles. Accordingly, both stripped off and towelled themselves dry. As the wireless operator remarked, *"It was a good job that the plane did not then fall apart. The wreckage would have contained the dead bodies of both pilots, stark naked. The News of the World would have had a field day"*.

Tony Spooner again: *'I am unable to describe the highly unpopular route to Takoradi and back. It took 23 days in all and passed through several unhygienic hell-holes where crew members had died of malaria or worse. I was tipped off that the way to avoid being rostered on this was to take to dinner the heavily made-up little blonde girl who was the secretary of the man who made out the rosters. I was not enamoured, but . . . Bristol boasted a restaurant, Horts, Where excellent food was served (at a price) despite the wartime limitations. Their wines were excellent, too. Moreover, the little blonde secretary turned out to be less fierce than she looked. It was all well worth never having to risk dropping dead on the Takoradi route.'*

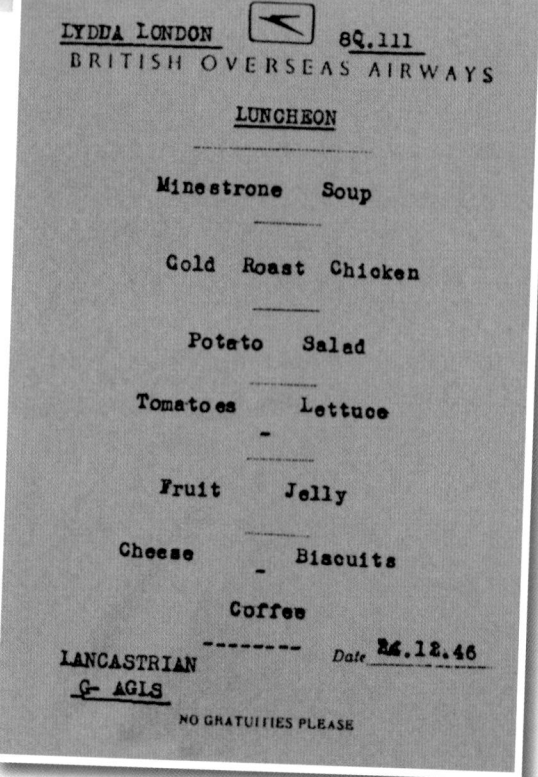

Much work was now being done by the Departments

Handley Page HP.70 Halton I G-AHDW *Falaise*. The aircraft was disposed of to Aviation Traders in 1948. *(PHT Green Collection)*

within BOAC to draw up tentative operating plans but there was still no firm indication by the Government as to the future shape of the industry, inevitably making it very difficult for the planners. The Chairman continued to do what he could to get plans moving by guiding the planners with assumptions and in December he followed this up with a list of questions about the future on the basis that BOAC must be prepared to produce its proposals to Government if asked. Early in January 1944, the Chairman presented a confidential memorandum to the Board setting out his views on a wide range of matters relating to post-war operations as policy guidelines. These were considered at the first meeting of the Board Development Policy Committee on 20 January at which the departmental heads were present and the guidelines were approved.

Apart from guidance as to what routes should be operated, the statement dealt with the difficult question of what shape British air transport should take and how it should be controlled.

The conclusion was that it would not be feasible for a private enterprise to be the sole management - unprofitable long-distance routes could be neglected, Government support would be needed in the fight against foreign competition and, as air transport is a form of public utility Government would inevitably have a considerable say.

On the issue of whether there should be a single chosen instrument or several, the Board's view supported, at least for the immediate post-war period, the setting-up of a central parent company for policy and overall direction with a series of subsidiary operating companies linked together regionally; it was considered that an organisation of this kind would allow for decentralisation and at the same time give the single direction needed to avoid wasteful use of resources. It was thought that Government influence could be secured through a 51% interest and that private organisations such as the shipping and railway companies could then participate in minority interest.

As to how the Government should exercise control, the Board suggested a Ministry of Transport or Communications with an Under-Secretary responsible to the Minister for air transport activities or alternatively these activities might form part of the Department of Overseas Trade in the Board of Trade with a Parliamentary Secretary for civil aviation. This demonstrated BOAC's anxiety that the sponsoring Minister should be of sufficient importance to protect and promote the interests of civil aviation in the Government as a whole. But whichever Ministry was in control, the Board emphasised that complete independence of Management must be assured, a situation which the company had never experienced due to its total

Lancastrian G-AGLT *Newcastle* is serviced between flights.
(BOAC via PHT Green Collection)

BOAC Dakota G-AGFX in the immediate post war colour scheme comes in to land at Croydon. *(PHT Green Collection)*

subservience in wartime circumstances to the Air Ministry. On the question of participation by railway and shipping companies, the Board took the view that there was no reason why there should not be co-operation in developing overseas airline operations, although the impression had been gained that both rail and shipping interests, whilst desirous of participating in some way, were not seeking physically to operate services themselves.

Two other aspects were dealt with at this January Board meeting. The first concerned future relations between BOAC and the Dominions and Colonies. The Board agreed to adopt a stance in meeting the post-war aspirations of the Dominions, accepting that they might well wish to operate reciprocal services to London or to follow a policy of sectionalisation on a route as had already happened with Australia, where QANTAS operated between Australia and Singapore and BOAC between Singapore and London. South Africa and Canada could be expected to operate parallel services to London. India, on the other hand, might present problems if the Indian Government were to insist on its local operators participating in the actual carriage of through traffic going beyond India. As to the Colonies, it would be a matter of local services, which BOAC would need to ensure fed traffic on to the trunk routes and that BOAC should assist such local airlines with maintenance and technical expertise, but, as Lord Knollys emphasized *'...what we should avoid is apparent responsibility in the eyes of the public Without effective control'*.

The Board accepted that Europe was a special case in the light of the military situation which would make operations by Transport Command inevitable in the early stages. Whether subsequently, this should be an international united European airline or a series of national companies was an open question but, whichever organisation were to be adopted, there ought to be close co-operation so that Europe could be regarded as one entity in the same way as the United States. It was clear from a broadcast he made on 17 January 1944 that Knollys was in favour of the private sector playing a part, when he said *'...the public interest is involved in the development of air transport together with the best features of private enterprise which we need in order to develop an industry which requires vision, initiative and enterprise'*, and this was in accord with his guidelines and his contacts with the

Avro York G-AGNO of BOAC seen at White Waltham in 1945. *(author's collection)*

Entrance to the Lancastrian's baggage compartment was via the upward-hinged nosecone. *(BOAC via PHT Green Collection)*

Minister and the DGCA. The signs were clear that shipping interests wanted to become actual airline operators. Many of them had changed their articles of association to permit this; in February 1944 five shipping companies, the Royal Mail, Blue Star, Pacific Steam Navigation, Booth Lines and Lamport & Holt registered a company, British Latin American Air Lines. This development caused the Government to make clear that no assurances had been given to the five companies and that no firm plans for post-war operations had yet been decided. But matters were undoubtedly coming to a head, and Critchley told his Executive Committee on 23 February 1944 that the appropriate Cabinet sub-committee was considering the shape of the future South American service. That BOAC felt some concern about its standing on a future South American operation is evident from a decision in April that a BOAC service to South America ought to start as soon as possible either with Boeing flying boats or with Yorks. On his return from an extended tour over Empire routes, Knollys Wrote on 28 April 1944 to the Secretary of State for Air, Sir Archibald Sinclair, on behalf of all the Board Members. The letter disclosed a genuine feeling of frustration in BOAC at the apparent lack of consultation by the Government about the future. In spite of the planning material available in BOAC and its up-to-date knowledge of overseas conditions and although it was the single chosen instrument for civil air transport overseas from the UK *'...neither the Board nor the Chairman have been brought into consultation on matters affecting the future development of British civil air transport, which they are now aware have been discussed with representatives of the Dominions and the United States'*.

Knollys submitted four papers on aspects of overseas operations, evidently to illustrate the kind of problems to be met and the extent of BOAC's studies and to stimulate some response from the Minister; these papers covered Feeder Lines, British air communications to and through India, Ceylon's place in Empire air routes and Egyptian Airways. It was not until 5 June 1944 that the Secretary of State replied with a number of comments on the papers but not of much consequence and he concluded his letter by saying *'I am sorry you felt you were not consulted, but we have been considering the broad international and not the British civil aviation scene, but we are now getting down to the latter'*.

Two days earlier, Lord Beaverbrook had written on 26 April asking for BOAC's views on a number of alternatives for post-war airline operations. Beaverbrook's first possibility was a single British chosen instrument to operate in parallel or in pool with Dominion chosen instruments; the second was an international operating agency for European trunk routes excluding internal services except in Axis territory; the third a joint Commonwealth Corporation to operate a round-the-world service stopping only in British and Commonwealth territory and assuming separate reciprocal British and Commonwealth chosen instruments for slower services and for routes to foreign countries; and finally the question of shipping and other interests joining in the control and operation of overseas air services.

Knollys' reply of 5 May agreed with the ideas in the Beaverbrook letter, but made it plain that BOAC considered there should be *'only one British Public Corporation for all British overseas airlines based in this country'* but that such a policy would not be incompatible with the inclusion of shipping, railway or other interests, although they should have only a minority holding. On the question of a joint Commonwealth Corporation operating a round-the-world service, this would be practicable but under such a plan the UK would be giving up for the benefit of the Commonwealth as a whole an operation which would have been its most outstanding service. With the agreement of the Secretary of State, Knollys included in this reply copies of the papers he had sent to Sir Archibald on 28 April. He also sent Beaverbrook a copy of a paper he had written, following his recent overseas tour, on a possible type of organisation for controlling air services throughout the Commonwealth. This paper, written on the assumption that BOAC would continue to be the only chosen Instrument for overseas operations, recommended the setting up of a Commonwealth Board composed of the relevant Government officials to discuss and decide matters of high policy. The Board would cover such matters as the routes to be operated, by whom and at what frequency and decide any subsidy contributions. Translating Government policy into action would be coordinated through a Board made up of the Chairman and Chief Executives of the operators concerned. The decisions of this Empire Air Board, interpreted by the Executives Board, would be put into effect by the respective UK and Dominion operators who would be operating reciprocal services between the UK and the Dominions.

Shipping and rail interests continued their pressure on Government for a share in post-war aviation and in the Lords debate on 10/11 May 1944 Lord Essendon put the case for the shipping companies owning British Latin American Air Lines, and Lord Kennet sought participation for railways. Lord Beaverbrook replied that BOAC would continue to be governed by the existing statute until repealed and asked the private interests to say straight out if they wanted the BOAC Act repealed or did they want to operate certain selected routes, that is, those with the highest level of profit, leaving the residue of unremunerative routes to BOAC. Lord Londonderry asked for the BOAC Act to be repealed.

The government was at last starting to think in depth about civil aviation in peacetime and was seeking opinions from those directly concerned. On 15 May 1944, Hildred sent a letter to Knollys setting out some ideas for a draft

Left: *'Fly to the Far East'* - A BOAC poster from the late 1940s depicting old and new modes of transport.

Below: Shorts Solent G-AHIU Solway making an appearance at the Farnborough Air Show. *(PHT Green Collection)*

York G-AGNV *Morville* was another Avro airliner used by BOAC in the early years after the war. It passed to Skyways in 1955.

paper to go to the Prime Minister. He floated three concepts, an Empire Corporation to operate all Empire trunk routes; an Empire Corporation but excluding routes to countries which might not wish to participate; and, finally, one or more Corporations to operate particular routes. He concluded that the first option was ruled out because it was known that Canada was not in favour of any joint operating body but the second could be a possibility if the South African route were excluded since South Africa also preferred to operate parallel services. Hildred suggested that there might be a Commonwealth Corporation on a limited basis, covering the one main trunk route between the UK and New Zealand, with other routes such the trans-Pacific service possibly coming under the same organisation. Hildred explored in some depth how the Commonwealth Corporation would function and what the control of the participating Governments might be. He proposed among other matters, the setting up of an Empire Council to be responsible for the policy of the Corporation and for seeing '...*that it was being run efficiently and economically*'.

Knollys replied on 18 May 1944 expressing doubts about the feasibility of such an Empire Corporation because of the inevitable problems - both political and in the sphere of human relationships - in a joint Board with appointees from several countries each owing allegiance to their own Governments. He thought that no Board could run an organisation if another body dealt with policy, efficiency and economy, as Hildred had suggested. To be constructive BOAC had given much thought to the whole question and a memorandum was attached to Knollys' letter setting out BOAC's views. These proposed the formation of an Empire Council to take decisions on all matters of high policy normally reserved to Governments, with the translation of Government policy into action coordinated through a committee composed of the Chairman and Chief Executives of the operators concerned. On the operating side BOAC felt there should be reciprocal operations between the UK and the Dominions carried out by the respective UK and Dominion operators.

This direct response from BOAC took the Air Ministry aback, and Hildred made this evident in his reply: '*We have been warned. It is clear your proposals represent a standpoint so different from the provisional inter-departmental concept that they merit quite separate consideration*'.

To make sure that Government were in no doubt about the thinking in BOAC a paper entitled '*Considerations on the future capital and organisational structure of BOAC*' was sent in July 1944 to both the Secretary of State and the Lord Privy Seal, Lord Beaverbrook. This was complementary to the earlier paper on '*A single British*

Just another Dak waiting for another revenue service. BOAC's G-AGKA. *(PHT Green Collection)*

chosen instrument' sent to them in May. This paper set down principles on the assumption that the BOAC Act 1939 would remain substantially unchanged with the State having enough control to protect public and national interests but independent management on commercial lines would have to be assured. As to the structure of the Corporation, BOAC would retain its rights and responsibilities as laid down in the 1939 Act. If subsidiaries to operate particular routes or services were formed, BOAC would be an active holding company. The majority of its Board would be appointed by the Minister and minority elected by private shareholders. BOAC should hold a majority of the capital of any such subsidiaries. Proposals as to the financial aspects were included in detail as well as the policy for dividend payment.

That same month BOAC submitted to the Air Ministry particulars of the routes it wanted to operate after the war covering all Empire territories and the North Atlantic in great detail, including all intermediate stops.

Meanwhile, discussions between BOAC and shipping and rail interests continued. The Chairman had pursued his talks with Elder Dempster Lines on a proposal for air services between the UK and West Africa, in which both organisations would be interested, and at its meeting, on 8 June the Board decided to lay down certain principles where association with outside interests took place. These covered the need for the State to have adequate supervisory control to protect the national interest, and, where the Government gave direct financial aid, there had to be means of safeguarding the taxpayer; independent commercial management within Government policy had to be maintained, air-mindedness had to be the ruling force in directing policy and finally that terms of participation should be adequate to attract private interests.

So by this time a policy of part public and part private participation had been virtually accepted in BOAC, and the Chairman widened his talks with the shipping companies; in June he had a discussion with Irvine Geddes in his capacity as representing all shipping companies likely to be interested, and he followed this up in September with a meeting with Sir Ernest H. Murrant MBE, the Chairman of Furness Withy & Co. Ltd., about co-operation on the South American route. Knollys also met Lord Balfour of Burleigh for talks about collaboration with British Railways on services to Continental Europe within the framework of whatever Government policy might be adopted.

Meanwhile, out in the real world the Stockholm trips were very popular. If one could arrive on schedule, via Goteborg, there was just time to get to the big department stores before they closed. The trick was to look around for a sales girl who resembled one's wife or girlfriend, and to hand her a fistful of notes to get herself things that she liked and which fitted her. Then to rush away to the hardware or other departments where kitchen gadgets, pots and pans, crockery and such useful things as bath plugs, coat-hangers etc were sold: articles which were next to impossible to get even after the war was over. Finally, just as the store closed, the sales girl would be waiting by the door with her choice of garments nicely wrapped and with the correct change!

The future of air services in Europe posed different problems to the Empire routes where services had been more or less maintained throughout the war. Continental Europe had been overrun by the enemy and had been a war zone so that inevitably it would fall to the military to restore air communications as and when territories were freed from enemy control and only later would it be possible to hand over to civil operators. Even before the Allied invasion had begun, a proposal to form a Group within RAF Transport Command to operate services into the mainland of Europe as soon as the war situation permitted was being discussed at the Air Ministry with Transport Command and BOAC. The DG was reporting progress made to the Board as early as April 1944, and in June his talks with the AOC-in-C Transport Command had reached the point where BOAC was to nominate and second key staff to operate services to Europe. The Group would be under the authority of the Command in view of the military situation, but the eventual intention was to transfer it to BOAC. By following this policy, it assured BOAC, as the DG explained to the staff, of a 'ready-made experienced organisation in Europe'. By the end of June Transport Command had formed No. 110 Wing to carry out this task and BOAC staff had been seconded; thus the future of British civil aviation within Europe was already taking shape.

Government plans for civil aviation's overall future

Shorts S.25 Sandringham G-AKCO *St George*. The aircraft was only in BOAC service for about a year. *(PHT Green Collection)*

were also progressing. On 8 October the Secretary of State for Air presented a White Paper to Parliament setting out views on the principles to govern international civil aviation post-war, thus preparing the ground for the British delegation attending the Commonwealth Air Transport Conference in Montreal and then the ICAO Conference in Chicago. On 12 October came the long-awaited Government announcement of the appointment of a Minister of Civil Aviation, the post going to Viscount Swinton who had been Secretary of State for Air from 1935 to 1938.

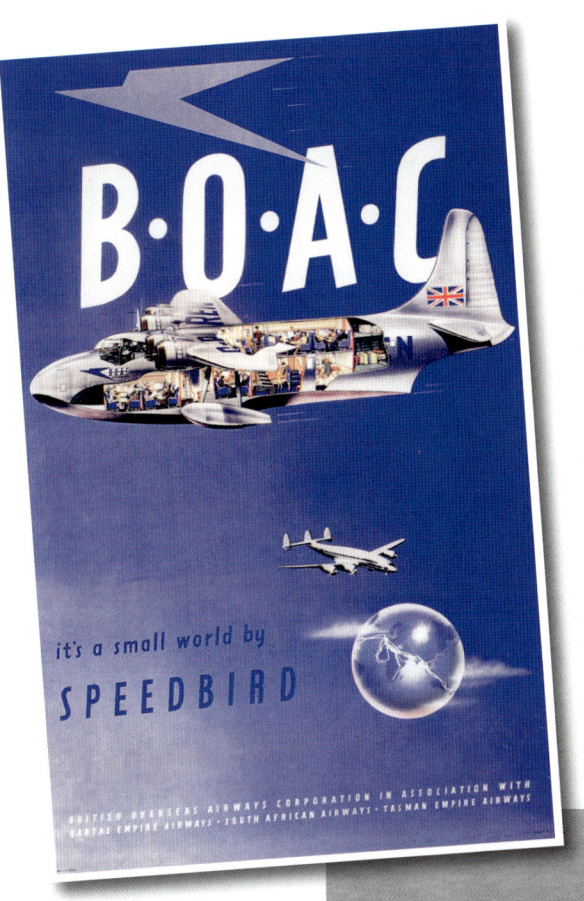

On the Minister's return from North America, he reported the results of these Conferences to the House of Lords in a Debate on 16 January 1945. The discussion centred on the outcome of the Conferences and did not deal with the Government's plans for the structure of British civil aviation. However, by the end of January, the Government's intentions became much clearer when they produced a secret document entitled *'Government Plan for the Organisation of Civil Aviation'*, a copy of which BOAC received and discussed at its Board meeting on 30 January. The Government plan finally removed the uncertainty of the past two years as to who the participants in British overseas air transport were to be and precisely what routes were to be operated. The Board were happy with the

BOAC's Shorts Solent 3 G-AKNO was flown onto the River Thames and then taxied upstream past Limehouse Reach to Tower Pier to be named *City of London*.

The aircraft was used on the BOAC-South African Airways Springbok route to South Africa carrying a maximum of 39 passengers.

proposals since in their view they were close to their own ideas already put to Government in July 1944.

The first item in the Government plan listed the routes to be operated, divided into five areas. Empire including the North Atlantic, South America, Egypt, Iraq, Europe and internal services within the UK, followed by a list of *'...the prerequisites of an efficient organisation for civil air transport',* with a comment that such prerequisites could only be obtained by drawing on the experience and personnel of the best undertakings in other transport fields. This was an indication of the success achieved by the shipping and railway lobbies: BOAC was to be a partner with these other organisations on routes it operated, and the door was left open to these outsiders *'...to be interested in some of BOAC's routes'.*

The route allocation in the plan was precise. BOAC would operate the main Empire routes and to Canada in parallel with the Dominions and India possibly forming joint operating companies later. The company would also operate to the United States, possibly with shipping companies having a minority interest. To West Africa, BOAC and shipping would fly in partnership - clearly with the BOAC/Elder Dempster discussions in mind. On the Pacific, a joint operating company was envisaged between BOAC, Australia and New Zealand in parallel with a Canadian service. Shipping was also to have a share in the Bermuda-USA and Bermuda-Canada routes, in a combination of BOAC, the Bermuda Government and Furness Withy.

As to local feeder services in Commonwealth territories, the plan allowed for these being operated by local companies, but it would expect BOAC to give at least technical assistance. Particular reference was made to Egypt and Iraq because of the negotiations proceeding between these countries and BOAC for the setting up of joint companies.

This left two major zones uncovered, the West Indies, Central America and South America area, and Europe. The South American route was to be operated by the shipping companies who would also take in the West Indies, and BOAC would be allowed only a minority interest.

The plan was less precise about Europe, suggesting that it might be sensible to combine the UK internal and Continental services and with interchangeable aircraft and tickets. Tickets should also be interchangeable with ground transport. Such a company would be a combination of shipping, railways, pre-war domestic airlines and BOAC, but the majority control should lie with shipping. It was suggested that travel agencies might take an interest as well.

It was made clear that Government would have broad control over policy and would allocate routes to operating companies and appoint members of the boards of directors, but that day-to-day management should rest with the

The Board of BOAC were well aware that they needed efficient designs if they were to compete with the Americans on the lucrative North American market - and the only way to initially do that, was to use similar equipment, not available from UK manufacturers.

G-AHEL was one of the early Model 049 Constellations originally built for the USAAF, but was never delivered, being put in service by BOAC in July 1946 and named *Bangor*. It could seat fifty-four passengers.

Avro Lancastrian I G-AGMD *Nairn* served with BOAC from 29 June 1945 until July 1947 when it passed to QANTAS.

companies. It was also clear the Government intended to set up a central Government-owned company to own or lease aircraft, which it would sell or lease aircraft to British and overseas companies and would cover a wide range of activities from aircrew training to staff welfare and from technical assistance to special responsibility for research work. Finally, there was going to be a change in the name of BOAC, the new title of 'British Commonwealth Air Services' being more indicative of its main functions.

BOAC made representations to the Minister about the plan, particularly in respect of the South American service where it was felt that if BOAC were to be a partner at all, it would inevitably be held responsible for the success of the operation; this being so, BOAC must have a say in the direction of the company. There was general approval at Board level for the Minister's plan, and on 5 February 1945 Knollys wrote to Viscount Swinton that it provided 'the most suitable tools for the job'. After all their hard work during the war years, the staff of BOAC were not at all happy at the prospect of the shipping interests reaping post-war benefits. The Board on the other hand having noted that in future the Corporation would be under the control of the Minister of Civil Aviation in place of the Secretary of State for Air except for certain purposes connected with the war effort, placed their appointments on 5 February at the new Minister's disposal at the proper time so that he

Chintz chairs, occasional tables and paper flowers did little to disguise the fact that the passenger terminal at London Airport in 1946 was ex-military marquees with electric cables strung up in the roof area.

might have a free hand.

On 13 March 1945, the Government produced its White Paper on British Air Transport that contained no surprises for BOAC. It was based on the premise that a single chosen instrument was unsuited to deal with future expansion. Consequently, it encouraged the intervention of surface transport especially shipping. In each of the three main route divisions. Empire and Atlantic, South America and Europe, surface companies were to play a part. BOAC would be the controlling operator in the first, and for the second a new company would be set up controlled by those shipping companies already associated together in British Latin American Air Lines Ltd. BOAC would have a minority interest and participate in the management; for the third a new company would also be established, to be run by a combination of the railways, the short sea shipping lines, travel agencies and BOAC plus any pre-war domestic airlines interested.

As it was thought by Government at the time that these short-haul routes would be more profitable than the long-haul Empire routes, BOAC would also participate and be allowed a substantial though not controlling holding. No subsidies were to be paid but, if an operating company was required by Government to undertake a new route in the national interest, a temporary subsidy could be granted.

The White Paper covered other aspects, such as the overhaul of aircraft which was to be carried out by a combined organisation of the three Companies, as also the training of aircrews and ground staff. Management would be allowed by the Minister to manage, but he was to control broad overall policy. Spelt out was the Government's intention that the new Corporations should use British

Like many other airlines, BOAC maintained a full Technical Publications Library - 'Tech Pubs'. Some of the aircraft handbooks were 'interestingly illustrated' by the manufacturers, which got the attention of the engineers! The page on the left is from *The Lockheed Constellation Pocket Handbook* from September 1951 covering the Model 649 and 749 aircraft and relates to the section on the aircraft's fuel system It is liberally stamped B.O.A.C. Central technical Library throughout!

This Constellation had an interesting start to its flying career. It was laid down for Eastern Airlines in the USA, but was completed for Aer Lingus as EI-ADD *St Kevin,* but was stored until sold to BOAC in June 1948 as G-ALAN. The aircraft was named *Beaufort* and served with BOAC for eleven years.
(all BOAC via Hugh Jampton Collection)

Constellation G-AHEM *Balmoral* was scheduled to become part of the USAAF as 42-94557 but was never taken up. Instead it became a -049 and was registered on 6 April 1946. *(BOAC via author's collection)*

Right: The Constellations operated the 'Mayflower' services on the North Atlantic. With the innauguration of the Boeing Stratocruisers on the route with their famed 'Monarch' service, as shown in this BOAC poster celebrating New York with a 'Strat' just visible top right, the 'Mayflower' services became almost second-class!

aircraft, and until new British types became available, the Government would lease aircraft to them. Although not stated in the White Paper, it was already accepted that the new Corporation for the short-haul routes would be named British European Airways and that there would, after all, be no change in the title of BOAC.

In all this BOAC made it very plain that they were willing partners in the Government's proposals, and Knollys issued a statement to the Press to that effect.

On 15 March 1945, a Lords Debate took place on the White Paper; it met with general approval except for Lord Morris, a Labour peer, who strongly criticised the inclusion of shipping and railway interests. Ten days later the Ministry of Civil Aviation Act formally set up the Ministry and defined the Minister's responsibilities and duties; thus another step in the transfer from military to civil control was completed.

In line with the White Paper, BOAC continued discussions with the shipping and railway companies, but progress was slow, made slower by the changing political scene. In May the wartime Coalition Government ended, and the Conservatives took over under Churchill, but at the General Election two months later a Labour Government was returned to power, known to have a different policy from the Conservatives for civil aviation. Lord Winster, a Labour peer, became the new Minister for Civil Aviation. At their meeting on 7 August, the Board decided that if their views were sought by the Minister, they would inform him that their July 1944 Paper still represented their thinking. This was perhaps surprising since this advocated the advantages of private enterprise participation in BOAC whereas the new Government was known to be in favour of complete nationalisation. However, BOAC then sent these thoughts to the Minister and to Sir Stafford Cripps with the comment *'that to ensure effective action, BOAC should remain the principal Government agent for civil aviation for the vitally important months during which Government will be in the process of firming its decisions'*.

These decisions had been taken by the autumn and were made clear in a House of Lords Debate on 1 November 1945. The Government announced they had decided that public ownership took overriding priority in air transport and that there should be no financial participation by existing surface transport interests. So there was to be nationalised monopoly control of scheduled services although for charter work the door was left open for private operators.

In a series of exchanges, Hildred told McCrindle that BOAC must assume there would be three Corporations and hinted at close control by Government even to the extent of having a Government appointee on the Boards and of the Minister chairing the three Corporations' meetings, plus a central Government purchasing agency.

Right: BOAC's Constellation *Bristol II* is seen undergoing service at a rather snowy Gander. *(BOAC via author's collection)*

Below: *'London - Sydney by Constellation QANTAS Empire Airways in Parallel with British Overseas Airways Corporation'*

On 6 November Critchley wrote to both Hildred and Cribbett at the Ministry of Civil Aviation advocating one Corporation with four separate divisions; Atlantic, Empire, South American and Europe, but the die was already cast. At the second Lords debate on 6 November 1945 it was not surprising to find Lord Reith strongly supporting the concept of three public Corporations. In response to concern from the Conservative peers regarding Governmental control, the Minister emphasised that *'it was not part of the Minister's duty to interfere with the day-to-day administration of the Corporations'*.

BOAC believed there was still time to influence Government decisions so on 8 November Knollys sent the Minister a statement of views. After making the point that any Board appointed by the Minister must be left free to manage the day-to-day affairs, the statement dealt with principles based on complete public ownership. Three alternative organisational systems were considered. The first proposed four separate divisions, North Atlantic, South Atlantic, Empire and Europe, all with their own staff and chief executives but reporting to a common Board; the second proposed autonomous subsidiaries formed for each of the four regions coordinated by a parent Corporation - really an extension of the first proposal. The third idea was for three separate and independent Corporations with some coordination through committees. In his covering letter the Chairman came down in favour of the second proposal namely four regional subsidiaries, each autonomous and each with its own Board reporting through to the central Board responsible to the Minister. Knollys suggested that on the subsidiaries' Boards there could be representatives of private industry.

On 20 December 1945, the Government issued its White Paper on British Air Services. Three separate statutory Corporations were to be set up under public ownership and control; BOAC was nominated for the Commonwealth routes and the routes to the United States and the Far East, the other two Corporations would cover respectively Europe and the domestic UK services, and South America. Maximum freedom of management was to be granted to the individual Boards. As to subsidies, air services must be made self-supporting as soon as possible, but it was accepted that State aid would have to be given to support essential but unremunerative routes. Capital would be provided from public funds, and profits or losses accrue to the Exchequer. Government policy would require the three Corporations to use British aircraft, and the White Paper referred to the fact the Government was doing all possible to accelerate the production of civil types to re-equip the new Corporations, but meanwhile, they would have to use modified military machines. The Corporations would be permitted to operate charter services as would private operators.

On 1 August 1946, the Act carried into effect the proposals of the White Paper and set up two new Corporations, British South American Airways and British

BOAC saw a desparate need for American airliners so as to be able to compete on the North American route.

Above: Boeing Stratocruiser G-AKGH *Caledonia* is seen here in flight, the left hand side of the flight deck with the radio operators position is shown here on the left.

European Airways which together with the existing BOAC formed the three Corporations of the Act. The Minister was given considerable powers in an Act which was itself imprecise, for instance, he was empowered to set out the geographical areas of operation for each of the Corporations. Failure to do so in the Act resulted in problems in the years ahead. He had general powers of direction; he approved all Board members (but not the Chief Executive), he had to be given each year the Corporations' plans and expected expenditure outturns. The Minister had the power to direct that aircraft on the British Registry should be used on the Corporations' services and during the Debate on the Bill the Parliamentary Secretary stated it was the general policy of the Government to require the Corporations to use British manufactured aircraft and the Minister's approval would be needed to purchase foreign aircraft. Excess expenditure was to be covered by Exchequer grants, and all regular services were reserved for the Corporations and their associates.

An Air Transport Advisory Council (ATAC), later used as a licensing Board, was set up to consider representations from the general public as to the adequacy of the facilities provided by a Corporation or any other matter referred to it by the Minister.

European operations which had been re-started by 110

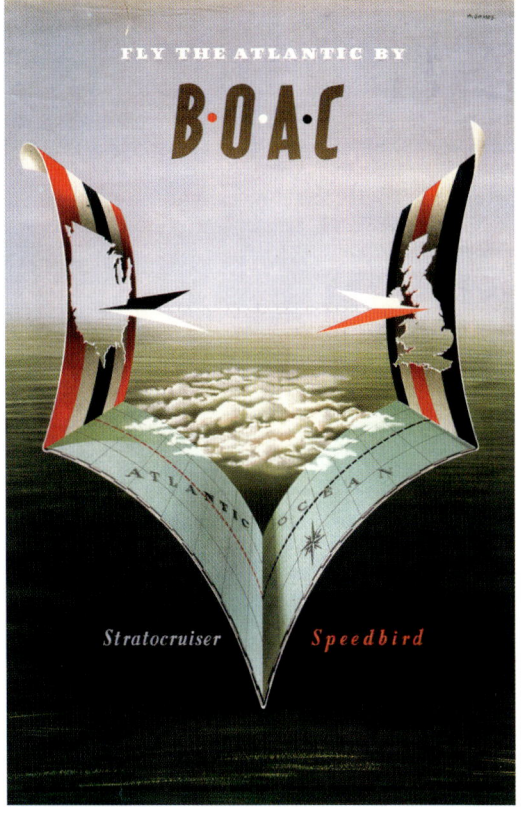

Wing of RAF Transport Command were handed over to a European Division of BOAC in February 1946 and from 4 March operated with civil aircraft and crews in civilian uniforms as a preliminary to the setting up of BEA. With the passage of the Act into law on 1 August, these operations were transferred by BOAC to the newly formed BEA. Thus the principle of publicly owned Corporations was established and remained the pattern throughout the rest of their existence.

On 6 August Knollys made a public statement on the Act drawing attention to the machinery it would provide for close co-operation between the three Corporations and reminding the public and the staff that *'this country is now feeling the effects of five years of non-production of civil aircraft. Even with the Tudors in service, there would be a gap when our competitors would be operating newer types now being built in America while we might temporarily be flying the world's No. 1 air route with aircraft which, through no fault of our industry, would not be so large, fast or economic as those of our competitors'*.

Re-equipment
There was never any doubt about the urgent need to re-equip BOAC when the war was over. Even if the new types projected in the years immediately prior to the outbreak of war had materialised, six years had passed, and rapid technical progress, especially in America, would have made them out of date; they were certainly not competitive with the products coming from American factories, the DC4, the DC6 and the Lockheed Constellation. The only British aircraft produced during the war for civil purposes was the Avro York, and even this was a development from the Lancaster bomber. BOAC would undoubtedly have been happy to deploy a full fleet of Yorks from 1944 even using the passenger-cum-freighter version used by Transport Command, but it entered service with BOAC too slowly with a fleet of only five aircraft by March 1945, and as was later demonstrated on the route to South Africa could not hope to compete with the pressurised Constellations.

By the end of January 1944, little progress had been made on the various Brabazon proposals but there was one encouraging development in that the proposed jet propelled aircraft, Type 4, had been handed over to De Havillands to design - the only experienced manufacturer in that field.

BOAC considered the right policy was to encourage the industry to produce entirely new transport aircraft in order to leapfrog American development rather than to dissipate too much of the industry's effort on aircraft for the interim. All these views were put to the Air Ministry and to the DGCA in September 1944.

On 14 September the Chairman of BOAC confirmed at the Board Meeting that the Government's policy was to use British aircraft and therefore obtained the Board's agreement to an intial requirement for Tudors, fourteen of Mark I and thirty of Mark II. By then a statement had been put to the Air Ministry bringing up to date BOAC's aircraft requirements for the post-war years. On the assumption that the war ended at the close of 1944, BOAC's needs in 1945 would be met by Lancasters and Yorks, and for the period from 1946 to 1949 it would rely on Tudor Is for the North Atlantic and Tudor IIs for the Empire routes and South America. BOAC's suggestion that Merlin-engined DC4s from Canada - Later called the Argonaut - might also be used was turned down by the Air Ministry on the grounds that only British aircraft should be operated by the chosen instrument. For the 1949-54 period it was hoped that the new Brabazon types would come into use, although there was concern in the Corporation at the lack of progress particularly as to the Brabazon type III for use on the wide-ranging Empire routes as a follow-on to the Tudor II. At the Air Ministry's request BOAC set out again its operating requirements for the Brabazon III and re-emphasized its great interest in the Brabazon Type IV - the jet or gas-turbine development.

On 1 November 1944, the Ministry of Supply (MoS) acting as agents for the Civil Aviation Department of the Air Ministry placed an order with A. V. Roe's for fourteen Tudor Is. It is important to note here that there was no direct contract between BOAC and the manufacturers as all such matters were dealt with by the Air Ministry under the Order made in April 1940.

Also by November, Avro had given dates for the Tudors, the first production Mark I to be ready in March 1945 and all fourteen aircraft by the following August. As to the Mark II, BOAC was told verbally by Avro the first aircraft was expected to be ready in May 1945, so in April BOAC increased its requirement for the Tudor I from fourteen to twenty and for the Tudor II from thirty to seventy-nine.

Britain was desperately short of manpower in what today would be called the service sectors of industry. So the senior management of BOAC addressed the problem of finding, if only on a temporary basis, cabin crews to man the new aircraft being acquired for the predicted rapid expansion of air travel following the war's end.

There would be flight crews in abundance with the winding down of the RAF bomber forces, but experienced catering-trained staff simply were not available. The solution to this problem must have come to someone in BOAC's head office in Stratton House, Piccadilly, as he gazed out over Green Park. Why not use Air Training Corps (ATC) cadets?

The answer was elegant in its simplicity. They were air-minded, and would jump at the chance of being paid to fly. They could be used as assistant stewards to supplement the small number of ex-Imperial Airways cabin crew who had been retained. Since the cadets would be working as helpers to these very experienced airborne staff, their lack of catering background would not be too important. The Air Ministry was consulted and convinced, the scheme approved, and signals were sent out to ATC squadrons and flights throughout the country, offering the opportunity for teenage boys to work in the aviation industry. So it was that in the latter half of 1945 large numbers of youths were travelling from all over Britain to attend selection boards held high up in the tower of Airways House near Victoria Station.

The successful candidates were sent across to Piccadilly to the BOAC Medical Department at Stratton

Above: The first Handley Page Hermes IV G-AKFP at the SBAC display at Radlett with the BOAC 'Speedbird' logo on the nose.

Right: The neat Hercules powerplants and undercarriage of the same airliner.

House to undergo a rigorous aircrew medical. Then it was back home to wait, hopefully, a letter of acceptance and the so-called 'worldwide' contract of employment.

It did not take long before the bright idea of using boy stewards began to go very wrong from the Corporation's viewpoint, as these teenagers gained a reputation for living the high life 'down the line' in areas with aviation connections where they were seen drinking and roistering with their mentors and social tutors, the ex-Bomber Command crew members on secondment from the RAF. It was a common sight in 1945-46 to see a 17-year old catering apprentice in his heavy greatcoat, often with gold-bar epaulettes, current girlfriend on his arm, entering a Bournemouth or Bristol super-cinema, a Singapore silk scarf draped casually under his turned-up collar and a Macropolis, Bombay, cigarette dangling from his lips. It must be said that they set quite a dash in the drab surroundings of that early post-War period.

The war ended with BOAC using a fleet of seven different types of flying boats and eleven different types of landplanes, giving a total of two hundred and seven aircraft at the beginning of 1946, thirty-five being on loan from the Air Ministry. None of these could compete with designs appearing in the United States. This situation was fully realised by BOAC and indeed, had been foreseen in the early years of the war. It was partly pressure from the Corporation which led the Government to set up as early as the end of 1942 the Committee with Lord Brabazon as Chairman to prepare outline specifications for the types of civil transport aircraft it considered would be needed post-war and to suggest those firms most suitable to tender for design work when circumstances permitted. It was also to indicate those military types which .might be modified to make them suitable for civilian flying immediately after the war and thus create work for a manufacturing industry badly needing support once the requirement for military aircraft fell away.

By September 1946 BOAC were at a point where they could offer something like a worldwide service, but it was a far cry from what is available today:

Route	Aircraft	Frequency
London - New York	Constellation	3 weekly
Prestwick - New York	Constellation	1 weekly
Prestwick - Montreal	Liberator	5 weekly
Bermuda - Baltimore	Boeing 314	3 weekly
Cairo - Karachi	C-Class	1 weekly
Durban - Calcutta	C-Class	1 weekly
Cairo - Kisumu	C-Class	1 weekly
Cairo - Nairobi	Lodestar	2 weekly
Cairo - Basra	Lodestar	2 weekly
London - Calcutta	York	1 weekly
Poole - Singapore	Hythe	2 weekly
London - Jo'Burg (With SAA)	York	4 weekly
London - Cairo	Dakota	9 weekly
London - Cairo	York	2 weekly
Poole - Cairo	C-Class	1 weekly
London - Beirut	Dakota	1 weekly
Cairo - Istanbul	Lodestar	3 weekly
Cairo - Athens	Lodestar	1 weekly
Athens - Teheran	Lodestar	1 weekly
Lydda - Terheran	Lodestar	1 weekly
London -Sydney (with QEA)	Lancastrian	3 weekly
Poole - Sydney	Hythe	3 weekly
Cairo - Aden	Lodestar	3 weekly
Cairo - Addis Ababa	Lodestar	2 weekly
Cairo - Karachi (via Aden)	Lodestar	1 weekly
Cairo - Port Sudan	Lodestar	1 weekly
London - Lagos	Dakota	3 weekly

The Tudor debacle.

Handley Page, which had a large order from RAF Transport Command for a four-engined aircraft developed from the Halifax – the Hastings - were promoting a civil version, the Hermes, pressurised and with a tricycle undercarriage. But in early 1945 BOAC was still looking to the Tudor II for Empire routes and considered there was no point in buying the Hermes. At the turn of the year, it was hoped deliveries would start in September for the Tudor I and April 1946 for the Tudor II.

The Tudor I prototype was the last of Avro's wartime aircraft to be assembled in the Experimental Department at Ringway Airport, and Bill Thorn and Jimmy Orrell - who was to be the co-pilot - had been studying all the intricacies of the new airliner's design and instrumentation, flying controls, and engine and fuel systems. The first flight ocurred on 14 June 1945 and was already behind schedule.

In an effort to make up some of the slippage, A V Roe and Co Ltd was gearing up production to roll all 20 out of the Woodford assembly sheds as rapidly as possible. Two prototypes had been ordered, and the second, G-AGST, although following behind G-AGPF, would be involved in the fitting of substantial modifications and not actually fly until June 19th, 1946 - after the first two production Tudor Is, and only a day before the third! Also at the time of first-flighting the Tudor I prototype. Bill knew that BOAC had agreed to buy a stretched version for its Commonwealth routes and that the MoS had now ordered 79 of these larger versions. With a potential order book of 99 brand new pressurised airliners, the 'heat' was certainly on at Avro's at Woodford to get the Tudor I prototype fully airworthy and to iron out all the likely new problems that would inevitably concern the brand new pressurisation system.

After the first few flights, however, Bill Thorn knew that there were serious problems in the flying characteristics of the Tudor I, and they were a lot more difficult to overcome than first thought. The problems soon became obvious:
1. Excessive take-off swing, due to poor rudder control.
2. Excessive buffeting of the aircraft on approach to the stall, but at relatively high speeds.
3. Early wing-tip stalling.
4. Rudder 'kicking' at moderate angles.
5. Tendency to 'bounce' on landing.
6. Loss in estimated performance.

The BOAC Development Flight received its first Tudor I on 5 September, but there were still problems with overweight, with pressurisation and in meeting the new international airworthiness standards. By then the latest BOAC estimates of availability were October 1945 for the first production Tudor I and July 1946 for the first production Tudor II.

By October there was an even more significant concern in BOAC about the aircraft situation for the North Atlantic. The Tudor I had aerodynamic problems, and the first fourteen aircraft would have to be taken unpressurized and carry heavy oxygen equipment; moreover, delivery dates had gone back again, and BOAC would not receive its second aircraft before February 1946 at the earliest. As for the Tudor II, delivery had gone back even further, and there was uncertainty as to the date of the first production aircraft. To help fill the gap caused by these delays, BOAC had agreed to use converted Halifax bombers, renamed Haltons, which were due to be ready for service between May and June 1946 for operation on Empire routes.

The Minister was equally concerned and wrote to the Chairman on 4 October disturbed that the Tudor I could not be in service before the end of 1946. He even suggested the possibility of leasing American types but agreed there needed to be real concentration on the Tudor I and urgent investigations as to the Tudor I replacement to compete with the coming DC-7 and the Boeing Stratocruiser.

The future aircraft situation was reviewed at the Board Development Policy Committee on 18 October regarding the North Atlantic. It had to be accepted that up to the end of summer 1946 there was nothing the Corporation could do to ease the difficulties and the British position would have to be held as best it could with Boeing 314As supplemented by Liberators to Canada. As for the Tudor, it had to be assumed it would remain unpressurized for up to two years.

In the face of all these uncertainties on the intensely competitive North Atlantic, the Board asked the Minister for authority to order five Lockheed Constellation 049 aircraft immediately with an option on a further five. As an insurance to cover the vital period up to 1950 for which

The prototype Avro 688 Tudor I pictured at Ringway before the first flight by Bill Thorn and Jimmy Orrell. It became airborne on 14 June 1945, still not bearing any registration letters, although it eventually became G-AGPF.

only the Tudor I seemed likely to be available, it was also decided to seek permission to secure an option on five Boeing Stratocruisers, an aircraft regarded by BOAC as likely to be the most dangerous competitor and deliveries of which were expected to start in twelve months' time. At this meeting one further important decision was taken in the light of the Tudor debacle, for all future orders for aircraft, the Corporation must be protected against failure to achieve delivery dates, weights and performance.

The decision to seek American aircraft having been taken, the next step was to secure Government approval, a difficult task in the light of the country's acute shortage of dollars and the Government's repeated policy to make use only of British types. Knollys wrote to the Minister on 20 October 1945, making a case for purchasing five Constellations at a cost around $1,200,000 for each aircraft including equipment and spares, with an option of a further five. The Chairman stressed the urgency if BOAC were to get on Lockheed's order list to obtain aircraft by the summer of 1946 - five aircraft would enable a daily London - New York service to be operated. He reminded the Minister that for the critical period 1946-50 BOAC's plans had been based on the Tudor I. Delays in the progress of that aircraft now meant BOAC could not start a scheduled service at the earliest before the second half of 1946. But that assumed that aerodynamic problems shown up by flight trials and problems with pressurisation, weight and the Rolls-Royce Merlin engines would be overcome. For good measure the Chairman, now asked permission to try to obtain an option on five Stratocruisers having a seating capacity more than five times greater than the Tudor I.

To further make the point, attached to the Chairman's letter was a statement of comparative performance over the London-Gander stage for the Tudor I and the known American types including the Constellation and the Stratocruiser. The comparisons showed how far behind the Tudor I would be in payload and performance and in the Chairman's words *'We cannot hope to compete effectively or maintain British airline reputation on the North Atlantic with the Tudor I alone'*.

This request for Constellations came just as the Chairman was approached by Hildred as DGCA on the subject of the Brabazon I. Hildred referred back to when BOAC promised support for the project but with reservations about being committed to a capital outlay forecast for two prototypes at about £3,000,000. Hildred reported that the current estimate was nearly £7,000,000 which took no account of the cost of fitting turbo-props to the aircraft; he asked whether the Corporation was still in favour of the venture proceeding? Without the Brabazon I the North Atlantic picture was bleak, and Hildred suggested several possibilities - flight refuelling for Tudors, increasing the range of the proposed Brabazon III, awaiting the arrival of the jet-propelled Brabazon IV, developing an entirely new landplane, and finally a greatly enlarged flying boat, the Saunders-Roe Princess. Not surprisingly Knollys reminded Hildred that from the start BOAC had disagreed with the Brabazon I as being too large and too great a step into the

Two views of Avro Tudor 2 G-AGRF *Elizabeth of England*. This aircraft had a somewhat cmplicated history. It was delivered to BOAC for trials on 21 January 1947, but rejected and returned to Avro on 11 April. It was converted to a Tudor 4B in 1948 and sold to British South American Airways, and then re-acquired by BOAC on the take-over of that company later that year.

Somewhat strangely the aircraft not only had its name painted in an unusual position, it also had the word 'Speedbird' painted alongside the logo. *(Peter V Clegg collection)*

Above: Tudor G-AGRE in flight

Right: a mock up of a Tudor 2 cabin.

unknown and added that his Board felt the project should lapse 'if other provision for suitable aircraft for the North Atlantic can be provided by 1950'.

Regarding Hildred's alternatives, BOAC did not favour flight refuelling since a great deal of development work would need to be carried out before it could be shown to be suitable for passenger carrying. On the Brabazon III, BOAC was adamant this important Empire type should be left unaltered; the Brabazon IV was of great interest, and BOAC would be willing to order a small number to be ready for 1950. It was opposed to any attempt to design another new type from scratch in the light of the American competition and preferred to look to the Brabazon IV for the North Atlantic in due course. The Saunders-Roe flying boat might be of interest, but its feasibility could well depend on adequate ground facilities.

It was clear that the Socialist Government would take no action on BOAC's request for American aircraft until after the impending agreement with the United States on dollar credits and lend-lease. By early December that agreement was signed, and Knollys was pressing the Minister for a decision, made more urgent by the announcement of PAA's order for twenty Stratocruisers. BOAC, alarmed by the PAA order, now asked for a firm purchase of five Stratocruisers in place of only options, in addition to the 1five Constellations. After further delay, until Congress had passed the loan and lend-lease agreement, BOAC was told on 24 January 1946 they could order the Constellations, but permission was withheld from going ahead on the Stratocruisers. So at last on 4 February a cable was sent to Lockheed ordering five Constellation 049s and twenty-eight engines. BOAC with relief felt it could now face up to American competition on the Atlantic at least in the short term. The Constellations for BOAC all came from the US Army Air Force and were delivered to BOAC in May, June and July 1946. Strenuous efforts enabled BOAC to start the first British London-New York passenger services on 1 July with Captain Oscar P. Jones in command of G-AHEJ *Bristol II* with seating for 42 passengers.

Chapter Four

Becoming Competitive

Without any doubt, BOAC was facing severe problems with the outbreak of peace. These were expressed by Viscount Knollys in the Annual Report for 1945/46, where he concluded his final summary with the gloomy phrase '*...it must be expected that financial results at least in the immediate future will be substantially worse than in the year under review*'.

Such an outcome was inevitable; during the war there had been only one objective without commercial considerations - to operate services following the requirements of the Government in the furtherance of the war effort. To do so, BOAC used converted military types in small numbers to the extent that early in 1946 it was operating a fleet of two hundred and seven aircraft of seventeen different types fitted with sixteen different kinds of engines from as many as nine bases. It was a complicated system that required an inflated number of staff to make it work.

It was going to take time to get this makeshift organisation and equipment into a semblance of a commercial airline; to do this would depend on the availability of aircraft designed specifically for civil operations since in a competitive world economics would be paramount.

BOAC entered the post-war period with an existing route network that offered immense potential for growth in the future. The USA and Canada were available to the West, the African Continent to the South, and the Far East and Australasia to the East. All of this was mainly an English-speaking area so far as potential travellers were concerned with close family and business ties to the United Kingdom. Moreover, except for the USA, almost all the countries were either in the Commonwealth or were Colonies. They housed large numbers of expatriate British personnel who would repeatedly be taking home leave.

This meant that BOAC had significant reserves of traffic to access and was able to operate during this period without restriction to a large number of overseas destinations. There were many countries which were later to prove difficult in those early days after the end of hostilities who welcomed BOAC operations and sought advice and technical assistance to get their own airlines off the ground.

No airline can operate for traffic purposes into and through foreign territory without the agreement of the country concerned, and although the UK made great efforts through ICAO to secure the acceptance of some form of multilateral cooperation, this proved unsuccessful with the result that bilateral agreements had perforce to be negotiated. At the time the Bermuda Agreement was reached between the UK and the USA in early 1946, the UK had only five such agreements in force, but by the end of the year the number had risen to twenty, and it continued to increase rapidly mainly as BEA spread its wings into Europe. In the early post-war days from 1946 to 1949 difficulties for BOAC over bilateral agreements were minimal, but they increased in later years.

Commonwealth co-operation was assured with the negotiation of a bilateral agreement between the UK and South Africa, but a meeting to this end was frustrated due to the calling by the United States on the UK initiative of a conference in Chicago in November 1944. This was intended to hammer out the shape of post-war civil aviation and was to be attended by the free nations of the world including the USSR which, however, declined to be present. Prior to the meeting the Commonwealth countries met again and adopted as their united aim the establishment of an international authority to regulate frequencies and capacity on all trunk routes in the interests of order in the air and the avoidance of wasteful competition, and Lord Swinton's opening speech to the fifty-four nations' representatives set this theme. The five-week meeting in Chicago had little difficulty in reaching agreement on technical issues but failed to reach accord on international control.

The USA, with a domestic philosophy of free enterprise, regarded civil aviation as just another form of trade; they ignored the catastrophic effect which this kind of trading would have on other international carriers who did not have the same resources and traffic generating ability. This was utterly antipathetical to the UK policy of order in the air with a particular aim at this time when the post-war scene was to be established, namely, to obtain protection at least for a period for the airlines of those countries which because of the war had for one reason or another not been able to develop their aircraft or their airlines.

The USA was fundamentally opposed to any form of international regulation. Though the USA and UK teams tried to work out some form of compromise, this proved impossible. It is worth noting that the USA insisted on concluding an International Air Transport Agreement, the adherents to which granted each other 3rd, 4th and 5th freedom rights; this was signed by the

US, some South American countries and Sweden but was found by the US to be so unsatisfactory that three years later they withdrew their adherence.

The Conference set up an institutional framework for post-war civil aviation - The International Civil Aviation Organisation (ICAO). This convention established a multilateral legal basis on which international air transport could be further developed by bilateral agreements between States. The Convention made a fundamental distinction between scheduled and unscheduled flights. While traffic rights for scheduled services were to be for bilateral arrangements, the Convention gave aircraft not engaged in scheduled flights the power to make trips in transit across each contracting State's territory and to make stops for non-traffic purposes. Under the Convention were established cardinal principles - that each State had sovereignty over its own airspace, that registration in a country established an aircraft's nationality, that uniformity of conditions govern aircraft engaged in international air navigation, and that there would be international co-operation in securing the maximum standardisation of aeronautical laws, procedures, codes and practices and in improving facilitation.

Establishment of these principles was of vital importance to BOAC with potential post-war routes stretching out worldwide. As important was the setting up of the International Air Services Transit Agreement (IASTA) under which its signatories granted the airlines of other contracting States the right to overfly and to land for non-traffic purposes; the problems which arose in later years with countries which did not sign this agreement underline its value to BOAC.

The Convention also established the right to withhold the right to carry cabotage traffic; this was of untold value to BOAC with British Colonies spread at convenient intervals around the world.

Thus the Chicago Conference had valuable results for BOAC's future trunk operations. An incidental domestic principle was established in that the BOAC team at the Conference was consulted at all stages of discussion and before any agreement was reached. This was after that accepted as standard practice at bilateral negotiations and at international meetings.

While the failure to reach agreement on the control of traffic rights on a multilateral basis forced the UK to obtain the rights that BOAC required under bilateral agreements, it is probable that BOAC obtained a better result in this way than would have been possible under any international arrangements if such had been agreed at the Chicago meeting.

Once the Government infrastructure had been set up

Christening of the first of the Handley Page Halton. Left to right: - Sir Victor Tait; Capt DD Haig, co-pilot; Eng/Off GA Battye; Capt WG Buchanan; Sir William Welsh; Sir F Handley Page; Provost James Strachan of Falkirk; and Robert Lyle, Town Clerk.

Air Vice-Marshal Donald 'Don' Bennett was the driving force behind British South American Airways. Above right: On 1 January, 1946, Lancastrian G-AGWA *Star Light* embarked on a route proving flight from Heathrow to Buenos Aires via Lisbon, Bathurst, Natal, Rio de Janeiro and Montevideo. From left to right are Radio Superintendent McGillavry, Donald Bennett, 'Star Girl' Mary Guthrie (first stewardess to join the airline) and, recently out of the RAF, Wing Commanders Cracknell and Alabaster, both of them Captains under training. Above left: Sir Alan Cobham (left) before the first major trial of his flight-refuelling system in which Air Vice Marshal D.C.T. Bennett (right) flew non-stop to Bermuda in 20 hours in a Lancaster III. Twenty-one further trans-Atlantic flights followed between May and August 1947. Above: *Star Light* was the first Lancastrian to enter service with BSAA. On I December, I 945 it became the first commercial aircraft to use the new London (Heathrow) Airport.

for the post-war years, it was necessary to have available an airline framework for international co-operation. Accordingly, the leading airlines met in Montreal in April 1945 and established IATA with its headquarters in Montreal — also the seat of ICAO. IATA was formed as a free association of scheduled airlines and was in effect a rebirth of the International Air Traffic Association in existence before the war. Eligibility for membership was determined by whether or not an airline's government was eligible for membership of ICAO. Besides an annual general meeting, traffic conferences on a geographical basis were set up which became responsible for the setting of airline tariffs on scheduled services; international acceptance of lATA's fare setting is reflected in that most bilateral agreements base tariffs on IATA rates, though subject to Governmental approval.

For BOAC its importance also lay in the inter-airline arrangements it sponsored — the IATA clearing house in London, the multilateral interline agreement, a worldwide policy as to travel agents with standard commissions and standard tickets and waybills.

BOAC played a leading part in setting up IATA with McCrindle, one of the original members of the Executive Committee.

Vital to the post-war scene was an air agreement between the United States and the United Kingdom. This was negotiated and signed in Bermuda in February 1946 - the UK giving up the idea of predetermination of capacity and the United States finally agreeing to the negotiation of fares by IATA subject to Government approval. Capacity was to be related to the requirements of traffic between the countries of origin and destination of the traffic and to the requirements of the regions through which the routes passed after taking account of local and regional services. Route schedules were drawn up covering all main routes between and beyond the UK and the USA including round-the-world routes for both sides; US and UK airlines had the same points available in the USA. Access was given to the US to virtually all British colonial points the world over, then including India, Singapore and West Africa.

The result was probably the best which could be obtained as the UK team were under instruction from the British Government to reach agreement. The agreement failed to give any real protection to BOAC against excessive carriage by US operators of 5th freedom traffic against their dumping of capacity. The real benefit to BOAC was probably that this type of agreement became that used virtually in all bilateral agreements between the UK and other nations so that what was a real benefit to the US was likewise available to the UK elsewhere in the world.

With the formation of ICAO, the rebirth of IATA and the Bermuda agreement, a framework had been created for the resumption of civil aviation.

Much of BOAC's Commercial Planning work was concerned with the setting up of a pattern of fares and rates for the deliberations at the IATA Traffic Conferences. A foretaste of the problems encountered in trying to work out an everyday fare basis between companies and subsequently between Governments was experienced following the North Atlantic Traffic Conference in March 1946 which agreed on a fare,
among others, of £90 from New York to London. But when the ticket was put to the CAB by the US Airlines, the Board turned it down on the ground that it was not based on operating costs; the IATA fare-fixing process had then to be repeated, a situation which was to recur several times in the future and involved the commercial staff of the IATA airlines over the years in an immense amount of work and endless negotiations.

BOAC's marketing effort had made a positive start, but the going was hard, for BOAC faced constant governmental meddling – for example a Minister took upon himself at a meeting with the three airline chairmen to lecture them on the form and content of their advertisements saying that it was *'a waste of effort to publicize the initials of BOAC, BEA and BSAA separately'!*

The difficulties met by the sales teams in the field were concerned with lack of capacity, with low frequency and uncompetitive aircraft. The travel agents were all competing for post-war traffic and felt they had to spread their business between the airlines, mainly if they could not precisely meet their clients' wishes with what BOAC had to offer them and this attitude affected not only passenger but also cargo traffic, where BOAC had repeatedly to refuse loads for lack of capacity. BOAC did the best it could to cater for the cargo market by the operation of Lancastrian freighters on both the South African and Australian routes, but its best was still inadequate. It was not until the spring of 1949 that conditions. began to improve and the Government eased the travel restrictions.

On all routes competition was increasing but was especially keen on the North Atlantic where BOAC was heavily out-gunned. By the spring of 1949, BOAC was

Boeing 377 Stratocruiser G-ALSC *Centaurus* with a Commer passenger coach alongside. 'LSC had originally been ordered as LN-LAF by Scandanavian Airlines System before being taken over by BOAC. The destination board on the coach states 'South Africa via Sicily, Egypt, The Sudan'.

The Boeing Stratocruiser was famed for its lower-deck lounge and bar area - something that BOAC's marketing department was quick to capitalise on, as shown here with a picture from one of their sales brochures.

still operating only three Constellations per week to New York of which two were routed over Prestwick. Over the same route but direct London-New York PAA was operating ten Constellation and two DC4 services while AOA had seven Constellations - a total of nineteen US airline services a week against BOAC's three.

Competition from the sea was also a factor. A high percentage of travellers were still using ocean liners even at the end of this period. For example, it was not until 1957 that air passengers over the North Atlantic equalled those by sea and the relative positions on the UK-Australia route were even more in favour of sea travel, but there the air fare was well above even the top of the first-class sea fare. The potential for building up the flow by air was clear enough but had to wait for exploitation until the air fares became more competitive in later years.

Corporate Organisation...
There had been little change in the Corporation's overall structure since its formation. Reith, Pearson and Knollys were all full-time Chairmen, and each had had a DG who was in effect a Managing Director. With the retirement of Critchley early in 1946, Knollys reviewed the organisation and decided that it would be inappropriate to appoint a Chief Executive from within the Corporation and prejudicial to the team spirit to bring someone in from the outside. So he concluded there was no satisfactory alternative to himself combining the tasks of Chairman and Chief Executive.

The Chairmen then had sixteen separate functions reporting to him, although in practice this was an understatement since one purpose consisted of the Regional Directors of which there were five each in charge of a mirror organisation of Head Office. At that time the five Regions were India and Burma under Gilbert Lee in Karachi, Middle East under Keith Granville in Cairo, Southern Africa under George Baldwin in Durban, West Atlantic under Vernon Crudge in New York and finally Home Stations and West Africa under A. J. Quin Harkin at Airways Terminal. These Regional Directors reported to the Chairman and were responsible for their Regions. No doubt this cumbersome set-up survived for so long because of the divisive effect of the war; communications were often tricky between the UK and the overseas areas which led to overseas bases being established and in some cases operations which did not even touch the home territory. The operating activities of the aircraft fleets were divided into Lines, with the Line managers reporting to their respective Regional Director.

Even before the introduction of the new organisation, the Chairman had raised the status of the heads of the Technical, Administration and Commercial Departments, so that the former Chief Accountant, G. T. Mellor, became ADG (Administration), the Technical Director, Air Vice-Marshal Sir Victor Tait, ADG (Technical), and the Commercial Director, R. D. Stewart, ADG (Commercial). At that time Campbell-Orde was Assistant to the Chairman dealing with Technical developments, including new aircraft types.

The new organisation which appeared at the May 1946 Board was intended to meet the requirements of the post-war period with the controlling executives based in London and with the creation of a new commercial organisation. It consisted of a Head Office with five functional departments plus a training organisation and the establishment of three new Divisions, Atlantic, African (later African and the Middle East) and Eastern, each under the command of a Divisional Manager responsible to the Chief Executive and based in the UK. The operating Lines were allocated to Divisional control in accordance with the territory where they mainly operated, so eventually, the Atlantic

The beginning of the journey for many BOAC passengers - Airways Terminal, London.

Division embraced No. 3 Line - Constellations - the African Division No. 1 Line (after the disappearance of No. 5 Line based on Cairo and No. 8 Line based on Durban) - Dakotas and later Haltons - and the Eastern Division No. 2 Line - Lancastrians and Yorks - and No. 4 Line - Sunderlands. The European Region was to be disbanded once the Divisions had taken over, whilst the Overseas Regions became areas within their respective Divisions and reported back to Divisional headquarters. Maintenance and overhaul of aircraft in the UK was to be placed under Head Office and, when conditions permitted to be centralised at Heathrow Airport. In addition to these three main divisions, one further Division played a vital part, namely the Repair Division very primarily centred on the Treforest overhaul and repair base although it also included the Croydon shops which were being run down as more work was diverted to the Line base at Hurn.

The Chairman in his dual capacity had a hefty load, which he hoped to lighten by the appointment of two senior assistants. McCrindle retained his title of DDG, although no DG as such existed, his responsibilities being confined to international affairs and to acting as central secretariat for IATA matters in BOAC. The three ADGs reverted to plain Directors of Technical, Administration and Commercial Departments respectively and Campbell-Orde was put in charge of a Technical Development Department.

Lord Knollys retired on 30 June 1947 after having done all that could be expected of him during a tough period to fight BOAC's battles with Government for equipment and bases, and in spite of communication, difficulties having shown his face world-wide to anxious staff cut off so long from home leave. His place was taken by Sir Harold Hartley, the former Chairman of Railway Air Services and former Vice-President of the London Midland and Scottish Railway. He had been a member of the BOAC Board from January 1946 while waiting to become the first Chairman of the newly created BEA on 1 August 1946.

BOAC gained a second experienced aviation man in Air Commodore Whitney Straight who also came in from BEA with Sir Harold, to the post of Managing Director (Chief Executive). He was an experienced pilot, had run his own aviation conglomerate before the war and finished the war in command of RAF Transport Command 46 Group in the UK after having built up the Command's 216 Group in the Middle East. Effective from Whitney Straight's appointment McCrindle was made Managing Director (External Affairs).

The new Divisional set-up came into being during the latter part of 1946, but within a year the new men at the top announced a revised organisation approved by the Board on 18 June 1947. The next month Whitney Straight outlined the proposed changes both to Management and to the staff under the auspices of the National Joint Council. The Divisional basis was maintained but streamlined by merging the African and Middle East Division to form an enlarged Eastern Division. The operating Lines would continue to function within each Division, and all en route stations would report direct or through area managers to the Divisional heads. The two men appointed to manage the Divisions were Vernon Crudge for Western and John Brancker for Eastern while Keith Granville was put in charge of the UK Region covering all sales, traffic and catering affairs within the UK and acting as agent for the Divisions in the home territory. Rather strangely overall day-to-day responsibility was withdrawn from Head Office, and Head Office chiefs were made advisory only. The Technical Department ceased to exist as such, but Sir Victor Tait up to then Technical Director became Deputy Chief Executive to Whitney Straight. Campbell-Orde remained responsible for technical development and Stewart for commercial development.

In this new set-up, executive engineering became the responsibility of the Divisions. The Corporation was to be controlled by a Management Committee comprising the heads of the Divisions together with the Head Office advisory chiefs all under the Chief Executive. There were other Board changes in 1948. Sir Harold Howitt who had been Deputy Chairman since May 1943 retired on 31 March, and his place was taken by a new appointment. Sir- Miles Thomas initially became Deputy Chairman of BOAC and later Chairman when Sir Harold Hartley retired fifteen months later.

It was not long before an even more radical reorganisation occurred. The event which triggered this off was the Government's declared intention to merge British South American Airways with BOAC, discussed secretly at a special Board meeting in BOAC on 25 February 1949. Miles Thomas led the discussion, having had a request from the Minister to explore the best means of amalgamating the two Corporations. He submitted to the Board Heads of Agreement arrived at between the Chairman of BSAA, John Booth, and himself after discussions between himself and Sir George Cribbett, the DGCA.

It was proposed that operationally the merger should start on 1 April 1949 although under the Airways Corporations Act 1949 the merger formally took place on 30 July. There were to be two Deputy Chairmen, Whitney Straight responsible for the technical side of the enlarged BOAC, and John Booth for the commercial team but in particular, it was proposed ultimately to move to a functional organisation in BOAC, to abolish Eastern and Western Divisions and to give much greater emphasis to the operating Lines, along with a preference for amalgamation with BEA.

Whitney Straight, who had been absent through illness from the special Board meeting, but had been sent a copy of the Heads of Agreement, reacted sharply, protesting that the decisions *'...have been arrived at in a most superficial manner'*. Whitney Straight's concern was over the plan once again to re-organise BOAC, pointing out that it was only a few months ago that the Board had agreed to structure the Corporation on a de-centralised and non-functional basis and that re-organisation had only just been completed. In his view to reverse existing policy accepted very recently by all concerned *'can only result in serious doubts being created in the minds of staff as to whether the Board of the Corporation knows its business'. He* made it clear that he was completely opposed to a further change to a functional set-up at that time. He also raised doubts about the wisdom of including any reference to a merger with BEA. Furthermore, he asked some a number of questions as to any resultant liability on BOAC for BSAA's fleet of Tudor IV aircraft and for BSAA's potential liabilities in connection with the Saunders-Roe SR.45 flying boat being built for them.

Whitney Straight's comments were realistic and cast a somewhat disparaging light on the Heads of Agreement as first submitted to the Board. His objections were supported by another member of the Board, Lord Burghley, who had not been able to attend the special Board meeting.

As a result of these objections, an improved and simplified Heads of Agreement was put to the regular Board meeting on 10 March when the offending reference to BEA and to a functional organisation was omitted although the ultimate disappearance of the Divisions was foreshadowed. Nevertheless, the new organisation came into effect on 1 January 1950 under the Chairmanship of Miles Thomas.

To achieve this, on 15 March 1949 the Minister, Lord Pakenham, announced the decision in Parliament, and Sir Harold Hartley issued a notice to all staff making it clear that staff assimilation would be handled through the National Joint Council. There was no reference to any further reorganisation, and in fact, at the April meeting, Miles Thomas submitted a memorandum which showed that wiser counsels had prevailed. His words were that '...*it would be quite suicidal to attempt at this date to make a material alteration to the scheme of re-organisation that only became effective from 1 January this year*'. After referring to the arguments for and against divisional and functional organisations, the recommendation was that BSAA would be absorbed into BOAC as a third Division, and the question of a further reorganisation should be left open for the time being. To all this, the Board agreed.

The Airways Corporations Act formally brought about the merger as from 30 July 1949. The reasons given to Parliament by the Minister were linked to the grounding of BSAA's Tudor IV aircraft following the loss of two aircraft, and the absence of any re-equipment possibilities in the near future. Furthermore, the Minister considered that operations to the Americas should be tackled as one entity. In December, the Air Corporations Act became law, a consolidating Act bringing under one roof all the enactments relating to the Corporations.

So the period from January 1946 to July 1949, brought many changes not only in the organisation of BOAC but also in its Management. During these formative years, BOAC built up a comprehensive range of activities necessary for running a global airline. The area where the war years had left the widest gap lay in the commercial field. It was early in 1946, in the expectation that the Government's priority control on aircraft space would be relaxed that the Corporation set about establishing significant selling and advertising organisations. In April of that year a comprehensive statement dealing with the policy of the Department, the ground it should cover and the relationship between the Department and the Publicity and Liaison Department. The emphasis was on selling and advertising both of which required building up from scratch. It was decided in April to open a sales office selling directly to the public in Regent Street where sufficient accommodation existed for a booking centre and counter and housing the Head Office of the Sales Branch. It was from these small beginnings that the worldwide marketing organisation grew up in later years.

Although a small Medical Department had been inherited from the pre-war years, there was a need to take advantage of all the experience gained in aviation medicine during the war. By June 1945 Sir William Tyrrell, in charge of medical services, had expanded the department to deal with staff health both at home and overseas, also covering the particular requirements of aircrew both as to health checks and matters of fatigue.

It was also planned to keep in touch with medical research on hygiene and especially to cater to the needs of staff in challenging climates overseas. With the extension of routes, the problems of quarantine of both aircraft and passengers were pressing and called for an expert - and diplomatic -handling by the Medical Department to ease matters. Sir William's proposals were approved by the Board in January 1946 and from these foundations grew up a medical and dental service second to none in the airline industry.

In the field of aircrew training, progress was made and, with the availability of Aldermaston, BOAC was able to establish a cental training school in March 1946 for landplane crews. This was not an ideal base as the buildings were scattered and unsuitable for training purposes, but the need was urgent for many pilots and engineers from the RAF needed to obtain civil licences. There was a considerable intake of new men due both to the multiplicity of aircraft types and to the departure of seconded RAF personnel now that the war was over. The uncertainty as to when the new aircraft types would be arriving, especially the Tudors, caused difficulties as the continued use of military conversions with limited carrying capacity required a higher number of crews.

In September 1946 after discussion between BOAC and BEA, a proposal was put up for a joint training organisation. Many doubted the wisdom of this, mainly for costs, and it was February 1947 before the scheme was approved and on 30 April Airways Training Limited was incorporated. In its first year of working the company incurred a loss of £44,600 and although it was fully occupied until the end of 1947 by September of that year, BOAC then concluded that training could then be decentralised back to the Lines at a considerable saving in expense! As a consequence Airways Training as a company was run down with the object of closing its activities and training at Aldermaston ceased as the end of September 1948.

1946 also saw the birth of a top-level committee which had considerable influence on policy. Although BEA was not formally set up until the passing of the Civil Aviation Act in August 1946, steps had been taken well before that to ensure there should be consultation between the two Corporations. To arrange for this, the Airlines' Chairmen's Committee (ACC) was formed in January 1946 with appropriate sub-committees. Initially, BSAA was also a member, but after the merger with BOAC the Committee was continued and remained in existence for the rest of the lives of BOAC and BEA. It proved an invaluable forum for the exchange of views, and for the hammering out of standard policies. In later years when the pressures of the competition were severe, and differences of opinion arose from time to time between the two Corporations at levels below the Chairman, the Committee served in effect as a final court of appeal and frequently proved its worth in settlement of complicated issues. In particular, it enabled the Chairmen and top executives to work out common policies where appropriate especially when an approach to the Minister was involved.

...Facilities...

For an international airline to function efficiently a whole range of ancillary facilities are necessary. The airline needed adequate headquarters and bases for maintenance and overhaul, airfields and marine bases with long enough runways and alighting areas, terminal facilities suitable for passenger and cargo handling, effective ground organisation, communication systems and navigational aids including approach and landing aids and reliable air traffic control. Not surprisingly after six years of war some of these requirements were found wanting or unsuitable for civil purposes. It is true that the war had been a forcing ground for great advance particularly in the availability of concrete runways and in the extensive use of radio and radar. Unfortunately, the great new runways built for military aircraft were for strategic reasons in locations often unsuitable for civil exploitation. There were important exceptions of which Gander in Newfoundland and Goose in Labrador were possibly the best examples. Until civil aircraft appeared which could fly the North Atlantic non-stop westbound as well as eastbound - and the DC-7 was the first of this breed - the Atlantic civil operation could not have been developed so rapidly after hostilities ceased if it had not been for these two vital refuelling stations.

The Atlantic route was reasonably well equipped with radio aids, meteorological services and radio communications as a result of the earlier Return Ferry Service operations, but this was far from the case on other routes where lack of adequate airfield lighting restricted night flying. This limited many operations to daylight hours with a repercussive effect on utilisation and so on costs. Ground aids overseas had been provided at many points by RAF personnel but as these were demobilised a vacuum was created, filled where possible by staff provided by BOAC. It was to meet these needs that International Aeradio (IAL) was set up and where local Governments were not equipped for the task, IAL gradually took over and was widely used by the MCA to meet ICAO recommended practices.

Through all the upheavals BOAC continued to provide their passengers with a wide range of information to assist their travel. One item that now seems almost archaic was advice on what to pack, and

The early days of Heathrow. Above: the original control tower.

Right: A line of telephone boxes and a post-box stand somewhat incongruiously outside the rudimentary passenger 'facilities' at London Airport in 1946.

Left: A policemans 'stands guard' outside the entrance to the passenger reception tent. Marquees were used before the erection of temporary buildings, which in turn were replaced by the dedicated passenger terminals.

The ground is covered with either wooden 'duck-boards' or Pierced Steel Planking' (PSP) to keep the travelling public's feet out of the mud!

The BOAC *'Before you Take Off'* booklet was in five sections, Booking, Planning, Packing, Clothing and Flying. There was also a supplimentaty leaflet to be completed regarding Heath, Passport, Departure, Baggage and Currency requirements.

how much things weighed. This was something of a hangover from the pre-war days of Imperial Airways, when in 1935 IA published a booklet 'Advice for your Maid or Manservant as to what should be packed or required for the journey'. About the only difference was the remove of the phrase 'Maid or Manservant'!

'The speed of air travel is such that the transition from mid-summer to mid-winter may be a matter of hours as far as ground temperatures are concerned, and on a long flight you may meet wide variations of temperature at your points of call. Your choice of clothing should be correspondingly flexible. Our staff will be glad to advise you on the most suitable choice for your particular journey.

In temperate climates, however, no special considerations in regard to clothing are necessary, but tight-fitting garments, including tight shoes or boots should be avoided. A warm overcoat in winter and a light coat in summer should always be carried.

When flying from temperate to tropical climates, start in your everyday clothes, but carry tropical clothes with you as a part of your luggage so that a change into a light-weight summer dress or suit can be easily made as soon as you feel it is necessary.

The temperature of the air falls quickly with increase in height and you may find conditions cooler than you

THE LIST ACCORDING TO BOAC - THINGS TO PACK

FOR LADIES	lb	oz
28 inch suitcase	6	8
22 inch case	3	13
3 pairs shoes: sports, promenade, walking	3	0
Beach slippers / mules	1	5
Evening shoes	0	6
Belt	0	4
2 slips and knickers	0	11
2 silk and wool vests and knickers	0	13½
2 camiknickers	0	6½
4 nightdresses	1	5½
Dressing gown	1	3
6 pairs stockings	0	8
Bathing dress	0	5
2 jumpers	0	9½
Cardigan	0	7½
Tweed suit	2	7
Flannel suit	1	7
2 afternoon dresses	2	14
Evening dress	1	12
2 dinner dresses	2	8
4 washing frocks	4	2
Crepe-de-chine blouse	0	6
Lightweight hat	0	2
2 pairs gloves	0	4
12 handkerchiefs	0	5
1 evening bag (empty)	0	9
Total	**38**	**5½**

FOR MEN	lbs	oz
28 inch suitcase	6	8
22 inch case	3	13
Tweed suit	5	2
Tropical suit	2	8
Dinner suit	4	12
Lightweight suit	4	0
Pair lightweight flannel trousers (or 2 pairs shorts and 2 pairs stockings)	1	8
Walking shoes	2	0
Brogues	2	12
Pumps	1	14
Bathing Trunks	0	12
Beach Sandals	1	2
3 pairs pants	0	12
3 vests	0	12
8 shirts	2	8
12 collars	0	4
8 ties	0	4
Braces	0	3
8 pairs socks	1	0
12 handkerchiefs	0	8
3 lightweight pyjamas	2	4
Pair suspenders	0	2
Pullover	0	6
Scarf	0	3
Dressing gown (light)	1	9
Total	**47**	**6**

expect. A warm overcoat is a good thing to have with you, even when flying into or through the tropics, but rugs and pillows are obtainable from the steward.

The clothing lists which follow are intended merely as guides, but they are based on practical experience and should help you in your choice - both from the point of view of the type of clothes you will need and also how much you can fit in to your 66 lb. free allowance.

A step forward was taken in 1949 in the UK with the agreement to establish 'airways' under control from the ground and with aircraft separated by agreed standards and with both military and civil aircraft brought under a common air traffic control (ATC) system. This result was a product of the Government's Inter-Departmental Aircraft Control Committee which had formulated such a policy with the object of setting up joint air traffic control centres where required.

In the immediate post-war years, it was hard to find accommodation in or near London where all the scattered office staff could be concentrated. Some staff were still in Bristol, and others were parcelled out in buildings around Airways Terminal. Only a limited number of more senior staff had been accommodated in Airways House, Berkeley Square, which had been in use since 1944. Space was acquired in Stratton House, Piccadilly, but this was inadequate to solve the problem. In November 1946 the Board authorised the purchase for £620,000 of the Simmonds Aerocessaries factory on the Great West Road at Brentford as a solution to accommodation shortages and also to satisfy the long term requirements. In just over a year around 1,200 staff were installed in the former Simmonds building and by December a plan was in place to house all the London staff in three buildings - Stratton House, Airways Terminal, and the new building at Brentford. This made it possible to give up Airways House, Berkeley Square, as the lease ended in March 1948. Extra accommodation had to be provided at Brentford by the erection of semi-permanent huts, and it was expected that space in Stratton House would be redundant.

Not only did BOAC had to find competitive aircraft they were also in desperate need of a single maintenance, overhaul and operating base for its landplanes. This was to be the new airfield at Heathrow. The development of Heathrow into a London airport was of enormous importance to BOAC. There had been an airfield at Heathrow since 1930 known as the Great West Aerodrome, used mainly for testing the products of Faircy Aviation. When the war broke out the RAF took it over but construction was slow and it never became fully operational. The Air Ministry were developing it for the RAF and when it became the responsibility of the MCA at the beginning of 1946 one runway, a control tower and a few buildings had been completed.

The opening of the new airport for regular use was linked to the provision of suitable buildings otherwise it could not operate; moreover, in BOAC's case alone, there was urgent need for accommodation for up to 3,000 staff. The same kind of problem existed at Hurn and Aldermaston, whilst the flying boat base at Poole could not be moved to Southampton unless essential works could be carried out.

BOAC entered 1946 with twelve maintenance and operating bases for its services - Whitchurch, Hurn, Northolt (earmarked for BEA), Croydon, Treforest, Aldermaston, Hythe, Poole, Montreal, Baltimore, Cairo and Durban. In 1947 these millstones round the airline's neck remained unchanged for although Northolt became BEA's responsibility and Durban was closed with the final departure of the C-Class boats, its place had been taken by Bovingdon and the early stages of Heathrow. The tide began to turn in 1948 with the closure of bases at Bovingdon, Cairo and Baltimore but even in the summer of 1949 there remained eight including the new complication of Filton in place of Montreal with all the added expense of a base eighty-odd miles from the operating point at Heathrow - and these numbers do not take into account the new flying-boat bases at Lake Mariut, Alexandria, in place of Red el Farag, and at Korangi, Karachi.

The effect of this widely dispersed organisation on efficiency and costs was incalculable; at one time BOAC estimated it was involved in an extra charge of around £1m per annum as compared with what would be the cost if all landplanes could be maintained and operated from one central base.

At bases such as Hurn and Bovingdon, there had to be provision for hangars and workshops and facilities for housing staff although these were known to be only temporary expedients until Heathrow became available.

There was much-wasted effort - Bovingdon being a case in point. This base was opened in the summer of 1946 at the airfield which had been in use by the US Army Air Force. No. 1 Line moved there from Whitchurch, still retaining their base at Whitchurch for the Dakota operations but making the move with the object of forming at Bovingdon a support for the Haltons. There were serious problems, and delays with the Halton but operations to West Africa took place from September 1946 but ceased a few weeks later due to failures in the Halton's hydraulic system, and they did not restart flying until July 1947; meanwhile, the base had to be kept manned.

By the summer of 1947 about a thousand BOAC staff were at Bovingdon with all the support facilities that were needed - and that at a time when conditions were difficult; in fact, in the bad winter of 1947 they were even more difficult than they had been during the war for fuel, food and housing. The Bovingdon base lasted for eighteen months and due to technical problems, the Haltons clocked up far fewer revenue flying hours than had been originally planned. Writing in January 1948, the Chief Accountant reported that in the financial year 1946/47 the aircraft were effective for a total of only 1,000 hours and that of the £193,000 costs at Bovingdon, as much as £160,000 was unproductive. It was only when temporary hangars became available at Heathrow that the Halton could be transferred there and enabled the Bovingdon base to be closed early in 1948.

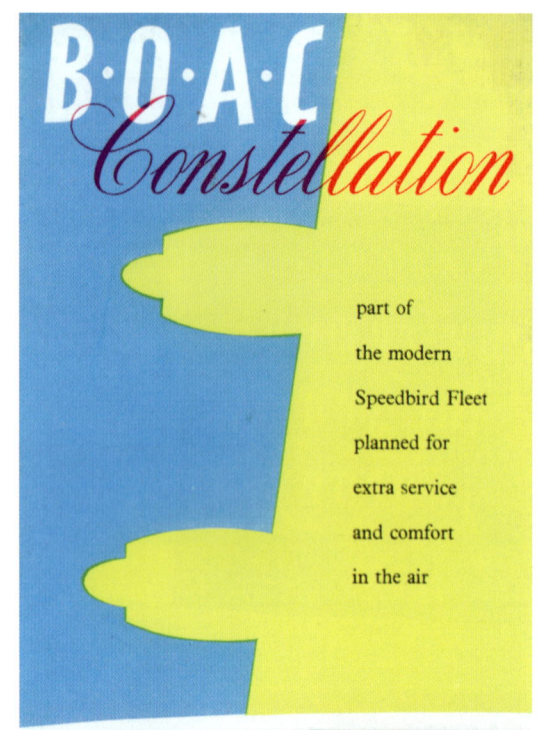

On June 1 1946 Heathrow was formally opened to international traffic with a bare minimum of facilities. There were some prefabricated buildings left by Transport Command and the BOAC staff took these over for working areas and additional accommodation was provided by caravans. Passenger reception and handling had to take place in tents. No hangars or workshops were available and BOAC continued to use Hurn and Bovingdon for maintenance. The staff rose to the occasion in these difficult conditions and under the newly appointed station manager, G. G. (Gerry) Hawtin, were able to overcome all the handicaps by their enthusiasm, but once again the Corporation was faced with the task of building up a station from scratch as it had already done at Whitchurch and Hurn. Not least of the problems was the need to position service aircraft

Flying is the simplest way of travelling with infants and B.O.A.C. makes special arrangements for their extra comfort. Our long experience, together with careful study of the needs of babies and young children in flight, enables us to anticipate any small problems likely to arise during the journey and the following information is accordingly issued for guidance.

So the quotation went in the BOAC *'Babies as Air Travellers'* information leaflet - One hopes that the addition to the overhead rack was waterproof!

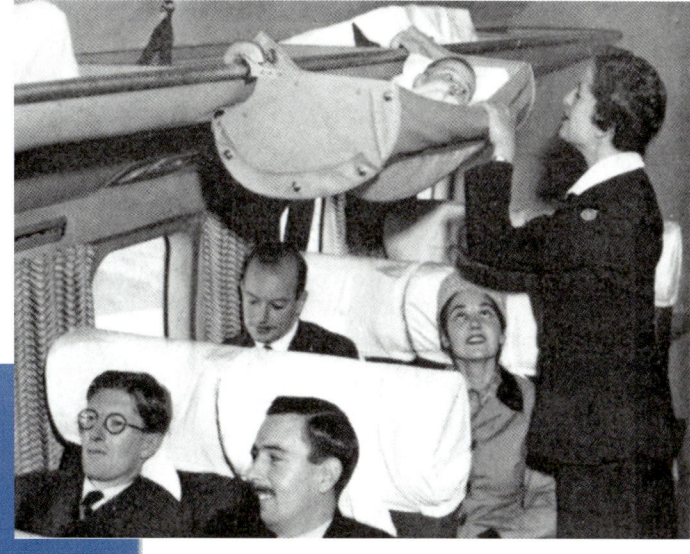

from maintenance bases into Heathrow to be ready for departures. To cap it all, staff housing had to be found in the neighbourhood of the new airport.

A comprehensive plan was devised for the new airport layout and made public at the end of 1946. This called for the construction of a tunnel from the North side into the central area, where it was planned to build the main passenger terminals and a new control tower. By early 1947 work was proceeding in levelling and preparing the ground in this central area and meanwhile temporary accommodation for offices, the passenger terminal, cargo warehouse etc. had been erected on the North side. Unfortunately the country's financial crisis in the autumn of 1947 caused work on the airport to be slowed down. However, by the summer of 1949 a new passenger hall and restaurant, also on the North side, had been opened and facilities for passengers, whilst certainly not ideal, were acceptable. Maintenance was to be established on the East side of the airport where the main maintenance bases for both BOAC and BEA were to be positioned, but apart from the Haltons no major maintenance and overhaul could be carried out due to the absence of the necessary shops. This remained the position until the DC4Ms, the Argonauts, came into service in late August 1949. In the meantime. Hurn and Filton had to soldier on, Whitchurch having closed at end of December 1948 and most of its engineering staff transferred to the new base at Filton.

The transfer of the engineering base at Dorval, Montreal, to the Bristol Aeroplane Company's works at Filton, near Bristol was not without difficulty, and the move reflected credit on all those BOAC personnel involved in organising and monitoring the progress particularly Charles Abell, Manager of No. 3 Line. Dorval had become the main base for the Liberators during the war and then subsequently for the Constellations in No. 3 Line. At that time it was an ideal base; conditions in Canada were easier than in England, spares could be more readily obtained and local staff had

Right: A BOAC traffic clerk checks a passengers's travel documents before he starts his flight.

Below: The front cover of a BOAC leaflet advertising their fully reclining Slumberette '...the chair that BOAC designed for the sky'.

BOAC Slumberette

FOR SUPERLATIVE COMFORT IN THE AIR

Air

become experienced in coping with American aircraft. Things were not so rosy back in 1947 Britain; the country was in the throes of a financial crisis with an acute shortage of dollars. There was growing criticism at the dollar expenditure incurred at the Dorval base although as BOAC had pointed out, there as no alternative in England which could take large aircraft. Hurn was already fully stretched and the buildings there were still inadequate to deal with existing operations. There was a serious housing shortage which would make it almost impossible to find accommodation for transferred staff.

Political pressure increased, and in September 1947 a meeting was held in London between the Managers of the Atlantic Division, No. 3 Line representatives, officials from the MCA and the MoS to discuss the problem. The MCA made it clear that accommodation for No. 3 Line at Heathrow could not be provided before mid-1950 and even then it was not certain to have enough housing for BOAC staff. A committee was set up that established Filton as the only possible solution. This would require hangars and workshops planned for use in the production of the Brabazon, to be allocated to BOAC. It was proposed to earmark one bay of the large three-bay hangars at Filton for BOAC's needs and although the MoS opposed this on the grounds of the possible effects on the Brabazon progress, MCA was prepared to accept any resultant delays and the Bristol Aeroplane Company gave their consent.

On 5 December 1947 the proposal to use Filton was announced in London and formally approved by the Board on 8 January 1948. Government consent was given on the understanding that BOAC would bear the expense of the move and of setting up the new base. This decision meant that BOAC would have to cover the cost of the move from Dorval, the construction of workshops and offices and the laying on of public services together with all the costs of bringing the airfield at Filton up to the standard required by regular day and night operations. This involved putting down landing aids, airfield lighting etc. and even bearing the cost of the salaries of the MCA airfield personnel.

The plan was to house and maintain at Filton the six Stratocruisers the Corporation would be acquiring, hopefully in the spring of 1948; Dorval would continue to look after the Liberators and the Constellations until later in that year. So in due time Filton was to be the main maintenance and overhaul base for all these aircraft until Heathrow was ready to take them.

The contractors started work in April 1948, meanwhile BOAC's Administration Department had the task of finding housing within a reasonable distance of Filton. It was thought that 350 houses were required and this in the midst of an acute shortage. Eventually a solution was found by setting up pre-cast concrete block houses, known as 'pre-fabs' built by the Bristol Corporation at a cost of £100,000.

By the end of November about one hundred staff had been installed and by February 1949 the move was almost complete, with one hundred and eighty-five of the houses were already occupied. By 31 March staff numbers in Dorval had dropped by nearly one thousand and further reduction was planned. It was expected that savings in dollar payments for salaries, wages and overseas allowances would amount to about $4m in a full year. As against this saving had to be put the loss in dollar revenue resulting from the reduction in transatlantic frequencies while the move was in progress. To BOAC it was not only an upheaval to one of its major routes but also a capital outlay of nearly £350,000 together with the actual cost of the move put at £361,583. Much of this would have been avoided had Heathrow been ready.

With the closure of Dorval as an overhaul base, somewhere had to be found for Liberator maintenance; the answer came in the use of Prestwick where the work was contracted out to Scottish Aviation.

In addition to the problems of achieving a central base for landplanes, BOAC was reviewing its flying-boat policy, with an urgent situation growing at Poole. The passenger terminal there was in the Carter's Tile Works building and had done sterling service but inevitably had meant inconveniences to embarking and disembarking passengers. Poole was also expensive, involving a considerable amount of aircraft positioning from the overhaul base at Hythe.

The Corporation was convinced it should return to Southampton. Plans were put to the MCA in 1946 for the development of Berth 50 there, but in February the Ministry experts raised objections to the proposed new building. BOAC had already spent over £30,000 for new floating docks and inability to develop Berth 50 would make the floating docks and inability to develop Berth 50 a pointless expenditure. The advantage of the scheme was the convenience it offered to passengers who could be brought by train right to the Marine Terminal on the waterside next to the flying boats. The Ministry suggested the possibility of an entirely new flying boat base at Langstone Harbour as recommended by Pakenham's committee. This idea was turned down by the Government as being against the national interest as it might interfere with the naval base at Portsmouth.

The new terminal was finally opened by the Minister, on 14 April 1948; in his address he somewhat surprisingly gave a hint as to the shortness of life which the flying boats might be facing when he said that Britain was *"the only country employing flying boats on an extensive scale and we faced a particular problem not to place ourselves at a disadvantage with competitors using landplanes"*.

Most staff moved over from Poole to Southampton on 29 March 1948 and the last flying boat operation at Poole took place on 30 March. On 31 March the first flying boat since 1939 arrived at Berth 50 Southampton.

Above: Boeing Stratocruiser G-AKGM *Castor* undergoes the pre-start-up procedure.

Below and right: BOAC made much of their famed 'Monarch' service offered to passengers, where on the Stratocruiser full sleeping berths were available.

...and aircraft.

By the beginning of 1946 BOAC's Technical Development staff were involved with an almost overwhelming number of airliner projects and were finding the existing government arrangements for design and procurement of civil aircraft cumbersome, complicated, inflexible and leading to unnecessary conflict between the airline and manufacturers. The system required the MoS to place orders with the manufacturer and the Corporation's requirements to be channelled through the MCA to the MoS and then to manufacturers, with the Treasury also involved. It was the MoS and not BOAC, the user, who placed contracts.

BOAC hoped to see new British aircraft types and so be confident that the international challenge could be met. Unfortunately, the reality was very different. The Tudor I which was to have filled the gap in 1946 and 1947 on the North Atlantic until the new types were available now seemed unlikely to be ready before late 1947; this had been accepted by Government and BOAC were permitted to acquire five Constellation 049s. Even if the Tudor Is were ready by 1947, the competition on the North Atlantic from the Americans - and especially those operating the Boeing Stratocruisers - expected late in 1947 or early 1948 would make seats on the Tudors unsaleable. What was disturbing was the lack of any aircraft to follow on for the years 1948 to 1950 after which it was hoped that the DH 106 Comet, the Bristol Brabazon, or the Saunders-Roe SR.45 - named the Princess - would be available to fight the North Atlantic battle.

BOAC had no alternative but to seek the Minister's authority to purchase five Stratocruisers, in addition to the five Constellations to close the gap at least up to 1950; the Stratocruiser would take over from the Constellations by 1948 as the latter lost its competitive edge. After an initial refusal, the Ministry permitted this purchase in August 1946, and an agreement was entered into with Boeing, the delivery dates to be spread between August and November 1947.

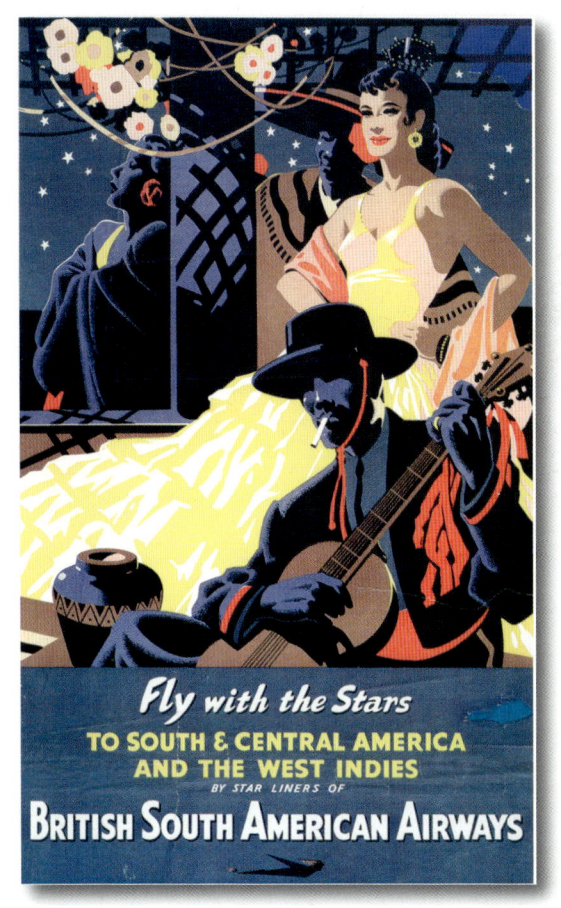

Flying to the sun!
Two posters, the right being a Spanish language version for BOAC's Stratocruiser service to the Caribbean, that replaced the BSAA 'Star Liner' service as advertised on the right.

In spite of feelings in favour of the flying boats, as the large long-range landplanes from the US factories came into use, a view surfaced in BOAC that the flying boat era was drawing to a close. By the spring of 1946 the writing was on the wall when the BOAC policy was defined as to continue flying-boat operations for three years or so until landplanes were available to meet the demand. Meanwhile, flying boats would be used where airfields for landplanes did not exist; moreover, at that time the SR.45 was still a possibility.

In January 1947 the Corporation again considered what the future of the flying boat was to be. Undoubtedly there were still advantages, particularly the availability of alighting areas reasonably near large cities around the world and at no great outlay of capital, but the disadvantages were telling - the extra cost of providing facilities, the fact that the Corporation would be the only major user of flying boats and so unable to get the benefit of the reciprocal use of facilities overseas, the problems of night take-offs in areas open to other surface craft and the time lost in moving passengers to and from the aircraft and city centres. However, BOAC had no alternative in the absence of suitable British landplanes but to carry on operating flying boats for several more years and the policy spelled out in the spring of 1946 was confirmed.

So by the autumn of 1946 BOAC was able to cover itself on the North Atlantic up to at least 1950 without falling behind its competitors in quality of aircraft - quantity was a very different matter. So in respect of the Atlantic operation, the 'Fly British' policy had had to give way for the time being as otherwise no British service at all could have been provided.

The Empire routes, with their less demanding range, the problem of a replacement for the converted military types was less acute. BOAC had depended on the Tudor II to hold its position until the Brabazon III appeared, and had told the MCA on 4 January 1946 that it would support the construction of three of this type, at that time designated the Avro 693. By April it was evident the Tudor II would not be ready for service before the spring of 1947, and there were much the same doubts about its design as were occurring with the Tudor I. The only other aircraft which could help to fill the interim gap until the Brabazon came along was the Solent. The MCA had ordered twelve of these from Shorts early in 1946. They were a civil variant of the Seaford able to carry up to 30 passengers. The first aircraft was launched in November 1946, but a full C of A was not granted until a year later, and it was not until 14 April 1948, the

As with the Saunders-Roe Princess, the Bristol Brabazon was intended for use with BOAC - only one was completed, and never entered airline service.

day the Minister opened Berth 50 at Southampton, when the first aircraft, ceremonially named *Southampton* on that occasion, left on a proving flight to Vaaldam near Johannesburg, This regular service to South Africa started on 4 May, but there were problems with the wing-tip floats and in July all eight aircraft in service with BOAC were withdrawn for modification. By mid-October, all the Solents were back in service, and they remained on the South African and East African routes until superseded by the Hermes in November 1950.

Progress was too slow for the Corporation's liking, but eventually the Minister of Civil Aviation announced in Parliament on 21 January 1948 that the Government was consulting *'a few men of wide business and administrative experience to assist in reviewing the procedure for ordering aircraft'*. True to form, the Government set up a committee of advisers known as the Hanbury-Williams Committee on Ordering Procedure which began its work in April 1948. At about the same time the three Corporations instituted the Corporation's Aircraft Requirements and Contracts Committee to coordinate and formulate requirements on new aircraft and to process orders through the agreed machinery with Government departments and manufacturers.

The British flag might well be driven off the world air routes which the British had pioneered. The Minister went on to tell the House of the authorisation to BOAC to buy Canadair Argonauts. He put this across as a Commonwealth product with Rolls-Royce engines, the Merlin. The Argonaut would carry forty passengers as against thirty-two in the Tudor IV, with a greater range.

From a Bristol Aeroplane Company sales document comes this stunning colour cutaway of how the interior of a Brabazon would have been fitted out - how much input came from BOAC is not known.

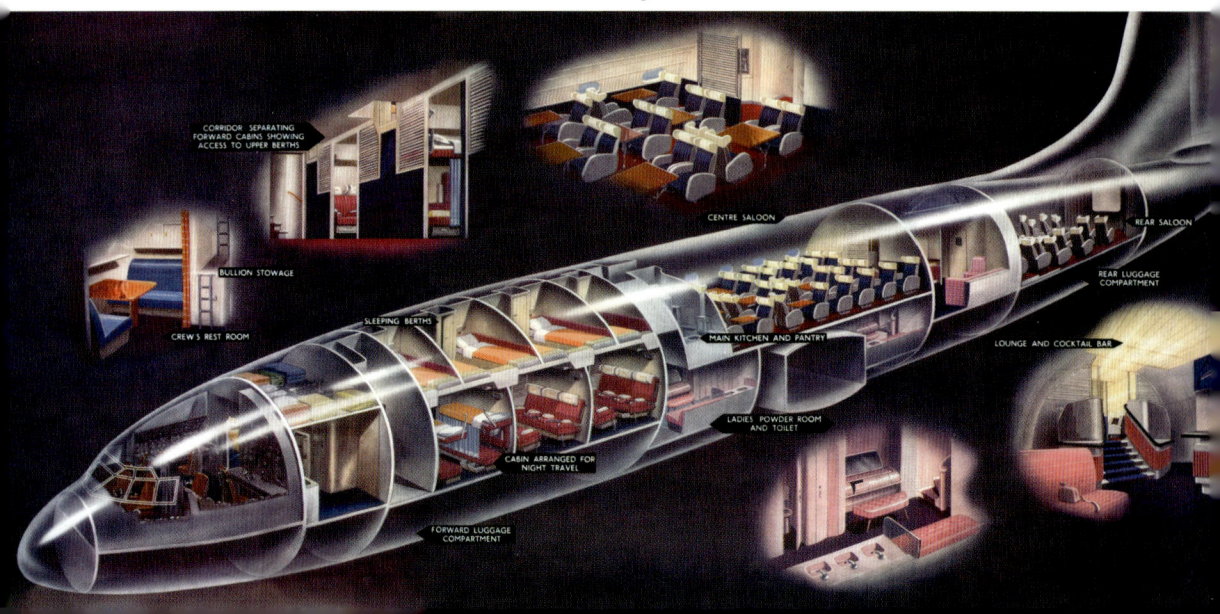

A BOAC travel poster aimed at the American public encouraging them to 'Fly to Britain'

As far back as 1944, BOAC had suggested the purchase of DC-4s from Canada, but the idea was rejected by the Air Ministry on the grounds that the chosen instrument must fly British. BOAC had kept in touch with the Canadian situation through its close association with Trans-Canada Airlines who had started a service from Montreal via Iceland to London with DC-4Ms in May 1947 and were fully committed to the Merlin-engined DC-4. With all the uncertainties surrounding the Tudors and the Hermes, here was an aircraft coming off the production line and successfully flown with British engines in passenger service. The type was pressurised with a tricycle undercarriage and able to carry forty passengers, even having a small horseshoe-shaped lounge at the rear of the cabin. In the autumn of that same year demonstrations of the aircraft's capabilities were given by the manufacturers to BOAC who were told that aircraft modified to make them suitable for the medium as well as long range could be available for service in 1948.

These were strong sales points and, and in addition to reminding the Minister of them, the Chairman attached to his letter of 8 January a convincing financial case for the DC4.Ms as compared with Tudors and Hermes. It is worth quoting verbatim Sir Harold's last paragraph to Lord Nathan because it sums up completely BOAC's attitudes as a worldwide operator in words which continued to apply long after the date they were written:-
'*We make this recommendation in the full belief that the ultimate success of British civil aviation depends not only on the ability of the British aircraft industry to produce the best civil aircraft in the world but also on the existence of strong British airlines with a worldwide reputation for efficiency. We recognise that there must be the closest collaboration between the industry and the operators, and we are now and have always been, most anxious to give the industry the full benefit of our experience. In our view, however, the surest way for the British aircraft industry to take the lead is to concentrate all resources on the rapid development and production of new, advanced and exportable types such as the Ambassador and the D.H. Comet rather than dissipate further effort in the production and modification of the interim aircraft.*'

It was clear that the field was wide open to an aircraft such as the Canadair with its British engines, available within a year or eighteen months and able to offer competitive service on the all Empire routes and moreover a tried success in commercial operation. The financial estimates produced by BOAC showed over a five-year period up to 1952/3 an accumulated surplus of over £10m on Empire routes as compared with a loss of over £10m if the Tudors and Hermes were used.

The arguments were well and truly joined. Whitney Straight for BOAC pressed the Canadair case with telling reasons in the face of considerable political and press opposition on the grounds (quite unjustified if the situation were regarded dispassionately) that once again BOAC was selling out the British manufacturers in favour of American products. In July Sir Harold Hartley gave an assurance to the Minister that the addition of Canadair aircraft to BOAC's fleet would not prejudice in any way the introduction by BOAC of the Hermes, the Medium Range Empire type and the DH 106 Comet. BOAC would continue to co-operate in respect of the so-called national experiments, the SR.45 and the Brabazon. To the credit of the MCA they not only recognized the validity of BOAC's case but acted swiftly; following the Minister's July decision, the Technical Development Director was able to report to the Board in September that a Letter of Intent had been signed with Canadair following receipt of authority from the Minister and that staff were already in Montreal finalising the DC-4M specification for BOAC's use.

The purchase agreement was signed on 30 September 1948, and the first Argonaut was delivered to BOAC at London Airport on 29 March 1949, six months ahead of the contractual date and two months ahead of the target date. On 23 August 1949, the Argonaut entered passenger service and quickly took over from Yorks to the Middle East, Yorks and Lancastrians to Pakistan, India, Ceylon and the Far East. At long last the turn of the tide had come; as the Annual Report for 1949/50 records, the Argonauts on the Far East route turned a deficit of £80,000 a quarter in 1948/49 into a contribution to overheads of £140,000 for the last quarter of 1949/50.

At the time the Government was considering the

The first prototype Saunder-Roe SR.45 Princess G-ALUN in flight along the English South Coast. Although intended for use with BOAC and BSAA, only three were built, and none entered service.

merger of BSAA with BOAC at the beginning of 1949, an immediate problem arose as to the availability of aircraft to fill the gap created by the withdrawal of all BSAA's Tudors. Some means had to be found to keep the services operating on both BSAA's main routes via the Caribbean to Central and West Coast of South America, and via Lisbon to Rio de Janeiro and down the East coast. The solution was in the acquisition by BOAC of four more Stratocruisers on offer for payment in sterling from the Svensk Interkontinental Lufttrafic (SILA), the Swedish forerunner of SAS. All four aircraft were for delivery by the end of 1949 and purchasing the aircraft in this manner was a way of circumventing the dollar deficit. The plan was to carry BSAA's former traffic to the Caribbean and the West Coast of South America on the regular BOAC Constellation services to New York but extending them down via Bermuda, Jamaica and Lima to Santiago; BSAA's former East Coast of South America service over Lisbon was to be operated by Argonauts released from the London-Montreal services by the new capacity to be provided on the North Atlantic by the four SILA Stratocruisers.

BOAC's Stratocruisers fleet was thus being increased from six to ten aircraft and the first Stratocruiser operation took place on 6 December 1949 from Heathrow Airport to New York.

Not only had the Stratocruisers fleet been enlarged in this manner but by 1948 BOAC had also been able to increase its number of Constellations from the original five up to eleven, consisting of six 049s and five 749s. It was in June 1947 that the MCA were formally approached by the Corporation for permission to buy one additional Constellation for use on its Atlantic services. The aircraft could be purchased for $435,000 and would enable BOAC to increase its Atlantic frequencies by three per week. The case made by the Corporation demonstrated that there would actually be a dollar surplus by reason of extra dollar-earning traffic carried. Permission was granted and the aircraft came into service early in 1948.

The next development concerned the Australian route; QANTAS had expressed dissatisfaction with BOAC's Lancastrian landplane and Hythe flying-boat services to Sydney. This became a difficult issue betwen the two partners when QANTAS began in December 1947 to operate to London with Constellations; they felt they were bearing an unreasonable share of the burden, especially since BOAC's Lancastrian flights had become solely mail and freight carriers once QANTAS' Constellations, with their far greater passenger appeal, were on the route. It happened about that time thatAerlinte, the Irish Airline, which had acquired five Constellation 749s for Atlantic services, wanted to dispose of them, and BOAC with Government approval was able to purchase them for use on the Australian route alongside QANTAS whose Constellations were also Type 749; in doing this BOAC was able not only to operate a far more economic and competitive service than previously with Lancastrians and Hythes, but also to ease the strains on the BOAC/QANTAS partnership. After the necessary modifications, the aircraft were put into service in December 1948 and until Heathrow Airport was ready were based at Filton. In fact one of the aircraft involved was hired out to QANTAS to operate on behalf of BOAC until BOAC had an adequate stock of spares to support all five.

Around the routes - the North Atlantic.
BOAC began 1946 knowing it would be late off the mark against its powerful competitors PAA and American Overseas Airlines (AOA). It was 1 July 1946 before BOAC could enter the competitive field with the introduction of Constellation 049s between London and New York. The service opened at a twice-weekly frequency with the Constellation schedules providing for a flight time of 17¼ hours via Rineanna with an elapsed time of 19¾ hours. The return flight with the prevailing wind was 14 hours or 16½ hours elapsed time. The aircraft was pressurised and had seating for forty-seven passengers although this included four seats forward customarily used for crew-rest positions and one place near the rear passenger entrance door not

necessarily used. The schedule involved a late evening departure from Heathrow and an arrival in New York about midday local time. Passengers were provided with meals on the ground at Rineanna and Gander although some hot and cold meals were carried on board in vacuum flasks. At that time the standard IATA fare London-New York was £87 one way and £156 return.

By May 1947 five services a week was operating to New York, two via Prestwick and three via Shannon in Eire. A new service, once weekly, was commenced to Montreal on 15 April 1947, BOAC's first commercial service to Canada. On the same day, the first service by TCA with DC4M aircraft left Montreal for London. It was still not possible for BOAC to operate a daily facility London-New York - a sore point with the Corporation's USA sales organisation who claimed that American travel agents were resistant to any operator offering less than a daily frequency. In fact BOAC's loads per aircraft were as good as or better than competitors' but the problem was frequency, something that was increased as fast as aircraft availability allowed; in July 1947 the London-New York frequency was increased from five to six per week and London-Montreal to twice weekly.

The Bermuda service was operated in spite of the competition from PAA with their aircraft based in New York and from the Furness Withy ships; all this capacity made operations unprofitable, but BOAC took the view the route was a showpiece for its Atlantic flights. Holiday trips by Americans on the BOAC Bermuda-bound aircraft were more often than not their first taste of British service, and BOAC was convinced this influenced bookings to its Atlantic services. Also, Bermuda was a valuable staging post likely to be of increasing importance on future trunk routes.

By the summer of 1949, both PAA and AOA were operating Stratocruisers into London Airport from New York PAA daily and AOA twice weekly; this was serious competition for BOAC with its five weekly Constellations.

The Liberators continued to operate Prestwick-Montreal, carrying the staff, stores and mail thus acting as a relief to the Constellations and avoiding any loss to foreign carriers of westbound British mail. One unusual development was the start in February 1948 of a once-weekly experimental operation to test out flight refuelling, but this was suspended at the end of May. Liberator operations finally came to a close on 27 September 1949, and so ceased to form part of BOAC's fleet after eight years in service.

Africa - UK-West Africa and the Trans-African route
BOAC's shareholding in Elders Colonial Airways Limited, which had ceased to trade, was virtually written off by 1947. Internal services within the four West African colonies were taken over by the West African Airways Corporation (WAAC) formed in 1946 by the four Governments of Nigeria, Gold Coast, Sierra Leone and Gambia. For some time BOAC operated local services for the new Corporation, but by March 1948 BOAC disposed of assets in West Africa not required for its trunk service.

On the trunk route Dakotas were operating between the UK and Accra at a three-times-a-week frequency, with an extension across Africa to Cairo at a reduced rate of two per week commensurate with the decline in importance of the trans-African link now that the more direct route to Cairo was open. This trans-African section was suspended on 9 June 1946 and shortly afterwards BOAC moved the West African terminal of its UK service from Accra to Lagos.

After that up to the summer of 1949 services from the UK were operated in turn by Dakotas, Haltons and finally Yorks. The Haltons took over from the Dakotas in July 1947 on a coastal route via Casablanca, Dakar and Accra to Lagos but the Governor of Nigeria complained at the length of the schedule and brought pressure to bear through the Colonial Office for a faster timing. This required the Haltons to fly on the Sahara route via Algiers or Castel Benito to Kano and then to Lagos but reduced their carrying capacity over the longer stages. In spite of this, the frequency was increased shortly to six per week although this just meant that the service lost more money. The Haltons were never an adequate aircraft for the route and were withdrawn in May 1948 when the Yorks took over at a frequency of five per week which was maintained until the Hermes relieved them in the summer of 1950.

UK to East Africa
Prior to the formation of the East African Airways Corporation (EAAC) on 1 January 1946 as a joint organisation of the three territories of Kenya, Uganda and Tanganyika, BOAC had undertaken at the Government's request to operate a series of local routes with De Havilland Rapides until the new Corporation was ready to take over the aircraft and run the services. In due course, EAAC was appointed BOAC's General Agents in the East African territories, and McCrindle was named to the EAAC Board in 1947. BOAC was to act as technical adviser and seconded some of its staff but had no financial interest in the new airline.

Nairobi was a calling point on the London-Johannesburg York services and continued to be served in that way throughout the 1946 to 1949 period. Additionally, from February 1948 York terminating services were introduced building up to four per week by October. After the Solents had been brought on to the UK-South Africa route, Nairobi was given a Solent terminating service in place of the Yorks, introduced at a frequency of three per week, used facilities on Lake Naivasha, near Nairobi.

With flags flying, Hermes G-ALDS *Hesperides* throws up clouds of sand during start-up at Khartoum.

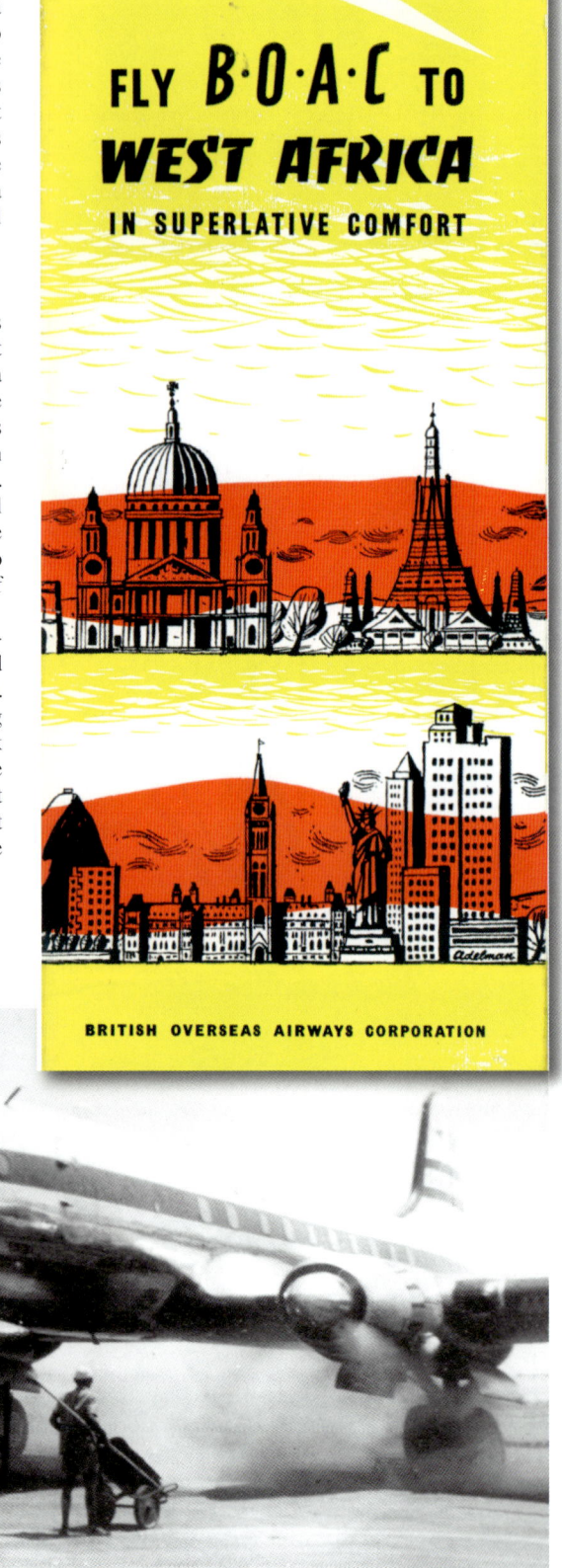

The post-war years saw activity in the field of Colonial development in the three East African states and especially in Tanganyika where an ambitious scheme to grow groundnuts for the production of food oil was underway. Support from the new Overseas Food Corporation enabled BOAC to justify the Solent frequencies, and in December 1947 a new service was begun to Dar-es-Salaam, initially with Yorks but with Solents from February 1949. Unfortunately, the high hopes of the groundnuts scheme were not realised and by late October of that year the Dar-es-Salaam service was withdrawn, and traffic between Nairobi and Dar was carried by East African Airways.

UK to South Africa
On 10 November 1945, the London-Johannesburg landplane service was inaugurated with Yorks, in partnership with SAA. The routeing was from Hurn via Castel Benito, Cairo, Khartoum and Nairobi to Johannesburg, Malta subsequently being substituted for Castel Benito. The elapsed time from Hurn to Johannesburg was 29 hours 20 minutes. This operation had been planned at the SAATC Conference held in South Africa in 1945. It had been agreed that the C-Class boats would be withdrawn at the end of December 1946 and that the Yorks would be replaced by Tudor IIs as soon as they became available. The York frequency had been built up to four per week by July 1946 and the accelerated schedule introduced. This fast timing was the outcome of an experimental sleeper York flight carried out in February 1946 from Hurn through Malta, Cairo, Khartoum and Nairobi to Palmietfontain Airport, Johannesburg. Apart from testing the reactions to the fast timings of the guinea-pig staff passengers, the objective was also to sell the concept of a sleeper York operation to SAA. The South Africans were not much impressed with the comfort of the York, considering it to be a particularly noisy aircraft which would bear heavily on passengers on the long flights between South Africa and the UK.

When the Chairman visited South Africa subsequently, he found there was intense opposition from SAA to any continuance of the Yorks on the route. They had their eyes on a fast service by DC4 Skymaster aircraft which they were already using between Johannesburg and Capetown. Discussions which between the two airlines and talks between BOAC and the MCA on the Chairman's return envisaged a Skymaster service by SAA starting in the summer of 1947 on a fast schedule, with BOAC operating a slower program using Solent aircraft until such time as both companies could work the same type, hopefully the Hermes IV when available. There was some talk of the Tudor I but that soon faded away following the aircraft's disappointing trials.

SAA started their Skymaster operation in July 1946 at a low frequency stepping it up to three per week fourteen months later. Meanwhile, BOAC had increased its York frequency to a maximum of six per week by March 1947. The Solents which had entered the route in May 1948 began to take over from the Yorks but, dogged by technical problems, had to be temporarily suspended in July and did not resume until mid-October, their capacity in the meantime being provided as best it could by Plymouth flying boats - as the Shorts Sandringhams were called by BOAC and by SAA Skymasters.

In July 1948 Sir Harold Hartley toured African stations with the specific purpose of studying the flying-boat route to South Africa and the future of the joint service with SAA. He was hugely impressed with the sightseeing possibilities of the route, and how these could be enjoyed from the promenade deck of the Plymouth flying boat with its excellent view of the terrain. Captains at that time gave passengers a good look at the main route attractions such as Luxor, Victoria Falls and Murchison Falls by low-level flying. But in spite of these attractions, Sir Harold realised, regretfully, that the financial burden of the flying boats was excessive with their use restricted to daytime only and with the need for repeated and expensive nightstops. It could not go on much longer.

Talks were again held with SAA in Johannesburg, and Sir Harold found General Venter, their Chief Airways Manager, concerned about the volume of traffic SAA were losing to KLM's DC6s operating between Amsterdam and Johannesburg on the West Coast route via Kano, offering a schedule of 27 hours as against 35½ hours by SAA's Skymasters. Eventually, SAA sought to purchase DC6s to compete, but permission was refused by the Minister of Finance. Shortly afterwards General Venter came to London and to BOAC's concern said that SAA was considering a partnership with KLM on the Kano routeing and appeared to feel that an arrangement of that kind was not incompatible with the BOAC partnership on the Springbok route nor with the 1945 UK-South Africa Air Services Agreement.

It is interesting to observe how dominant sea travel still was to South Africa at that time and it would have been a brave forecaster who would have foretold the eventual disappearance of the long-established Union Castle Southampton-Capetown service as a result of air competition. In the six months period from October 1948 to March 1949 the percentage of total passenger traffic from the UK to South Africa carried on the ships fluctuated month by month from 88% to 93%; air travel had scarcely dented the hold sea travel had on the route.

Services to and radiating from the Middle East
The Middle East presented a complex and ever-changing political and commercial picture. Airfields were as inadequate as the ground facilities; and services were liable to interruption from outbursts of cholera: nor had demand for services settled into an enduring pattern, with the oil companies in particular uncertain of their continuing needs. All these factors led to repeated changes in BOAC's operating model to such a degree that any attempt to record each and every move would result in a confused jumble of service alterations.

The Middle East had three areas of service - transit operations from the UK to India and beyond; the service from the UK was terminating in the Middle East area; the local facilities were radiating from Cairo and Aden.

The transit services at the beginning of 1946 consisted of the three-weekly C- Class flying boats on the Horseshoe route, the three-weekly Sunderland service to Calcutta and Singapore, the three-weekly Yorks to Calcutta and three-weekly Lancastrians to Australia.

As to terminating services in 1946, Cairo at that time was the sole Middle East point served and was provided with a high-frequency Dakota operation of ten flights per week.

The services radiating from Cairo continued to rely on the fleet of nineteen locally based Lockheed Lodestars and were operated to the requirements of the MCA through the Middle East Air Transport Board. At the start of 1946, the pattern was becoming less intensive, and flights were being regularly operated to eight points - Baghdad, Khartoum, Nairobi, Istanbul, Aden, Karachi, Addis Ababa and Port Sudan. At that time the demand for travel from the oil companies working in the Persian Gulf had not yet exploded, but it would not be long before they would have a considerable influence on BOAC's Middle East route pattern.

By early 1947 the overall pattern of operations had not radically changed. As to transit services, the Horseshoe route had closed, and the C-Class boats were withdrawn in March 1947, but the Hythe Class boats were now operating to Hong Kong and to Sydney; the

Argonaut G-ALHC *Ariadne* on a pre-delivery flight over the St Lawrence river in Canada. *(BOAC via Simon Peters Collection)*

A BOAC poster showing the interior layout of Argonaut G-ALHC *Ariadne*.

Lancastrian service to Sydney was unchanged, and the Yorks to Calcutta had been stepped up in frequency.

There were innovations in BOAC's terminators in the Middle East. A once-weekly Dakota service started to Beirut on 1 July 1946 and to Lydda on 8 April 1946 but shortly afterwards the latter was suspended and not resumed until 2 May 1947. A reduced Dakota frequency served Cairo supported by the three-weekly York service and by the reappearance of the *Golden Hind*, the G Class flying boat, for a year from 2 September 1946.

In 1947 the oil companies requirements began to make their influence felt on BOAC's aircraft and route planning for the Middle East area. Damascus, Baghdad, Basra, Bahrain, Kuwait and Abadan were all points of interest to the companies, and the problem for BOAC was how to provide a service with a limited fleet in the face of all the demands for space coming from so many quarters. The problem was accentuated by the withdrawal of the C-Class routine operation which aroused a great deal of concern in the Gulf area. The Political Resident in the Gulf protested to the India Office at the withdrawal of facilities especially from Bahrain, and the Sheikh of Bahrain complained that the local pearl trade would be severely affected primarily by the loss of the C-Class connection to the Indian sub-continent.

The UK Sales Branch who were in regular contact with the Anglo-Iranian Oil Company (AIOC) recommended as early as November 1946 that a weekly service should be operated to the Gulf to meet their needs for staff movements and these taken with the staff flows from the Iraq Petroleum Company and the Bahrain Petroleum Company would justify a service terminating in Bahrein.

Local services from Cairo began to diminish in 1947, and only six points were served, Baghdad and Port Sudan being dropped and also the Hadramaut coast route to Karachi; Teheran was introduced by means of a service Cairo-Athens and Athens-Teheran; this was at the request of the Foreign Office who wanted a direct connection from the UK to Teheran which was obtained by linking the BOAC operation at Athens with the BEA UK-Athens flights.

Two events in 1947 demonstrated once again the value of BOAC's experience in an emergency. In August 1947 the Air Ministry received an appeal for BOAC's help from Air Headquarters in New Delhi to move 7,000 Pakistanis and their families from Delhi to Karachi following riots after the partition of India. BOAC provided two Yorks, a Lancastrian and twelve Dakotas and three charter companies were called in by BOAC to help; BOAC managed the whole operation under the command of Air Commodore H. G. Brackley, Assistant to the Chairman. The first load left Delhi on 1 September, and the evacuation was completed on 15 September, the aircraft involved having flown in all over 300,000 miles and moved 7,000 people from Delhi and 1,500 from Karachi. As many as eighty passengers were carried in a York and up to fifty-seven in a Dakota.

A second emergency evacuation was organised by BOAC at the request of the Indian Government. It was a much more significant undertaking with over 35,000 refugees airlifted between India and Pakistan. The majority of the fleet of aircraft used were supplied by charter companies, but all were under the overall control of BOAC again through Air Commodore Brackley.

'The operations on which the BOAC have been engaged for some time on behalf of the Government of India for the transport of refugees across the borders of India and Pakistand have been concluded today'.

So said a message from the Prime Minister of India in his introduction to 'Operation India' BOAC produced two booklets on the refugee crisis that included a photograph below right captioned 'Loading stores at Karachi - 3,103 items of bedding, towels, table cloths; 2,900 items of crockery and cutlery, amongs other things'.

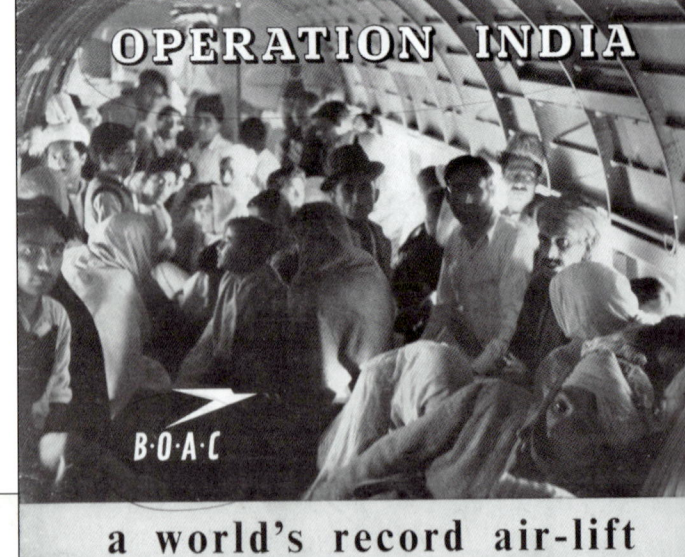

Pressure from the frightened refugees was such that aircraft had to be protected against the rush to get on board and again, aircraft were loaded up to the hilt - Dakotas taking sixty or more per trip. Both these operations were carried out successfully without any casualties, and both Indian and Pakistani Governments expressed their appreciation to BOAC.

In 1948 and 1949 there was a build-up of services terminating in the Middle East. It was these experiences which subsequently led BOAC to adopt as an active policy plan which included Beirut, Lydda, Cairo and Teheran on its route pattern, since the chances were that at least one of these points would be open even in emergencies. This policy paid off well in future years.

In 1948 developments also occurred in the oil business. The Bahrain terminator had been built up to thrice weekly by February 1948, but lengthy discussions with the oil companies led to a changed pattern to give more comprehensive coverage of stations. On 18 March a twice-weekly Hythe service was begun routed UK-Marseilles-Augusta-Cairo-Basra-Kuwait-Bahrain and after some delay due to the need to obtain Iranian permission a York service was started UK-Castel Benito-Cairo-Abadan-Basra on 13 April. A somewhat surprising development then occurred. BOAC's Commercial Director, R. D. Stewart, completed arrangements for Skyways to take over all Gulf services on charter to BOAC for one year using DC4 Skymaster aircraft. The two Hythe services were withdrawn at the end of April and the York to Abadan and Basra early in May and were replaced eventually by Skymaster services to Bahrain, Damascus and Abadan, the first service to Bahrain leaving the UK on 9 May 1948. Scheduling problems to suit all the varying needs of the oil companies and to match the ground facilities were a constant headache; at Kuwait, for example, the strip was in open desert and at night livestock, and local people tended to stray over the landing area.

Dakotas continued to operate from the UK to Cairo and were joined by Yorks in October 1948. The Dakota service from the UK to Lydda was forced to cease operation because of the political situation in Palestine and the violence which ensued following partition. By April 1948 it was clear BOAC would have to evacuate its staff and the last Dakota left Lydda for the UK on 28 April with the airport deserted except for British troops. It was fourteen months before service could be resumed. The Dakota service from the UK to Teheran which had replaced the Cairo-Teheran service in August 1947 was gradually taken over by Yorks from September 1948.

Purely local services in the Middle East in 1948 and 1949 tended to centre on Aden as BOAC was interested in setting up a domestic airline, Aden Airways Limited, which was incorporated on 1 March 1949 as a wholly owned subsidiary. The new airline used Dakotas to run services to Addis Ababa, Riyan, Nairobi and Cairo. From Cairo itself, only two services were operated, to Addis Ababa and to Nairobi but these were withdrawn at the end of November.

Services to the East
Here also the story was complicated by the frequent changes of aircraft type. Following the withdrawal of the C-Class boats in January 1947, services to the East consisted of Yorks to Calcutta, Delhi and Ceylon and of Haltons which operated briefly to Karachi and Ceylon but were withdrawn in September 1947 and replaced by Yorks once more. To the Far East Hythes operated to Singapore and Hong Kong although Plymouths took over the Hong Kong service in June 1947. An extension from Hong Kong to Shanghai lasted only one week at the end of 1948 as the advancing Communist troops made it no longer possible to operate. Plymouths also extended to Iwakuni, Japan, and to Yokohama in November 1948. By January 1949 all Hythe flying boats had been withdrawn.

Services to Australia
The Lancastrians which began through services from the UK to Sydney on 31 May 1945 in conjunction with QANTAS continued to operate in 1946 but were becoming less and less acceptable as a passenger aircraft. By December 1947 the frequency was reduced from three to one per week, and only mails and freight were carried. However, BOAC began a twice-weekly Hythe service to Sydney on 12 May 1946, increased to thrice weekly by early July on a schedule which was lengthy and slow, leaving the UK on Day 1 and not arriving in Sydney until Day 7. BOAC's partners, QANTAS, and the Australian Government became very restive; with the failure of the Tudors and the unsuitability of the Hermes there was no British aircraft in sight capable of providing a competitive service on the route. QANTAS themselves had therefore decided in July 1946 to order four Constellations and had suggested that BOAC should order three, which together would then make it possible to provide four services weekly between Australia and the UK with BOAC still operating its three slow services by Solents.

An urgent partnership meeting was held in Cairo on 23 October 1946 to resolve these issues, at which QANTAS, whilst regretting it could not use British aircraft were determined to press ahead with Constellations, claiming that at the Canberra Conference in March 1946 it had been agreed that new types of aircraft would replace the Lancastrians and Hythes as early as the autumn of 1946. BOAC had to explain that the UK Government were opposed to its using non-British aircraft and to the expenditure of dollars which a purchase of more American aircraft would involve; all BOAC could offer was a continuation of Lancastrians and flying boats.

BOAC reported the situation to the MCA at a meeting in London on 18 December 1946. The existing operations to Australia were losing around £1m a year, whereas QANTAS claimed its planned Constellation service would break even but that BOAC's flying-boat operations would by 1948 be losing nearly £1.25m a year, a loss which QANTAS would have to share under the partnership arrangements. Knollys emphasised to the Minister that a competitive service could be provided only by American aircraft and that the Australians would not accept a profoundly inferior operation by the British side of the partnership; hence the continued existence of

The Australian flag on the central fin of this Lockheed Constellation is a give-away that this was a QANTAS flight that has just arrived in the UK.

the partnership might be in real danger.

The outcome of this meeting was a realisation that the only practical solution lay in the purchase of more Constellations for BOAC, but there were Treasury problems on dollar spending to overcome. The Chairman of QANTAS, Hudson Fysh, repeatedly stressed the urgency of the matter. The inferiority of BOAC's Hythe and Lancastrian operations became even more glaring when QANTAS introduced its Constellations on the route in December 1947 cutting the journey time down to four days.

The Irish airline, Aer Linte, the trans-Atlantic arm of Aer Lingus, was offering its five Constellation 749 aircraft for sale in sterling, so providing a way around the dollar problem, just had been done with the former Swedish Stratocruisers. The British Government gave permission for BOAC to acquire them and at last, on 1 December 1948 BOAC joined QANTAS on the Kangaroo route as a Constellation operator but at a lower frequency, and the Hythes were withdrawn in February 1949; the Lancastrian service continued but only as a freight and mail carrier. The 'Fly British' policy adhered to even when there was no adequate British aircraft to use had created a crisis, which although finally solved, continued to affect BOAC/QANTAS relations for some time to come.

Chapter Five

Entering a New Age

Although BOAC had consistently regarded the development of a Brabazon III type aircraft for Empire routes as being of high priority, there was an air of unreality about the designing of A. V. Roe's Avro 693, which seemed to have been given a low priority by the manufacturers. Back in January 1946 BOAC had asked the MCA to instruct the manufacturers to proceed with the design and manufacture of three Avro 693 aircraft, Brabazon III type. However, in July of that year Lord Rothschild, a Board member, paid a visit to the A. V. Roe factory and came away with the impression that the Brabazon III was not even on the drawing board. By October the Corporation had re-thought its requirements for Empire types which it considered fell under two heads - a fast mail and passenger aircraft for maximum stages of 2,000 miles and a slower aircraft carrying 30 to 40 passengers over stages of 900 to 1,200 miles with flexibility of operation as to airfield and, capable of conversion to tourist standards. To crystallize these ideas, the Chairman produced a paper in November which cleared the air. It was evident that the original concept for the Brabazon III was no longer relevant in the changed competitive circumstances, and in any case its purpose of longer range could be met by the Brabazon IV and DH.106.

What was wanted was a design which would give a still air range of 2,500 statutory miles rather than the 3,500 miles of the first concept, able to accommodate up to thirty-two passengers and around 2,750 lbs of mail and freight over a maximum stage of 1,350 miles, pressurized and air-conditioned and able to operate from second-class airfields as to runway lengths and strengths with a cruising speed of 320 mph - initially if necessary with reciprocating engines.

These requirements were submitted to the Self Committee for passing on to the manufacturers. Some

The end of an era! Shorts Solent 2 G-AHIN *Southampton* in flight above the Cunard liner *Queen Elizabeth*. Both were soon to be replaced by the jet age.

delay then occurred before the firms were asked to tender for the new design but BOAC was advised by the Ministry in February 1947 that ten firms had been approached by the Ministry of Supply; of these firms five submitted tenders including the Bristol Aeroplane Company Limited and A. V. Roe & Co. Ltd.

On 23 June 1947 the MoS called a meeting at which the MCA and BOAC were present to examine the technical aspects of the tenders. Subsequent to this meeting the Technical Development Director put a memorandum to the Technical Committee of the Board recommending selection of the design submitted by the Bristol Aeroplane Company, known as the Bristol 175, later to be given the type name 'Britannia'. This was for an aircraft of around 94,000 lbs gross weight using four Bristol Centaurus engines with normal seating for 32 passengers and a cruising speed at 20,000 feet of around 316 mph. The Board on 7 August 1947 agreed to this recommendation and both the MoS and the Self Committee were prepared to support BOAC's selection, Bristols forecasted deliveries starting in 1951.

Discussions with Bristols got under way quickly including early consideration of the aircraft fitted with Bristol Proteus and Napier Nomad engines. In October agreement had been reached with all concerned as to the changes to be made to improve the tender design. These included an increase in the maximum seating to forty-two at 48 inch seat pitch and the ability to increase to forty-eight at 42 inch pitch. Design gross weight had increased to 103,000 lbs enabling full volumetric payload to be carried over Empire medium stages. Direct operating costs were to be slightly below those for the Constellation but with better take-off performance. The revisions made to the original design had a dramatic effect on forecast operating costs. For example, over the Nairobi-Johannesburg sector the cost in pence per capacity ton mile came down from 16.7 pence to 8.5 pence.

The MoS sought design studies for a long-range Empire type. Although BOAC had not been consulted when the MoS drew up a statement of requirements, BOAC said it was prepared to examine any ideas put forward. Discussions with Bristols continued on the 175, including provision in the design for the aircraft if required to be developed for operations on Empire routes over longer stages. The importance of the 175 in BOAC's planning was evident; it would be the only new type other than the advanced DH.106 and BOAC considered it essential that provision should be made in the airframe design for conversion to propeller turbines if desired. BOAC also thought it would require great development and production effort to have the aircraft in orthodox reciprocating engine design ready by 1954.

By March 1948 the provisional specification had been amended to include the possibility of a later version with Proteus or Nomad engines and the proposed contract was being discussed when difficulties arose between Bristols, the MoS and BOAC over questions of price and guarantees. On price the manufacturers were asking £400,000 per aircraft whereas BOAC took the view that £300,000 for a conventional aircraft like the 175 was a fair price. BOAC was also unhappy about the proposed guarantee for failure to meet performance flights and to cover late deliveries. Some progress had

For a number of years BOAC had its passenger terminal at London Heathrow at an area known as 'The North Side', or to give it its formal name 'London Airport North' as seen here. To the left is the A4 Bath Road with the roundabout and access tunnel to the central area in the foreground. *(author's collection)*

The Captain and crew of BOAC Hermes *Hengist* on the inaugural UK to West Africa service - 6 August 1950.

While so much was going on with Bristols and Handley Page, one manufacturer had been quietly proceeding with the pure jet design originally listed as the Brabazon IV; this was De Havillands and the aircraft the DH.106. In January 1946 BOAC had shown its faith in the project by informing the DGCA that it would support an order for eight aircraft and in August confirmed to Sir Henry Self the requirement and added that it would like a been made by the end of 1948 when the MoS told BOAC they had placed an order with Bristols for three prototypes, although it was not until July that the company received written confirmation.

Use of the aircraft on the North Atlantic began to appear in contacts between BOAC and Bristols, but Sir Harold Hartley was anxious that the two primary purposes for the Bristol design should be maintained, namely as an insurance against any failure by the DH.106, and as a workhorse aircraft for the medium-range Empire operations. Consequently BOAC's technical staff were instructed to ensure in effecting any improvements to the aircraft that nothing should be included in the contract which might impair its effectiveness over the medium ranges.

On 8 July 1948 the Board authorised the placing of a production order for twenty-five Bristol 175 aircraft subject to further price negotiations but no action was taken with Bristols as BOAC was awaiting production proposals from them.

Suddenly in November the MCA asked BOAC to engage with them in a study to see how the twenty-five aircraft should be operated in relation to the rest of the fleet and particularly how they might affect the number of DH.106 aircraft BOAC would require. By February 1949 justification for the order had been given and accepted by the Ministry but it was June before they gave their approval. There remained negotiation over price to be settled with Bristols who were still asking £400,000 per aircraft. On 28 July the contract with Bristols was signed at a price of £445,000 for each Centaurus-engined aircraft including the levy payable to the MoS. One issue was still outstanding, the speed of development of the Proteus II, which was wanted to follow on from the Centaurus-engined version.

conditional order placed for a further eight aircraft to come in the production line after the four which BSAA was likely to order. At the same time BOAC reported it was thinking of an axial flow engine on the grounds that the planned Ghost engines might not be adequate to achieve transatlantic operation. The reply received by BOAC to this approach was only partly encouraging since it raised doubts about the order of priority as between BOAC and BSAA for the initial batch of aircraft. BOAC protested vigorously as it had supported the aircraft's development enthusiastically since its inception and was prepared to use it on the North Atlantic and Empire routes or the Empire routes alone if the aircraft were unsuitable for Atlantic conditions.

The existing system of placing contracts via the MCA to the MoS which then dealt with the manufacturers was a breeding ground of anxieties and difficulties for the actual users. As to the DH.106, the Deputy Chairman complained to Sir Henry Self that although he had seen the draft contract, *'BOAC was hardly mentioned and everything was in the name of the MoS'*.

In November 1946 BOAC wrote to Sir Henry Self as Permanent Secretary to the MCA, again confirming the order for eight aircraft and asking for first refusal on a further batch. Additionally, the Ministry was requested to place a development contract for a DH.106 version and axial flow engines.

Some progress was being made in official circles as the MoS on 17 January 1947 agreed to instruct De Havillands to proceed with the manufacture of sixteen DH.106 aircraft, of which eight were for BOAC and six for BSAA. Price per aircraft was not to exceed £250,000. BOAC was doing all it could to urge along the DH.106 and showed its confidence in the project by

The BOAC Operations Room - the control point for the movement of BOAC aircraft. The facility was staffed day and night.

now seeking an option on a further seventeen aircraft over and above the initial eight, making twenty-five in all, and this was confirmed by the Board on 7 August 1947 subject to satisfactory agreement as to price, delivery and performance. The total cost of the order was assessed at about £16.25m. De Havillands accepted the order but stipulated that Ghost engines should be fitted to the first twenty-five aircraft to be built.

By June of the following year the MoS had placed a firm order for eight aircraft and the contract for the remaining seventeen was under discussion, BOAC having expressly left open the engine position for the last eight of the additional seventeen so that more thought could be given to a more efficient type of engine. At that time De Havillands were saying that the first aircraft should be flying by May 1949. The first flight was in fact made on 27 July 1949 and that and the following flights were completed without any mechanical or other problems. So the future of the project seemed bright and BOAC looked forward to being the first operator of a civil jet aircraft, hopefully in early 1952.

In the summer of 1948 the MCA officially announced that BSAA would operate the SR45 instead of BOAC and contacts between BOAC and Saunders-Roe ceased; but BOAC were not to be so easily released from this problem aircraft. With the subsequent disappearance of BSAA into BOAC, Whitney Straight was having to ask his technical staff *'to look carefully at the SR45 position having regard to the possibility that this contraption will be back on our plate'*.

An important meeting was held on 29 March 1949 at the MCA at which Sir Miles Thomas for BOAC and John Booth for BSAA were present. The meeting was to decide what should be done with the boats - by that time BOAC had made a domestic decision to abandon all flying-boat operations in the future. Sir George Cribbett, the DGCA, pointed out that his Ministry had

Stratocruiser G-ANUC *Clio* rests between flights.

already recommended cancellation when BSAA came along and decided they wanted to take the boats and the Minister then obtained Government endorsement for three and possibly seven aircraft. On these grounds now that BSAA was merging with BOAC, Sir George considered BOAC had a moral obligation to take over the order.

Sir Miles Thomas for BOAC was prepared to take only three boats with Proteus II engines and only on the basis of a subsidy being available, although at that time the estimated price per aircraft had increased to £850,000. In reporting back to the Board Sir Miles, by then Chairman, said the Corporation had agreed with the MCA for the purchase of three boats equipped initially with the Proteus II engines and the cost of the airframe alone would be £700,000, but the Proteus II engines would be hired from the MoS, who would also bear the cost of the developed Proteus III engines which were essential in BOAC's view to give the aircraft any chance of adequate operation on the North Atlantic.

So the long story dragged on with further delays in engine deliveries. However, with more performance figures, it became clear that BOAC would be unable to use the aircraft on the North Atlantic unless the Proteus engines were available but by 1950 the on-service date for the aircraft had gone back to 1955 at the earliest; the problem of its deployment continued to exercise the minds of Government and of BOAC well into the later years of Sir Miles' Chairmanship.

The Royal Flights

The reign of Queen Elizabeth I is recorded in our history books as the age of seamen and seamanship. Historians of the future may well look back on the second Elizabethan era as the real beginning of Britain's air age. For it was by aeroplane - first in a Dakota of East African Airways, then in a BOAC Argonaut - that Queen Elizabeth II journeyed to England for the first time after her accession. And it was at London Airport, on a misty February evening in 1952, that she first stood on British soil as Queen.

The first Royal flight undertaken by BOAC was as far back as October 1951, when Princess Elizabeth and the Duke of Edinburgh flew to Montreal by Stratocruiser. In the following year they took off again in an Argonaut for South Africa via Nairobi, but with the sudden death of the King on 6 February the Princess, now Queen, hurried back on the same Argonaut arriving at London Airport on 7 February. Royal flights succeeded each other each year and in the House of Commons on 18 April 1956 it was confirmed that the Corporation would continue to provide aircraft for the long-distance flights. The most extensive flight with the most significant number of off-route calls was over the period from 20 January to 6 March 1961 when the Queen and the Duke of Edinburgh travelled in a Britannia 312 throughout India and Pakistan, to Iran and Turkey and Cyprus - over 14,300 miles in all, involving an immense amount of planning, including the operation over the same complicated routeing of the second Britannia with members of the Royal Household. A unique occasion took place on board the Queen's Britannia before it landed back at Heathrow, when Her Majesty held an informal investiture and made Captain Meagher, CVO, his second-in command, Captain Morrell, MVO: John Gorman, Chief of Security, CVO and 'Buck' Buchanan, his Deputy, MVO.

In the years since that history-making flight, the Queen and other members of the Royal Family have millions of miles in BOAC airliners, on tours through the Commonwealth and to most of the major countries of the world. As the United Kingdom's 'Chosen Instrument' it was only right and proper that BOAC was tasked with carriage of the Royal family in long distance trip overseas.

Though these royal flights are of course tremendously important, they are not, as many people imagine, something entirely apart from the airline's standard operations. They are really what may be called 'special charters', and they are dealt with in basically the same way as charter flights booked by travel agents or other organisations.

A contemporary report from BOAC files gives a flavour of the times: *'An aircraft used on a royal tour is not kept solely for this purpose. It is merely taken off scheduled service and used for the 'charter'. Of course, its interior has to be altered to meet the unique requirements. For instance, the forward cabin may be made into a sitting-room, with comfortable divans. Next to it will be a dining-room furnished to seat eight people. There will also be dressing-rooms; the remainder of the aircraft is fitted out for the Household Staff.*

The crews on a royal flight, though carefully chosen, are not given any special training outside their regular

The scene is set for a royal departure. On board Boeing Stratocruiser G-AKGK *Canopus* is Her Majesty the Queen (then Princess Elizabeth) making her first international flight in October 1951. *(BOAC via author's collection)*

working routine. And for the personnel in BOAC's Control Centre at London Airport, the royal aircraft, once it is airborne, becomes another chalked number on the board - to be watched and plotted with the same care as all the other aircraft making their way across the world. In other ways, of course, royal flights are very different from those of ordinary passengers. The aircraft used throughout the journey is a vital part of any royal tour, and its schedule must be planned to fit in with the elaborate ceremony and protocol which surrounds state visits overseas.

The planning usually begins many weeks before the actual departure. At meetings of all the people concerned, the flight schedule is prepared; the type of aircraft best suited to the job is decided on; the crew is chosen. In some cases, it is necessary for the captain to visit the airports on the route, to study the technical facilities available and to make the arrangements required for the handling of his aircraft. Before the Queen's Indian tour, for example, Captain Meagher who was to pilot the royal airliner (a Britannia) made a reconnaissance along the entire route. With his second-in-command, Captain Morrell, he flew in a Heron of the Royal Flight through Persia, Pakistan and India, to familiarise himself with landing conditions at all the scheduled stopping-places.

Other details to be settled include the menus, which are chosen at Buckingham Palace and prepared in

Above: On 31 January 1952, Princess Elizabeth and the Duke of Edinburgh flew out of London airport in BOAC Argonaut G-AJHK *Atalanta* to begin a tour of East Africa.

Right: While the Royal couple were in East Africa, the King died. The Argonaut conveyed Queen Elizabeth II and the Duke of Edinburgh from Entebbe to England on 7 February 1952, where she was met by bare-headed Winston Churchill and Clement Attlee.

Above right: the royal cabin interior of a BOAC airliner, in this case a DC-7C. *(all via AVM J de M Severne, The Queens Flight)*

From a slightly later time, the dining room 'conversion' of a BOAC Comet 4 designed for Prince Philip.
(via AVM J de M Severne, The Queens Flight)

BOAC's London Airport kitchens; the handling of the large amount of luggage which the Queen has to take with her; the provision of a spare aircraft in case a technical fault should delay the flight; and the planning of any engineering work to be carried out on the royal airliner.

On the day of departure, the gleaming jetliner is towed on to the tarmac an hour or two before take-off, and the last stages of fuelling are completed. Meanwhile, the captain and his crew have reported to the Station Operations department, and have begun the job of planning the flight and calculating the fuel. The most up-to-the-minute weather forecast is supplied by Air Ministry experts, and when all the details are complete and approved by the captain, the crew go to the aircraft.

The flying staff then begin the series of checks and cross-checks which precede every BOAC departure. The stewards and stewardesses put the final touches to the cabin, and make sure that everything is in perfect order for the Queen's arrival.

Not far from the aircraft, the crowds have been waiting for some time to watch the Queen depart. Among them are press photographers and reporters. Huge cameras are mounted and set carefully to obtain close-up pictures of the departure. Newsreel and television units fit their film cameras on top of their motor cars, which line one side of the Queen's approach to the aircraft (marked by a length of red carpet).

With about ten minutes to go, a line of gleaming limousines, headed by the Queen's special glass-sided' car, moves smoothly past the policemen at the airport entrance and stops outside the lounge used solely for distinguished travellers of royal or ambassadorial rank. The Queen and her husband are greeted by officials, and after a brief pause, they begin to walk towards the aircraft.

On both sides, as flashbulbs pop and cameras whirr, while radio commentators from many countries talk purposefully into their microphones. Uniformed airport staff salute the Queen as she walks up the aircraft steps, accompanied by members of her family. A few moments later the family re-appears, the Queen waves goodbye from the aircraft doorway and then takes her seat.

Exactly to the minute, the door is pulled shut, the engines begin to whine, and the aircraft rolls towards the beginning of the runway.

Another royal tour has begun'.

A New Chairman

On his appointment as BOAC's Deputy Chairman on 1 April, 1948 Sir Miles Thomas DFC, MI.MechE, MSAE, (later Lord Thomas of Remenham) looked forward to taking over the Chairmanship from 1 July in accordance

The Bristol Britannia in BOAC colours is seen here at Bromma Airport Stockholm during a sales tour. (Bristol Aircraft)

with the arrangement made with him by the Minister, Lord Nathan. He succeeded Sir Harold Hartley and so found himself in charge of a nationalised service industry, a complete contrast to his former post as Deputy Chairman and Managing Director of Morris Motors.

His seven-year 'tour of duty' was full of incident, and his powers of resilience were tested to the extreme. He took up his post determined to streamline BOAC and to carry out the Minister's original directive to turn losses into profits. His time as Chairman fell into two eras, the first from 1949 to 1953 and then from 1954 to 1956.

The first period was one of exciting achievement; the financial year 1951/52 saw the first operating profit, the first surplus after remuneration of capital and by 1952/53 no requirement for any Exchequer grant: by 1953/54 the operating surplus was over £2m as compared with a loss of over £6m in 1949/50, and the accumulated deficit was brought down below £1m. The passenger fleet had been rationalised to four types all pressurised, and by 1952 included the first pure jet civil aircraft in the world, the Comet 1. The reorganisation of the Corporation's internal structure - a task specifically given to him by the Minister - was completed by the early summer of 1950 and by drastic pruning up to senior levels the new Chairman had been able to reduce staff numbers down to 15,811 by December 1950 as compared with nearly 19,000 when he took over the chair and with almost 20,000 after the merger with BSAA in September 1949.

In spite of the feeling engendered among the staff who survived the cuts, the new Chairman had some success in raising staff morale. He was a natural communicator and believed in his own spoken and written words as a means of getting a message across to the outside world and to the staff. For the first time staff were reading favourable comments about their airline in the Press, and Sir Miles made the most of the coming Comet as a world beater using it for a platform to get the BOAC story across to the public and incidentally get pleasing publicity for himself. This was something new to the staff; previous Chairmen had shunned publicity, but he used it to the full with the expert help of his newly appointed Press and Public Relations Officer, Freddie Gillman, a former journalist. BOAC was now becoming a household name in the country.

Following his remit from the Minister, the new Chairman set about his task of reducing BOAC's payroll with a will, particularly among middle and senior management. He claimed to have dispensed with the services of over eighty executives and commented in later years after his departure from the airline that *'when pruning in BOAC really got rolling, we began to shed our fat at an almost visible rate'*. It was inevitably a painful process, never forgotten within the airline and deeply disturbing to staff, not only because of redundancies but also because of the new re-organisation which had been introduced on 1 November 1949 and which was to be made fully effective early in 1950. The 1948 re-organisation had scarcely had time to settle down before the new changes were announced.

On the day Sir Miles took over the Chairmanship, BOAC was still in the transitional stage in respect of aircraft. Out-dated types such as Lancastrians, Yorks, Plymouths and Solents continued to operate passenger services, but there was a massive programme of new types ahead, although not all of these met their promised delivery dates. The Corporation had to prepare staff for the introduction shortly of four, possibly five, new models, with the phasing out of the older types within a year or so. The task facing the Technical Development and Engineering staff was formidable. They were involved in preparations for over sixty aircraft, Hermes, Argonauts, Stratocruisers and the new dimension, the Comets. Also, the SR.45 remained a possibility and beyond that the Bristol 175.

The policy for BOAC's aircraft purchase laid down by Government was still 'fly British' unless a cast-iron case could be made out otherwise such as to bridge the gap on the North Atlantic. It was a policy which for the operator contained disadvantages; BOAC was inevitably the innovator with the risks that entailed, and, moreover, the sales of British aircraft being limited resulted in high unit prices from the manufacturers. It seemed to BOAC to be unreasonable that it should have to bear the cost of pre-operational development which meant in essence that the airline was subsidising the manufacturers. At the very start of the new Chairman's term of office, Whitney Straight as Deputy Chairman put these thoughts in writing to the Board and also to the Select Committee on Nationalised Industries.

Whitney Straight was responsible for making recommendations about future aircraft types, and part of his job was to review the forward programme. One such review he carried out at the end of 1949. All twenty-two Argonauts had been delivered by Canadair, and five Stratocruisers, with five more about to be delivered together with eleven Constellations were in the fleet. Whitney Straight's concern centred on the new British types still in the developmental stage, the Princess, the Britannia and the Comet.

One of the first acts of the new Chairman was to sign a production order for twenty-four Bristol 175s fitted with Centaurus engines at a price of £445,000 each, including the Ministry of Supply's fee of £45,000 airframe development levy. There was to be accommodation for fifty passengers, with provision for an increase to sixty-two and capacity built into the design for a high-density version at maximum seating with a seat pitch around 39 inches. There was also a requirement for the structure to be suitable for the installation of some sleeperettes at a greater pitch. The design had to be such as to permit the fitting of Bristol Proteus engines at a later date.

This was the position on the B.175 at the time of Whitney Straight's December 1949 Review to the Board, with one important difference, namely that the development of the Proteus engine had suffered a serious setback which could well cause a delay in the Proteus-engined version. In contrast to these B.175 problems, the Comet 1 was showing great promise and could be in operation at least three years before the B.175. Somewhat surprisingly Whitney Straight recommended that the MCA be informed BOAC no longer considered the B.175 as an essential part of its future aircraft programme, mainly on the grounds of delay and the promising outlook for the Comet. In making this recommendation Whitney Straight also commented on the likely developments in the United States where Boeing were poised to produce a jet transport aircraft quickly once an initial order had been received.

So the MCA were informed but the Cabinet Air Transport Committee decided that the purchase contract for the B.175 should only be reviewed so as to reduce capital expenditure. Meanwhile in March 1950, following this decision, the MoS cancelled the development contract for the Centaurus-engined version, and subsequently the Corporation was required to reduce the aircraft order from twenty-five to fifteen all of the Proteus-engined version, and one of the prototypes to be ordered by the MoS was to become eventually the sixteenth aircraft. It was at this time that the MoS allotted the name Britannia to the B.175.

Agreement in July 1950 was reached between Bristols and BOAC under which capital expenditure on the aircraft was to be reduced. The contract gave BOAC the option of purchasing only fifteen aircraft at the higher price of £567,000 per aircraft or, if twenty-five aircraft were purchased then at £483,000 per aircraft; the Board approved capital expenditure of £12.5m, for the project consisting of fifteen aircraft from Bristols plus one aircraft from the MoS together with spare engines and spares.

Bristols were able to give some good news a year later, having found it possible to increase the maximum gross weight of the aircraft from 130,000 lbs to 140,000

Above: Sir Miles Thomas poses in the First Officer's seat of the first Britannia 100 handed over to the company on 30 December 1955.

Right: Sir Miles receiving the log-book of the first Britannia.
(both BOAC via Simon Peters Collection)

lbs with increased tankage thus making it feasible to consider it as a possible North Atlantic type. Later that year BOAC reviewed its capacity needs, particularly in the light of IATA decisions about the development of tourist-class traffic. It was expected this would lead to a requirement for more aircraft and, in addition, the Corporation's Commonwealth partners were showing interest in the Britannia. At that time BOAC plans assumed five Britannia on passenger operations by the beginning of 1956. After taking these factors into account the Chairman decided to tell the MCA and Bristols that BOAC would not exercise its option to reduce the Britannia order from twenty-five to fifteen aircraft and in January 1952 the MCA finally confirmed approval to a total order of twenty-five aircraft excluding the purchase of the second prototype to which BOAC was committed by a separate deal. The revised interior layout of the aircraft to suit tourist traffic provided for a total of ninety-one seats.

There were still worries about the reliability of the Proteus III engine which was run up for the first time in April 1952 and on 16 August the first prototype Britannia but with Proteus II engines completed its maiden flight. In October the Board decided that negotiations should be started with Bristols for the purchase of five B.175 freighters but without any firm commitment, and early in 1953 Bristols were able to tell BOAC that the last ten aircraft out of the order for twenty-five could be converted into the lengthened version at an estimated cost of around £100,000 per aircraft. A study was completed in May by BOAC which clearly showed direct cost advantage to the lengthened version, now called the Britannia 300, as compared with the standard version, the Britannia 100 although it is noteworthy that the study indicated both types would need a two-stop operation on the North Atlantic westbound to give a reasonable payload. Accordingly, Bristols were instructed to deliver the last ten aircraft in the Mark 300 version.

On 24 April 1953 Sir Miles Thomas visited the Bristol Aeroplane Company and confirmed an order for three freighter aircraft and an option on a further two, these latter to depend on whether the ATAC gave a decision in BOAC's favour on its North Atlantic freight application (in fact the requirement was altered later in the year down to only two freighters). He also took an option on a further five of the Mark 300. In October 1953 Bristols announced that the price of the Mark 100 had increased from the date of the original contract

Left: Sir George Cribett on the left, and Sir Basil Smallpeice make speeches on the departure of the inaugural Britannia 102 to Johannesburg - the date was 1 February 1957.

A stunning picture of 'the whispering giant' BOAC Bristol Britannia G-ANBA *Britannia*. *(both BOAC via Simon Peters Collection)*

Part of the BOAC passenger check-in area at London Airport. *(both BOAC via author's Collection)*

(July 1949) from £483,000 to £680,000 per aircraft because of escalation and change orders.

At long last the second Britannia prototype fitted with Proteus III engines had its maiden flight in December 1953 but on 4 February 1954 after only 51 hours flying a failure of the reduction gear permitted the compressor turbine of the starboard inner engine to over-speed and disintegrated. The aircraft had to be put down quickly on the mud of the Severn estuary.

The time spent in finding a solution to this failure and troubles during the first flight of the resumed flying programme put back the commencement of the first production aircraft's flying and the tropical trials. In July 1954 a revised delivery programme to BOAC was being offered by Bristols, eight aircraft between May and December 1955, with the fifteenth aircraft not before June 1956, about a year later than Bristol's previous indication; BOAC's own view was that there might well be a further year's delay before all fifteen could be handed over to them.

At this stage, BOAC began to be gravely concerned about their future competitive position, particularly on the North Atlantic. The future of the Comet was uncertain, and the Court of Inquiry had not yet completed its findings on the Comet disasters; as for the Britannia, delays in delivery were building up, and in any case, there was as yet no Britannia type which could do the London-New York stage non-stop. Although the Britannia contract had been signed as far back as July 1949, it had taken five years before the aircraft's first overseas development flight, and it seemed almost inevitable there would have to be more prolonged testing as a result of the Comet accidents. This inevitably meant that deliveries would be further delayed.

The period of most concerns to the Corporation was from 1956 to 1958, and by September 1954 it had begun a serious examination of possible American aircraft as insurance. The most likely contender was the DC-7 series from the Douglas Corporation, mainly as discussions were in progress to fit Rolls-Royce RB109 engines to this aircraft. It would offer non-stop westbound flight London-New York with a payload much higher than the Britannia 300. The Board approved an approach to the Minister for authority to purchase up to fifteen DC-7Ds and a letter from the Chairman to him in October set out all the reasons for BOAC's request, at the same time being careful to stress the Corporation's continuing support for the British aircraft industry. Involved in the overall equipment situation was also the question of a commitment by BOAC to a revised Comet 2 aircraft, acceptance of which was being represented by the Minister as of national importance. Important exchanges of letters between the Minister and the Chairman took place in November 1954, and it became clear that Government was likely to approve a DC-7 purchase of some kind and at the same time it took note of BOAC's intention to introduce the Comet 2 on its routes subject to the findings of the Comet Inquiry and to the usual technical considerations. To complete the DC-7 story, Whitney Straight and Campbell-Orde made a quick visit to Boeing, Douglas and Lockheed in December 1954 and reported back to the Board that the right aircraft to go for was the DC-7C with its more comfortable delivery dates than a possible DC-7D. This was accepted by Government and approval for the purchase but for ten aircraft only was given to BOAC early in 1955 on the understanding that these aircraft be resold when long-range Britannias were introduced into regular service on the North Atlantic. The cost of the DC-7C purchase including associated equipment came to over £13m; deliveries took place without any delays between October 1956 and April 1957.

At the time of BOAC's discussions with the Minister in November 1954 about the possible American purchases, the National Press came out with a hostile story that *'BOAC is buying American again'* and the Chairman felt it necessary to issue a public statement in reply. This emphasised BOAC's commitment to the Britannia project but drew attention to the delays - the 1949 contract delivery dates for the first 25 Britannias were from May 1954 to October 1955 and to date, five years later, no individual Britannia had flown pressurised for more than a total of 120 hours. It did not deny BOAC's interest in the DC-7 series but concluded with these words: *'You cannot carry passengers on promises - and as BOAC is engaged in a vigorously competitive international business it is vital that as a British airline we should be in a position to offer present and future passengers throughout the world aircraft which are as up to date as those of any of our competitors... ...Today BOAC is supporting the British aircraft industry with orders to the value of more than £50m and has on order two types of aircraft, both of them British, de Havilland Comet and Bristol Britannias. It is our firm policy to buy British wherever possible, but it is also our statutory responsibility to operate an airline. To do so, we must have the right aircraft. Moreover, it is essential that we should have them at the right time if we are not to lose trade to the airlines of other nations.'*

In December there was a meeting between BOAC and Bristols at which their Chairman indicated that all the fifteen Mark 100 aircraft on order should be delivered by April 1956 and the ten Mark 300 by March 1957 - providing there was no delay

A iconic landmark at London Heathrow for many years was the control tower in the central island.

Above: A view inside 'the cab' when the tower was new, showing a controller handing over to another.

Air traffic control officers occupied the glazed cupola on the roof, working in close cooperation with their colleagues of approach control, located two floors below, guiding aircraft to the runways.

Left: BOAC's passenger terminal at Victoria.

Below: Queens Building. In the 1950s its primary function was to provide entertainment for the thousands of visitors arriving daily to see a variety of aircraft and to watch the multiplicity of activities of an international airport at close quarters. To quote a BOAC brochure of the day: *The young ones find delight in wandering over the spacious roof gardens with their parents, or in sailing their toy boats on the blue circular pond. Those of riper years, mostly males with technical leanings, invariably sit with pads and pencils and aircraft identification charts, feverishly writing down registration numbers called out to them by their friends manning the roof garden's telescopes. On the gardens there is a variety of shops, some selling souvenirs, books and films, others devoted to the needs of the inner man by the provisions of light snacks, drinks and ice-cream.*

Such was 'Queens' . . . a place of wonder and enjoyment for all ages.

in the granting of the Certificates of Airworthiness. Bristols were given a clear indication by Sir Miles that BOAC was also looking elsewhere for capacity - a veiled reference to the DC-7 possibility as an insurance for the North Atlantic route to cover the 1956 to 1958 period.

In spite of this, Bristols were happy to receive an increased order from BOAC to take thirty-three aircraft made up of fifteen Mark 100, eight Mark 300 and ten of a long-range Mark 300, with a firm intention to order sixty Bristol BE25 engines eventually to re-engine all the 300 and 300 LR series, in order to achieve better fuel consumption and greater flexibility of operation. The new requirements were advised to the Minister on 3 February 1955 for his approval. On the assumption that the C of A for the Mark 100 would be granted by August 1955, Bristols guaranteed delivery of four Mark 100s by the end of 1955 and the balance of eleven by August 1956; all the Mark 300s were to be delivered by the end of 1956 and the long-range 300s between October 1956 and April 1957. The long-range version of the Mark 300 would be capable of operating non-stop between London and New York with sixty-four first-class or ninety-three tourist-class passengers.

After negotiations, the price to be paid for the Britannias was rising, and by April 1956 the Mark 100 stood at £768,000 each, the Mark 300 at £895,000 and the Mark 300LR at £945,000, the last two prices being subject to escalation. At the request of the MoS, BOAC agreed to let it take the first of its 300 series, receiving in replacement another of the 300LR series thus altering the make-up of the Britannia fleet to fifteen Mark 100, seven Mark 300 and eleven Mark 300LR. The Chairman signed the agreement with Bristols for these aircraft on 10 August 1955, but by October he was having to tell the Board that following receipt of later delivery dates from Bristols, the Mark 100 would suffer a delay of about three months in its introduction into service.

The route-proving trials of the Mark 100 had not surprisingly thrown up specific defects, and the ARB would not give full clearance until they were satisfied these had been rectified. Two Mark 100 aircraft were finally accepted by the Corporation on 30 December 1955 but with reservations and without a full C of A, with the delivery of four more promised by May 1956 but without any definite dates for the balance of the Mark 100 order. It was hoped that all eleven of the long-range series (now the 312) would have been delivered by December 1957. These delays in the Britannia deliveries resulted in the Corporation having to continue to operate Constellations, Stratocruisers and Argonauts beyond the planned retirement dates.

The first proving flight to Johannesburg with a Britannia 100 left London Airport on 29 March 1956 and at that time it was planned to start passenger services on that route in July.

It was at this point that a further delay in service introduction occurred, due to the problems experienced on the return of the first proving flight from Johannesburg. A power failure had occurred an all four engines when the aircraft met the inter-tropical front between Livingstone and Nairobi. This was the first warning of the problematic 'flame-out' problem which was to extend over an extended period and delayed the first Britannia service with passengers to 1 February 1957 well into the time of the next Chairman.

Britannia in service - for the passengers
BOAC and Bristols were keen to see that the passengers percieved the Britannia in the best possible light - and did this by making sure that the press both reported and recorded what today would be termed a series of trip reports correctly.

Thousands of people see the Britannia every day, as they wait with varying degrees of impatience in airport departure lounges in Zurich, Tokyo, Khartoum or Sydney. To them, the Britannia probably looks much the same as any other airliner - perhaps a little more elegant and attractive than most, but still fairly ordinary. It certainly doesn't cause the stir of excitement that still accompanies the screaming arrival of the occasional rakish-looking jet. Until someone notices that the men working on the Britannia somehow look rather small that is!

For so graceful are its lines, so pleasantly proportioned, that it is often difficult to realise just how large an aeroplane it is! To give some idea of scale, it measures more than 125 feet from the point of its radar nose fairing to the tip of its tail, compared with the DC-7C's 105 feet, weighs nearly ten tons more than the next heaviest airliner, the Super Constellation, and exceeds the dimensions of the Stratocruiser in its wingspan.

So much for sheer size. The next striking thing about the Britannia is the impeccable surface finish, although you cannot fully appreciate it until you are close enough to see your face mirrored in the glittering undersurface of the wing, or pause at the top of the entrance stairs to look along the huge shining cylinder of the fuselage and see the airport lights reflected there with hardly a ripple. Here, you will say to yourself, is precision quality on a truly massive scale.

There are two main doors conventionally located in the port side, the forward one up near the flight deck being customarily used by the crew and the other, about three-quarters of the way back towards the towering tail fin, leading into the passenger entry vestibule. The most important point about the Britannia's cabin is that it is roomier than that of any other airliner flying today. It starts off in its unfurnished state with an unobstructed length of ninety-nine feet, stretching from the forward pressure bulkhead of the flight deck to the rear pressure bulkhead which is roughly in line with the leading edge of the tailplane. Combined with a floor level which, except for the aftermost ten feet, is constant throughout its whole length, this has allowed the layout and accommodation specialists a very free hand in catering for the individual requirements of different airlines, without reference to the fixed structural members that usually seem to pop up just where they are not wanted.

The cabin is also no less than twelve feet wide, as compared with about ten feet in most contemporary airliners, which gives a gangway wide enough to allow natural movement from one end of the aircraft to the other. It also means that six-abreast seating in two units of three, usually considered to make the middle passengers feel like the filling in a sandwich, becomes comparatively comfortable, and this is the arrangement

Big tails outside the base hangars - BOAC Britannias G-ANBE and 'NBJ at Heathrow.

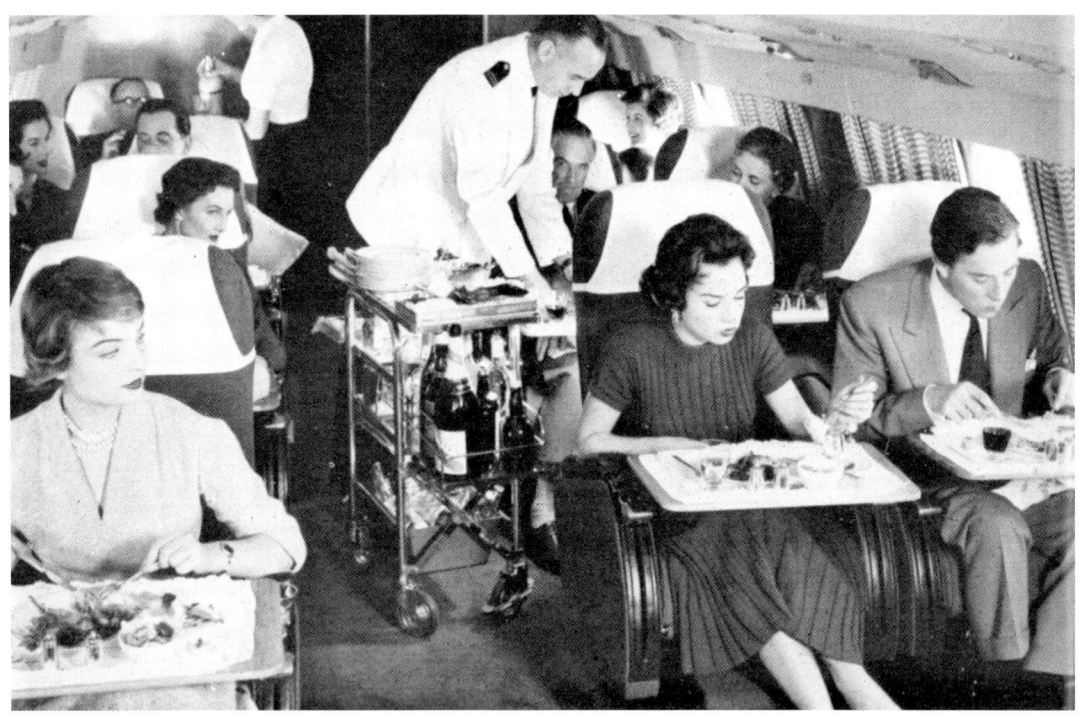

Dinner is served in a BOAC Britannia at 30,000 feet in the aft first class cabin. So claimed the caption for this obviously posed picture in a BOAC travel brochure.

The food was prepared in an all-electic galley that could produce anything from a quick cup of coffee to a full six course meal. *(BOAC via author's collection)*

adopted for the standard all-tourist version. It seats ninety-three passengers with a standard of leg room usually only associated with first-class travel, the seating being pitched at forty-one inches so that every row coincides with one of the enormous elliptical windows.

The claustrophobic 'tube effect' that would result from the use of a single 77-foot cabin was broken up by partitioning it into three smaller cabins, the main one seating sixty-six passengers beginning in line with the front of the engine nacelles and extending aft, slightly beyond the wing trailing edge, to the entry vestibule. This area is defined by an arrangement of partial bulkheads containing the ship's library and either a cocktail bar, or wardrobe space for hanging the usual hats, coats and cameras, and for the winter dress or tropical suit that some passengers like to keep handy for later in the journey. In the aft cabin are a further twenty-seven passengers, seated five-abreast as the fuselage begins to taper towards the tail which houses, down two steps, the elaborately equipped toilet compartments.

On a long journey, it breaks the monotony to be able to spend an inordinately long time having a 'wash and brush-up', and great care has been taken to make the surroundings as pleasant as possible. Thus the ladies' powder-room contains a dressing table and both wing and full-length mirrors, the men's dressing-room a selection of electric shavers and plenty of light in the right places, and each toilet compartment an electrically operated flushing toilet - unusual aboard an aircraft. There are three of them, the centre one having an ingenious movable partition so that it can be made available to either men or women, at the steward's discretion, according to the look of the passenger list. Two more toilets occupy the unavoidably noisy position up forward abreast of the propellers and serve as a division between the main cabin and a small front compartment fitted out as a crew rest-room with two bunks and two seats.

Finally, opposite the crew entry door comes the all-electric galley, where packaged meals, pre-cooked and loaded at the airports, are kept refrigerated or quickly heated according to requirements, and next to it the steward°s pantry unit where the trays are laid and dishes washed. A further advantage of the large diameter fuselage is that it has made it possible to combine ample headroom in the cabin with a comparatively high floor level. This leaves space for two very roomy cargo and luggage holds under the floor, one ahead of the wing and one aft, and loaded through big doors in the flanks of the fuselage. Each hold is just over 3 feet high, 4 feet 6 inches wide and some 30 feet long, with a reinforced floor to support loadings of up to 75 lb./sq.ft, and with built-in floodlights in the ceiling for use during loading operations. They are ventilated, heated and pressurized by the cabin air-conditioning system so that livestock can be freighted and, for example, liquids will not freeze or leak in the passengers' baggage. In an emergency, the air supply can be cut off, and carbon dioxide injected as a protection against fire.

As well as the four tons or so of baggage in the holds, the cabin luggage racks can carry another 1,350 pounds, and if necessary, a section of seating can be removed to make room for more. All the seats are

So quiet you could hear the fizz in your drink!

So went the advertsing slogan for BOACs Britannias, brought about by the smooth, quiet, Bristol Proteus turboprops. *(BOAC via author's collection)*

carried on four main rails which run the whole length of the cabin floor and which are perforated at one-inch intervals to provide the attachment points. Thus, for example, the small forward cabin can be quickly converted to additional passenger accommodation or to provide extra space for, say, Christmas mails. In addition to allowing the seating to be easily rearranged, even if necessary in the course of a flight, the system of rail-mounted seats makes it possible for the operator to use the same aircraft in a variety of different roles. They range from a high-density 130-seat coach to a luxury first-class 64-seat sleeper suitable for transatlantic non-stop services, with seventeen single upper berths which stow up into the roof and fifteen double-width lower berths, formed by the ingenious folding of some of the seats. This is one of the layouts used by BOAC on their transatlantic runs; for their mixed-class services, the rear cabin is given over to sleeping berths and first-class seating for twenty-eight, with fifty-four tourists occupying the main and forward compartments.

The characteristic quality shared by all these layouts is a most pleasant sense of space and elegance. Even in the high-density 130-seater, where a certain amount of cramping is only to be expected at coach-class fares, the seat pitch of thirty-six inches does not restrict legroom unduly. However, there is of course correspondingly more breathing space in the tourist accommodation, while the first-class arrangement of a mere pair of double seats across the twelve-foot cabin affords a degree of freedom and spaciousness undoubtedly unequalled in any other airliners.

As you sink into your seat, two other factors strike you as adding to the pleasing atmosphere: the soft even lighting from the hidden fluorescent strips in the ceiling panels, and the unusually broad field of view from the windows. There is the usual slight loss of visibility from those seats in line with the centre of the wing, but despite the size of the Proteus nacelles, which look enormous from here, you are seated high enough to be able to see over them to the wing-tips, and to take in the view in front or behind them. Turbo-prop airliners, in the shape of the Viscount, have already established themselves as a favourite with air travellers by their inherent quietness, comfort and almost complete lack of vibration. However, in the sudden numbing silence that follows the muffled bump as the main door closes, you realise how undesirable would be a completely quiet aeroplane. Every wheeze and sniff and cough, the cries of a child, the private conversation three rows behind you, the clatter of plates in the galley and all the tiny whines and groans of motors and pumps drift up and down the cabin, until they are gradually absorbed into the smooth rushing of the air-conditioning system as the engines are started.

That, and the sight of the four great propellers beginning to pick up speed, are the only signs that the turbines are running at idling revolutions. Unlike piston engines, they need no warming-up period, and a moment later you are taxiing fast towards the take-off position. A brief engine check and then the seat presses you firmly in the back as seventeen thousand horsepower take you tearing down the runway, accompanied by a heavy rumbling from the huge paddle blades of the propellers as they beat the air at a majestic thousand revolutions per minute.

There seems to be virtually no sound from the turbines at all, and when power is pulled off to begin the climb, the noise level throughout the cabin becomes low and entirely unobtrusive. Only in the forward toilets is there any real volume of sound from the thudding of the propeller tips, but such is the effectiveness of the soundproofing that even in the front two rows of seats propeller noise is barely noticeable and elsewhere you might be riding in a jet. The muted thunder from the efflux pipes can, in fact, just be sensed from the seats near the tail, but anywhere in the cabin, you can chat to *another passenger several paces away without raising your voice.*

This was the era of the catchy advertising slogan, and so it was with the Britannia; The unofficial name for the aircraft was 'Whispering Giant' - it was big, and it was quiet for the time, so it was an acceptable phrase to use. However, it has been recorded - although unproven - that for a short time at one long-haul destination through the imaginative efforts of a translator, Malay-speaking passengers embarking at Kuala Lumpur airport were apt to get rather the wrong impression from the brochure, which described the Britannia not as a 'Whispering Giant', but as a 'Thundering Elephant'.

Then there was the claim allegedly attributed to Bristols that on a Britannia passengers could 'hear their drinks fizz'. This probably depended on how acute your sense of hearing was. However, the story goes that at least one passenger attempted to prove or disprove the claim, and caused a scene by lifting his glass to his ear just as the plane hit an air pocket. It happened to be his first drink of the day, but his eager assurances to anyone who would listen were greeted with raised eyebrows and the stewardess, who helped dry his clothes, politely declined to bring him further experimental material! But back to the brochures:

Physical impressions of the Britannia have a sort of dreamlike quality - you feel almost as if you might be travelling around in some thickly padded and heavily muffled hotel. One of the operating characteristics of the Proteus engines is that their revs seldom need to be altered during the approach and landing; there are no sudden brazen outbursts from the exhausts, no alarming changes in the note of the engines to convince the habitual armrest grabbers that the pilot must be struggling desperately to save their lives! Vibration, one of the leading causes of travel fatigue, is so slight in the Britannia that the surface of your drink remains unruffled; Pressurisation keeps the cabin altitude' down to the equivalent of about 6,000 feet even when cruising in the stratosphere at 35,000, and plenty of fresh air, essential in preventing nausea, is continuously pumped into the cabin via concealed ducts and through the adjustable louvres above the seats. The ventilating supply is not only heated or cooled according to requirements but is automatically humidified at high altitudes, where the extremely dry air is otherwise liable to make your eyes smart and to irritate your nose and throat. These are some of the factors that contribute towards a superb level of travel and enable one to arrive after a twenty-four-hour flight feeling fresh and rested, instead of wanting to crawl into bed with eyes tingling and ears singing. This is what, from a passenger viewpoint, the long years of research and development have led to. Nowadays an airliner is judged by the traveller solely on the standard of comfort it can provide, since it is reasonable to assume that beyond all doubt he will arrive safely and even, with luck, on time. Yet although such things may justifiably be taken for granted, a glimpse of some of the Britannia's engineering highlights serves to broaden the perspective of Bristol's magnificent achievement.

Enter the Comet!

The aircraft which made the most significant impact on BOAC during Sir Miles Thomas' Chairmanship was undoubtedly the De Havilland Comet, an impact which was initially highly advantageous and subsequently

'At eight o'clock on the misty morning of Saturday April 2nd 1949, the Comet was pushed out of its hangar for the first time, so that engine runs could be made...' So said the caption that went with this colour picture that appeared in the *DH Gazette* of August 1949. At the time only the port pair of Ghosts had been fitted. The long wooden building in the background is the pilots hut which later housed the library and tech reps offices. *(DH Hatfield)*

The warm greetings at Ciampino Airport, Rome. The Comet carried eleven passengers, including Sir Archibald Russell, GCB, MBE, Permanent Secretary of the Ministry of Supply, and his two colleagues at the Ministry, Air Marshal Sir Alec Coryton, KBE, CB, MVO, DFC, Controller of Supplies (Air), and Mr. Musgrave, Under Secretary (Air), the official entrant for the record, also the editors of 'The Aeroplane,', 'Flight' and 'Aeronautics', and several De Havilland executives were carried. DH Hatfield)

The Esso refueller at Ciampino airport at the completion of the refuelling of the Comet, during which about 1,400 gallons of aviation kerosene were pumped aboard in ten minutes. By now the BOAC 'Speedbird' emblem had been painted large on both sides of the aircraft's nose. (DH Hatfield)

deeply disappointing. At the time he took over, the Comet was just about to make its first test flight. Subsequent test flying produced encouraging results, so encouraging that the Board agreed in December 1949 a letter could go to the Minister advising him that the Corporation would prefer to dispense with the Britannia plans and concentrate instead on later developments of Comet 1, thus meeting the Government's request for reduced capital expenditure and at the same time putting BOAC in the strongest possible competitive position.

For the full story of the development of this airliner, see the author's *Comet! The world's first jet airliner* by (Pen & Sword Aviation 2013)

This proposal came to nothing but BOAC was showing increasing interest in a Comet 2, holding its orders for the Comet 1 at fourteen aircraft. The need for a jet aircraft larger than the Comet 1 was apparent in the light of signs that the American manufacturers - especially Boeing - were studying pure jet projects able to fly the North Atlantic non-stop in either direction.

In September 1950 the nucleus of a Comet Flight was set up in BOAC with Captain M. J. R. (Roly) Alderson in charge with the object of developing flying techniques for this radical new type and of doing all possible to speed up the introduction of the aircraft into service once deliveries had been made.

Early in 1951 De Havilland's released information about the expected performance of the Comet 2 with Avon engines and with a promised delivery of the first aircraft by October 1953. The aircraft was planned to provide an increase in range over the Ghost-engined Comet I of about 750 miles for a given payload, but a one-stop North Atlantic operation westbound would still be necessary. BOAC, therefore, decided to ask the manufacturers to convert five of the fourteen Comet Is on order into Comet 2s. Also, six more Comet 2s were held on option for BOAC. At that time deliveries of nine Comet Is ranged from June 1951 to October 1952 with four Comet 2s from October 1953 to April 1954 and the remaining Comet 2s to be delivered in October 1954. If taken, the six Comet 2s on option could be delivered between December 1954 and July 1955.

Towards the end of 1951, the Corporation's operational development flying programme with the second prototype Comet I was nearly complete and much had been learnt about new operating techniques for jet-powered aircraft. Tentative plans were made to introduce the aircraft by March 1952 on the London-

Johannesburg and London-India routes. An estimate of the operating costs on the London-Johannesburg route showed that the Comet would be in the same kind of bracket as the Argonaut or the Stratocruiser but cheaper than the Hermes.

Given how the Comet was a quantum leap above every other airliner that flew, it was not surprising that wherever it went, new records were set.

Many milestones in flight testing were achieved in a remarkably short period - on 8 August 1949 a Mach number above 0.8 was attained in a shallow dive, and on 10 November a height of 43,000ft was reached. During the same flight the 375 miles from Edinburgh to Brighton was covered in 42 minutes, averaging about 530 mph, and during a 5 hr 35 min flight at 35,000 - 40,000 feet on 14 November the 590 miles from the Shetland Isles to Hatfield were clocked in 60 minutes. A noteworthy feature of the early trials was the high degree of serviceability attained, and the aircraft averaged well over one hour's flying a day during its first 110 days. The ease of engine changing and other servicing, the quick turn-round between flights, and even the simplicity of the cockpit check - which had brought a favourable comment from London Airport control tower - was noticeable.

On 25 October 1949, John Cunningham took G-ALVG from London to Castel Benito in Libya and back in under twelve hours. This was the Comet's first flight to an overseas airport. The trip was one of a series of flights to measure the economy of the aircraft and to ascertain the optimum cruising conditions. For the tests to achieve their purpose of providing reliable operational data with the conditions of the flight were strictly controlled with pre-set figures for cruising altitude, and engine speed and no attempt was made to achieve spectacular times at the sacrifice of accurate data. Up to the end of October, these flights were made on a medium loading corresponding to full tanks and partial payload. Gradually the loading was increased, and in mid-November, it was near to operational weight. In other respects, the flights were fairly representative of commercial conditions except in the matter of the pressurising of the fuselage. Up to the Castel Benito flight, only about two pounds per square inch of pressure differential had been used - higher pressures would be soon employed.

As a preliminary to the flight to Africa John Cunningham flew the Comet to London Airport on 22 and 23 October to carry out a series of night landings and Ground Controlled Approaches in mixed weather to familiarise himself with the airport and the control procedure. Towards the end of this series of landings, the Comet was operating in very poor visibility.

On 24 October the Comet left Hatfield at about 21.30 and touched down at London Airport some ten minutes later. Cunningham and his crew retired to bed leaving the aircraft in the hands of the ground crew for refuelling and a final check.

At 0400 weather reports were examined, navigational and radio briefings checked, and after a hasty cup of coffee and to the accompaniment of flash-bulbs Cunningham and his crew of Peter Bugge, second officer. E. Brackstone Brown, flight engineer, and G. Blackett, navigator and radar-operator, climbed aboard.

Taking off at 06.33 BST in darkness and light rain the Comet climbed on course to its required height of 35,000 feet. Eyebrows were raised at London Airport when twenty-seven minutes after take-off Cunningham sent a routine message reporting passing through tenuous cloud at 31,000 ft. At 09.56 BST the Comet touched down at Castel Benito, Tripoli, having accomplished the flight in three hours twenty-three minutes, representing a block-to-block speed including the climb and landing circuit of 440 mph. After a late breakfast, during which the Comet was refuelled with Shell Aviation Turbine Fuel, Cunningham left the ground again at 12.04 on the return flight. During the homeward trip an accurate 500-mile stretch, in which light head- and tail-winds alternated, was covered in sixty-one minutes. As on the outward journey, the altitude was 35,000 ft., the cruising speed being allowed to increase as the weight of the fuel lessened. Over northern Europe, cloud and strong crosswinds were encountered, but over the Mediterranean the winds were light and the sky almost clear. At London Airport the weather improved during the morning, and there was sunshine with occasional showers when Cunningham circled and landed at 15.19. The return flight had been made in three hours fifteen minutes at an average block speed of 458 mph.

The Ministry of Civil Aviation, who were naturally most interested in studying traffic control problems for jet aircraft, gave the fullest assistance in handling the Comet at London Airport. At Castel Benito, the airfield was in the hands of International Aeradio Ltd., and all the facilities at that place worked with smooth efficiency. Underwing pressure refuelling was used at both London Airport and at Castel Benito with fuel and mobile equipment provided by the Shell Company. BOAC, who as potential operators of the Comet were watching the progress of the tests with the keenest interest, gave their full co-operation particularly in matters of meteorological and route information, engineering facilities and communications.

The BOAC Flight Development Unit

There were concerns about operational aspects, understandably so, since the design represented a substantial advance in airliner performance without calling for radical changes in operational procedure. The Comet was an aircraft of moderate wing loading - far less in fact that some of the latest propeller-driven airliners of its time. Thus in all ordinary handling respects, the Comet was orthodox.

Considering handling in the broader sense, the whole problem of operating the Comet demonstrated itself to be – as it was designed - a practical proposition from the outset. Traffic control in all its many aspects was acknowledged to be in need of development, even for propeller-driven airliners; as such development

Every now and then a picture appears that just screams to be reproduced as large as possible - this striking picture - taken in the circuit over London Airport, is of the second Comet, G-ALZK, in BOAC livery. (DH Hatfield)

materialised so the economic and practical superiority of the jet aircraft would show up to a better advantage.

De Havillands had designed the Comet 1 to best suit stage lengths of up to about 1,750 miles, and down to as low as 600 miles. These figures had full regard to the fuel reserves and all other operational features of the aircraft, which had been well demonstrated, in over fifteen hundred hours of test flying accomplished in the two and a half years since 27 July 1949. Useful experience had been gained, not only as a result of the manufacturer's trials but also by the Comet Unit of BOAC which in the course of a six month period of proving trials during 1951 flew over 92,000 miles on the Empire air routes with the second Comet prototype, G-ALZK.

Sir Basil Smallpeice (with the 'e' before the 'i') had a number of jobs within BOAC, including that of Managing Director. He recalls the establishment of their own dedicated Flight Development Unit for the Comet: 'Meanwhile great progress on the de Havilland Comet had been made in our Flight Development Unit led by Captain 'Rowley' Alderson. Captains Majendie and Rodley were the two pilots selected to work with him to develop the operating procedures and techniques for getting the best out of these first generation pure-jet aircraft. This was an entirely new and untried field, and they had no one else's experience to draw on. Never before had there been a passenger airliner which would fly at speeds of over 450 mph and at altitudes of up to 40,000 feet. Moreover, the nature of jet engines was such that aircraft consumed fuel at a higher rate, causing real problems if for some unforeseen reason it had to go into the normal holding stack at a much lower altitude while waiting to come in to land. This - and much else besides - required the mastering of new operating techniques. The best methods of using jet aircraft on civil air routes had to be tested and proved and proved again. It was all done with the quiet competence and thoroughness which I had come to expect and was so typical of BOAC's flight operations management.

The Comet 1 was due to go into passenger service on the route to South Africa at the beginning of May 1952. The planned schedule of training flights was well on time, and in April we took the press down to Rome for lunch and back the same day. That may not sound much of an achievement these days. But in 1952 it took the most advanced piston-engined aeroplane some four-and-a-half to five hours, depending on the wind, to cover the 900 miles between London and Rome. But in the Comet, with a following wind, we reached Ciampino Airport near Rome in under two-and-a-half-hours. Distances between places, as measured in time, were on the point of being halved.

Plans were already in hand for the Series 2 Comet

Above: the 'waving off' party as Comet Yoke Peter leaves London Airport for the inaugural flight to Johannesburg.

Left: Captain Alastair M A Majendie, MA, FRAeS boards G-ALYP for the world's first jet passenger service.
(both DH Hatfield)

Captain Geoffrey de Havilland (left), John Cunningham and members of the BOAC and De Havilland Boards 'see away' the world's first jet airliner passenger service. *(DH Hatfield)*

Above: Charles Sims, who had been photographing aircraft for 'The Aeroplane' for more than 25 years, records the scene at London Airport as the Comet is prepared for its inaugural flight at lunchtime on May 2 1952. *(DH6283Y photo)*

Right: passengers' luggage is manhandled into the pressurised freight compartment of the Comet before departure. *(DH6291J photo)*.

with Rolls-Royce Avon engines which would come into operation later. This was a long range aircraft capable of carrying its capacity payload on stages of up to 2,500 miles, and, having the advantages of a greater thrust and a lower specific consumption, its operational aspects would be less critical than those of the Comet 1. Moreover, the entry of the Comet 2 into service would be facilitated by the widespread commercial experience which the Comet 1 would have obtained, as well as by developments which should have by then taken place in air traffic control.

Whilst from the start of the flight tests De Havillands naturally concentrated on evaluating the performance and handling characteristics of the Comet, it was realised, even in the early design stage, that the high speed of the Comet and the basic characteristics of the jet engine would call for a close study to determine the best operating procedure to enable the aircraft to function smoothly on the established airways, and at the same time to take full advantage of the speed, comfort and economy which it could offer. It was always evident that the many novel characteristics of the Comet would call for a special study of the already acute problems of air traffic control, and, whilst this was primarily a matter for the airline operators and controlling authorities, Comet G-ALZK landed at London Airport on 2 April prior to being handed over on loan to BOAC. This aircraft was the second of the two Comets ordered by the Ministry of Supply for development purposes. A comprehensive programme of flying had been drawn up in collaboration with the Ministry of Civil Aviation and the Ministry of Supply.

After preliminary flight trials in the United Kingdom which were largely concerned with a series of performance measurements for the purpose of producing a BOAC Cruising Control Manual, experimental flights were being made between the UK, Cairo and Calcutta, with extensions to other routes as experience was gained No passengers were carried on these flights, but up to 28 May, the Comet flew 147 hours.

'LZK differed in several respects from the machines being built for BOAC. The interior layout was not standard and was not fully soundproofed or furnished.

The first Comet 1 to be delivered to the Corporation, G-ALYS, arrived at London Airport on 15 January 1952 and on 22 January the Minister signed the C of A; at about the same time the RAE at Farnborough were conducting a series of tests into metal fatigue.

The original order for fourteen Comet Is and six Comet 2s had been altered to eight Comet Is and twelve Comet 2s, the latter with Rolls Avon engines and lengthened fuselage. Seating in the Comet 1 was limited to thirty-six, but the Comet 2 could take a total of forty-four passengers. With a cruising speed of nearly 500 mph and flying at the height of around 40,000 feet, the Comet offered passengers an entirely new flying experience at a speed which far outstripped any competitive piston-engined aircraft. Deliveries to BOAC had kept near enough to forecasts to enable a start to be made on scheduled passenger services as planned, and the first-ever jet operation began from London Airport on 2 May 1952 to Johannesburg flown by Comet 1 G-ALYP, completing the journey in 23 hours 40 minutes compared with thirty-two hours by the Hermes it replaced.

De Havillands were maintaining the momentum of deliveries and by June 1952 six Comet 1s had been accepted by the Corporation and by 31 March 1953, all eight Comet 1s were in service. It was planned to bring the Comet 2s into service in 1954, and the pilot manpower was being stepped up to enable a start to passenger operations by June 1954.

The Corporation had already been in touch in late 1952 with De Havillands about guarantees of performance for a Series III Comet as the airline's forward plans had a requirement for twenty of this type. A start was made preparing to build up a Comet 3 fleet when in April 1953 the Board gave approval to the purchase of a maximum of six Comet IIIs at a price of £715,000 each, subject to performance guarantees adequate for North Atlantic operations. The Comet 3 was designed for uprated Avon engines and with an all-up weight of 145,000 lbs as compared with the 105,000 lbs of the Ghost-engine Comet 1 and 120,000 lbs of the Comet 2 with its lower-powered Avon engines. Cruising speed of all three Marks was about the same, 490 mph, but passenger seating in the Comet 3 had risen to fifty-eight. The stage lengths over which the optimum results might be achieved varied considerably. With the short-range Comet 1 it was only 1,200 miles, with the Comet 2 1,950 and with the Comet 3 2,150, still well short of economic non-stop Atlantic operation.

Above: the world's first fare-paying jet-liner passengers board the Comet at London Airport for Rome and points south to Johannesburg *(DH6291M photo)*.

Left: amongst the thirty-six passengers on the 2 May flight were Miss T Coleridge-Taylor, daughter of the famous musician, who was herself a noted conductor and composer, and Miss D. Hannaford, a 22-year-old London policewoman. *(DH6291I photo)*

> 'In view of the already heavy demands for seats on our Comet flights, please let us know your requirements as far in advance as possible'.
>
> An invitation to fly on the world's first passenger jet service from BOAC dated 24 March 1952.

A World's first!

In the afternoon of 2 May 1952, the British Overseas Airways Corporation Comet G-ALYP - 'Yoke Peter' in the parlance of the day - took off from London's Heathrow Airport bound for Johannesburg, on the world's first jet-airliner service. Operating as Flight number BA113, the aircraft arrived at its destination 23 hours 37 minutes later, at twenty-three minutes to four local time on 3 May, three minutes ahead of the scheduled time. It left again at nine o'clock local time on 5 May on the first north-bound service and arrived at London Airport two minutes early at 07.48 on the morning of 6 May.

An indication of the special operating technique employed by high-speed jet airliners was apparent even at the start from London Airport on 2 May. At a 14.45 the last passenger embarked and the door was closed; at 14.58 the engine-starting routine commenced. Promptly at 15.00 the Comet moved off from the apron and taxied direct to the duty runway, a distance of nearly three miles. At 15.10 the aircraft was lined up on the runway, the four Ghost engines were opened up to take-off power, the brakes were released and twenty seconds later, at exactly 15.12, the Comet was airborne en route for Rome.

The crew that afternoon were Captain A M Majendie, First Officer J G Woodill, Engineer Officer W I Bennett and Radio Officer R W Chandler. In the cabin was Steward E W Charlwood, and Stewardess J P Nourse.

The passenger list for the Heathrow departure was Mr. Alex Henshaw, Mr J S Crossley, Mr D Carter, Brigadier G Ross, Mr O A Cookman, Mr P F Knight, Mr R D Gwyther, Mr E T Pinkney, Mr J Garlick, Mr W K Peters, Mr P Sraffa, Mr D P Bertlin, Miss D Hannaford, Mr S C Brealey, Dr J M Brown, Colonel E P J Ryan, Mr W E Lawson, Mr G N Wright, Mr B I D Jackson, Mr L Morris, Mr R Brook, Mr C H Brigish, Alderman J H Wemsley, Mr D Willis, Mr Fraser Wighton, Mr G W Pearson, Mr W A Walker, Mr G Movshon, Mr E C Bailey, Mr O Garlick, Miss A Coleridge-Taylor, Mr T West, Mr A C Hales, Mr B Hardy, Mr L Orton and Mr S Naude.

At Beirut, there was a crew change: Captain J T A Marsden, First Officer K Emmott, Engineer Officer T W Taylor, Radio Officer G L Coutts, In the cabin was Steward A C McCormack and Stewardess E P Courtney.

Khartoum saw another crew change for the flight to Johannesburg : Captain R C Alabaster, First Officer D T Whitham, First Officer B A Arterton, (flying as supernumerary crew) Engineer Officer J A Johnson and Radio Officer R J Dolman. In the cabin was Steward T D Irwin and Stewardess A Cartmell.

Sir Miles Thomas had gone on ahead with the last training flight and joined the service at Livingstone - there was no trunk-line airport at Salisbury at that time - and went on with the aircraft to its destination at Palmietfontein airport as Jan Smuts airport was not then built. Two days later he left in the Comet on the first northbound jet service from South Africa to London, arriving on schedule - the

Right: refuelling at Livingstone Airport, Rhodesia.

Above: arrival at Johannesburg.

culmination of countless hours of devoted effort by thousands of people in Britain and other parts of the world.

Reading through the documentation of the time, it is impossible not the hear in your mind the clipped, precise tones of the BBC newreader explaining to his listeners the dawning of a new Elizabethan age as he described the scene at London Airport and the following journey down to Johannesburg. It was all so 'frightfully British old chap':

'Airliners of many nations throng the busy parking apron, their cabin tops white against a threatening sky, and the spectators crowding the terraces and public enclosures are a jostle of precautionary raincoats and umbrellas. But the weather flaunts British ceremonial tradition with a brilliant dazzle of sunshine as, promptly at three o'clock, the shrill scream of turbines cuts suddenly across the heavy background hum of the airport.

All eyes turn to where Yoke Peter, its full complement of thirty-six waving, chattering passengers embarked amid a fusillade of flash-bulbs, is swinging out on to the taxiway. Standing with the little send-off party of BOAC executives are Sir Geoffrey de Havilland, Ronald Bishop and John Cunningham: their feelings can well be imagined as a few minutes later a wash of thunderous sound comes rolling back from the runway and the Comet climbs swiftly away, heading east on the world's first commercial jet service.

In command for the first part of that historic trip was Fleet Captain Alistair Majendie, whose flair for mathematics had enabled him to contribute much towards devising BOAC's new jet operating technique. Strong headwinds over the Alps made them nine minutes late arriving at Rome to refuel, but they made up time on the next leg, and in the gathering dusk swept in low across the surfbound coast of the Lebanon, to land fifteen minutes ahead of schedule at Beirut.

While the clattering kerosene bowsers once more disgorged their thousands of gallons into the great wing tanks. Captain Majendie handed over to the slip crew who were to take the service on through the night. And so to Khartoum, silvery in the waning moonlight. More fuel, another fresh crew. a stroll for the yawning passengers, and they were off again, the brazen sun tipping the rim of the desert as the Comet winged southwards to Entebbe on the edge of Lake Victoria; then on to a last refuelling stop in the midday furnace heat of Livingstone. Climbing away over the Victoria Falls, there were still a few minutes in hand and they dawdled along the last leg of the journey, making wide turns to kill time. When Yoke Peter finally touched down at Johannesburg, twenty three and a half

Her Majesty Queen Elizabeth The Queen Mother and Her Royal Highness The Princess Margaret alighting from the Comet G-ALYR after a four-hour flight from Hatfield on 23 May.

The aircraft was piloted by John Cunningham (extreme left), and Mr. Peter Bugge, Accompanying the Royal visitors were Lord and Lady Salisbury (right) Air Commodore Edward H. 'Mouse' Fielden, Captain of the Kings Flight, Group Captain Peter Townsend, Sir Miles Thomas, Sir Geoffrey and Lady de Havilland and Mr F. E. N. St. Barbe.
(DH Hatfield)

hours after leaving London, it was just two minutes ahead of schedule'.

The clockwork precision of this 6,774-mile flight did more than thrill the crowd of 20,000 who were waiting to greet the Comet on its inaugural schedule; it impressed the whole aviation world. Long after the South African service had settled down into the normality of simple routine, there remained a widespread general interest in this revolutionary mode of travel.

During May the Comet service was to operate once weekly in each direction, leaving London on Fridays and Johannesburg on Mondays, but because of the great demand for seats an extra service was operated, leaving London on 24 May. The frequency increased to three services a week from June.

The Comet service was operating on the Springbok route in conjunction with the Constellation services of BOAC's partners, South African Airways. The Corporation's Hermes services, which flew three times a week between London and Johannesburg on the west side of Africa, would be withdrawn progressively as the Comet frequencies are increased.

In the original time-table the actual flying time of the service was eighteen hours forty minutes southbound and eighteen hours fifty-five minutes northbound. Later, when formalities at the transit points were accelerated, the journey time was reduced. Plans were in hand, depending on the political situation in Egypt, to re-route the service through Cairo instead of Beirut, which would bring a saving of 450 miles, thus cutting nearly an hour off the schedule time.

Sir Miles Thomas, was understandably proud: '*The introduction of the Comet into regular airline service as the world's first civil jet- propelled aircraft is indeed a memorable and historic event. Moreover, it is an event in which Britain can well take pride. Not only does it mark the fulfilment of the aims which inspired British initiative and enterprise at the close of the last war; it justifies the faith which the manufacturers and the British Overseas Airways Corporation placed in this remarkable product of the De Havilland factory, where the labours of skilled workers have been crowned with such striking success.*

When the war was over we were convinced that a jet airliner was practicable both for its flying qualities and from the maintenance point of view. One of the problems that exercised us most was the question of operating cost. As a result of exhaustive investigation, however, it became apparent that this new type of air transport should be competitive economically and from that moment we had no doubt but that the pure-jet aircraft would become a British achievement.

The inauguration of the Comet into passenger service on a great British trunk route in little more than five years from the decision which set in motion the whole machinery of detailed design and manufacture, test and development, is deeply gratifying to everyone who has been engaged in the task - and the confidence felt at the inception of the project has gained force during the intervening period.

The Comet certainly heralds a new era in international travel. It will in effect halve the size of the world, and passengers will appreciate in particular the high standard of comfort and the smooth-riding qualities of an aircraft in which the lack of vibration is so notable a characteristic. Furthermore, the Comet introduces a form of powerplant which is basically simple and unmistakably offers much scope for development.

The British manufacturers and the British airline operator, in their close working harmony, together have the ability to retain the ascendency that has now been established for our country in the world of civil aviation.

The Comet was instantly fashionable, exclusive and futuristic. Its essential up-to-dateness was reflected in a load

BOAC never missed a trick in the publicity stakes. This special Air Mail envelope is one of a series that was sent to different sales managers 'down the route' on the occasion of the first London to Singapore service - this one being from the Cairo - Karachi leg.

factor of 89% on the South African route - 32 seats filled out of 36 on every single flight - throughout the first few months of operation. On 11 August, it had begun flying a regular schedule between London and Colombo, and on October 14 the service was taken on from Karachi through to Singapore.

Again the load factors jumped to record levels while slower competitors lost business, and travel agents' waiting lists for each new service became longer and longer as the fashion spread.

More royal attention
John Cunningham borrowed brand new Comet G-ALYR from BOAC to take the Queen Mother and Princess Margaret on a 1.800-mile jaunt during the afternoon of 23 May. Luncheon and tea were served to the Royal Party during the flight, which extended down to Switzerland and northern Italy, embraced the south coast of France and the Pyrenees to Bordeaux and thence home. Before embarking The Queen Mother and The Princess Margaret spent a little time at Hatfield, where they saw Comet manufacturing operations. In the style of the day it was reported that *'Her Majesty and Her Royal Highness expressed great pleasure and enjoyment in the flight, in the course of which each of them spent considerable periods in the control cabins'*.

Not to be outdone, Prince Philip flew back from the Helsinki Olympic Games in one; Pandit Nehru looked down on Mount Everest from the flight deck of another which was making a series of route-proving runs to Ceylon, Singapore and Tokyo.

BOAC scheduled services had started modestly enough with a once-weekly service to Johannesburg, a route mileage of 6,724 miles. In June, following the delivery of the fifth aircraft the previous month, the Johannesburg schedule was increased to three services per week. By August the ground organisation on the route was beginning to run so smoothly that at Khartoum and Livingstone the transit time was cut from sixty to forty minutes - the first time that a BOAC aircraft had been scheduled for so brief a stop. This reduction in ground time and the re-routing of the flight through Cairo enabled elapsed time for the journey between London and South Africa to be reduced from 23 hours 40 minutes to 21 hours 20 minutes. The delivery of the sixth and seventh aircraft by July was followed on 11 August by the introduction of the service between London and Ceylon to be flown once weekly, 5,961 miles in an elapsed time of 20 hours 35 minutes. Until April, 1953, this service worked under a severe handicap because the Ceylonese Government refused permission to uplift passengers from Colombo, which made it necessary for the homeward bound aircraft to fly empty as far as Bombay.

On 14 October a weekly service between London and Singapore was started covering 7,833 miles in 25 hours 30 minutes, and on the last day of October this was stepped-up to twice weekly. On April 3, 1953, London was linked with Tokyo by once-a-week Comet service scheduled to cover the 10,380 miles, reaching halfway round the world, in 36 hours 20 min. elapsed time. Ten days later, on April 13, the frequency of the service was doubled.

Thus within eleven months of the inauguration of the world's first jet service, Comets were flying 122,000 miles a week on a network of routes with an unduplicated mileage of 20,780. During the year the Comets flew 9,443 revenue hours and carried nearly 28,000 passengers thus achieving 104,600,000 revenue passenger miles.

So much for the bare record. In assessing the scale of the achievement it must be remembered that the Comet in use by BOAC during the inaugural year, the Series 1, was the undeveloped Comet and that the fleet of eight consisted of the first Comets off the production line.

As Sir Miles Thomas said: *'If there was scepticism about the practicability of a jet air service within the industry, the public did not share it. As was to be expected the early services were fully booked for months ahead and bookings for the first service itself were being received by BOAC over a year before the start and before the applicants knew where the inaugural flight was to be routed. There is always a large number of people anxious to be 'first' and it was recognised that the novelty factor which led to the heavy bookings achieved during the first weeks of operation could not be considered as necessarily representative of the normal demand. In the event, however,* there has been no detectable falling off in the high load factors, and, if anything can be said to have been proved by 12 months of service, it is that the Comet has become firmly established as a favourite medium of travel with the public, who have quickly come to appreciate not only the value of the speed but the unprecedented degree of leisureliness and comfort, which enables them to reach their destination unfatigued and even refreshed by the experience'.

'Statistically the figures are remarkable: from the start of the service until the beginning of January, 1953, the average passenger load factor over the entire network was no less than 87%. and the overall load factors including passengers, mail and freight have been nearly 80%. Individual route figures show up even better. During February, 1953, for instance, a total of 1,116 stage-seats were offered outbound between London and Singapore and of these seats 1,028 were occupied, representing a passenger load factor of over 92%. In practice it is hardly possible to improve on this figure. On a long trunk route some concession must be made to passengers who require to travel only part of the way and it is not always possible to fill the seats thus vacated. Another reason which prevents every seat from being occupied is the passenger who having booked a seat just does not turn up. This 'no-show' problem, the solution for which has defied the best brains in the air-line business, can affect load factors by as much as six percent, although for the Comet the figures are not so high'.

Passenger appeal

In common with all major airlines, BOAC provided each passenger with forms on which they were invited to make comments about the quality of the service. This facility was widely used on the Comet services and, as was expected, the vast majority of passengers who turned in comments were full of praise for the comforts of jet travel — however, the gentleman from across the Atlantic who stated that he could detect no improvement in noise and vibration must be considered as having been 'got at' by the opposition! A lady passenger evidently more friendly disposed towards the Comet, wrote that from her own experience she was able to assure the Corporation that there was no truth in the rumour that the high cruising altitude disintegrated nylon stockings and underwear. For the most part, however, the comments were eulogistic, and sensible.

Sir Miles Thomas: 'Notable among the few criticisms which have been made is a general dislike of the seats in the forward cabin. Unlike the seats in the main cabin of the Comet 1, which are adjustable through a range of positions from upright to reclining, the fore-cabin seats are fixed and thus do not provide the same degree of comfort for sleeping. This feature was, in fact, dictated by the position of two structural bulkheads which limit the length of the cabin. Because of the justifiable criticism alterations have been made in the Series 2 Comets which will permit the eight seats in this compartment to be of the same adjustable type as in the main cabin. In the Series 3 Comet, too, all the seats will be adjustable. Another criticism, and more difficult to correct, concerns the toilet facilities. With passenger loads which are predominantly male the equal provision of toilets for both sexes inevitably leads to congestion in the men's room, particularly at shaving time. Until the British traveller becomes more continental in his outlook there does not seem to be any answer to this complaint in an aircraft the size of the Comets 1 and 2'.

During the year's operation a great deal of flying time had been occupied in crew training and proving flights. Sir Miles Thomas went into some detail. 'Since the regular services began in May, 24 proving and training flights covering 381,700 miles have been completed prior to the opening of new routes and this is additional to the 248,600 miles covered during 28 similar flights before May 2, 1953. These figures take no account of the local training flights covering general familiarisation, landing practice including three-engine wave-off approaches and approach-pattern procedure.

Each BOAC captain, before taking a Comet on regular service, has completed at least 60 hours of flying on the aircraft, about 10 hours of which consists of local flying followed by at least two trips under supervision over the route he is to operate. In addition all pilots undergo a nine-weeks' course of instruction covering the technicalities of the airframe, the engine and the theory of cruise control. Under Captain Rodley, who is in charge of Comet flying training, there are five of the fleet captains who are qualified as instructors and after qualification each captain has a check-flight of two hours or more every six months. The training syllabus of first officers, flight engineers (both of these groups must complete at least three route trips before qualification), radio operators and the cabin crew varies in detail according to their duties, but is equally thorough.

At the year's end the target of 40 operational Comet crews was achieved, but even before then five captains and ten first officers were undergoing ground courses and initial flying training in anticipation of the introduction of the Comet Series 2 into service.

Operational Lessons

After a complete year of operation the overriding lesson was that many of the expected troubles did not materialise and that the Comet had a much greater degree of operational flexibility than was indicated by theoretical considerations.

Sir Miles Thomas again: 'As is well known, the jet's fuel consumption is high during taxying and ground manoeuvring, although a considerable saving can be achieved if stopping and restarting can be avoided and the aircraft kept moving fairly fast. For this reason a novel take-off routine has been adopted at London Airport and other points where traffic congestion might cause delay.

In the case of the Comet clearance to take off is obtained from traffic control before the engines are started so that it is possible to taxy without stopping to the duty runway and then take-off without interruption. This procedure, which is made possible by the rapidity with which the Comet engines can be started and the fact that no warming-up or full-throttle checks are required, has worked very smoothly in practice and has not in itself led to any serious take-off delays.

After take-off it is desirable, for fuel-economy reasons, to make an uninterrupted climb to cruising altitude. The Comet's climb is more rapid and the air speed is higher than in the case of a piston-engined aircraft and the resultant climbing angle is rather better.

Although since the start of the Comet service there has been no case of an engine failure during take-off, such a condition is simulated frequently during training and it is well established that the three-engine rate of climb with full load leaves an ample margin of safety and is much superior to that of a comparable piston-engined aircraft.

The cruise technique adopted by BOAC consists of flying at a constant angle of incidence which is achieved in practice by flying at an indicated air speed selected in relation to the weight, alterations in speed being made at half-hourly intervals as the weight decreases.

This procedure results in a gradual climb. Depending on load and the length of the stage and to some extent on the ambient temperature, the cruising height will vary from about 35,000 ft. to 42,000 ft'.

On the score of punctuality the Comet service has acquitted itself well, and has indeed already put in a record which compares favourably with that of the well-established types of the BOAC fleet—no mean achievement for the first year of operation of a new type of aircraft.

During the course of the year the standard of punctuality has steadily improved, mainly due to an improvement in mechanical reliability resulting from the increasing knowledge of the aircraft and the growing experience of the engineering staff. The mechanical *reliability of the Comet is considered below in greater detail and here it is enough to say that the troubles experienced have been of a nature such as might be expected when a new aircraft is put into service. The proportion of time lost due to mechanical trouble during the period showed a marked improvement, but the overall picture showing a reduction of about 44%. in time lost was not so favourable because of the exceptionally high incidence of fog in London from October through to the end of February. During this period London Airport was on many occasions closed, sometimes for days at a time, with a consequent disruption of all air services. Other factors which contributed to delay, although only to a small extent, were concerned with traffic control and ground services, passenger service and late connections. In the final reckoning the record shows that 33% of the services arrived on time - or early - and 61%. were within 2 hours of schedule.*

A glance at the BOAC route network operated by eight Comets gives an indication of the increasing rate of utilisation being achieved. During May, 1952, when the service started, the total utilisation was in the order of three hours per day, of which only just over one hour was revenue earning. With the gradual introduction of new services and the stepping-up of service frequencies the rate of utilisation steadily improved. By October, 1952, the figure for the total had reached 4 hours and, with the start of the service to Tokyo, five hours. The increase in the Tokyo schedule to two services a week, ten days later, raised it to 6.5 hours per day.

As already mentioned, the normal basis of statistical reckoning tends to penalise the Comet and this is particularly so in the case of utilisation because the great reduction in stage time which has been accomplished cannot be matched by a proportionate reduction in transit time at intermediate stops. If one considers a future aircraft in which, for the sake of argument, the stage time is reduced to one hour and the transit time is also one hour it then becomes obvious that the maximum possible utilisation, without taking maintenance into account, which could be achieved with such an aircraft, could never exceed 12 hours a day. Thus it can be appreciated that a figure of 6.5 hours' daily utilisation is one that can be considered as a good beginning for the Series 1 Comet, with its comparatively short stage lengths, when the work-done factor is taken into account. More scope for improvement will be available with both the Series 2 and 3 Comets which will offer progressively longer operating stages.

Serviceability

Closely linked with the rate of utilisation was the degree of mechanical reliability achieved. Some degree of mechanical trouble was to be expected in a new type of aircraft and, in addition, however efficient the pre-delivery instruction of engineers may be it is inevitable that in the early stages of service with a new aeroplane the factor of unfamiliarity and even, perhaps, over-caution, would affect the serviceability standard. As was expected, mechanical

A cross-section of the Comet Fleet personnel on parade at London Airport. In the centre of the front row is Captain M J R Alderson, Manager of the Fleet. On the left is Captain E E Rodley, Officer in Charge of Training and on the right is Mr R A V Dismore, the Fleet Maintenance engineer. Immediately behind Captain Alderson is a Comet crew headed (from left to right) by Captain T B Stoney, First Officer P A Wilson, Engineer Officer J H Kingston, Radio Officer I M Clark, Steward E A Johnson and Stewardess Miss J Todd. Flanking the crew to the left are representative members of the administrative staff and to the right representative members of the engineering staff. In the back row are representatives of the ground services including an MCA Air Traffic controller and a Tarmac controller. Captain A M A Majendie in charge of Comet operations was abroad when this photograph was taken. *(DH Hatfield)*

delays were fairly prevalent at the start of the Comet service and during the first month the average arrival time at the terminals was behind schedule to an amount equal to nearly 40% percent of the journey time. The standard of serviceability quickly improved and the final figure showed that time lost by mechanical delays was reduced to 9%. Most of the troubles which were encountered at first were of a type which could quickly be eliminated as each aircraft became 'bedded-down' in service.

One of the most serious faults and one which nearly led to the temporary grounding of the fleet was the misting of the windscreen, which without any previous warning suddenly started to occur in June 1952, when the aircraft were on the final approach to Beirut and Khartoum airfields. This phenomenon occurred only in certain conditions of temperature and humidity The cure, simple enough in principle but involving a fair amount of engineering, was to increase the flow of warm air across the inner face of the screen. Needless to say this modification had a high priority and was incorporated in every aircraft within a very short period.

Of a similar but less serious nature was the trouble encountered with the early type of windscreen wiper, which having given every satisfaction during the test-flying period proved to be not up to the job on service. Its replacement by a more satisfactory type - hydraulically driven - was achieved only after a considerable amount of development work. In spite of the many hours of flying, many of them fruitless, which were spent in searching-out icing conditions, it was again not until the aircraft encountered service conditions that some maldistribution of the hot-air in the de-icing system was discovered.

Sir Miles Thomas: *'In summing up the record of serviceability for the first year of operation it can be stated that the Comet and its Ghost engines have earned unstinted praise from the engineering department of BOAC. The incidence of faults with the first production aircraft and engines has been considerably below that which was expected by the engineers, a point which is fully borne out by the unprecedented increase in hours which has been found possible between the check 4 inspections. Starting off at 200 hours - admittedly representing a cautious approach - the check 4 period has been progressively extended to 1,040 hours. This rate of increase represents a record, which has not been approached by any other aircraft in the history of the Corporation'.*

The first full year of operation has shown that the Comet is a mechanically sound aircraft capable of operating in a normal manner with regularity at unprecedented speeds on the world's air routes, but what of the economic picture? In spite of authoritative statements to the contrary there are those who without grounds assert that the Comet 1 has proved uneconomic and that the BOAC services are being run at a loss. Even if this were true the operation of the small fleet of Series 1 Comets by BOAC would still have proved worthwhile because of the invaluable experience which has been gained in jet operation, experience which can be obtained in no other

Above: BOAC Comet G-ALYP at Naneda Airport Tokyo on 8 July 1952 at the end of a proving flight to Japan.

Left: Captain A.M. A, Majendie of the Comet Fleet, who piloted the Comet on its flight to Japan, and A. C. Campbell-Orde, Operations Development Director of BOAC are here seen with the ladies of the Takarazuka Opera Company who greeted the aircraft at Tokyo airport. BOAC planned to start a Comet service between London and Tokyo early in 1953. *(both DH Hatfield)*

Comet G-ALYP on the apron at Kadena Field, Okinawa, Saturday 28 February on the last of the proving flights. The Okinawans in the foreground are gathering painful pebbles and packing-case nails in order to prevent damage to aircraft tyres but unfortunately for the Comet they missed one and a puncture resulted.(*U.S. Air Force photograph by Robert Coiner*)

way and which has enabled BOAC to establish the commanding lead which it enjoys today among the world's airlines. But it is definitely untrue.

In considering the matter of airline economics and the break-even loads for a given aircraft, a large number of factors which vary route by route must be taken into account. First, the rate of revenue for each route is fixed by the fare structure and the differing proportion between passengers and mail loads over different air routes and different stages has a great influence on the revenue figures. The cost of fuel varies considerably from place to place mostly because of local duties and taxes, whilst the cost of maintaining the line stations naturally differs considerably according to location. All these important considerations greatly influence the break-even load factors which consequently vary considerably from route to route irrespective of the type of aircraft employed. Taking all these factors into account it has been established that on the routes on which BOAC have been operating Comets (connecting London with Johannesburg, Colombo, Singapore and Tokyo) an average load factor of 75% shows a profit.

When considering the profits made it must be borne in mind that eight Series 1 aircraft operated by BOAC represent only the first stage in Comet development. No more of this type will be built and already the Comet 1A with an improved performance is in service. At the present time the Series 1 aircraft are benefiting by high load factors, and this advantage can be expected to last for some years, as there will be no other jet airliners during this time to challenge their monopoly of the travelling public's preference. On this basis, therefore, it is reasonable to suppose that the Series 1 Comets will have gone far to pay for themselves by the time the progressively more profitable Comets 2 and 3 are put into service by BOAC and this takes no account of the value of accumulated experience.

In the final analysis the year's operating figures have shown that the eight Series 1 Comets, which have been operating at an overall load factor of nearly 80%., have made sufficient profit during the period to cover the interest on the capital expenditure. This calculation is on a realistic costing basis, exactly as is applied to all the Corporation's fleets, and includes the cost of route proving and training flights spread proportionately over the life of the aircraft.

The Corporation can thus look back with justifiable satisfaction on the record of their first year of Comet operation in which they have shown the world that the spirit of the British merchant adventurer is vigorously alive today, and in which they have gained a lead that their competitors will be hard put to recover. But once having assimilated the varied experience gained they do not look back. The Members of the Board of the British Overseas Airways Corporation and the small team, under Captain Alderson, who bear the responsibilities for the day-to-day operation of the Comet fleet are farsighted men - actuated by the enthusiasm of crusaders - who more than most can visualise the future in store and who even now are laying plans for the long-range operation of the bigger and better Comets to come.

By Jet Airliner to Japan

Sir Miles made mention of the BOAC service to Japan - the opening of the Comet service of BOAC between London and Tokyo, with a weekly departure each way from 2 April- that became twice weekly from 13 April - demonstrated with fresh force the facility of jet travel. Even the Series 1 Comet, working relatively short stages, was able to cut the timetable on this line from the 86 hours taken

G-ALYZ at the end of the runway at Rome Ciampino. The damage caused by the broken undercarriage is clearly visible. *(Hugh Jampton Collection)*

by the Corporation's twice-weekly Argonaut service to a mere 36 hours.

The eastbound traveller left London at 09.00 and, flying against the sun, passed the first short night between Beirut and Karachi, and the second between Bangkok and Tokyo, by way of Manila and Okinawa, reaching Tokyo for breakfast - only 36 hours after departing from England.

Westbound, with the sun, take-off from Haneda Airport, on Tokyo Bay, was at about midnight and after a long day flight between Manila and Karachi, arrived in London for early breakfast the next morning.

One weekly service was via Rome, Beirut, Bahrein Karachi, Calcutta, Rangoon, Bangkok, Manila, Okinawa and Tokyo. The other weekly service called at Cairo instead of Beirut, at New Delhi both ways instead of homeward only. One service called at Rangoon outward, the other homeward. The Comet flying time for the whole journey was scheduled at 28½ hours, against 44 hours for the Argonaut.

With typical thoroughness BOAC spent nine months proving the Comet route to Tokyo before opening it to traffic. Between 4 July 1952, and 9 March 1953, eleven trial flights were made with Comets between Bangkok (on their Comet twice-weekly line to Singapore) and Tokyo, and many captains and crews were thus familiarised with every feature of the natural and the contrived aids on the last three legs. Hong Kong, served by the Argonauts, was not available for the Comets because of the lack of a suitable alternative airport, so the jets flew from Bangkok to Manila and then to Okinawa and Tokyo.

The editor of the *De Havilland Gazette* was on the last proving flight: '...*we confess to being surprised at the size and resources of Okinawa. It is 70 miles long and has 600,000 inhabitants, and the American forces have constructed two fine air bases, Naha for civil and Kadena for military operations, both comprehensively equipped. We must admit also to a suppressed flutter at the turnout of bomber crews with their trousered wives, jean-clad juniors and bright Cadillacs, and at the footage of Kodachrome which they expended on the Comet as it taxied in. They are enthusiastic and generously co-operative, and manage their heavy air traffic well. Operating along with jet fighters, as well as many lumbering transports and bombers, the Comets ask for no favours, only efficient handling, and that they are accorded.*

The same may be said of Tokyo, where the military and civil air traffic can only be described as intense - it is up to 600 in-or-out movements a day. The BOAC crews have experienced tremendous winds in this region - 180 and 200 knots at Comet height - and believe that the velocities off the Asiatic seaboard at high latitudes are the greatest in the world, even 400 knots and above in rare jet streams. But with seasoned experience they remark that it is all a matter of fuel reserves and frequent fixes, and readiness to take advantage of every favourable element.

Arriving over Honshu in moonlight and four-eighths

Comet G-ALYV had been in service with BOAC for less than a year when it crashed near Calcutta.

G-ALYS in the short-lived 'large' BOAC titles seen at Farnborough after it was requisitioned by the RAE. The aircraft was subjected to fuel overfilling and leakage testing using a Dorset fuel bowser.

cloud late in February, circling Tokyo's seven-million population and landing at Haneda to be greeted by cheerful ground crews, gloved and muffled against the wintry weather, is an experience for an occidental.

Then Comes Disaster...

The Comet was the focus of the hopes of many in the aviation industry - it had been delivered on time and lived up to its specification. Then a series of disasters that rivalled the loss of the RMS Titanic struck the design – disasters that had worldwide implications. Again, the full story of these accidents and incidents are told in the author's *Comet! The world's first jet airliner* by (Pen & Sword Aviation 2013)

The first occurrence happened in Rome: on 26 October 1952 BOAC's G-ALYZ, operating on a London-Johannesburg flight with intermediate stops, was ready for take-off from Rome-Ciampino Airport. It was taxied to runway 16 and lined up; all pre-take-off checks were made, and the elevator, aileron and rudder trim were set at the neutral position. The Captain's estimation of runway visibility was five miles but with no horizon. The flaps were lowered to 15 degrees, and the windscreen wipers were operating. The engines were opened up to full power; RPMs were checked at 10,250 on all engines; fuel flows, engine temperatures and pressures were all correct.

The brakes were released, and the aircraft accelerated normally. At an airspeed of 75-80 knots, the nose wheel was lifted off and a slight tendency to swing to starboard was corrected. At 112 knots the aircraft lifted from the ground by a backward movement of the control column, and when he considered that the aircraft had reached a safe height Captain Foote called for 'undercarriage up'. At that moment the port wing violently dropped, and the aircraft swung to port; the controls gave the normal response, and lateral level was regained. Foote realised that the aircraft's speed was not building up, although he made no reference to the airspeed indicator. A pronounced buffeting was felt suggesting the onset of a stall, and in spite of two corrective movements of the controls, the buffeting continued. Before the First Officer had time to select undercarriage up, the aircraft came down on its main landing wheels and bounced. It was now evident that the aircraft's speed was not increasing and Foote was convinced that there was a considerable loss of engine thrust. He was also aware that the aircraft was approaching the end of the runway and so the take-off was abandoned. The undercarriage struck a mound of earth as he was closing the throttles and the aircraft slid for some 270 yards over rough ground. The main undercarriage units were wrenched off, the port wing and tailplane smashed into two of the airport's boundary lights, and the aircraft finally slid to a stop on the muddy

G-ALYP - the aircraft that disappeared off Elba on 10th January 1954 - it had previously suffered a heavy landing in August 1952.

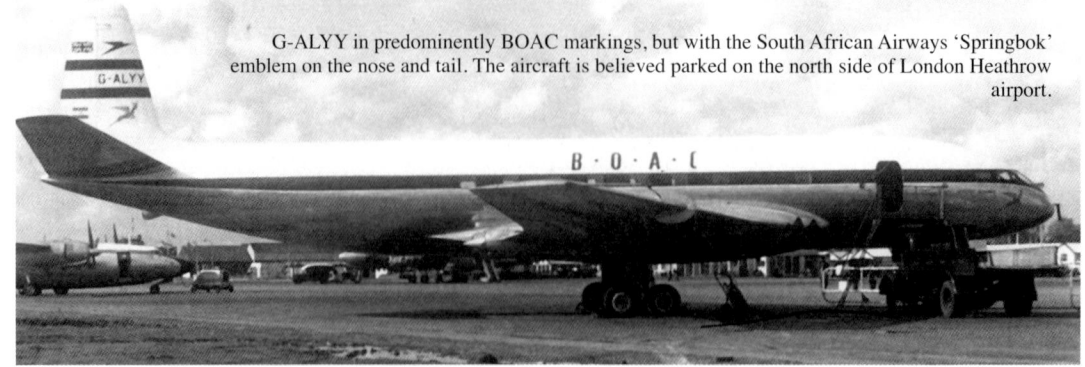

G-ALYY in predominantly BOAC markings, but with the South African Airways 'Springbok' emblem on the nose and tail. The aircraft is believed parked on the north side of London Heathrow airport.

ground. Mercifully, though fuel poured from a ruptured tank, no fire broke out and the occupants, though severely shaken, all escaped injury.

At this point there occurred a curious change of emphasis in BOAC's aircraft policy regarding the development of the Comet and Britannia. In spite of Whitney Straight's enthusiasm in favour of the pure jet as compared with the turboprop, individual Board members, possibly influenced by the accident to a Comet I at Rome, and after comparing operating costs of the Comet, already in operation, with the Britannia which had not yet even entered service, caused the proposed order for twenty Comet 3s to be reduced to five.

Then on 3 March 1953, Canadian Pacific Airways lost CF-CUN on a delivery flight, led by Captain Charles R Pentland, the company's Director of Overseas Flight Operations. Alongside Captain Pentland during the take-off for Rangoon at 03.00 hrs after a quick refuelling stop at Karachi, Pakistan was Captain North Sawle, CPA's chief overseas pilot.

Superficially, it appeared exactly the same as the Rome incident, but there were significant differences. Captain Pentland was short on Comet experience, having flown one for less than ten hours. CF-CUN was heavier than G-ALYZ, and the night air at Karachi was hotter – and therefore thinner than the air at Rome.

Again there was an inquiry, this time conducted by the government of Pakistan, with representatives present from Canada, the Air Investigations Branch and De Havillands. For whatever reason, the lesson from the Italian incident had not been learned. It seems that a new phenomenon, determined as 'ground-stall', had been discovered in the Comet. The report was not published by the Pakistan government, which issued only a press release. Neither the Canadian nor the British government published a report, and again IFALPA disputed the findings.

On 27 August 1953, the first flight of the Comet 2 occurred, with expected C of A flying completed by the end of the year. In September the required manufacturer's guarantees of performance were received, so the Corporation placed a firm order for five Comet IIIs with options on five more. A significant improvement in performance of the Comet 2 by the adoption of a droop leading edge modification was found to be achievable with a minimal delay in the date of service introduction.

On 10 January 1954 Comet I G-ALYP en route from Singapore to London was lost in the sea shortly after leaving Rome in a position about ten miles south of Elba. As the reason for the accident was unknown at the time, the Corporation itself took immediate action to ground all Comet flying as from midnight 11 January. A rigorous technical examination of the Comet followed in BOAC, and after an exhaustive inquiry into all possibilities together with de Havillands, the component manufacturers, the ARB, the Accidents Investigation Branch and the RAE, sixty-two modifications to the aircraft were incorporated before any return to service. By 23 March it was accepted that all possible remedial action had been taken and Comet services to Johannesburg restarted. All the modifications incorporated in the Comet 1s which were relevant to the Comet 2 were included in the first two Comet 2 aircraft, and it was still hoped that these two aircraft would be delivered to BOAC by mid-April.

But 1954 still had more disasters to come. Shortly after

G-ALYS seen at London Airport in June 1952, wearing the early BOAC titles

the resumption of Comet flying the Corporation was to suffer another and even more grievous blow, and on this occasion, BOAC's partners were also directly affected. On 8 April G-ALYY - a BOAC Comet operated by South African Airways - disappeared into the sea south of Naples. Once again Comets were immediately grounded, and the MCA subsequently withdrew the aircraft's C of A.

This was a bitter blow to BOAC. For two years the airline had pioneered civil jet flying, had operated successfully worldwide in spite of the forecasts of impracticability due to airport control and a host of other possible problems, and had outdistanced all its powerful competitors. As Sir Miles said, *'my aspirations and expectations had fallen out of the sky'*.

With the salvage operation in full swing, arrangements were made to test a complete Comet fuselage in a huge water tank at Farnborough, putting it through simulated flight loads and actual pressurisation cycles in the hope that some area of weakness or failure would be located. The desirability of carrying out such tests had in fact been mooted in the early 1950s as more and more significance was attached to the phenomenon of metal fatigue by British and American manufacturers. There was an increasing perception that complete fuselages would need to be pressure- tested to the point of failure whilst being subjected to dynamic loading. De Havilland had tested sections of fuselage and such areas as cabin window and aerial cutouts, but the suspicion was growing that such accepted 'best practice' was inadequate for heavily pressurised high-speed aircraft.

There were feelings of deep shock within De Havilland's when the Farnborough fuselage failed at 1,830 simulated 'flights'. A large section of fuselage had begun to detach from the line of a skin crack starting at an ADF aerial cutout in the top of the forward fuselage. This was an incomparably earlier failure than that recorded in the original Hatfield tests, which had included pressurisation up to two times the design pressure.

From shock and disbelief grew a gradual understanding that the company test specimens had not been representative of production actuality. In the first place they were hand-made without recourse to press tools, and within limits and tolerances inevitably above the standard hitherto acceptable production specification. Secondly, and more importantly, the test sections had been subjected to pressure more than twice the projected working level in the course of the long programme to prove structural integrity. Ironically these higher pressures had themselves made the tests unrepresentative by bedding or firming down the various component parts and actually altering the molecular structure of the material to the extent of improving its

BOAC's advertising material from this period was much more educational and informative than just simply offering the airlines products. They made use of a lot of so-called 'colourised' black and white images that today look strange but are remarkably evocative of the time.

Right: certainly colourised, this picture shows three types in use with BOAC at the time - the Douglas DC-7C, the Boeing Stratocruiser and the Bristol Britannia. *(BOAC via author's collection)*

'Step inside the luxury cabin of a B.O.A.C. airliner and immediately you warm to the atmosphere of elegance and good taste. Experts have designed every detail for your comfort and enjoyment-the harmonious colour schemes, thick pile carpets, restful lighting, wide-vision windows. The limited number of seats ensures maximum spaciousness, and on all de Luxe Services you travel in a full-length sleeper-seat, easily adjustable to give complete relaxation all the time. Not the least of your pleasures are the food and drink. You wine and dine from menus to tempt the most fastidious, enjoying cocktails and canapés, the most expert cuisine, the choicest wines, the best liqueurs. Light refreshments may be ordered at any time, and the well-stocked bar is always open. The Steward will give you a list of the wide variety of amenities for your entertainment and use during flight, down to such details as the toilet preparations and cosmetics provided in the dressing rooms. Throughout your journey, the cabin staff study to please you, giving the efficient, courteous, personal attention which has made B.O.A.C. service world-renowned. For unsurpassed luxury in the sky - fly by B.O.A.C.!'

So said A BOAC brochure from 1956 (BOAC via author's collection0

The scene at Ciampino Airport, Rome, during the course of route-proving trials carried out by BOAC with the second Comet in June, 1951. In March, 1950, the Comet set up a new international speed record by flying the 900 miles from London to Rome in just over two hours. It was from Rome Ciampino that most of the Comet disasters transpired. The Shell refueller in the foreground is a Viberti which came under suspicion as to being a contributory factor in causing the accidents. *(De Havilland Hatfield)*

fatigue performance. This phenomenon, not then appreciated, is now widely understood.

Thus De Havilland's painstaking research programme in this area had yielded wildly optimistic results. The eventual enquiry into the loss of the two Comets totally cleared the company of blame, but a terrible price had been paid for progress. It was also clearly established and that no culpable shortcomings had been revealed within the aircraft's development programme conducted by De Havilland's and BOAC.

Two tasks faced the Corporation; to find capacity to replace the grounded Comet fleet and so to keep as many operations going as possible and to do all possible to assist in the public inquiry ordered by the Prime Minister and at some stage to decide what future any modified Comet might have in its fleet.

As to the public inquiry. Sir Arnold Hall, Director of the RAE, led the investigation which was immensely helped as the result of work by the Royal Navy in locating and recovering much of the wrecked Comet off Elba. The salvaged pieces were sent to Farnborough, and Sir Arnold and his team set to work to re-build as far as they could the Comet fuselage. At the same time, another Comet fuselage was undergoing pressure-testing in a huge water tank by continually raising and lowering the pressure whilst the wings were flexed by powerful hydraulic rams. In this way, it was demonstrated that the cause of the accidents lay in the incidence of metal fatigue in the fuselage panelling owing to frequent pressure cycles. The Court of Inquiry blamed no-one; the aircraft had been constructed in accordance with the known state of the art at the time. Out of disaster came some good in that the findings were of immense value in the development of civil jet aircraft worldwide. BOAC paid the penalty of being the first in the van of progress.

The reaction of BOAC's Chairman to the eventual findings of the Inquiry was predictable. In his memoirs, he wrote: *'To say that we were relieved was the world's understatement. At last, we knew what was wrong with the Comet; it was an easily correctible fault. All that was needed was a thicker skin and a system of skeleton construction... ... This system of fuselage construction has since been widely adopted both in this country, in France and America for both civil and military jet aircraft, so that the sacrifices made in the history of the first Comet did provide a valuable contribution to the future of all jet aircraft.*

It was a measure of BOAC's faith in the Comet that

Canadair DC-4M Argonaut G-ALHO *Althea* in the later 'blue tail' colour scheme.

Above: The prototype Bristol Britannia Series 101 G-ALBO, owned by the Ministry of Supply, but in the colours and markings of BOAC.

Left: Prince Phillip visits BOAC's headquarters on 20 May 1959. One aspect was the Britannia Hangar, and the Prince is seen here with Charle Abell, Basil Smallpeice, Ted Chandler, the Prince's equerry Squadron leader John de M Severne and Harry Pearn.

scarcely a month after the Naples accident the Board approved the Comet 2 order for twelve aircraft subject to the Court of Inquiry's findings still to come, and work on the Comet 3 was continued but on a single shift basis.

By September 1954 there was still no clear indication as to the cause of the accidents, and in consequence, BOAC lacked guidance as to any future Comet commitments, but it was decided that whatever the inquiry outcome the Corporation would not operate Comet Is again. The inquiry did not, in fact, commence its work until 19 October, and its findings appeared in February 1955. Meanwhile, De Havillands had proposed to BOAC a complete re-building of the Comet 2 fuselages with thicker gauge material, a re-designing of the windows and strengthening of the wings. All this would involve greater weight, and so the loss of payload and similar changes to the Comet 3 design would mean it could no longer be considered competitive for North Atlantic work, although it might prove attractive for Eastern operations. Consequently, BOAC began to think of the substitution of Comet 3s for Comet 2s.

The Corporation was never able to estimate precisely the financial effect of the Comet withdrawal. Not only was there a direct capital loss but also a severe loss of revenue which would have been attracted to the Corporation's services; moreover, replacement aircraft had to be acquired in a seller's market at very high prices.

It might have been thought that the great question in BOAC would have been whether or not to press on with the Comet project. In his memoirs, Sir Miles said *'various suggestions were made that we should change the name of the Comets so that the new series would start without stigma. I took the stand that that was an entirely unacceptable suggestion. I took the view to soldier on, retain the name Comet and prove by deed that it was staunch. It was a decision of which I have never been ashamed'.*

By February 1955, although the Cohen Report into the disasters had not yet appeared, the Chairman wrote to the Minister, John Boyd-Carpenter, informing him of the Corporation's decision to concentrate on the Comet 3 to secure a unified fleet and to take advantage of the Comet 3s greater range capability which should enable it to operate with only two stops to South Africa. Consequently, BOAC's existing order for twelve Comet 2s and five Comet 3s should be changed to seventeen Comet 3s; in fact, the total order was increased to twenty aircraft. Shortly afterwards in April 1955, BOAC altered the order once more, this time to nineteen Comet IVs with a greater all-up-weight and seating capacity and capable of a one-stop westbound operation on the North Atlantic. The aircraft was to be equipped with Rolls-Royce Avon RA 29/1 engines of over double the thrust of the Ghost engines in the Comet I. At the same time, BOAC would acquire one Comet 2 aircraft fitted with Avon RA 29 engines for purposes of testing before the introduction of the Comet IV. It was unfortunate that Government insisted priority should be given to finalising the Comet 2 for the RAF, thus preventing De Havillands from immediately concentrating on the Comet IV series. The Comet IV could not in consequence enter BOAC's fleet until the second half of 1958, and it was well into the next Chairman's term of office before the contract with De Havillands was signed.

There was one further chapter to the Comet story before Sir Miles resigned in April 1956. This was the possibility of a design developed from the Comet IV, to be

A pair of Stratocruisers - G-ALSD *Cassiopea* furthest from the camera - undergo maintenance at Heathrow.

called the Comet V, suitable for the installation of Rolls-Royce Conway engines. An aircraft of this kind would give transatlantic range and a revenue payload of around 20,000 lbs. Before he left Sir Miles pressed the Minister hard for a decision as to whether this aircraft should go ahead but no decision was taken, and the proposal came to nothing. How it would have matched up to the Boeing 707 had it been proceeded with, is still a matter for speculation.

With its Britannia and Comet IV orders, BOAC considered it would be adequately equipped with capacity in the years immediately ahead, and this was one reason why the Corporation showed little interest in the Vickers' 1000 project, a development from the military Valiant. RAF Transport Command ordered the V.1000, but difficulties arose over the specification and serious doubts about the aircraft's range capability; in the House of Commons on 28 November 1955 the then Minister of Supply, Reginald Maudling, announced that contracts for one prototype and six production aircraft for Transport Command had been cancelled. The view not only of BOAC but also of the Transport Aircraft Requirements Committee was that the proposed aircraft was not a competitive proposition in the face of American progress in jet design for the North Atlantic.

The removal from the Corporation's fleet of all the Comets meant a loss of capacity of about 21%, and it became an urgent task to find ways of filling the gap. There were three possibilities; first to see to what extent the

Argonaut G-ALHG *Aurora* taxiing past spectators at Heathow. Security, what security?

With the loss of capacity from the Comet fleet, it fell to other airliners serving with the Corporation to carry the load. The Bristol Britannia had it's share of teething problems on introduction to service, but started the inaugural Johannesburg service on 1 February 1957.

utilisation of the remaining fleet could be stepped up; second to seek capacity assistance from the two main partners, QANTAS and South African Airways, and, if needed, charter from outside sources and third to acquire if possible more capacity of the right type.

It was one of the benefits from the crisis that preconceived notions about the limits to utilisation had to give way to the current needs. Operational, commercial and engineering planners worked urgently together to extract any wasted hours in the existing schedules. Time on the ground was reduced, turn-round time speeded up and aircraft check times were shortened by speedier maintenance. The results were surprising; in the year following the Comet withdrawal an increase of 12% in Argonaut utilisation was achieved and for most of the year each aircraft was providing over 9½ hours per day of revenue flying, a considerable contribution equivalent to an additional two aircraft in the fleet; similar increases were obtained by the Constellation 749 fleet where a record level of utilisation at 10½ hours per aircraft per day was reached for a limited period.

But useful as all this was, it was far from filling the capacity gap. There was no alternative to some service withdrawals as any additional aircraft acquired would undoubtedly have to be modified to BOAC's requirements and time would be needed to carry out this work before aircraft could be put on service. The choice for withdrawal fell on the South American services; they were wasteful of Argonaut capacity, were a long low-frequency operation and were at that time uneconomic, made worse by currency problems within South America, rate undercutting and restrictive local Government attitudes. Additionally, in April 1955 it was decided to withdraw the UK-Israel service where Argonaut capacity had not been earning its keep and could be put to better use elsewhere.

The number of aircraft in the Corporation's fleet was thus reduced for most of 1954 to forty-three; twenty-two Argonauts, ten Stratocruisers and eleven Constellations. For a short time four Hermes aircraft were brought out of retirement and operated the East African services between July and December 1954, but this was an awkward business as the aircraft had been out of commission since 1953 and crews dispersed to other types.

This was one of those times when the real value of BOAC's partnership arrangements was demonstrated. As a matter, of course, discussions were held without delay with QANTAS and South African Airways to see how they could help. Both were able to assist to some degree on the partnership routes - but still, this was not enough, and there would have to be aircraft purchases. In addition to the Comet lost capacity, further losses had been suffered by the accident to Constellation G-ALAM at Singapore in March 1954 and to Stratocruiser G-ALSA at Prestwick in December, so that an urgent programme of replacement with quickly obtainable secondhand aircraft was put in train. The Singapore loss was replaced by the purchase of a new Constellation 749A which Howard Hughes had acquired for TWA, and a further four 749s were obtained from QANTAS who were in the process of getting Constellation 1049s; the opportunity also arose of doing a swop with Capital Airlines of the seven BOAC Constellation 049s for seven of Capital's 749As with a financial adjustment in favour of Capital. This meant BOAC would have a fleet of sixteen Constellations, all 749s, standardised as soon as modification to BOAC's requirements had been done. It was, in fact, mid-summer 1955 before all the QANTAS and Capital Constellations had been modified and were in service.

On the North Atlantic BOAC was looking for more Stratocruiser capacity and was able to buy one from PAA and six from United Airlines; allowing for the aircraft lost at Prestwick, this increased the total fleet to sixteen.

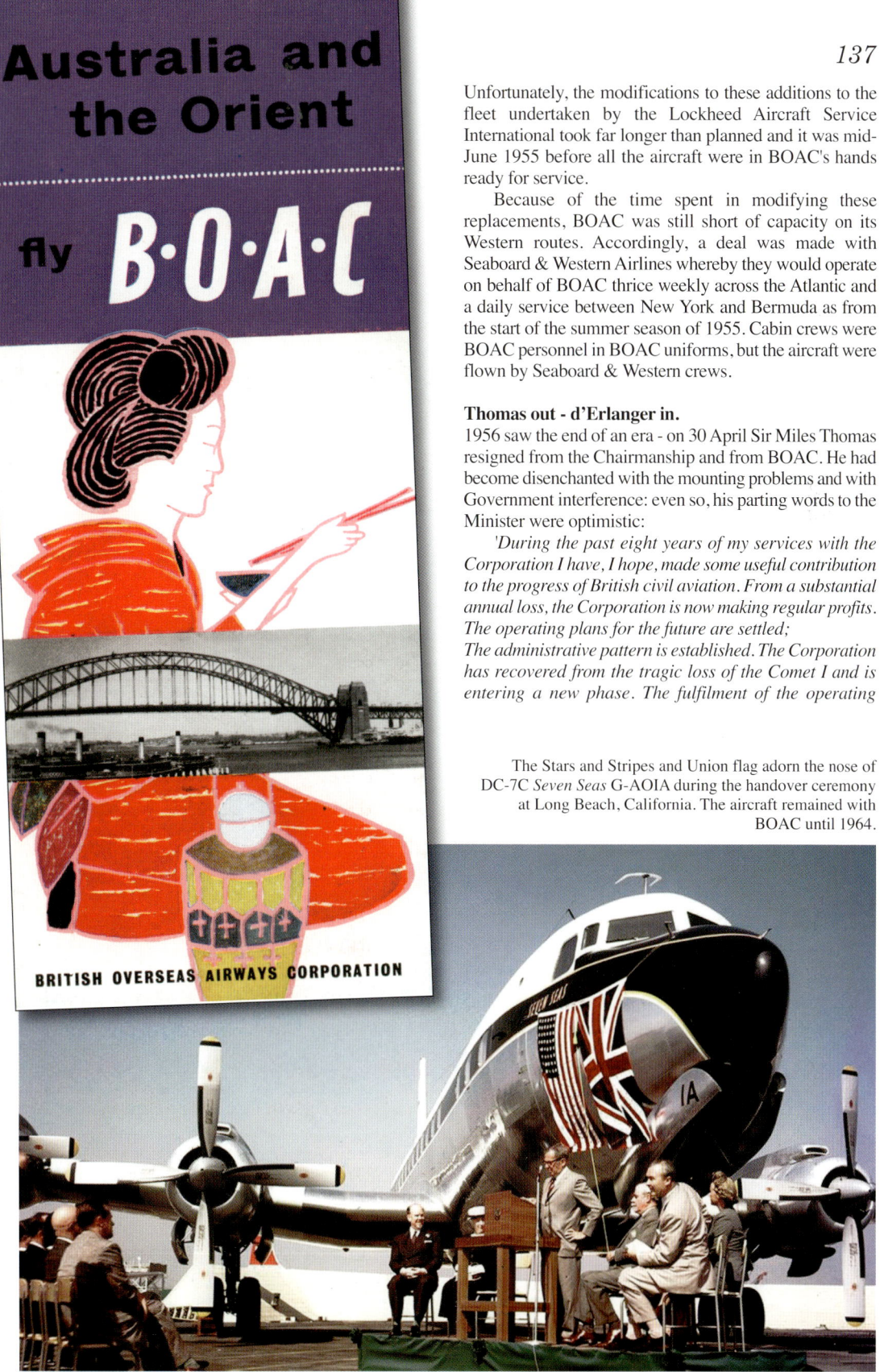

Unfortunately, the modifications to these additions to the fleet undertaken by the Lockheed Aircraft Service International took far longer than planned and it was mid-June 1955 before all the aircraft were in BOAC's hands ready for service.

Because of the time spent in modifying these replacements, BOAC was still short of capacity on its Western routes. Accordingly, a deal was made with Seaboard & Western Airlines whereby they would operate on behalf of BOAC thrice weekly across the Atlantic and a daily service between New York and Bermuda as from the start of the summer season of 1955. Cabin crews were BOAC personnel in BOAC uniforms, but the aircraft were flown by Seaboard & Western crews.

Thomas out - d'Erlanger in.
1956 saw the end of an era - on 30 April Sir Miles Thomas resigned from the Chairmanship and from BOAC. He had become disenchanted with the mounting problems and with Government interference: even so, his parting words to the Minister were optimistic:

'During the past eight years of my services with the Corporation I have, I hope, made some useful contribution to the progress of British civil aviation. From a substantial annual loss, the Corporation is now making regular profits. The operating plans for the future are settled;
The administrative pattern is established. The Corporation has recovered from the tragic loss of the Comet I and is entering a new phase. The fulfilment of the operating

The Stars and Stripes and Union flag adorn the nose of DC-7C *Seven Seas* G-AOIA during the handover ceremony at Long Beach, California. The aircraft remained with BOAC until 1964.

DC-7C *Seven Seas* G-AOIA in flight.

programme with the Britannia depends only on the deliveries of the aircraft, which have now started'.

Gerard d'Erlanger took over on 1 May 1956. He had had a long connection with flying, and at the formation of BOAC, he was one of the early Board Members, being appointed in March 1940. While remaining on the Board, he was put in charge of the Air Transport Auxiliary until it was wound up in November 1945. At the time of the Board's resignation en block in 1943, he was the one Board Member not prepared to resign. In 1946 he was made Managing Director of BEA although his tenure of office was short as he stepped down early in 1949 but remained on the Board. Since 1952 he had been a member of the ATAC. According to the Minister, Harold Watkinson, he was most reluctant to take over the BOAC chair and then only on a part-time basis.

His style of management was more reserved and in distinct contrast to the open type favoured by Sir Miles. He did not project his personality as Sir Miles had done to the staff and the outside world. His task was not made any easier by the arrangement of the appointments immediately under him. The Minister had decided that his Deputy Secretary at the Ministry, Sir George Cribbett should be appointed not only to BOAC's Board but should also be the Corporation's, Chief Executive. It seems extraordinary that in doing so, he had overlooked or ignored the Air Corporations Act 1949 which limited the Minister's powers to the appointment only of the Board and the Chairman and that it was for the Board to nominate the Chief Executive. What is more, the position of Chief Executive had assumed even greater importance for the reason that d'Erlanger was only part-time Chairman. In fact, the Board appointed the Deputy Chief Executive and Financial Comptroller, Basil Smallpeice, as Chief Executive and Cribbett became Deputy Chairman in place of Lord Rennell (who had resigned the Deputy Chairmanship and from the Board but who was almost immediately reappointed as a Board Member). The Minister's action over the position of Chief Executive was not a happy augury for smooth working among the Corporation's top executives.

With a part-time Chairman, a full-time Deputy Chairman and a full-time Chief Executive, it was essential that their respective responsibilities should be clearly defined notably as there was some clash of personalities. This was in fact done at the new Chairman's first Board meeting on 1 May 1956. Apart from the Chairman who would look after policy, planning and overall financial control, Cribbett was allocated responsibility for liaison with Government Departments (as an ex-civil servant he was apparently well fitted for that task), for BOAC's Associated and Subsidiary Companies and relations with the Independent air companies; in the absence of the Chairman he would be in overall charge of the Corporation. The Chief Executive, as Managing Director, was given the task of running the Corporation on a day-to-day basis and was responsible for its efficiency, economy, and safety of operations with executive authority over all departments.

In the view of the Corbett Report, commissioned by the Government in 1962 some of BOAC's difficulties during the d'Erlanger chairmanship were due to the unsatisfactory top level arrangements which did not make for clear and decisive leadership. In hindsight, Corbett also considered the Board as a whole was not strong enough, did not pursue the interests of the Corporation solely and paid over-much regard to the Minister's wishes, wishes too easily reinforced by reference to the 'national interest'.

Be that as it may, the Chairman took over the reins at a time when vital decisions had to be made about future aircraft equipment and when aircraft already on order had to be brought as quickly as possible into a satisfactory condition for entry into service by constant discussion and argument with the manufacturers. There were, of course, many other important issues - organisation, commercial and technical, problems with Associated and Subsidiary Companies, relations with independent operators and so on, but the overriding consideration during d'Erlanger's chairmanship concerned aircraft and decisions about aircraft which would affect BOAC for years to come.

The fleet in the early part of 1956 consisted of twenty-one Argonauts, sixteen Constellation 749s and sixteen Stratocruisers; four Britannia 102s had been delivered by Bristol's, but these were still on development and training.

Orders were in hand for a further eleven Britannia 102s, seven Britannia 302s and eleven Britannia 312s, nineteen de Havilland Comet 4s, one Comet 2E, ten Douglas DC-7Cs plus fourteen Vickers Viscounts for the Subsidiary Companies. It was fully understood that the DC-7Cs were stop-gap aircraft as insurance against severe delays in the deliveries of the longer range Britannias for the North Atlantic. It will be recalled that the Minister had required BOAC to dispose of the DC-7Cs when the Britannias came on the North Atlantic.

These provisions for capacity in the immediate years ahead seemed entirely adequate and, at the time when the new Chairman took over, his primary concern was to see that BOAC remained competitive on the Atlantic routes. The danger signals were already flying following PAA's order for twenty Boeing 707s and twenty-five Douglas DC8s. BOAC needed not only insurance against problems with the Comet 4 but also a turbojet aircraft able to stand up in size and range with the American contenders on the Atlantic crossing. There would also be a need for a replacement for the Comet 4 by 1963 or so for Eastern and Southern routes.

Although progress was being made with the Britannia, there were still problems; especially with the aircraft was experiencing severe difficulties with engine flame-outs in flight, and the programme of proving flights had had to be suspended. BOAC had received five aircraft by that time but had had to decide to delay introduction into service to 1 August 1956. The fifth Britannia was accepted on 9 May 1956, but BOAC increased its retention fund from £100,000 to £250,000 per aircraft in the expectation of necessary modifications. In June 1956 Bristol's conducted 'flame-out' trials at Entebbe and subsequently claimed that these had disclosed the cause of the problem - the passing into the engine of ice during flight in wet and dry icing conditions - and that modification to the Proteus engine would be installed without delay.

Unfortunately, the Managing Director had to tell the BOAC Board on 26 July 1956 that Bristol's modifications had not cured flame-outs and what made matters more serious, these were not confined to flying in the inter-tropical front. BOAC flew one of its Britannias, G-ANBC, at the end of August into conditions likely to cause problems to prove the worth of the modifications but with disappointing results; the Corporation had no alternative but to declare the Proteus engine unacceptable in its existing form on any permanent basis. An event occurred on 13 September 1956 which brought matters to a head. The Chief of Flight Operations, Captain J. N. Weir, telephoned Headquarters that while he was flying a Britannia on trials over Malaya, all four engines had suffered flame-outs. Bristol's were therefore told that the Corporation would not proceed with any commercial operation by Britannia 102s, would not take delivery of any more Britannia types until the malfunctioning of the aircraft had been cured, and reserved all its rights under the contract, and the Chairman informed the Minister of this decision. On that same day, 13 September 1956 the Chairman had a meeting with Bristols who then decided to obtain the assistance of the National Gas Turbine Establishment and the RAE to find a solution. A Committee was set up under the chairmanship of Dr Gardner of the RAE on which those interests concerned including BOAC were represented.

Britannia G-AOVN is prepared for an early morning start from London Heathrow.

Droning across the skies - DC-7C G-AOIB

In spite of these problems and following a demonstration tour in the United States with a Britannia 102, flown to avoid any risk of icing conditions, Bristol's were hopeful of securing an order from Northeast Airlines of Boston for the aircraft provided they could supply Britannia capacity for Northeast to use on trial from December 1956 to April 1957; BOAC was asked to agree to postpone deliveries of three aircraft to enable Bristol's to meet the requirement which could lead to an order for ten aircraft. After some heart-searching, the Corporation said it was willing to help in view of the dollar prospects, which it would then claim as some offset to the dollar expenditure it had asked for in its Boeing 707 requirement.

The first outcome of the Gardner Committee's work was a plan to fit glow-plugs in the engines which would cause automatic re-lighting on flame-out and in November 1956 further tests at Entebbe with glow-plugs fitted yielded encouraging results, although this method was not a necessary cure.

Whilst the trials had shown the glow-plug modification to be acceptable, there was no evidence so far to show that the risk of damage to engines caused by ice had been overcome. Further trials did, however, show such cost to be considerably reduced by the use of shields on the air intake tunnels. Although the aircraft could not be said to meet BOAC's full requirements as it could not be operated in icing conditions unless an automatic re-lighting system was installed, the Corporation was now prepared to accept

Britannia 102 G-ANBH on a test flight just before delivery to BOAC. The aircraft appears to be in full company livery - apart from the passenger door! (BOAC via PHT Green Collection.

further deliveries, with the safeguard that if Bristol's modifications did not reduce the risk of engine damage to reasonable proportions, BOAC would be entitled to return the engines and receive back all payments made less depreciation to be agreed.

The extra cost of the glow-plugs was estimated at around £180,000 per annum for all fifteen aircraft, and although there would be engine 'bumps' on re-lighting, these would not be detectable in the passenger cabin. As there was now no question of unsafe operations, BOAC decided at long last to introduce the first Britannia 102 passenger service on 1 February 1957 to Johannesburg.

Meanwhile, BOAC had come to the conclusion that no benefits would be obtained by fitting Bristol BE25 engines to the Britannias and asked Bristol's not to go ahead with the manufacture of the sixty engines it had ordered. Not surprisingly Bristol's considered BOAC to be committed, a view also taken by the MoS which pointed out that the purchase of the BE25 was a condition of its approval to the DC-7C order. BOAC agreed that although no formal contractual commitment had been made, it was under a moral obligation, and so informed the MoS but pointed out at the same time that it had never received any draft specification. Later in 1956 BOAC finally told Bristol's the BE25 engines were definitely not required, fortunately as it turned out since subsequent information showed that the engines could not have been installed without significant modifications to the aircraft. In the eventual settlement between Bristol's and BOAC over the whole Britannia issue concerning delayed deliveries and performance, the BE25 matter dropped out of sight.

By the end of 1956, Bristol's had delivered ten B.102s on which BOAC held back £550,000 retention money on account of deficiencies. There were still five more to come to complete the order, and on these, it was the Corporation's intention to retain a further £375,000. At the same time, BOAC took counsel's opinion as to the Bristol liability under the contract for damages for their failure to deliver aircraft in accordance with the contract specification. Counsel took the view that Bristol's were liable under the contract terms for damages amounting to £500,000 in respect of performance deficiencies and delivery delays. Bristol Aircraft, on the other hand, were claiming payment from BOAC for the extra intensive flying incurred in overseas test flying.

Further delay in completing the order for all fifteen B.102s occurred due to industrial problems at Bristol's, and the fifteenth and final aircraft was not handed over to BOAC until 12 August 1957.

In the meantime flame-outs had again become more dangerous, occurring on the B.102 proving flight to Australia in April 1957, and then on 1 May a passenger flight out of Entebbe had experienced eleven flame-outs within eight minutes, and on three occasions two engines were out of action at the same

Above: Basil Smallpeice tries out the first officer's seat on board this Britannia 102 flight deck and is shown the controls by Captain B E P Bone.

Right: the then newley-appointed Chief of Flight Operations James 'Jim' Wier in December 1956.

(both BOAC Press Office via Simon Peters Collection)

The location looks like New York - Britannia G-AOVF waiting for another load of passengers, with a company DC-7C in the background, in front of KLM Constelation PH-LDE and a pair of British West Indian Island Airways Viscounts behind that.

time, and manual re-lighting was necessary. There was no hazard to the aircraft, but the psychological effect on the crews was worrying. BOAC had to impose a maximum altitude limitation while these problems remained and this was a restriction on flexibility. In September 1957 the Chairman, the Chief Engineer and the Chief of Flight Operations called on the Minister and brought him up to date on the severe position regarding the Proteus engine performance. The Minister undertook to have the matter considered at Cabinet level to ensure adequate priority for research to solve the technical problem. During the time BOAC was continuing to fly passenger services with the B.102 but under a system of operating procedures avoiding the worst conditions, thus reducing flame-out incidents.

The technical difficulties affecting the B.102s were relevant also to the B312s, as both had Proteus engines. The order at the time of Sir Miles' resignation was for seven B302s and eleven B.312s, but early in 1957, BOAC decided to replace the seven B302s ordered by seven B.312s, making a total B.312 fleet of eighteen aircraft and involving an increase in the overall cost of £1,732,00 caused by the switch. The first flight by the prototype 300 Series took place on 31 July 1956, and by January 1957 the first B.312 flew to Winnipeg for cold weather trials.

Delivery of the first two B.312s to BOAC was scheduled for end June 1957, but a month later the delivery position was still obscure as the aircraft's performance was not up to specification. At this point in the midst of all the uncertainty and in the face of BOAC's urgent need for aircraft on the North Atlantic route, Bristol's asked BOAC to give up seven B.312 delivery positions to facilitate negotiations for a prospective sale to TWA. Once again the Board accepted the sacrifice in the national interest, although "if given a free choice the proposal would not be acceptable". In the event, the proposed sale came to nothing, but it is indicative of the attitude of the BOAC Management at that time towards their assumed responsibilities where the national interest was concerned.

By now the Minister had decided in conjunction with the MoS to present a paper to the Cabinet and asked for an up-to-date statement of the

The presentation ceremony on the occasion of Sir Miles Thomas leaving BOAC on 27 April 1956. From the left; Sir Victor Tait, Ken Staple, Sir Miles Thomas, Sir Basil Smallpeice, Ronald McCrindle and Freddie Gillman.

Comet 3 G-ANLO in flight, photographed against a typically impressive cloudscape. (DH Hatfield)

Corporation's views. These were set out in a letter from the Chairman on 11 October 1957 which dealt mainly with the B.312s. This vital letter pointed out that if the B.312s were not to enter service, the Corporation would lose £18m gross revenue in a year; moreover, it had already paid Bristol's about £12m in progress payments. The Chairman went on to state that *'if the Corporation had felt free to disregard considerations of national importance which they have hitherto regarded as precluding them from dealing with this problem on a strictly commercial basis, they could not but refuse to accept the remainder of the B.312s and claim damages against Bristols'*. The Corporation was feeling particularly alarmed at that point following a serious incident on 30 September 1957 when on a B.312 flight from New York to Miami two engines failed shortly after the aircraft had entered the cloud. The letter went on to point out the severe consequences of all these delays to the planned B.312 programme for the North Atlantic route. The B.312 was initially planned for delivery in October 1956 but if it were not to come into service until 1958 its competitive life on the North Atlantic would be extremely short in the face of the arrival of the turbojet aircraft, the Comets and Boeings, in late 1958 and 1959. There were also unanswered questions as to the aircraft's operating limitations creating severe traffic control problems, and the effect on crew morale. The Chairman closed his letter by listing possible alternative policies for BOAC - to refuse acceptance of the B.312 and claim against Bristol's, to seek Government compensation if national interests required acceptance of the aircraft, to request Government to take over the aircraft and hire it back to BOAC, or finally if all else failed for BOAC to acquire ten further DC-7Cs to tide over the gap. As the Chairman pointed out, his comments referred to the B.312, but similar defects afflicted the B.102s which the Corporation already had in service and which were being coped with by extraordinary measures.

By the end of October 1957 a Cabinet Committee chaired by Lord Mills had considered the whole matter in detail and the RAE had come up with a new anti-icing device; this consisted of directing jets of compressed air bled from the engine compressor into the air intake to provide a film of high-speed air over that part of the surface where ice accumulated. It was expected that the success or failure of this new modification would be known by December 1957.

Having completed its work on the technical issue, the Cabinet Committee then considered how best to sustain Bristol's financial stability. BOAC had to tell the Minister that there could be no commercial justification for further payments to help Bristol's and that severe consequences for its profitability would result if the B.312 were not on the North Atlantic by April 1958 in the face of American competition with non-stop services; even so the Chairman said he would recommend to the BOAC Board that help should be given to Bristol's since he understood national interests were at stake. Under pressure from the MTCA, BOAC agreed to accept further B.312 deliveries and make payments accordingly, to halve the retention monies and to pay half of the £1m cost of the extra flying caused by the flame-out trials (although as the result of a contribution by the MOS, the actual amount paid by BOAC was reduced to around £285,000). The Minister also requested BOAC to defer the release date to Boeing of the fourteen Stratocruisers, even though this would involve it in a loss of $352,000 on the trade-in price.

Under these arrangements, BOAC would be accepting deliveries of B.312s even though the aircraft fell short of its guaranteed performance. In these circumstances, the Board instructed the Chairman to ask the Government to protect the Corporation against the consequences of seeming to accept aircraft below guarantee and thus jeopardising its legal rights against Bristol's. When the Minister's reply was received, it was found to contain no specific protection. BOAC, however, made a payment to Bristol's of £915,000 on 6 November 1957 which included an amount of £285,000 for flame-out trials. The balance of the £915,000 was an advance payment made to help Bristol's financial situation and was to be put against the cost of future aircraft to be delivered after 31 December 1957. At the same time, the Corporation made it clear to the Minister that this action had been taken on the basis that it would expect the Government to indemnify it in the event of payments becoming 'in certain circumstances' irretrievable. In his reply, the Minister gave no specific assurance but simply stated that whatever happened the Corporation's problems would be 'sympathetically considered.

As the modifications proposed by the Gardner Committee, had proved themselves on trials, the first B.312 passenger service left London Airport on 19 December 1957 and successfully completed a non-stop flight to Idlewild Airport, New York.

By mid-March 1958 Bristol's claimed that the flame-out problem on the Proteus engines was solved and that the plan was to modify all the B.312s by April 1958 and all the B.102s by July 1958. In these circumstances, they asked BOAC to release the retention monies on both types. Taking the view it was early days, and the Corporation agreed to reduce retentions for flame-out by 50%.

The performance of the B.312 was still below guarantee and BOAC had agreed with Bristol's that they should be given until 30 April to rectify this; by early April the first B.312, G-AOVK, had been brought up to only 1% below the guaranteed figure and in view of this BOAC agreed to a general settlement negotiated between Sir Matthew Slattery for Bristol's and Gerard d'Erlanger for BOAC.

By May 1958 Bristol's had established that the performance was up to minimum guarantee and so retention monies were to be released on each aircraft as it became modified to the G-AOVK standard. BOAC's own review of the results of the flame-out modifications was completed by November 1958, and the conclusion reached that flame-outs had been virtually eliminated. The Gardner Committee submitted its final report to the MOS, and BOAC's Chairman wrote to the Minister expressing the Corporation's appreciation of the work the Committee had done, and also to Sir Reginald Verdon Smith, Chairman of Bristol's, for his company's efforts to secure clearance for the power plants.

Unfortunately, the B.102 on African and Eastern routes experienced persistent technical troubles in service from leaking fuel tanks and from undercarriage torque link failures, so that regularity suffered severely; even so, by the end of 1958 B.102s had taken over all Eastern routes except for one Argonaut service to Kuwait. By January 1959 all seventeen B.312s had been delivered to BOAC (one aircraft had been lost in an accident at Hum in December 1958) and were deployed progressively on Western routes and took over the South African services from the B.102s.

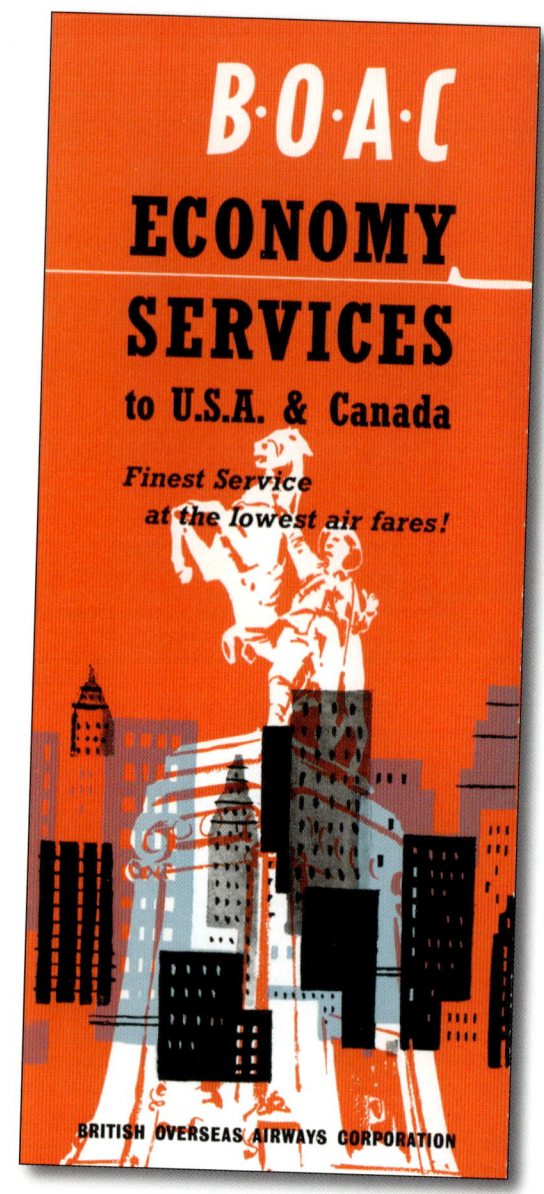

In sharp contrast to the troubled birth of the Britannias and the delays in deliveries, the DC-7C was ordered in March 1955 for delivery over the period October 1956 to April 1957 as per the contract of purchase. All ten aircraft were delivered without delays with the result that BOAC began route operations London-New York in January 1957 and within six months was operating eleven tourist and nine mixed class services per week including a twice-weekly service over New York to San Francisco. But the aircraft had a short life as was expected; after all, its purchase had been in effect an insurance policy, and it became outmoded in the face of the jet competition. By 1960 it was practically out of service on BOAC except for charter work and for freighter operations on the North Atlantic for which two of the fleet had been converted by Douglas.

Although a condition of the purchase of the DC-7C had been that they must be sold when the B.312 entered service, the Minister agreed subsequently that BOAC could retain the aircraft until the arrival of its 707s. By then the DC-7Cs' sale value would be massively reduced, and BOAC realised it would have a problem over possible massive capital loss.

Fares and Class battles

In late 1957 at a conference in Paris, France, the airlines constituting IATA decided to introduce a new cut-rate class, of travel on the North Atlantic. This decision took effect on 1 April 1958 on which date fourteen airlines introduced the new economy service between Europe and North America. It was widely expected that this would sound the death-knell of tourist-class travel over the North Atlantic.

With economy service, the airlines sought to reduce Atlantic air travel to its cheapest - and therefore simplest - level, with none of the trappings which had been associated in the past with long-distance air travel. The tourist-class concept, which since its introduction on the Atlantic on 1 May 1952, has been widely adopted throughout the world, was the first move in this direction and produced the desired results. In reducing still further the degree of service and amount of space to offer a 20% reduction in fare, it seemed to some that the airlines were going too far.

At the time there were concerns in some quarters as to just how tolerable - or intolerable - a twelve to fourteen-hour flight would be under the conditions offered. One example quoted a trip on a 103-seat, all-economy class Super-G Constellation - with only two toilets.

The limitations agreed by the Conferences for this service were that the seat-pitch should not exceed thirty-four inches, that the seats should have a restricted recline angle and that the only service in the cabin would comprise of simple, cold, inexpensive sandwich meals together with

Whereever in the world you are flying, it will pay you to consult your B.O.A.C. Appointed Travel Agent first. In 1001 different ways, he will save you time and money. His job only begins with booking your passage by B.O.A.C. and its associates to any part of the world. He'll see to all the formalities . . . book you the right hotels . . . give you full information about routes, timetables, connections. This comprehensive world-wide service is yours for the asking-just book where you see the sign of a B.O.A.C. Appointed Travel Agent.

(BOAC via author's collection)

Original caption: *'Sydney; 24 hours 24 minutes' flying time from Hatfield. Twenty thousand happy Australians greeted the Comet at Mascot (Kingsford Smith Airport) on Sunday afternoon, December 4. Enthusiastic children led a break-through onto the field and the Comet, when on finals, was asked to circle Sydney again while the runway was safely cleared. The crowd surged around the aircraft as it came to rest. To get a gangway and vehicles through, a water hose was used as a last resort, but the uniformed operator really hadn't the heart to spray those pretty frocks. Everyone laughed and cheered. Australians want jets, and would like British ones - but on merit only'.*

coffee, tea, milk or mineral water. No hot food would be served, and no alcoholic drinks should be offered, free or for sale. No 'give-aways' were permitted other than sweets at take-off and landing, and the usual toilet accessories.

By comparison with this, tourist-class passengers, with seats pitched at thirty-six to thirty-eight inches, with hot meals, a pay-bar service and so-called 'give-aways' to a - value of 25 cents a head - were travelling in luxury.

The operators had been influenced by the success of tourist traffic on the Atlantic, which by 1958 accounted for about 70% of the total. A very large proportion of these travellers, obviously, were paying for themselves - as opposed to governmental or corporate-funded tickets - so it followed that a reduction in fare would make it possible for still more self-paying passengers to travel.

The IATA members also suggested that most passengers who had previously flown tourist would fly economy in the future. This fact was supported by contemporary booking trends, which demonstrated that four to five economy reservations were being made for every tourist reservation, an overall 20% increase in traffic.

At the time of the fare launch, it appeared that the travelling public welcomed the economy-class plans but doubts remained. As one travel trade magazine said in an article at the time of its commencement: *'But the public is buying, at present, an unknown quantity. How many passengers who make an economy-class crossing will wish, we wonder, to repeat the experience, and how many will prefer, if and when a second occasion arises, to save up for a little longer and to travel tourist. Time alone will tell; some airlines certainly have had misgivings about economy service - although none can now afford to ignore it. It may be, on the other hand, that the majority of economy-class passengers will be folk who in any case would only make one trip of this kind in their lifetime; in which case their reactions are perhaps of less consequence'.*

Economy-class fares applied to journeys between the port of entry on the Eastern seaboard of North America and the first port of call in Western Europe. In fact, economy travel will be offered on flights starting and finishing beyond these two boundaries, but on such flights, the fares would be a combination of economy rates for the Atlantic crossing and modified tourist rates for the additional sector - although the entire flight would be to economy standards. The basic fare in 1958 US dollars from which all other economy rates were calculated was $252 (one-way) and $453.60 (return) between London and New York. These fares were approximately 20% lower than the new tourist rate which anyway was being increased on 1 April by about 8%.

In practice, the introduction of an economy class has presented difficulties, more especially for operators with small fleets. There was considerable uncertainty over the best proportions of the various classes, and the best configuration to use. Many of the airlines were favouring three-class configurations, with a large economy section and small numbers of tourist and first-class or de-luxe passengers. This arrangement made it possible to offer maximum frequency in all classes, but complicates bookings and could lead to empty seats in one section and loss of traffic in another. Few airlines, however, had a sufficient number of aircraft to be able to devote some to all-economy layouts, and those who did were still mixing two or three classes on some services. Most airlines were also retaining all-first and de-luxe services at a frequency of one or two a week.

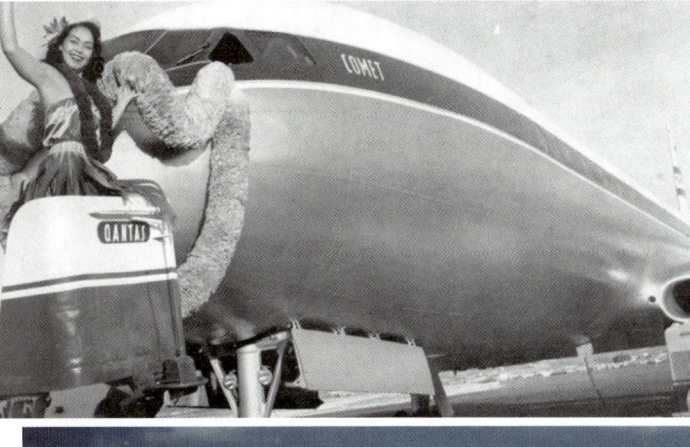

The Comet gets lei'd!
The South Sea interlude saw leis showered on aircraft and crew alike, and there was time for a swim on Waikiki beach, Honolulu during refuelling before a dawn take-off to cold Vancouver, Canada.

At the time of the fares launch, two airlines did not offer economy service. These were Icelandic Airlines, a non-IATA carrier already offering tourist standard service below IATA rates; and QANTAS, who could do so only by maintaining special aircraft for the London-New York sector of their round-the-World service. Because of their small fleets and the difficulty of modifying aircraft without serious loss of revenue, El Al and Iberia both had permission from IATA to offer tourist accommodation at economy fares for a month or two beyond the 1 April deadline.

In addition to the operators already flying on the Atlantic, Irish Airlines proposed to start their Dublin-New York service on 28 April, and so became the only operator offering nothing but economy travel. SAS, El Al and KLM declared their intention of dropping tourist-class completely, while retaining first and de-luxe classes.

Published schedules and plans indicated that at the height of the summer season at least 18,000 economy seats would be available each week in each direction between the Old World and the New. In all, about 200,000 more seats will be offered in 1958 to allow for the anticipated increase of 20% over the 1,003,000 passengers who crossed the Atlantic in 1957.

To give the reader some idea as to how traffic was building on the prestigious North Atlantic route, it is worth comparing what BOAC, and it's competitors were facing by looking at aircraft configurations and frequencies in March 1958. In these statistics, the weekly total of economy-class seats in each case is calculated on a one-way basis. It is also worth reminding the reader that airliners during this period were usually divided into the front or forward, main and rear cabins.

Air France: Daily L-1049G Super Constellation service with 12 tourist and 75 economy seats. Daily service with L-1649 Starliner with four bunks, 16 first-class and 49 economy seats; twice-weekly Starliner service with four bunks, 12 sleeper-seats, 15 tourist and 34 economy seats; once-weekly Starliner service with four bunks, 15 tourist and 60 economy seats. Total, 996 economy seats per week.

Alitalia: Nine services a week at summer peak by DC-7Cs with 16 first (rear), 6 tourist (forward) and 55 economy seats. Peak total, 495 economy seats per week.

BOAC: At summer peak, 21 services a week between London and New York by DC-7C with 16 first-class (rear), 17 tourist (main) and 37 economy seats. Five services a week London-Montreal by DC-7C with 16 first-class (rear cabin, with four bunks) or 18 tourist (rear) and 63 economy seats (main and forward). Seven flights a week London - Montreal - Detroit - Chicago by Britannia 312 with 12 deluxe or 16 first (rear), 15 tourist (main) and 54 economy seats. Two services a week London - New York - San Francisco by Britannia 312 with 12 deluxe (rear), 27 tourist (main) and 45 economy seats. Peak total, 1,560 economy seats per week.

Canadian Pacific: Two flights a week Vancouver - Amsterdam (originating in Australia) by DC-6B with 12 deluxe (rear), 10 tourist (centre) and 44 economy seats. One flight a week Vancouver- Amsterdam by DC-6B with 12 deluxe and 54 economy seats. One flight a week Montreal-Lisbon- Madrid by DC-6B with 12 deluxe, 10 tourist and 44 economy seats. Weekly offering, 142 economy seats. Commencing in June, Britannia 314s replace the DC6Bs. Configuration not yet decided: two basic layouts planned are 16 deluxe (rear), 48 tourist (front) and 24 economy (centre); or 22 deluxe, 12 tourist and 54 economy seats.

El Al: Four services a week (increasing to five) by Britannia 313 with 18 first and 72 economy-class seats. Peak total 360 economy- seats per week.

Iberia: Four services a week by L-l049G Super Constellation with 19 first-class (rear), 15 tourist (forward) and 45 economy (centre) seats. Peak total, 180 economy seats per week.

Irish Air Lines: Commencing April 28, three services a week (increasing to seven) by L-1049H Super Constellation (leased from Seaboard and Western) with 95 economy-class seats. Total, 665 economy seats per week.

KLM: Fourteen flights a week by L-1049G Super Constellation with 95-98 economy seats. Six flights a week (to Montreal) by DC-7C with 12 deluxe (rear), 8 first-class and 35 economy (main) and 6 economy (forward) seats. Total, 1,576 economy seats per week.

Lufthansa:-Twelve services a week by L-1049G Super Constellation and L-1649 Starliner. Configuration of Starliner was 6-10 first-class or deluxe (rear), 15 tourist (forward) and 59 economy seats. Peak total, about 700 economy seats per week.

Pan American: Company planned to split total capacity in proportion of 5% deluxe, 23% first, 9% tourist and 63% economy. Published summer schedules indicated 12 services a week (including seven on Arctic route to West Coast) by DC-7C with first and economy -seats; 14 services a week by DC-7C with_ tourist and economy seats; seven services a week by DC-6B with first and economy seats and 14 services a week by DC-6B with all economy seats. Peak total, about 3,500 economy seats per week.

SAS: At summer peak, 12 flights a week by all-economy DC-6B seating 81; four flights a week by all-economy DC-7C seating 85; and 14 flights a week by DC-7C with 28 first (rear) and 34 economy seats. Peak total, 1,788 economy seats a week.

Sabena: Fourteen flights a week by DC-7C with 16 first, 10 tourist and 48 economy seats, and seven all-economy flights by L-1049 G Super Constellation seating 97. Total, 1,367 economy seats per week .

Swissair: Eight services a week by DC-7C with 14 first (rear), four tourist and 57 economy seats. Three services a week by DC-6B with 16kfirst and 52 economy seats. Peak total, 612 economy seats per week.

TCA: At summer peak, 15 services a week by L-1049 Super G Constellation with 7 deluxe (rear), 4 first-class (rear), 15 tourist (forward) and 44 economy (main cabin) seats. Peak total, 660 economy seats per week.

TWA: Commencing 1 April, 11 flights a week by Starliner (including two on Arctic route to West Coast) with six deluxe (rear), eight first-class and 60 economy seats. Seven flights a week by L-1049 Super Constellation with 103 economy seats. Total initial offering, 1,381 economy seats per week. Peak total offering. about 3,500 economy seats per week.

Rebirth

Even before the disasters struck, De Havilland's were already working on the Comet 3 design. By the middle of 1952 the Series 3 Comet was in an advanced stage of design - to a point where it was publically revealed at the SBAC airshow at Farnborough in September. It was an enlarged version of the original configuration, incorporating all the lessons of experience gained with the earlier versions.

Initially, the Series 3 Comet was to have an all-up weight of about 145,000 lb. The four Rolls-Royce Avons of 9,000 lb. static thrust provided a cruising speed of at least 500 m.p.h.

Interior accommodation was to provide alternative seating arrangements for up to 58 first-class passengers or for up to 78 passengers in the tourist class.

The practical stage length, taking into account fuel reserves for climb, descent, headwinds, stand-off and diversions, meant that it was about 60% greater than that of the Comet 1 and the specific cost of operation was appreciably lower.

the aircraft for these flights.

28 December 1955 saw the return home of G-ANLO, the Comet 3 development aircraft, at the completion of a round-the-world proving flight. The purpose of the flight, as conceived, planned and executed by De Havillands, was to operate and prove the aeroplane in a wide variety of atmospheric conditions simulating as far as possible actual airline operations and to check the performance of the aircraft against the basis used for calculating the brochure performance figures of the Comet 4.

The outward configuration of the Comet 3 closely resembled that of the Comet 4 but the Rolls-Royce 10,000 lb. thrust Avon R.A.26 engines fitted were not so powerful as the 10,500lb. thrust Avon R.A.29 that were to be installed in the Comet 4. The total fuel capacity of the aircraft was less but the specific consumption would be

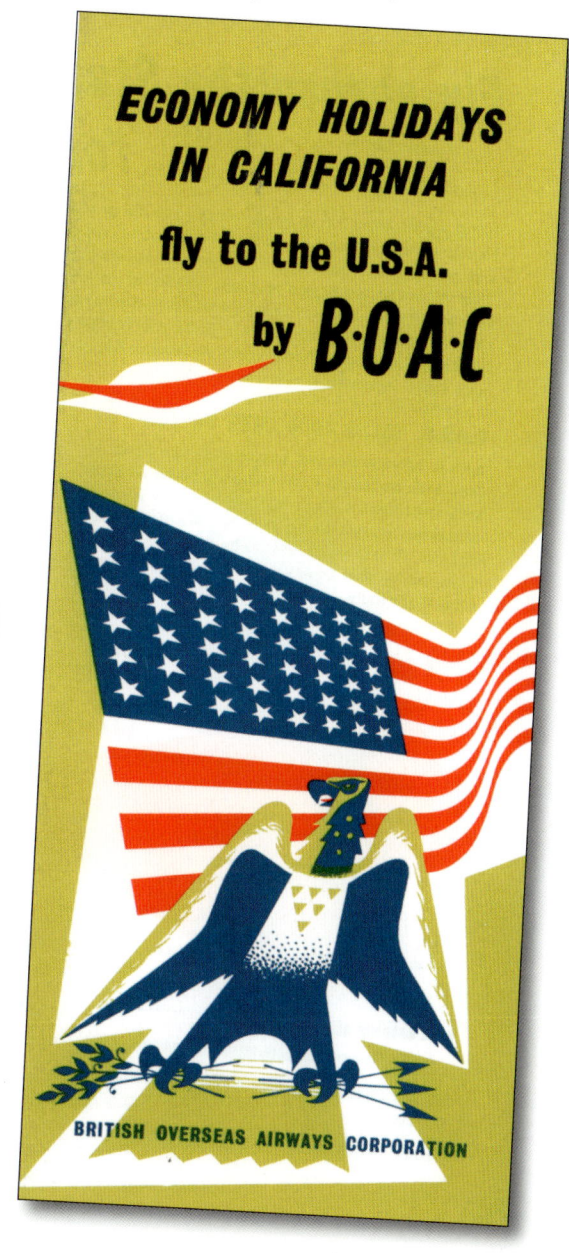

The prototype Series 3 was expected to fly early in 1954, and the first production aircraft was scheduled to appear late in 1956. In December 1953, BOAC announced that it was to order five Comet 3s to form the airliners initial express transatlantic fleet. A contract was signed on 1 February 1954. However, just as with the Comet 2, the incidents involving the Series 1 aircraft had a great impact on the Comet 3.

Only a single Comet 3 was built - G-ANLO, actually owned by the Ministry of Supply - and this aircraft served as a development machine for the Comet 4. Rolled out in May 1954, it took to the skies for the first time on 19 July. The first public display was at the 1955 SBAC show, when it appeared in BOAC colours. After Farnborough the aircraft was readied for a series of development flights, even though the investigations into the Comet crashes were still ongoing. The Series 3 was a completely different machine and so the authorities had no hesitation in clearing

Right: Notables on a notable occasion! On September 30, when the first two Comets were formally delivered, Sir Gerard d'Erlanger, Chairman of BOAC, shakes hands with Sir Geoffrey de Havilland, as he alights at London Airport. To the right of Sir Geoffrey are Sir Basil Smallpeice, Managing Director of BOAC, Mr. W. E. Nixon, Chairman of Havilland Holdings Ltd., and The de Havilland Aircraft Co. Ltd., and Mr. Aubrey Burke, Deputy Chairman and Managing Director. Mr. A. S. Kennedy, de Havilland Financial Director, is between Sir Gerard and Sir Geoffrey. On the left are Sir George Cribbett, Deputy Chairman of BOAC, Mr. John Cunningham, de Havilland Chief Test Pilot, and Mr. C. T. Wilkins, de Havilland Chief Designer. *(both DH Hatfield)*

appreciably improved. In spite of these limitations it was possible throughout the flight to simulate very closely the Comet 4 operating technique.

Frank Lloyd, Commercial Sales Manager and Contracts Manager of The De Havilland Aircraft Co. Ltd., was the executive in charge of the business aspects of the flight and John Cunningham, Chief Test Pilot of the Company, commanded the aircraft. Peter Buggé was second pilot.

As the *De Havilland Gazette* said '...*John Cunningham and all aboard were very pleased to fly in BOAC colours and greatly appreciated the decision of the Corporation to send Captain Peter Cane, who headed the pilots of the Comet fleet in 1952-54, to accompany the aircraft as a member of the crew throughout the tour.*

The Company is most grateful for the assistance given by BOAC at ports of call, and likewise by QANTAS Empire Airways, Trans-Australia Airlines, Australian National Airways, Tasman Empire Airways, Canadian Pacific Air Lines, Trans-Canada Airlines, the Shell Company and its associates, the Royal Air Force and the Royal Australian, New Zealand and Canadian Air Forces, as well as the airport and airway authorities throughout the world. Efficient help all round was received from the airway organisation, and the many who contribute to it with ground handling, maintenance, refuelling, meteorology, radio and communications, customs, immigration and other services'.

The Company took fullest advantage of the opportunity to demonstrate the aircraft to operators, technical authorities, the Press and public, and this aspect of the flight proved of the utmost value. Apart from the circuiting of cities to give as many people as possible a good view of the Comet, the aircraft was flown strictly in accordance with airline technique. The Sydney-Melbourne stage (circling Canberra) and the Toronto-Montreal stage were too short for representative flying, and the Melbourne-Perth was made unrepresentative by circling Adelaide.

Airline pilots, familiar with their respective respective sectors, who were carried as supernumerary members were as follows:- BOAC pilot, Captain Peter Cane who flew all round the world; QANTAS pilot, Captain I. D. V. Ralfe, who flew London-Sydney QANTAS Pacific pilot, Captain W. A. Edwards TAA, ANA, and Australian Department of Civil Aviation pilots CPA pilot, Captain W. S. Roxburgh CPA Director of Flight Operations, Captain B. Rawson TCA Flight Superintendent of Western Division, Captain A. Rankin who flew Sydney-Honolulu Melbourne - Perth - Sydney Honolulu - Vancouver and Vancouver-Toronto.

The aircraft was equipped with elaborate recording instruments and, in addition to flight and engineering crew, carried observers including a senior aerodynamicist from Hatfield.

For the world flight it was necessary to lay down special supplies of fuel at most of the calling points. In a few cases turbine fuel was available at diversion aerodromes, but on most of the sectors it was necessary to carry sufficient reserve fuel in the aircraft to enable a return flight from the alternative to the destination to be made should diversion become necessary.

Every flight was planned on a take-off weight with tanks full and this gave a maximum take-off weight available of from 140,000 lb to 142,000 lb depending upon the number of people carried.

The world voyage gave the chance at an early stage in the marketing phase of the Comet 4 to verify the capabilities publicly along world routes for which it was suited, for which it was designed, and with the great benefit of having on board skilled senior pilots familiar with the stages flown.

At the end of the Cairo-Bombay stage of 2,710 statute miles, 2,900 Imperial gallons of fuel remained in the tanks,

Three of the new Comet 4s at BOACs maintenance facility at London Heathrow.

sufficient for a further three hours flying. At the end of the Fiji-Honolulu 3,210 mile stage, 1,250 gallons of fuel remained, sufficient to fly a further 1.25 hours. At the end of the Montreal-London 3,350 mile stage on December 28 enough fuel remained in the tanks to circle London for an hour at low level and then, if diverted, to fly to Prestwick, 330 miles away, and there safely circuit and land.

One aspect of the Comet which had at last been made plain to the world as a direct outcome of the flight was its airfield behaviour.

The crew on this flight were surprised to find that in

Comet 4s at BOACs maintenance facility at London Heathrow. In the middle is G-APDB. *(DH Hatfield)*

some parts of the world, notably Hawaii, the public and even the aviation community seriously thought that a jet airliner would need a long runway, would climb at a flattish gradient over the city of departure, causing a noise nuisance, would circuit, approach and land rather fast, would be unbearably noisy when standing and manoeuvring in front of the terminal building, and might even scorch the paving! Technical people in Honolulu really thought that special regulations might be needed to keep people a considerable distance away from the jet engine intakes and effluxes during ground running. One newspaper referred to the efflux as *'...a blast as hot as a blow torch'*, and said that *'...the danger-point behind a jet engine is 100 yards'*.

The De Havilland Company had always stated that the Comet was designed to use runways and airport facilities as they exised, and had emphasised that a low wing loading was specified so as to be sure of a short take-off and steep climb and a low landing speed with a short landing run. The sight of the Comet landing on the short 7,000 foot runway at Honolulu - this was in the days before the reef runway - pulling up within 3,000 feet and turning off at the intersection, impressed the observers of the Hawaiian Aeronautics Commission. They were also horrified to see ground staff standing and walking quite close behind the aircraft's tail while all four engines were running fast enough to start the wheels rolling away from the parking bay. They were also astonished to learn that the fully loaded Comet 3 or 4 could climb out after take off far more steeply than modern piston-engined airliners.

It seemed to those on board the Comet that many had been misled deliberately or otherwise by Press accounts, especially in USA, during the previous two or three years, and they were pleased to dispel some of the myths and legends that were starting to grow.

Another misunderstanding which the world voyage tried to eliminate concerned the purpose for which the aircraft had been designed - a 'misapprehension' that is still around to this day. As the *De Havilland Gazette* reported: *'At every port of call the crew have had to explain that the Comet 4 was complementary to rather than competitive with the American conception of a jet airliner. It was smaller than the American jet airliners which are promised. Those aircraft, arising from a US Air Force requirement for a flight-refuelling tanker, are aimed at long-range operations with high traffic density, especially the coast-to-coast service across the United States and the non-stop connection between New York and European capitals.*

De Havilland designers had as their main aim the other trade routes around the world. On these the traffic was less dense and most of the centres of trade and industry are 2,000 to 3,000 miles apart. Designed to fit these conditions, the Comet has a world-wide suitability. It is, in fact, universally useful on world routes except for the one case of the non-stop service between New York and European capitals, on which it will need to make one halt on the westbound flight.

This North Atlantic route is exceptional in two respects: it has exceedingly heavy traffic, especially in the summer months, and it experiences strong westerly winds, especially in the winter months. These conditions call for an exceptionally large aircraft if the flight is to be made without stopping between European capitals and New York.

The Comet has had six years of continuous development flying, including 30,000 hours or 13 million miles of airline duty. There is no short cut to such invaluable knowledge as De Havilland and the British Overseas Airways Corporation have gained during these years, and all of it is built into the new Comet 4.

Without doubt jet propulsion is about to revolutionise the standards of world travel. Its two outstanding qualities are speed and comfort. Its speed of over 500 miles an hour virtually halves the journey time. Its comfort arises from the fact that the jet airliner flies extremely high, in the region of the smoothest passage, and it employs virtually vibrationless power. A long inter-continental journey is cut down to a very few hours, and the comfort and quietude and perfect air-conditioning of the aircraft make the journey seem even shorter still. One arrives without the sense of having travelled.

Arrival of the Comet 4

The intercontinental Comet 4 was designed for stages up to about 3,000 statute miles carrying, typically, about 67 mixed-class passengers. It was this version that was employed by BOAC on the North-Atlantic route with 52 mixed class and chosen as the express airliner – operating with 67 mixed class pax for all their other world routes.

The Continental Comet 4B, with longer fuselage and clipped wings was intended for inter-city networks with stages from about 400 to 2,500 statute miles, carrying about 86 mixed-class passengers. It had a higher maximum speed than the Comet 4 and less fuel tankage. This version was selected by, amongst others, British European Airways and Olympic Airways for their fast services across the continent of Europe.

The Intermediate Comet 4C combined the longer fuselage of the Comet 4B with the larger wing and fuel capacity of the Comet 4. It carried substantially more payload than the Comet 4 at the cost of a small reduction in maximum range, and was most suitable where neither very long nor very short stages were the main consideration.

Thus the three versions met the requirements of almost every airway system, typical exceptions being the few domestic routes within the United States of America where the traffic was many times heavier and called for a larger vehicle.

De Havillands freely admitted that they thought that outside the United States only the North Atlantic route could justify the large jet. Here the traffic was considerably lighter and more than a dozen airlines competed for it, but an increased frequency would not of itself yield further traffic growth, so the large jet could be filled. Also, because the stage was exceptionally

Crew of the first BOAC Comet 4 crossing: According to airline custom a flight with Press representatives and others was made a couple of days before the frst regular London-New York passenger service. Here are the crew at London Airport, Thursday 2 October, before departure on the pre-inaugural crossing in G-APDB. Left to right: Captain Ernest E Rodley, Captain Tom B Stoney, Captain C Farndell, Steward J Miller, Steward A C J McCormack, Stewardess Barbara Jubb and Stewardess Peggy Thorne. *(BOAC Press Office via P H T Green Collection)*

long - about 3,000 statute miles - and because of strong westerly winds, an airliner with a somewhat longer range than the Comet would be better suited - although the Comet 4 had long operated an excellent and profitable service on the route.

It was Sir Miles Thomas who in February 1955 had given De Havilland an Instruction to Proceed with the manufacture of nineteen Comet 4s for delivery in 1958 and 1959 but it was December 1956 before Basil Smallpeice was able to outline the main features of a draft contract with de Havilland. The price subject to escalation was to be £1.16m per aircraft and delivery dates were to be phased between September 1958 and December 1959; BOAC was to receive the first four aircraft delivered to any airline and De Havilland would be entitled to a bonus if they were able to give the Corporation eight months notice that delivery dates of all the nineteen aircraft would be advanced. On the other hand BOAC would be entitled to damages if the aircraft had not been delivered within six months after the planned delivery dates. Also included in the draft was a 'most favoured customer' clause which provided that if De Havilland sold Comets at less than £1.16m per aircraft or on more favourable terms than those in the BOAC contract within two years after delivery of the nineteenth aircraft, the price to BOAC was to be reduced accordingly and any more favourable terms substituted for those in the BOAC contract.

In February 1957, the Board approved a sum of $35,437,566 to cover the nineteen Comet 4s and one Comet 2E, with spare RA29 Avon engines, provision for escalation and change orders, and spares, equipment etc, and on 4 April 1957 the contract was formally signed.

In contrast to BOAC's experience over the Bristol Britannia - which was plagued with problems, mostly from the Proteus turbo-props - progress of the Comet 4 was unhindered by any serious problems.

Sir Basil Smallpeice: *'We had started Britannia services with the 102 to Australia in*

BOAC took great advantage of the publicity to be gained from the first transatlantic jet service - and rightly so!

Right: Captain Roy Millichap below the nose of G-APDC.

Below: G-APDC on the North side of London Airport. *(both BOAC/DH Hatfield)*

March, only a month after putting it into service to South Africa. On the route to Sydney it earned a very bad name for irregularity and unpunctuality due to repeated mechanical troubles. By August, Charles Abell reported that certain provisos we had made in accepting the aircraft from Bristol at the end of 1956 had not been met.

We had stipulated that we should never encounter engine damage through ice ingestion so that it was necessary to shut down engines in the air and that that the engine relighting system provided by the glowplugs would never require manual operation by the crew. We also stipulated that the aircraft could be used over all our routes without temperature or altitude limitations.

Capt Trevor Marsden,[Deputy Britannia Fleet Manager] held a meeting of pilots to consider the matter. It was agreed that, until mid-October, we must accept a height limitation and operate at low level between Karachi and Hong Kong, and between Karachi and Darwin; after mid-October, operations on all sectors would have to be reviewed in the light of experience and information gained.

Meanwhile, Captain Rendall [Britannia Fleet Manager] reassured his pilots that 'the whole matter is being pursued at the highest possible level. The Prime Minister has asked to be kept informed at short intervals of how things develop, and to this end a combined report by MOS, MTCA, BACI and BOAC is submitted twice every week to the Minister of Transport and Civil Aviation and the Minister of Supply.'

The long-range Britannia 312 also ran into

Right: British journalists arriving at Idlewild: well-known aviation writers of the British daily and Sunday papers and the news agencies, together with Mr Freddie Gillman, the efficient public relations officer of BOAC sfter the pre-inaugural flight from London on 2 October.

trouble, but of a different kind, after a promising start. At the end of June 1957 our first 312, G-AOWA, made the first ever non-stop flight from London to Vancouver, taking 14 hours 40 minutes to cover the distance of 5,100 miles. But on a proving flight which took off at 24.00 hours on Friday, 27 September, our second aircraft, G-AOVB, had two engines fail almost simultaneously after flying for about four minutes in cloud at altitude. Bristols thought that the quenching effect of the moisture in cloud caused a contraction in the compressor casing which resulted in the stator blades rubbing on the spaces between the rotor blades and causing destruction. This had not happened in the 102 engine; so what had gone wrong? It transpired that the stator blades in the 312 engine were made of a different material, which expanded more under heat.

No other airline in the world was liable to encounter such snags in aircraft recently delivered by the manufacturers. The matter was so serious - particularly against the background of the 102's history - that I felt I had to report personally to the Minister, Harold Watkinson.

The first production Comet 4, G-APDA, flew on 27 April 1958 The crew for this flight was John Cunningham as captain, assisted by Pat Fillingham, E. Brackstone-Brown, J Johnston and J Marshall. This aircraft was then used for extensive flight testing between 12 June and 10 July. The pilots involved with these flights included John Cunningham, Peter Bugge, Pat Fillingham and John Wilson, and they accumulated sixty-nine hours during twenty-five flights. During these sorties the complete flight envelope was explored and confirmed, while the

Left: Captain Roy Millichap and the crew of G-APDC shortly before departure from London Airport on the morning of 4 October 1958

Below: First regular transatlantic jet travellers: the departure from London Airport of the first-ever jet airliner service across the Atlantic Ocean. Official sanction from the Port of New York Authority had been announced the previous evening. Bookings, eastbound and westbound, some having been made with BOAC years beforehand, were confirmed overnight. This picture shows the passengers boarding G-APDC for the flight to New York.

onboard systems were given a thorough check-out.

A full C of A was issued to 'DA on 29 September 1958. The first aircraft to be delivered to BOAC was G-APDB on 12 September 1958 with a limited C of A but together with G-APDC the two aircraft were officially accepted by the Corporation on 30 September, the actual date provided in the contract. From then on deliveries proceeded according to plan with the nineteenth and last aircraft handed over on 11 January 1960. De Havilland had done an exemplary job in meeting contract dates.

Entry into service was smoothed as a result of the extensive trials the Corporation had carried out with Comet 2Es fitted with the Rolls-Royce Avon engines which were to be in the Comet 4s. These trial flights took place over a period of eight months between September 1957 and October 1958 over the London-Beirut stage and subsequently across the North Atlantic to gain experience of engine performance and with the object of introducing the Comet 4 first on the North Atlantic to compete with the Americans.

Sir Basil Smallpeice: *'As we moved through the first half of 1958, reports on the Comet 4 from Hatfield were more and more encouraging. In the light of the Comet 1 explosions, its airframe had been submitted to rigorous static tests under the supervision of the RAE at Farnborough. It passed satisfactorily. In other respects, too, production was well up to schedule and it looked as though de Havilland would repeat its 1952 performance and deliver the aircraft on time.*

But then we ran into difficulty over the Comet 4 with the authorities in New York, who would not authorise us to land the aircraft because of the noise it was alleged to create. We produced evidence that the noise level in 'perceived noise decibels' was not materially higher than that of large piston-engined aircraft and was lower than that of the French Caravelle. But the Caravelle had never flown in New York, which was unimpressed by that particular evidence. We brought as much pressure to bear on the Port of New York Authority as we could, through the Ministry in London and the British Embassy in Washington. As a first stage we succeeded in getting approval for Comet 4 training flights but not - repeat not, they emphasised - for commercial service flights'.

It was thought that these 'objections' on the grounds of noise from the Port of New York Authority were both a political and commercial excuse – just as they were years later when the Europeans tried to introduce Concorde into the USA. The Americans were still smarting that the United Kingdom had been the first to fly at jet airliner in 1949 and the first to put a jet airliner into service in 1952. They were damned certain that the Brits were not going to be the first to operate passenger services into and out of the USA!

Sir Basil again: *'The Port of New York Authority's willingness to meet us to this extent was, I feel, greatly influenced by the fact that Juan Trippe's Pan American Airways were also hoping to put the new Boeing 707 jet into service in the autumn of 1958.*

Clearly Americans would not stand in the way of Pan Am or a Boeing product. The Port Authority in New York also feared that, if they were too difficult, Pan American might encounter retaliatory action from London. The New York authorities were also aware that the jets - both Comet and Boeing - would be welcome at Boston.

Thinking about New York's reluctance to admit the Comet 4 on noise grounds, I was certain they would never keep the Boeing out. They would be bound to give way on our Comet 4 application in due course - but probably not until the fifty-ninth minute of the eleventh hour. So we went full speed ahead with our Comet 4 pre-service flight-training and other preparations as though there were no obstacles in our path.

Aubrey Burke from de Havilland's told me that they would be delivering our first aircraft on Tuesday 30 September, ahead of schedule. When the day arrived, de Havilland delivered not just one aircraft but two - a great achievement.

I was already determined to get to New York in our first aircraft as soon as possible. I wanted to see what I could achieve with the Port Authority with a BOAC Comet 4 actually sitting on the ground in New York - previous Comet 4 visits having been made with aircraft still belonging to de Havilland. The Ministry did not rate very high my chances of success in getting early Port of New York Authority permission for scheduled public services, but I decided to go all the same. Aubrey Burke and Captain James Weir of BOAC would come with me.

On the Wednesday we alerted the press and took them across on Thursday, 2 October. For their purpose we dubbed our westbound training flight to New York that day a pre-inaugural flight. The next morning, I met the Port Authority. After the meeting, there was nothing to do but return to our office on Fifth Avenue, and try to possess ourselves in patience until we heard the results.

On the Thursday before leaving my home in Esher, I conceived the idea that, as we now had two aircraft, we could inaugurate the world's first transatlantic jet service with a flight in both directions on the same day, passing one another in mid-Atlantic. Gerard 'Pops' d'Erlanger liked the idea, and said he would accompany the westbound flight.

While waiting in New York on the Friday for the Port Authority decision I was turning over in my mind the organising complexities of a double inaugural, when suddenly, about 5 o'clock, word came that a letter was on the way. It was 10 o'clock at night in London. There was still time to alert them for a possible flight next day.

Shortly afterwards, waving the Authority's letter in my hand, I stood on a chair in the Speedbird Club and told everybody that we could at last start New York services with the Comet 4 - and would do so next morning.

At midnight British Summer Time on 3 October - the Port of New York Authority publicly authorised jetliners to use the International Airport at Idlewild (now John F Kennedy International). Permission was subject to a number of restrictions, all of them designed to safeguard the amenities of the neighbourhood.

That the Comet was able to easily meet all the restrictions imposed by the Port of New York Authority - which represented one of the most noise-conscious communities in the world - and the fact that it did so without loss of range or payload were obvious sources of satisfaction to potential Comet operators throughout the world; most of them faced similar problems to a greater or less extent.

The actual wordage of the restrictions placed upon jet airliner operations by the Port of New York Authority was thus:

1. *The Comet would use Runway 25 as a mandatory preferential runway for take-offs in zero wind conditions and for all wind conditions which will produce a headwind component during take-offs on this runway, provided the crosswind component does not exceed 20 knots.*
2. *Runway 22 may be used for take-offs in lieu of Runway 25, in which case the pilot will make a right turn as soon as practicable after take-off sufficient to avoid flying directly over the communities which are in a direct line with the runway.*
3. *In the event that take-offs could be made on Runways 25 or 22 under the conditions set forth in (1) and (2), then, and only in that event, Runways 13R, 3lL and 07 will be used under the following conditions:*
    (a) For take-offs on Runway 13R, the pilot will make a turn to the right as soon after take-off as practicable, such turn to be made with approximately 15° bank. In addition, taking into account wind and temperature conditions, take-offs from Runway 13R will be so planned and conducted that the aircraft will not fly over any community underlying the flight path at an altitude of less than 1,200 feet and the pilot will observe the piloting procedures set forth in (5) below.
    (b) Take-offs on Runway 31L will be so planned and conducted, taking into account wind and temperature conditions, that the aircraft will not fly over any community underlying the flight path at an altitude of less than 1,200 feet and the pilot will observe the piloting procedures set forth in (5) below.
    (c) For take-off's on Runway 07, the pilot will make a turn to the right as soon after take-offs as practicable, such turn to be made with approximately 15° bank. In addition, taking into account wind and temperature conditions, take-off's from Runway 07 will be so planned and conducted that the aircraft will not fly over any community underlying the flight path at an altitude of less than 1,200 feet and the pilot will observe the piloting procedures set forth in (5) below.
4. *No take-offs will be made on any other runways without specific permission.*

Comet pictures were always 'firsts' including this view of the Andes en route to Santiago. *(DH Hatfield)*

5. All take-offs in 3 (a), (b) and (c) above will be made using the following piloting procedures:
   Initial take-off will be made with a power setting of 8,000 r.p.m. and 20° flap.
   Aircraft will be allowed to accelerate to V2 +15 knots during climb, and the pilot will maintain this speed to the best of his ability until he has reached the communities adjoining the airport.
   Just prior to or upon reaching the nearer boundaries of communities adjacent to the airport, as defined on Chart No. NYA-5967, the pilot will elfect a power reduction to 7,350 r.p.m.
6. Take-offs during the hours between 10 p.m. and 7 a.m. will be made on Runways 25 or 22 only.

In reality, what these regulations meant to the Comet could best be understood by considering the aircraft's take-off and climb performance at maximum all-up weight of 158,000 lbs in relation to each of the runways available for jet airliner operations. The map shows the layout of the runways and the proximity of dense urban

One of BOAC's Comet 4s undergoing servicing in the maintenance facility at London Airport before re-entering passenger service. *(DH Hatfield)*

BOAC's Comet 4 flagship G-APDA is seen low over Hatfield, with Hatfield House in the lower right corner *(DH Hatfield)*

areas.

For Runways 22 and 25 - with zero or light winds, or when the wind was in the south-west quarter, the Comet would use runway 25. The map showed the line of this runway to be over the sea and there are thus no noise problems. Alternatively runway 22, which wass in line with the Rockway community could be used and it was then a simple matter for the pilot to make a gentle turn of 30° to the right to clear the area.

Each of these runways was 8,000 feet long - more than enough for a maximum-weight take-off under any conditions likely to be encountered.

For runways 13R and 31L - if the cross-wind component on runways 25 and 22 exceeded 20 knots, then runways 13R or 31L were used, the choice being dependent on the direction of cross-wind.

If runway 13R was used then the line of flight was towards the Inwood and Cedarhurst communities. A steady 15° banked turn to the right, through 120°, however, enabled the aircraft to avoid the area completely. In any event the nearest houses were some 3.5 miles from the start of take-off and a straight-ahead climb on a hot day (80°F. -into wind component 20 knots) would by then have enabled the Comet to attain 1,800 feet, using the procedure laid down by the Port of New York Authority which called for a reduction to climbing power before over-flying built-up areas. This altitude was half as high again as the Authority's minimum of 1,200 feet and had been laid down to allow the Boeing 707 to operate out of the airport and was a clear example of the difference in take-off performance!

If runway 31L was used - when a right-hand circuit was in force - the aircraft had to pass over the Lindenwood community - approximtely 2.5 from the start of take-off. With a 20-knot headwind and maximum all-up weight of 158,000 lb. the Comet would, in theory, be at 1,100 feet at this position on a hot day (80°F.); on a standard day (60°F.) it would have reached 1,300 feet. However, practical operating experience at Idlewild showed that the combination of an 80°F. air temperature and a strong down- or cross-wind component on runway 25 were unlikely eventualities.

Runway 07 had to be used if the cross-wind component over runways 13R or 31L exceeded 20 knots and its direction was downwind on runway 25. Once again there was a community - Rosedale - at a distance of approximately 2.5 miles. The Comet would thus be at the altitude indicated in the previous paragraph for runway 31L. It was, however, entirely practicable to make a steady 15° banked turn to the right through 180° and this made it possible to avoid over-flying the built-up area.

The conclusion was that one remote contingency alone - a strong downwind component on runways 22 or 25, between the hours of 22.00 and 07.00 - would not affect the Comet. The Port of New York Authority ruled that jetliners must use runways 22 or 25 between these hours which could affect operators. BOAC, however, scheduled their New York-London Comet 4 flights outside this period. Cancellations or delays on this account were, therefore, most unlikely.

The Comet's good power-to-weight ratio and low wing loading provided a runway and climb performance unequalled among jet airliners and better than many contemporary piston-engined airliners. These qualities enabled it easily to meet the Port of New York Authority`s restrictions without loss of range or payload, a fact which was proved in practice by daily Comet 4 operations

Basil Smallpeice again: *'After the initial excitement we had to get down to work preparing for the service to leave New York in the morning. The DC-7C and the Britannia 312 still had to operate overnight eastbound because of their long flight times. Only the jets made it possible to schedule daytime flights from west to east. If we were to leave early from Idlewild (as Kennedy Airport was then called) and if the flight took no more than six and a half hours, we could reach London shortly before dark.*

Operationally, there was no problem. The aircraft in New York was fully serviceable, as was its sister aircraft in London, and both crews were on standby. So, on Saturday, 4 October, 1958, BOAC made aviation history by operating the first transatlantic jet service ever - and, to cap it, both ways on the same day. Capt Tom Stoney, our Comet flight manager, in command eastbound, took the aircraft up to 1,850 ft while still inside the perimeter fence of the airport, at which point he throttled back to reduce the noise level within limits acceptable to the authorities.

Out over the Atlantic we passed the other aircraft, out of sight, with Pops d'Erlanger on board and Capt Roy Millichap in command. Our eastbound flight to London took only 6 hours 12 minutes, thanks to a tailwind of 92 mph and the priority given us by Air Traffic Control over the UK. The aircraft glided in to a smooth touch-down at London Airport. A warm welcome was given us on the tarmac, and it gave me a particular glow of pleasure to find Miles Thomas among those who had come to greet us.

Not until nearly three weeks later were Pan American able to introduce their own transatlantic jet service with their newly delivered Boeing 707. Our team had scored another BOAC first.

BOAC's important 'first' was very nearly spoiled by an Engineering strike by BOAC's staff at London Airport and subsequent Comet services were held up until 14 November when the aircraft took over the daily Monarch services between London and New York from the Britannia 312s. In 1959 the three organisations concerned with getting the Comet back in international air service, De Havilland, Rolls-Royce and BOAC, were given the Hulton award for the most outstanding contribution to British prestige.

Not everyone enjoyed flying in the Comet however. In Ian Fleming's *For Your Eyes Only*, - the original book, not the movie James Bond laments that he isn't able to take the Stratocruiser: *'Two days later, Bond took the Friday Comet to Montreal. He did not care for it. It flew too high and too fast and there were too many passengers. He regretted the days of the old Stratocruiser — that fine lumbering old plane that took ten hours to cross the Atlantic. Then one had been able to have dinner in peace, sleep for seven hours in a comfortable bunk, and get up in time to wander down to the lower deck and have that ridiculous BOAC 'country house' breakfast while the dawn came up and flooded the cabin with the first bright gold of the Western hemisphere.*

It seems that James Bond thought that everything was too rushed on the Comet!

```
LONDON OCTOBER 4 - REUTER - BRITAIN TODAY MADE AIR HISTORY

BY WINNING THE RACE TO BEGIN ALL-JET COMMERCIAL AIR SERVICES

ACROSS THE ATLANTIC .

    TWO GLITTERING COMET FOURS CROSSED THE ATLANTIC IN OPPOSITE

DIRECTIONS - THE NEW YORK-LONDON PLANE MAKING THE TRIP IN SIX

HOURS 12 MINUTES - THE FASTEST CROSSING EVER MADE BY A CIVIL

AIRLINER.
```

Cut to the bone.

To say that American pride was hurt with the arrival of the transatlantic Comets was an understatement. Legend has it that the first passengers were booed as they stepped off the aircraft in New York by some. None were more incensed at being beaten than the Boeing Aircraft Company.

Boeing targetted Pan American World Airlines, which had traditionally 'bought Boeing' and which the U.S. government considered its 'chosen instrument' to represent the American commercial air fleet abroad. Undoubtedly a pioneer in embracing jet aviation, Juan Trippe, the airline's legendary chief executive officer, had early on expressed a keen interest in operating a passenger jet service capable of flying nonstop across the North Atlantic. Having seen the bright promise of the British Comet fade, Trippe played off two of the biggest domestic airplane builders, Boeing and Douglas. Both companies vied to appeal to Pan American's needs and offered the Boeing 707 and DC-8 respectively.

At first BOAC tried hard to compete. With its rapidly expanding use of the Boeing 707, especially on the transatlantic route, Pan American began a period of almost unchallenged success in the international airline industry. The airline, for example, was the first to recognize the importance to passengers of non-stop flights on long trips; it negotiated with Boeing for a version of the 707 that could fly for a longer time without refuelling, known as the 707-320. This allowed the airline to introduce true intercontinental service with non-stop London-to-New York flights on 26 August 1959. This was a perfect case of a dominant air carrier playing the lead role in defining the characteristics of a new class of

Left: BOAC's promotional material advertising their 'Monarch Luxury Service' to New York on the Comet 4.

Below: One of BOAC's Comets is caught by the camera on a proving flight through Gander in Newfoundland.

jets that the industry would produce. The 707-320 was eventually adopted by as many as eleven other airlines within a year.

Comet 4 commercial operations across the Atlantic were notably trouble-free and BOAC justifiably thought that, as a result of Comet 4 experience, and because of their previous Comet knowledge amounting to almost 30,000 hours, that the Comet had quickly passed through the early stage during which all aircraft minor operating difficulties were encountered and no defects in the various systems could be expected to cause delays in service.

Practical experience confirmed that there were no unusual crew problems associated with Comet 4 operations, pilots having no difficulty with conversion. From the navigator's point of view work might in theory be at a rather higher intensity because of the compression in time of flight; in practice, however, trans-Atlantic Comet navigation proved easy. Good Loran coverage was available for most of the flight and radio contact with either side of the Atlantic or the two weather ships, Charlie or Juliet, was maintainable through all but fifteen minutes or so of the flight; the Comet provided a smooth platform for Astro if required.

Passenger reaction was unanimously favourable; the four-abreast standard-class seating was deservedly popular and there was plenty of room for the cabin staff to provide the usual incomparable BOAC Monarch service.

From the public address system, over which the Captain briefed his passengers on the new and exhilarating experience of jet travel, to the serving of a luxury meal in conditions as steady and as smooth as (and somewhat quieter than) a West End restaurant, BOAC certainly 'took good care of you'.

As the publicity material said '...the size and arrangement of the passenger accommodation enable the operator to provide a personal element in the cabin service which travellers particularly appreciate. The Comet has proved by practical demonstration to leading airline people the world over that it goes anywhere using to-day's runways without extension or restriction. Now the passenger will decide'.

Basil Smallpeice, the Managing Director had even more to say: *If one had been asked, only 30 years ago, to forecast the changes which civil aviation would work in the field of international relations, one might have hazarded a number of guesses which have proved substantially correct, Short and medium-haul aircraft, plying between cities hundreds of miles apart, have very largely replaced ground transport in many parts of the world, with Australia as a prime example; regular inter—continental air services are an everyday reality; mail, in the carriage of which speed is paramount, is carried by air as a matter of course. In all these things, the foresight of those who founded the aviation industry has been more than vindicated.*
What I do not think the boldest pioneer could have visualised two or three decades ago was the astonishing growth of high speed, very long distance commercial air services, linking continents thousands of miles apart in a matter of hours, and carrying, moreover, tens and even hundreds of thousands of passengers a year, easily and regularly.

Two things have made this possible; the superb technical capacity of the modern airliner and its ability to fly at speeds which a few years ago would have been considered quite unattainable. It is this speed which, in the last analysis, has made the advent of the true international express air service a practical proposition. No product of human ingenuity can ever physically reduce the enormous distances which separate the continents of the world; the jet airliner alone has the means to bring the ends of the earth together in terms of time. The benefits which this shrinking of time can bring to millions of people throughout the world are incalculable.

For these reasons, we are especially proud of the

B.O.A.C COMET SERVICES - FEBRUARY 1960

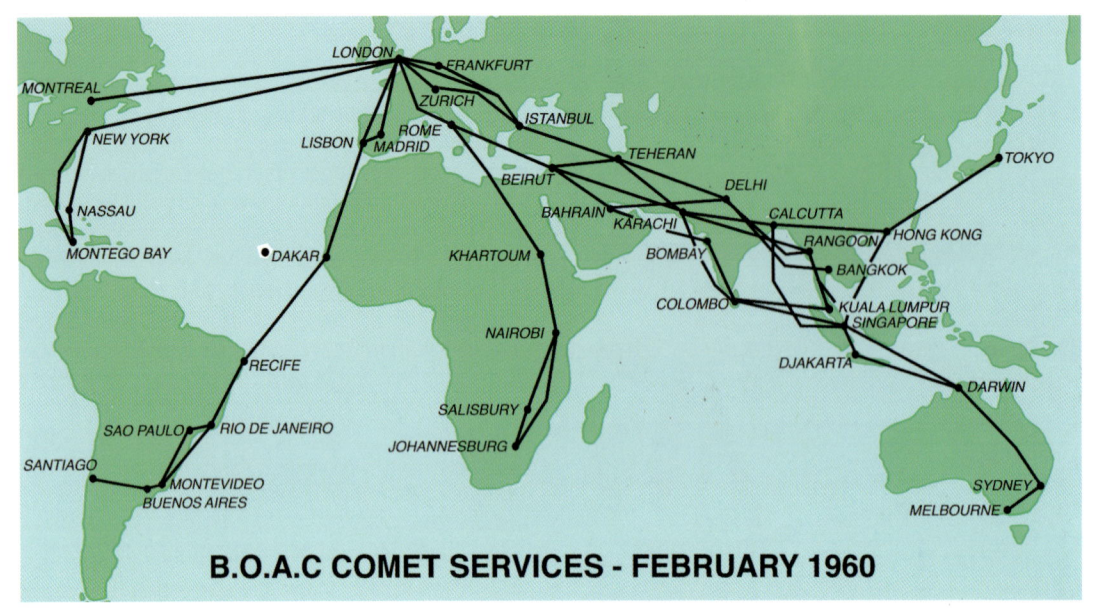

advent of the Comet 4 jetliner on services from Australia to Europe, for we believe that no other aircraft so fully expresses the revolution that the jet age has brought to commercial transport. By any standards, an airliner which can link two continents 12,000 miles apart in a day-and-a-half is an outstanding technical achievement. Its introduction, we believe, is a vindication of the faith which we have held in the Comet throughout its career and of the long years of exhaustive testing, development and training which lie behind its introduction into regular service on the Kangaroo route to Australia, the longest and most difficult of the many routes on which it has so far appeared.

With the introduction of jets on the Kangaroo partnership services, we may perhaps look at some of the possibilities of high-speed, intercontinental air - at some of the problems. With Australia- Europe jet services firmly established from the technical viewpoint, we must, I think, recognise that the next major development must lie in the field of airfares. Of this, I would say only that despite the difficulties which have become apparent recently in the task of recasting the structure of international airfares, BOAC reaffirms its belief that the cost of air travel can and must be reduced if we are to take advantage of the enormous work capacity of the modem jetliner and make air travel available to many millions of people who cannot afford the present level of fares. The great expansion of popular air travel in Australia, America and other parts of the world is some indication of the size of the market which long-distance air travel, too, must reach if civil aviation is to fulfil the promise of the jet age.

Another problem which is very much in our minds is that of reducing the time which the long-distance air passenger must at present spend on the ground. Constant increases in flying speed lose much of their point if the passenger is condemned to long hours on the ground, both at his destination and at calling places on his route. From the technical standpoint, this means ensuring that large and complex modern airliners are handled on the ground as fast and efficiently as possible. From the administrative angle, airlines, governments and many other organisations 'must pool their common knowledge and goodwill to ensure that the formalities of international travel are as few and as simple as possible.

These and many other considerations face the airlines of the Commonwealth as we begin to feel the full implications of the revolution which the jet age is bringing about. There is, however, no reason why we should feel alarmed at the problems ahead- It is, after all, only just forty years since the first aeroplane flew from London to Australia; and when we consider the enormous strides which civil aviation has made since then, and the many problems which have been met and overcome at every stage in the task of turning a lonely and hazardous enterprise into an everyday commercial undertaking, we should, I think, feel a sense of gratitude to those many pioneers, Australian and British, who laid such firm foundations for our generation to build on.

I believe we have every reason to feel pride in what has been achieved and faith in our ability to match those achievements in the years ahead. For us in BOAC, there could be no finer aircraft than the Comet 4 in which to express that faith.

Chapter Six

Associates, Mail, Cargo and the Independents.

BOAC's involvement in its Associated and Subsidiary Companies forms a vital part of the history of the airline. It is a complicated story which stretches right through the life of the Corporation. BOAC emerged from the war with a financial interest effectively in only two companies, QANTAS and Tasman Empire Airways (TEA), with a shareholding in the former of 50% and the latter of 38%. The shareholding in QANTAS was disposed of in 1946 when the Australian Government decided to take full control into their own hands and nationalised QANTAS. A BOAC investment in TEA remained for a time but, with an increase in the airline's capital, BOAC's shareholding was reduced to 20%.

From 1946 BOAC's interest in mainly, feeder line companies steadily increased to a point where in June 1957 the Board approved the concept of a holding company, BOAC Associated Companies Limited (BOAC-AC), specifically to look after these interests. The Corporation had always recognised the value of local services feeding traffic from out-stations into its main trunk line and realised that it was in its interest to develop close working arrangements in such operations by offering technical and managerial help. It was a natural step then to become financially involved where a good case for it could be made, mainly to obtain a useful measure of control and to ensure that BOAC's particular interests were protected.

The reasons for BOAC to become associated with certain established companies and in some cases to form new local companies were set out by the Corporation on some occasions, in one such they were defined as:
- To obtain feeder line traffic for the Corporation's trunk services and to keep to a minimum the traffic flowing from the area to competitors.
- To exert influence beneficial to the Corporation's operations and to minimise competitors' influence
- To assist the strategy of the Corporation's future development such as enlarging and protecting BOAC's traffic rights in the area.
- To hope of financial profit in a field not open directly to the Corporation.
- To protect and promote the national interest when requested by Government.

In the early days of this policy it was left mainly to McCrindle in his position as Adviser International Affairs to monitor the activities of the Associated Companies and in fact he became the BOAC representative on the Boards of a number of these companies, notably West African Airways and East African Airways, although BOAC had no financial interest in either. In the case of Malayan Airways where the Corporation had a minimal investment of 10%, John Linstead whilst he was Manager Far East was nominated as the BOAC Board Member. This kind of control could only be intermittent and by early 1951 BOAC's Chairman was becoming uneasy at the looseness of the arrangements from BOAC's point of view. He felt that too much time and effort by senior Corporation executives were being applied to these associated and subsidiary enterprises to the detriment of BOAC's primary task. He asked McCrindle to produce an appreciation of their value to BOAC.

With this study, the Chairman took the line that the Corporation's policy should be to exercise more direct control over wholly owned subsidiaries to ensure that BOAC's interests were not swamped by purely local

A stunning picture of QANTAS' Super Constellation VH-EAM in flight. At one time BOAC had a large share in the airline.

British Commonwealth Pacific Airlines' DC6 VH-BPF
(author's collection)

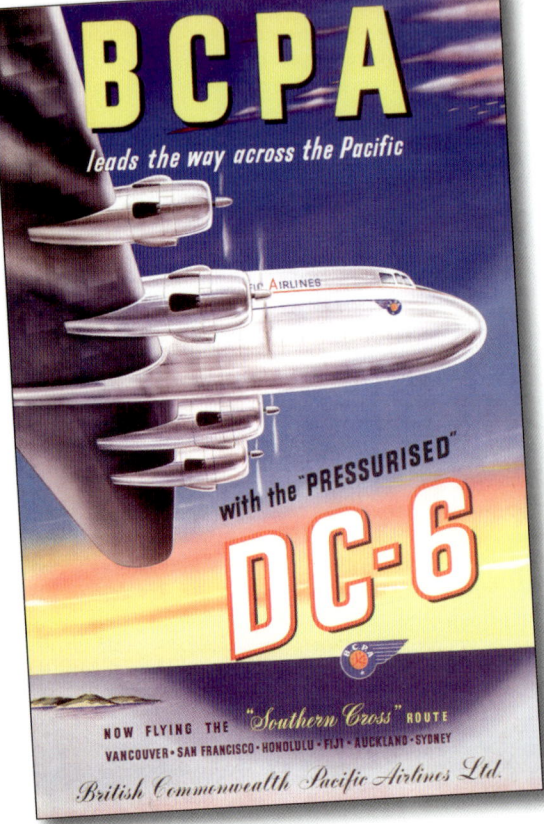

considerations. In the face of the Chairman's views, it seems somewhat inconsistent for the Board then to pronounce that the best organisational pattern was for the Chairman to be also Chairman of each subsidiary, with the Deputy Chairman and Financial Controller as members of the Boards; if this had been fully implemented even more time for these top-level executives would have had to be spent on these affairs. It was clear there was a need for a full-time executive at Head Office to ease the burden and act as a watchdog for the Corporation's interests. Accordingly, in February 1952 Captain V. Wolfson was appointed as General Manager, Subsidiaries, and it became his responsibility to report in future to the BOAC Board on both subsidiary and associated companies after consultation with Adviser International Affairs. This arrangement continued until his death in a Comet crash in 1954. His post was not filled as such, but instead, Gilbert Lee was appointed to the new position of London Manager, Subsidiaries, to hold a watching brief over the activities of the subsidiaries and the associates.

Up to the time of Lee's appointment, the results achieved by the subsidiary companies fluctuated considerably. In BOAC's financial year 1949/50 a radical change took place in the extent of BOAC's subsidiaries when following the BSAA merger BOAC acquired British West Indian Airways (BWIA) and Bahamas Airways Ltd. (BAL). In that year the losses of all the subsidiary companies totalled £287,490, increased to £301,039 in the following year but then declined to £30,612 in 1952/53 as the result of offsetting profits and by 1954/55 remained at the reasonable level of £45,192. In 1955 a further expansion took place concentrated mainly in the Middle East when BOAC acquired from Pan American Airways a company called Middle East Airlines (MEA) and then set up a holding company. Associated British Airlines (Middle East) Ltd. (ABAMEL), to cover the Corporation's investments in the group of Middle East companies. To manage the new holding company, a managing director, Sir Duncan Cumming, was appointed in June 1955. That year saw the losses mount steeply to £401,224 and alarmingly to £767,895 in 1956/57. It was at that point when BOAC-AC was set up under the Chairmanship of Sir George Cribbett, the newly appointed Deputy Chairman of BOAC, with the object of reversing the trend of mounting losses.

To understand why BOAC regarded the policy of involvement in local companies as essential it is necessary to go back to the immediate post-war years. British airline interests in colonial territories had been well preserved because of the international acceptance of the cabotage principle. But the move towards independence for the Colonies was under way which was to result in their becoming sovereign territories able to develop their airlines and to grant traffic rights to foreign operators if they so desired. BOAC was anxious to preserve an advantageous position locally not only to hold on to its long-haul on-line traffic but also to obtain the benefit of any local traffic fed into its main line rather than to its competitors.

One of the first and most successful of BOAC's subsidiaries was outside the field of direct airline operation. This was International Aeradio Limited set up at the end of 1946 with BOAC and BEA as the main shareholders joined later by other international airlines. The purpose was to provide communications, flying control and navigational aids for civil aviation at a time when many countries had neither the expertise nor the means to undertake this specialised work for themselves. The Company established itself quickly, found a ready market in the undeveloped countries, particularly in the Middle East and except for the initial years was consistently profitable.

In the same year, 1946, as BOAC relinquished its shareholding in QANTAS, a decision was taken by the three Governments of Australia, New Zealand and the United Kingdom to set up a new airline to operate between the Dominions of Australia and New Zealand and North America. The new company, British Commonwealth Pacific Airlines Ltd. (BCPA), was founded in June 1946 by the three Governments with shareholding as to 50% subscribed by Australia, 30% by New Zealand and 20% by the United Kingdom. BOAC, in fact, held the 20% British share as a nominee of the British Government. The new Company, having no aircraft of its own, relied on Australian National Airways to operate its services, which began in September 1946. This arrangement lasted until April 1948 when BCPA took over its flying on acquiring DC4 aircraft. During 1947 the Company increased its capital, and at British Government instigation, BOAC increased its investment to £160,000. By 1949 the Company had replaced its DC-4s by DC-6s offering a greatly improved service with these pressurised aircraft.

In October 1953 a full meeting of the South Pacific Air Transport Council took place in Christchurch, New Zealand, and a restructuring of the Pacific operation was agreed by the Australian, New Zealand and UK Governments' representatives under which QANTAS would absorb BCPA and thus become the sole Commonwealth operator for the trans-Pacific route. BOAC disposed of its financial interest in BCPA at that time, and the idea of a Commonwealth round-the-world service was to be promoted by the extension in due course of BOAC's Atlantic service to San Francisco to link there with QANTAS on the Pacific. At the same time, BOAC gave up its shareholding in TEA at the request of the New Zealand and Australian Governments to make that Company wholly owned by them.

BOAC owned a small shareholding in the Irish airline Aer Lingus which also operated for a short time a trans-Atlantic operation called Aerlínte Éireann using a number of Lockheed Constellations. *(author's collection)*

In the 1950s BOAC has financial involvement in Rhodesian & Nyasaland Airways Ltd, whose operation went well back into the 1930s, as can be seen by this 'first day cover' from July 1935.

Another 1946 investment was in the Irish Airline, Aer Lingus, in which BOAC took a small holding of 10% with BEA holding 30%. BOAC became disenchanted, partly due to the airline's losses in the early days, and somewhat as it became recognised that Aer Lingus was essentially concerned with intra-European traffic not within BOAC's sphere of interest. By early 1950 BOAC was negotiating with BEA to sell its holding which was completed in the summer of 1952.

Just before the commencement of Sir Miles' Chairmanship, and in addition to the interests referred to above, BOAC was involved to a lesser or greater extent in a wide range of overseas airlines in pursuit of a policy encouraged by Government, of assisting mainly Colonial territories to establish their own air transport companies. BOAC's activity in this respect was two-fold; on the one hand, to invest capital to secure complete control, and on the other hand to assist existing companies with technical and operational advice and in some cases to second management staff and even to grant loans. The purpose behind all these steps was mainly commercial to develop traffic locally which could feed into the Corporation's trunk routes.

BOAC's subsidiary and associated interest lay not surprisingly in the four main areas of its trunk route operations, Africa, the Middle East, the Far East, and lastly the West Atlantic/Caribbean.

In Africa there had been long-standing relationships in East Africa between Imperial Airways and Wilson Airways of Nairobi, in central Africa with the Rhodesian and Nyasaland Airways of Rhodesia and in West Africa with Elders Colonial Airways owned jointly by Imperial Airways and the Elder Dempster Shipping Company, so that when in 1946 the Labour Government approved the formation as public Corporations of EAAC and WAAC, it was natural that BOAC should assist with technical aid and advice through providing a Board member to each Corporation but without any financial commitment. Much the same kind of development occurred in central Africa with the establishment of Central African Airways Corporation (CAA) based on Salisbury, Southern Rhodesia, but financed by the three territories of Northern Rhodesia, Southern Rhodesia and Nyasaland. Here BOAC was asked by the UK Government to undertake an investigation into the affairs of CAA which subsequently led to the

In clearly what is a colour scheme very much baded on the BOAC 'blue tail' design, here is Aden Airways DC-3 VR-AAA. *(Simon Peters Collection)*

Aden Airways Vickers Viscount VP-YNB 'sometime in the 1960s'. *(Simon Peters Collection)*

appointment of a Manager provided by BOAC. By 1953 the Company had grown sufficiently to be able to operate a low-fare service from Salisbury to London alongside BOAC's services.

In 1958 BOAC took further steps to protect its interest in Africa by taking financial stakes in the new Ghana Airways Limited and WAAC (Nigeria) Ltd. These companies had sprung from the break-up of the WAAC which had ceased to exist from 30 September 1958 as a consequence of Ghana (formerly the Gold Coast) having attained independence. Agreement in principle was reached in May 1958 with the Ghana Government for BOAC to subscribe 40% of the capital of the new airline, the Ghana Government accepting 60%. The new airline's international services under the agreement were to be operated by aircraft and crews chartered from BOAC, but the intention was for Ghana Airways eventually to take over its international operations. The agreement provided that the Ghana Government would meet any loss sustained by the Company. The new Company was registered on 4 July 1958, and the agreement with BOAC was signed on 5 July. This arrangement lasted until February 1961 when BOAC's shareholding of 40% was taken over by the Ghana Government at par; the pooling arrangements between BOAC and Ghana Airways continued in effect.

The intention of Ghana to form its airline sparked off similar action in Nigeria and the BOAC Board in June 1958 approved in principle BOAC-AC participation in a new Nigerian airline, WAAC (Nigeria) Ltd. The capital was to be provided by the Nigerian Government as to 51% and as to 49% by outside interests. Elder Dempster with their long-standing interests in the area decided to take up a holding and the split agreed of the 49% was 16% to BOAC and the balance to Elder Dempster. BOAC was to charter aircraft to the Company for its international operations but specially between Lagos and London. BOAC-AC was entitled to 50% of any profit on the international services; the new Company was to be fifinancially responsible itself for such international operations, thus relieving BOAC-AC from making any provision as to the trading results of the Company.

In the same fashion as occurred in Ghana, Nigeria decided to take over complete ownership of the airline in March 1961, and consequently, BOAC-AC transferred its holdings to the Federal Government of Nigeria and Elder Dempster did likewise. The commercial agreement on pooling over the international operations between BOAC and Nigeria Airways continued in force.

As to the Middle East, at the close of the war British interests in Iraq were dominant and when the Iraq State Railways formed Iraqi Airways in December 1945, they turned to the UK for help. BOAC in due course supplied both technical, financial and commercial assistance through seconded staff but had no direct investment in the Company. BOAC, however, made loans to Iraqi Airways and after some delays repayment of the final debt of £160,000 was eventually established in 1950 when the Iraq State Railways secured a considerable investment from the City of London and the Bank of England authorised the settlement of BOAC's debt out of it. BOAC continued to help with seconded staff but was not subsequently financially involved.

In Iran, BOAC was briefly concerned with a local outfit called Eagle Airways and assisted them with seconded staff and stores, but the Company had a very brief life from June to November 1948 when operations were suspended, and the Corporation experienced difficulty in securing a final settlement of outstanding bills. In 1950 Eagle was absorbed into the new Iranian national airline, Iranair, and BOAC finally negotiated an agreement with Iranair on the best terms it could.

Reference has already been made in dealing with BOAC's relations with BEA to the difficult issue of Cyprus where BEA held a shareholding in the local company Cyprus Airways, and BOAC sought traffic rights in the island for its transitting services. By 1950 an initial agreement had been reached between BEA and BOAC under which BOAC would withdraw its services through Cyprus in return for inter-line arrangement with BEA and Cyprus Airways and a BOAC shareholding in the local company by taking over 50% of BEA's interest. This amounted to an investment of £42,523 by BOAC, representing a 23% holding. BOAC and BEA, both shareholders, continued to jockey with each other in extending their respective influence in Cyprus Airways by putting forward proposals particularly in 1955 for the hiring of Viscount aircraft to operate local services.

Another Middle East company in which for a time BOAC had a limited investment of about £10,000 was the Egyptian Engineering Company S.A.E. set up in 1949 to undertake overhaul and maintenance work at the Almaza base, Cairo. In November of that year BOAC decided to transfer its Dakota overhaul and maintenance to this Company as it was thought worthwhile to collaborate with an influential Egyptian group. Some progress was made by the Company for in its financial year 1950/51 it made a small profit. But by the following year the political upheaval in Egypt resulted in a complete lack of work, many British staff were withdrawn, and the Company put on a care and maintenance basis. Its future under the new Egyptian regime was unclear, and by the end of 1953, the decision was taken to liquidate it.

During the war, BOAC had operated a network of services from Asmara and Aden, and some of these operations continued after the war with Cairo-based BOAC Dakotas. It was felt that this type of local activity required the setting up of a local Company and on 1 March 1949 Aden Airways was formed as a wholly owned subsidiary. Its area of operation was broadly as far as Cairo to the North, to Addis Ababa to the West, to Nairobi to the South and into the Hadramaut to the East. BOAC's investment in the new Company represented by fully paid shares amounted to £85,770. A small profit was earned in the first six months of operation, and for 1950/51, the first clear year, the advantage was £12,118. In his report to the Chairman of BOAC in May 1951, McCrindle supported the continuance of BOAC's interest because if the Corporation withdrew, BOAC's US competitors would almost certainly step in, as TWA were already established in nearby Ethiopian Airlines. Useful revenue came to Aden Airways from pilgrim charters, and this contributed largely to profitable results. Over the next two years, 1952 and 1953, there was a proposal from BOAC to amalgamate Aden Airways and Cyprus Airways which received the blessing of the MCA, but difficulties arose with BEA as a shareholder in Cyprus Airways, and eventually, the Government of Cyprus preferred to operate its airline. In 1953 Aden Airways, at the request of Arab Airlines (Jerusalem), began to operate on their behalf with two of its Dakotas and in due course, Aden Airways held 49% of Arab Airways, based in Amman, Jordan.

1952/53 saw a further profit of £22,817 by Aden Airways, but in the next year a serious loss of £140,959 occurred, far more than the budgeted amount. As a result new Management was installed by BOAC and the Company's working overhauled, action which reduced the loss in 1954/55 to £6,381 and for the first six months of 1955, April to September, the Company was back in profit. All the shares were taken over by BOAC's new holding company in the Middle East, the Associated British Airlines (Middle East) Ltd.

The Corporation found Bahrain an invaluable staging post on the way to India and the East and it had every intention of maintaining its influence in the area and as far as possible keeping out the American airline interests who were already close at hand at Dhahran on the Arabian mainland. Accordingly, when the small local airline in Bahrain ran into difficulties, BOAC stepped in and purchased a majority holding in the Company, Gulf Aviation Ltd., in October 1951, the balance of the shares being held by residents. The Company at that time operated De Havilland Doves on scheduled services within the Gulf area and also undertook charter work for the oil companies. BOAC appointed an experienced General Manager, Captain M C P Mostert, and assisted the Company by the purchase of additional Dove aircraft. It was evident from the start that the Company had the makings of a profitable airline. Its first deficit of over £11,000 in the six months before April 1952 had been reduced to around £4,000 by 1952/53. That year, the General Manager, Subsidiaries, in reporting to the Board was able to confirm that the Company was on a sound basis and was being efficiently run. By the end of the financial year 1953/54 the Company had made a small profit enough to pay a dividend on the preference shares held by local Bahrainis, and in the year following had increased its profits more than fourfold. It continued to make profits until it was taken over by BOAC's new Middle East holding company in 1955. Events had already proved by then the wisdom of BOAC's involvement, and subsequent immense development in the Gulf area in the wake of the growth of the oil companies made this even more apparent.

There was another small company in the Gulf controlled by the Corporation. This had the rather grandiose title of British International Airlines Ltd.

Not the best of images it is true, but interesting nevertheless, DC-3 G-AMSM in the colours of Kuwait National Airways. The aircraft was acquired from Eagle Aircraft Services in 1954, being registered to British International Airlines at the BOAC address in July and placed on the Kuwaiti register. It returned to the UK and Skyways in 1955. *(Simon Peters Collection)*

Gulf Aviation started off with De Havilland Doves, but soon moved up to the unbiqitious DC-3, as demonstrated by G-ALVZ, and registered to BOAC. *(author's collection)*

(BIAL) and started its existence in 1951 as a BOAC subsidiary. It was based on Kuwait, and its purpose was to undertake charter work and in particular to act as a charter company for the Kuwait Oil Company. When local interests in Kuwait started up their airline, Kuwait National Airlines, in 1954, BOAC gave technical help and provided some flying staff, and the new airline's Dakotas were then maintained by BIAL.

BOAC had had a long connection with Kuwait National Airlines 'and had supplied it with technical assistance. In 1955 Kuwait National Airlines changed its name to Kuwait Airways but BOAC retained a direct interest in the area through its subsidiary BIAL which carried out charter work in the surrounding territories. The British Government was anxious that British interests should continue to play a leading part and when there was a risk of foreign involvement in Kuwait Airways, it encouraged Hunting Clan and BOAC to take positive steps to assist Kuwait Airways to avoid this: Hunting Clan dropped out of the negotiation but on 23 May 1958 BOAC and Kuwait Airways signed a five year agreement under which BOAC-AC was to have sole responsibility for management and for administrative, operational, commercial, and technical control of Kuwait Airways and would provide a Chief Executive. The Agreement was to commence on 1 June 1958 and BOAC-AC undertook to provide the airline with a Viscount on that date and a second Viscount a month later. BOAC-AC was to guarantee to Kuwait Airways shareholders a return of not less than Rs300,000 (£22,000) per annum and any profit over and above that was to be shared equally between BOAC-AC and Kuwait Airways, but BOAC-AC assumed full responsibility for any losses. In its Annual Report for 1958/59 BOAC, with the Minister's approval, said in respect of Kuwait Airways that *'the responsibility for this Company and for the consequential re-equipment and expansion of the scope of its operations was undertaken by the Corporation at the express request of the Minister of Transport and Civil Aviation'*. Following the BOAC-AC/Kuwait Agreement, BIAL was transferred to Kuwait Airways as from 1 April 1959.

Heavy losses were incurred under the Agreement amounting over the five years of its life to around £1.5m, and at the end of that time with the change in status of the Kuwait Government there was no longer any pressure from the British Government on BOAC to continue a loss-making arrangement. The whole issue received considerable airing during the proceedings of the Select Committee on Nationalised Industries in late 1958 and early 1959, and the Committee expressed the view that in the light of what occurred it was *'wrong that BOAC should have to carry this loss unaided'*

The last commitments entered into by BOAC in the Middle East during Sir Miles' Chairmanship arose from a decision to acquire a holding in Middle East Airlines (MEA), a decision which led to unpleasant financial consequences for the Corporation. The initial proposal was put to the Board by the Deputy Chief Executive, Basil Smallpeice, on 11 November 1954. It was based on the premise that Pan American Airways which held 36% of the shares would be prepared to sell them and that further shares could be acquired to give BOAC at least a 59% holding, necessary to secure Management control.

MEA was registered in Lebanon and owned a fleet of Dakotas operating local routes around the Middle East and the Gulf. Since 1950 the Company had been profitable but on a declining level. In fact, BOAC had been technical advisers to the airline shortly after the war, but this arrangement had been terminated in 1947 as at that time BOAC was unhappy about the Company's operating methods. The justification for BOAC's purchase as put to the Board rested on three main aspects - first it made good sense commercially, second, it would improve prospects for the sale of British aircraft in the Middle East and third it could increase British influence in the Arab world.

By April 1955 MEA became an associate company of BOAC and by May arrangements had been completed for the sale of just under 75% of the Company's shares to BOAC at a cost of around £335,000, although it was BOAC's intention to reduce its holding to only under 50% provided adequate powers of management remained with the Corporation. In any case it was found that under the Articles of Association no more than 49% might be held by foreign shareholders. At the time of these developments, it was the Corporation's view that MEA *was the key to the solution of the problems confronting BOAC in the area'*. BOAC recognised that it would not be easy to reconcile the divergent interests of MEA and the other local companies under BOAC's umbrella.

One of the conditions of purchase was that BOAC should appoint the General Manager or Managing Director and this it did by calling on Sheik Najib Alamuddin to fill the post; Sheik Najib had already been appointed previously as BOAC's Middle East advisor.

In May 1955 BOAC placed an order for six Vickers Viscounts for use by its Associated Companies and took an option on a further six of this type at the same time. Deliveries could not begin until early 1957 so arrangements were made for MEA to lease Viscount time from Hunting Clan and it subsequently purchased Viscounts from that Company; this enabled MEA to operate Viscount services with its aircraft from June 1956. In this first year under BOAC control, MEA incurred a loss of £132,414, mainly due to fundamental reorganisation to gear up to new routeings suitable for Viscounts to operate.

BOAC thus found itself in control of some local airlines based on Arab lands in the Middle East and by this means had avoided any threat from American quarters to obtain a dominant position there. A list of these local companies demonstrates how extensively BOAC had spread its influence - in Lebanon MEA, in Jordan Arab Airways (Jerusalem), in Kuwait BIAL, in Bahrain Gulf Aviation and in Aden, Aden Airways. What is more, BOAC although without financial commitments had influence either through Management Agreements or close working arrangements with Kuwait National Airlines and Iraqi Airways; BOAC also had a 23% interest in Cyprus Airways.

It was going to be difficult for each of these small companies to re-equip itself with modern aircraft and there was realistically no possibility of their combining to form a large unit, for reasons of national pride. BOAC, therefore, concluded there was a case for an aircraft service company to be set up in the area (and Beirut seemed the most convenient point) to acquire aircraft and hire them out to the companies and to undertake maintenance work for them and any other local operators who might desire it. A case was therefore put to the BOAC Board in January 1955 for setting up such a company under the title of the Mideast Aircraft Service Company (MASCO), using the added argument that it would provide a market in the Middle East for British aircraft manufacturers. It was initially hoped that BOAC would be joined in the venture by Hunting Clan and Skyways - BOAC thus demonstrating its willingness to work with the Independent companies - and also by Air Finance; the company would act for British aircraft and engine manufacturers generally.

BOAC accepted that it would mean in effect that the British would be helping the Arabs to establish an airline industry, but the hope was British products would be used. BOAC was careful to stress that *'it would be a mistake for any British interest to embark on the venture under the impression that it can be other than Arab in the long run'*.

A useful start was made with the agreement of Hunting Clan to join the MASCO venture and to charter to it three of their Viscounts. In due course, BOAC intended to allocate four out of the six Viscounts on order to MASCO, but deliveries were not expected before 1957.

MEA's DC-3 OD-AAG. The airline painted a *Cedrus libani*, commonly known as the Cedar of Lebanon or Lebanon cedar, is a species of cedar native to the mountains of the Eastern Mediterranean basin. *(Simon Peters Collection)*

Avro York OD-ACZ of MEA.*(Simon Peters Collection)*

As the setting up of MASCO progressed, it became apparent that there were two distinct functions involved, first a straightforward servicing company providing aircraft and maintenance and second a requirement for the supervision of control of BOAC's Middle East Companies. MASCO was 65% owned by UK money, with 35% coming from the Arabs. BOAC, therefore, decided to confine MASCO's responsibilities to the first function and to set up a separate holding company to deal with the second function. One of the advantages of an independent holding company lay in its ability to decentralise away from BOAC's Management the supervision of the local companies and thus free the top echelon of BOAC from much time-consuming work.

At its June 1955 meeting, the main BOAC Board approved the formation of the holding company, subsequently known as Associated British Airlines (Middle East) Ltd. (ABAMEL), with participation by BEA, Hunting Clan and Skyways. The appointment of Sir Duncan Cumming to act as Managing Director of the new Company followed. As part of ABAMEL, Aden Airways and British International Airways (Kuwait) were 100% UK-owned; Gulf Aviation Ltd was 51% UK-owned and Arab Airways and Middle East Airlines were 49% UK owned.

It was the intention that Hunting Clan and Skyways would take shareholdings in ABAMEL and MASCO, but Skyways found themselves unable to subscribe and BOAC took up their proportion. In November 1955 the BOAC Board approved the transfer of shares in Aden Airways, British International Airlines, Gulf Aviation and Middle East Airlines to ABAMEL. Thus by the time, Sir Miles resigned in April 1956 all was set for the restructuring of BOAC's Middle East interests.

The last of the big losers for BOAC was MEA, and it is convenient to consider at the same time the Mid East Aircraft Service Co (MASCO) both based in Beirut. Both Companies were formerly in the group of Companies coming within Associated British Airlines (Middle East) Ltd (ABAMEL), but ABAMEL's existence ceased in March 1959 and from 1 April 1959 MEA and MASCO both came directly under BOAC-AC. Both Companies during this period had a troubled life; MEA suffered heavily from the disturbed conditions in Lebanon and the Middle East generally and consequently made a substantial loss in 1958/59. A considerable proportion of the losses between 1957 and early 1959 arose from the airline's York freighter operation and the Yorks were withdrawn from service in March 1959.

1959/60 saw a marked improvement -and BOAC-AC received a share of a small profit. The Company was looking for re-equipment to improve its competitiveness - its fleet being made up of Viscounts and Dakotas - and in 1959 closed a contract for four Comet 4Cs for use in 1961 on which BOAC guaranteed payment of the instalments. But losses continued, and it was not until the new Chairman of BOAC was in office that BOAC disengaged itself from its commitment in MEA.

The story of MASCO is also an unhappy one. The purposes for which the Company was established were only partially fulfilled. It carried out engineering work for MEA but in doing so incurred massive expenditure in setting up adequate facilities; it never succeeded in acting as a hiring company for aircraft in the area nor in working in any supervisory capacity as had been originally intended. Consequently, it ran into financial difficulties and had to be rescued by BOAC, as otherwise, MEA would have been in serious trouble over its engineering work. Eventually, in 1960 it was arranged for MBA to do its maintenance work by taking over the engineering activities of MASCO.

In the Far East BOAC had two regional airlines, Malayan Airways and Hong Kong Airways. Malayan Airways was managed by the local British shipping agents, Mansfield & Company of Singapore. The Company was well run, BOAC had provided the General Manager of the airline and a 10% shareholding, and profits were made primarily during the political troubles in 1952 when surface transport became affected by communist guerilas. The Company used Dakotas on local operations, and a fleet of eleven aircraft had been built up by 1953. After the emergency was over, political developments were imminent, and Mansfield's felt reluctant to continue their management of the Company. BOAC took the view that the importance of the area astride the route to Australia called for a continuance of British control, and it was ready to increase its shareholding; otherwise, there would be a danger that control would pass to nominees of one of BOAC's trunk

Cathay Pacific's DC-3 VR-HDJ. *(via Sir Adrian Swire/Cathay Pacific)*

line competitors. The local Governments of Singapore, the Federation of Malaya, British North Borneo and Sarawak were all prepared to renew the existing agreement which was due to expire on 30 April 1947 for a further ten years provided Malayan Airways became a public company with limited liability and registered in one of the territories, and contributed 49% of the shareholding was offered to them. Accordingly, in April 1956 BOAC approved an investment of not exceeding £625,000 in the reorganised company. Malayan Airways was about to enter a new phase of expansion.

The second Far East company, Hong Kong Airways (HKA), was set up by BOAC in March 1947 as a wholly owned subsidiary, beginning operations in December to Shanghai and in January 1948 to Canton using Dakotas chartered from BOAC. Although traffic was good especially to and from Canton, the depreciating Chinese dollar had effects on the revenue collected in China and there was a loss of nearly £45,000 in the first full year ending March 1948; little improvement occurred in the following year when a service to Manila was introduced and a Plymouth flying-boat was time chartered from BOAC on the Hong Kong-Shanghai route.

With the future outlook in China uncertain due to the advance of the communist troops, it was essential that the Company should obtain new routes, but there was a local complication. Another British airline, Cathay Pacific Airlines (CPA), was also based on Hong Kong, and the Hong Kong Government delayed in making up its mind as to how routes should be allocated. It had been expected that HKA would be given routes North of Hong Kong and CPA routes South. Discussions to see what solution could be found were held in Hong Kong by Thornton and McCrindle with Jardine Matheson & Co., the powerful managing agents in Hong Kong who were already acting as BOAC's general agents. Eventually by the end of 1949 agreement had been reached between BOAC and Jardines under which BOAC agreed to sell its interest in Hong Kong Airways to Jardines, together with the three BOAC Dakotas; Jardines would continue to act as general agents for BOAC in the area.

With the fall of Canton to the Communists, the airline's services were suspended, and all that was left to the Company were minimal operations, including service from Hong Kong to Taipeh. By the end of 1950, the Company decided to dispose of its remaining aircraft and to charter as needed from Cathay Pacific. The British Government was concerned to see a strong regional airline in the Far East and began in 1951 to urge BOAC to participate in a such a venture jointly with Jardine Matheson and Butterfield & Swire - who managed CPA. The idea was that BOAC should have a 51% holding and would contribute spare Argonaut capacity. Negotiations dragged on but eventually in 1954 BOAC declined to become involved on the grounds that the proposed financial arrangements were unsatisfactory. This was not the end of the matter; the UK Government continued to press the importance of the issue, and by December 1954 BOAC had come round to the idea of participation provided Jardines and Swires also took part and provided the new Company's services did not extend beyond Calcutta westbound and Tokyo northbound, and BOAC was not called on to manage. Discussions continued with Swires, but they were reluctant to commit CPA to any new, untried company particularly in view of the lack of growth of HKA under Jardines' management.

Such slow progress was made that in September 1955 BOAC agreed to co-operate with Jardines in developing the northern routes from Hong Kong as a first step towards creating a more widely based regional company which Swire's might subsequently join. By October 1955 BOAC had reached an agreement with Jardines to provide Viscounts for HKA and to subscribe for some shares in the Company. The outcome in January 1956 was that BOAC signed up with Jardines to hold 50% of the shares and to invest a sum of £976,500 including the cost of the two Viscount aircraft.

Fortunately the story of HKA took a different turn in that after losses of over £167,000 in 1957/58 and £105,000 in 1958/59, BOAC-AC made arrangements with Cathay Pacific to take over the management of HKA and subsequently by July 1959 to acquire its entire capital; and so at long last the British Government's desire for one powerful British regional airline based in Hong Kong was achieved. BOAC-AC acquired as a result a 15% shareholding in Cathay Pacific which subsequently proved itself to be a very worthwhile investment.

The third region where BOAC was committed to local airline growth was in the West Atlantic and Caribbean islands, and the two airline companies concerned, British West Indian Airlines (BWIA) and Bahamas Airways Ltd. (BAL), both were acquired as a result of the merger between BSAA and BOAC in 1949.

BWIA then became a wholly owned subsidiary of BOAC flying inter-island services within the Caribbean and also between Kingston and Nassau and Kingston and Miami with Vickers Viking and Lockheed Lodestar aircraft. The first year, 1949/50, in BOAC's control was disappointing with a loss of £218,425. Whitney Straight and Booth were already on the BWIA Board, and at the Company's Board meeting on 24 August 1950 Smallpeice was also elected a Director as reported to the Annual General Meeting on 15 September. This had been recommended at the main BOAC Board on 24 July

The Cathay Pacific office at the Penninsula Hotel, Kowloon around 1955. *(via Sir Adrian Swire/Cathay Pacific)*

1950 which also decided that Smallpeice as BOAC's Financial Controller would assume the responsibility for co-ordinating BWIA's financial structure.

Results for 1950/51 were even worse, a loss of £267,000. It was clear by October 1950 that the year would show poor returns and in November Sir Miles visited Trinidad to see for himself what the prospects were. In his view, there was an urgent need to overhaul BWIA's route planning and commercial activities, and a shake-up occurred at the top level. Emphasis was to be put on trading profitably, and if politically motivated services had to be operated, these should be subsidised by local Governments. BOAC had agreed to use its Exchequer grant to meet the BWIA's losses and any deficit still left uncovered would be achieved by interest-bearing loans from BOAC.

At the beginning of 1951 Sir Miles asked McCrindle for an appreciation of the value of the Corporation's investments in its subsidiary and associated enterprises, and it was apparent the Chairman was becoming uneasy at the way things were going, for as he wrote at the time *'the amount of time and effort that senior executives of the Corporation have to apply to these subsidiary enterprises is a great burden and is apt to take our eye off the central target'*. The McCrindle report took the same view and in respect of BWIA considered that as inter-island communications were a public requirement, then the islands should finance them; if that proved impossible, then the UK Government should step in and BOAC should do all possible to get local interests and Governments to take financial stakes in the Company.

McCrindle concluded his report by stating a view which had some support within BOAC. He felt that the policy of having wholly owned subsidiaries with a majority of local directors, as was the case with BWIA, was wrong for this resulted in BOAC's Financial Comptroller being diverted from his main task of reducing the Corporation's deficit by reason of having to attend Board meetings of a small subsidiary in the West Indies, and that BOAC's business was the operation of worldwide trunk routes.

The outcome of the McCrindle report was a decision that BOAC should exercise more direct control of wholly owned subsidiaries and paradoxically with more representation on the local Boards, for this would mean more time spent by top-level BOAC executives on subsidiary company affairs.

With the changes in Management and greater concentration on selling its services, BWIA was able to finish the year 1951/52 with a considerable reduction in its loss, at £94,102, but there was an urgent need for re-equipment, and the decision was taken to order Viscounts as the fleet replacement. By December 1952 an offer of three Viscounts had been made by Vickers for delivery in March, April and May 1955 and this had been accepted, but meant that BWIA would have to soldier on for another two years or more with the existing Vikings and DC3s. In spite of this, the Company's results showed a steady improvement, with a reduction in loss each year until by the end of the financial year 1954/55 the figure was down to £11,931. By March 1955 BOAC's outstanding advances to BWIA totalled £435,298, and by that date, in the following year, the figure had risen to £1,862,500. These large amounts were linked with the Viscount re-equipment programme. It was planned that these aircraft would take over the denser traffic routes of BWIA itself and would operate on behalf of BOAC between Nassau and Miami, New York and Jamaica and New York and Bermuda. Some delay occurred in Viscount deliveries but the first aircraft was flying on service in October 1955, and by December regular operations began between Trinidad and San Juan. By March 1956 four Viscounts had been delivered to the Company and on a visit at that time Sir Miles had been impressed by the way the Viscount operations were going and said of the Company *'its future was promising'*.

Results for 1955/56 showed an increase in losses to £306,959. This was expected following the introduction of the large Viscounts and until traffic caught up with the greater capacity offered, but it was not long before doubts arose about the competitive strength and range of the Viscounts on the routes from the US to the West Atlantic and the Caribbean.

Early in 1957, the survey team led by Bampfylde and including a planner from BEA made their report on ways to improve aircraft utilisation and fleet deployment over BWIA's network. The plan called for an eventual fleet of eight Viscounts and four Dakotas, and an estimate of financial results when these aircraft were in full operation showed that apart from the routes then operated by BWIA as charters to BOAC on New York - Bermuda - New York - Nassau - Jamaica and Miami - Nassau, the remaining Viscount and Dakota routes could be flown at a profit. But the losses on the chartered routes would amount to over £350,000 in a year so that the allocation of routes as between BWIA and BOAC was of considerable financial importance. The decision taken was that the arrangements on the three routes flown by BWIA on behalf of BOAC should continue and the interim plan put forward by the survey team should be the basis of BWIA's operations in the immediate future. But further doubts were expressed both in BWIA and BOAC about the future competitiveness of Viscounts for the long distance services between the Caribbean and New York.

Up to 1958, BWIA had been operating its Viscounts on charter services for BOAC between New York-Bermuda, New York-Jamaica and Miami-Nassau, the Viscounts being provided by BOAC on the lease. Increasing concern in BOAC about the ability of the Viscount to compete with the large American jets on these routes led BOAC in April 1958 to discontinue these charter arrangements and operate its aircraft instead, and the Viscounts were withdrawn. The backing organisation for these Viscount operations was not then required, but this left BWIA with underused workforce leading to increased losses. In July 1958 the Board of BOAC-AC arranged for Air Commodore Powell to produce a report on BWIA, and he submitted this to the

Board in November 1958. It called for increased sales efforts, greater flexibility in the use of the fleet and a reorganisation of top Management. Although the BOAC-AC Board had intended that Air Commodore Powell should become the Chief Executive of BWIA, this was not welcomed locally, and in the outcome, he was appointed to the BWIA Board as the BOAC representative, but events showed that his influence locally was somewhat confined. BWIA's losses mounted, reaching £592,444 in 1958/59 and £618,335 in 1959/60, arising from a fleet of five Viscounts and three Dakotas. Such losses could not continue to be borne by BOAC. Attempts were made to get local and outside investment, but these failed including an attempt to interest TCA. Increasing concern was expressed in government circles especially in the Federal Government in the West Indies, and in July 1960 it appointed Sir Frederick Tymms to look into the whole question of air communications in the West Indies.

BOAC itself now realised it could not overcome the political and managerial problems inherent in an airline operating in the West Indies, and sought to disengage itself from the commitment but protecting as best it could the main interests of BOAC in developing traffic for its long-haul operations. By 1961 the Federal Government in the West Indies wished to acquire BWIA as the Federal airline on the approach of independence, but it was eventually the Trinidad and Tobago Government which took over in 1961 the total shareholding except for a 10% holding which BOAC-AC retained. As against net assets at book value of just over £1m, BOAC-AC received about half that amount from the sale.

As in the case of BWIA so with BAL, BOAC attempted to secure local participation in the airline but without success. Losses continued to rise, reaching over £112,000 in 1958/59. In that year BOAC through BOAC-AC decided on a new policy of securing experienced short haul Management for the airline and arrangements were completed with Skyways (Bahamas) Ltd as from April 1959 for them to acquire 80% of the shares and to take over the Management, BOAC-AC retaining a 20% holding. At the BOAC Board meeting of 11 September, 1958 confidence was expressed at the outcome of the plan which according to the then Minute *'fully protected the Corporation's position and provided that in certain circumstances we could re-establish our position in the Company if it would be in our wider interests to do so. This would not interfere with the network of services operated by the Company within the Bahamas and possibly to Florida but might involve our taking over services operated by the Company outside*

The Airline of the Caribbean' - British West Indian Airways, as represented by their Vicker Viscount 5Y-TBT

this network. The result of the Agreement would be that we should substantially reduce our losses and the Company would have a better prospect of improving its financial position with the removal of competition from Skyways (Bahamas) Ltd'.

The second company BOAC inherited from BSAA was Bahamas Airways (BAL) operating local services between the Bahamas Islands and also to the coast of Florida. In 1949/50, the first year of BOAC's ownership, this airline incurred a loss of £17,549 on its amphibian operations and also on its Dakota flying between Grand Bahama and Miami and West Palm Beach. These operations continued throughout the following year at an increased loss of £28,228. BOAC found itself in a difficult situation between its desire to reduce the losses by curtailing international operations and by fare increases and the need to satisfy strong local interests, who were concerned to promote the Bahamas as a major tourist attraction for the visitors from the United States. Moreover, BOAC could see that Nassau could grow in importance as a transit station for services not only into the Caribbean but also to Central and South America.

In February 1952 Sir Miles visited Nassau to see what could be done to bring the airline into profit. Already it was evident that failures by the end of March 1952 would exceed £100,000 which was out of all proportion to the size of the undertaking. On his return, Sir Miles recommended to the Board that BAL should be confined to inter-island services and that the international services should be taken over by BOAC. Following the reorganisation of the route pattern, together with certain Management changes, the result for 1952/53 was a reduced loss of £14,452 as against £101,658 for 1951/52.

In 1953 the Company was able to standardise its small fleet of Grumman Goose aircraft, but although a profit had been budgeted, the final result for 1953/54 was another loss, this time of £19,050 although by then the Bahamas Legislature had approved an increase in the subsidy of £10,000 per annum which was subsequently increased by a further £4,000 per annum.

Re-equipping became a problem largely due to the lack of landing strips in the outer islands. It was proposed that the Company should be supplied with DH Herons, but this could not be done until the Bahamian Government had constructed adequate landing grounds; meanwhile, the Company had to persevere with its four Grumman Goose amphibians. There was no improvement in financial results for 1954/55, and it was late in 1955 before the Herons were delivered and operations with them commenced; unfortunately, they could not be put to full use as the promised landing strips had still not been constructed. So the Goose aircraft had to continue with inevitably rise in costs and a loss for the year 1955/56 at £46,127, more than double that in the previous year.

At the close of Sir Miles' Chairmanship BOAC top Management remained convinced of the political and commercial value of its subsidiaries and of their eventual profitability, in spite of a huge total loss of over £400,000 in 1955/56.

The Corporation was also considering extending its interests in Associates, for in January 1957 it took a favourable view of giving financial help to Turk Hava Yollari (THY) following an indication from the MCA that the airline was desirous of purchasing British Viscounts and required technical and financial assistance. Obviously in taking this view BOAC was influenced by the need to have the Turkish authorities sympathetic to any re-routeings of Corporation services through Istanbul or Ankara, a highly desirable facility in the light of the disturbed Middle East situation. Later in 1957 BOAC agreed to provide five Viscounts against which THY would issue debentures for £1.5m bearing interest; additionally BOAC would acquire an equity holding of £0.5m to complete the Viscount purchase. In the same month a decision was taken to invest up to £59,500 in a Borneo Airways Company which the Government of British North Borneo intended to set up to operate local services in the Borneo territories. Although not referred to in the case put to the BOAC Board for the investment, it was undoubtedly the attraction of securing feeder traffic from a prosperous area with a sound future based on oil revenue which led the Corporation to become financially involved.

On 22 May 1964, Granville as Chairman of BOAC-AC presented the results of the Company to the BOAC Board for the last time before his appointment as Deputy Chairman of BOAC. He told them that BOAC-AC had made a profit in 1963/64 for the first time in its history, the tidy sum of £158,211 after tax and interest, a great stride forward from the loss of £654,806 in the previous year. Profits continued to be made in each of the next five years. When, early in 1966, the BOAC Board asked for. a review of Associated Companies including an assessment of the value to BOAC of its involvement in each of the separate companies concerned, Lee set out the reasons why BOAC had invested in each company. These were based not only on the benefit of a cash return on capital funded but also on the intangible benefits flowing from strategic or tactical considerations, such as securing a favourable traffic rights position for BOAC, using surplus BOAC aircraft capacity, forestalling the setting up of rival services in the area by BOAC's competitors and so on.

BOAC-AC was supervising BOAC's investments in twelve associated airline companies, including BCL's investment in Bahamas Airways and BWIA, but changes inevitably occurred. In April 1965 Borneo Airways had been absorbed by Malaysian Airways, and in 1966 the name of Malaysian Airways was altered to Malaysian-Singapore Airlines, each Government having a 29.5% shareholding to give local interests the majority holding. In 1966 EAAC repaid its loan of £250,000 and in consequence, BOAC-AC no longer had any financial benefit. Also in 1966 Air Jamaica, actually formed in 1963, began operations with BOAC-AC holding 33% of the ordinary shares. BOAC operated services on behalf of Air Jamaica until March 1969, the end of a three-year agreement. In 1967 Aden Airways, a 100%

subsidiary, was forced to cease operations by the severe political disturbances in the area and after losing a DC3 by an explosion in mid-air. Also in 1967, the Trinidad Government bought out BOAC-AC's remaining holding of 10% in BWIA; thus at long last BOAC was free of a commitment which had cost it dear over the years. On 1 June a new Company, Air Mauritius was formed with local interests, Air France, and BOAC-AC, BOAC-AC's shareholding being 27%.

Following the purchase by BOAC of Cunard's 30% holding in BOAC-Cunard Ltd in 1966, BAL again became a 100% subsidiary. There was no let up in its losses now borne again by BOAC in full. In 1967/68 the first full year after BOAC re-acquired a 100% interest, the damage was £382,000, and early in 1968, Lee was estimating the loss of 1968/69 at the considerable sum of £400,000.

In May 1968 Lee accompanied Adrian Swire, Deputy Chairman of John Swire & Sons Limited, London, and John Brembridge, Managing Director of CPA, visited the Bahamas in connection with the future ownership of Bahama Airways. Swires were contemplating the purchase from BOAC of a majority holding in the Company and meetings were held with the Bahamas Government. After negotiations, arrangements were completed in September 1968 for the sale of 85% of BOAC's interest to Swires, BOAC thus retaining a 15% holding and an obligation to continue its Subvention at decreasing amounts for three years.

Financial results of BAL under its new ownership continued to show heavy losses, and Swires finally decided that the Company should cease operations on 9 October 1970, and the Company was put into liquidation. Thus ended BOAC's long involvement with Bahama Airways Ltd going back to 1949. The Company had been a subsidiary of BSAA, and at the time of the BSAA/BOAC merger, it came under the control of BOAC. It had never been profitable, and it is difficult in retrospect to understand why BOAC persisted so long in paying its losses. In the 1966 review BOAC-AC had attempted the difficult task of putting an annual figure against incursion into the Bahamas by competing airlines, but this was inevitably a nebulous figure to set against the actual yearly losses in the more recent years of around £200,000 to £230,000, and in 1967/68, BOAC's last year of full control, a loss of £380,000.

In 1965 BOAC entered into an agreement with the catering firm of Trust Houses Forte (THF) under which a joint company was formed, Bofort Ltd., and incorporated on 1 October 1965 to carry out catering work for BOAC at Heathrow and elsewhere. It was also intended to use the company to develop hotel projects, a subject of concern to the Corporation. It was felt there was a pressing need for more hotel beds at some BOAC destinations and that lack of them would inhibit BOAC's efforts to promote tourist traffic and to fill its aircraft. Sir Giles initially took a cautious line; he made it clear he did not want to see BOAC saddled with losing enterprises and in February 1966 he laid down the principles '...which should guide BOAC and BOAC Directors on the Board of Bofort in establishing our policy towards hotels'. His main stipulations were that hotels should be built where if possible the local Government would assist by the provision of finance or the exemption from taxes, Where there was likely to be a year-round requirement to ensure profits, that every project must be shown to be likely to benefit BOAC and that the MoA must be advised about every project. By 1967 BOAC began to invest in new hotels, initially by supporting a decision of its subsidiary Gulf Aviation to build a hotel in Bahrein where BOAC had had a rest-house for many years. By March 1967 the Corporation was committed to a total investment of £435,000 covering the Panafric Hotel, Nairobi, Kenya Safari

Turk Hava Yollari Viscount TC-SEL rests between services. *(author's collection)*

Hotels, New Mauritius Hotels in addition to that in Bahrein. At the same time in conjunction with THF BOAC was engaged in leasing land in Guyana for hotel construction. A decision was taken to give a family name - Pegasus - to BOAC hotels where appropriate. By June 1967, with investments in hotel development building up, BOAC decided to set up a permanent organisation within BOAC-AC to administer its hotel interests and Craig was put on the BOAC-AC Board specifically to advise on hotel matters, and he became Chairman of the BOAC/Forte Hotel development committee. The first substantial commitment entered into under this new structure was the construction of a hotel in Ceylon with finance provided by BOAC-AC and by local interests. Other obligations followed, and by the end of 1968 a significant new project was underway in Kingston Jamaica) for completion in 1971. Also in June 1967, BOAC had agreed with a Pan Am subsidiary Inter Continental Hotels Corporation (IHC), to finance the construction of hotels in London, and a company was formed, BIH Ltd., to develop promising sites. IHC would provide the management of any such hotels. The first venture was the Portman Hotel which subsequently turned out to be a profitable investment. BIH Ltd endeavoured to develop a site at the head of Piccadilly as a first-class hotel, but after many planning problems BOAC decided that the greater need was for, the cheaper type of accommodation linked to tourist class travel; accordingly, it dropped out of this project and eventually terminated its arrangements generally with IHC. At the end of 1968 Lee as Chairman of BOAC—AC was reporting to the BOAC Board a proposal to form a European hotel corporation to build economy class hotels in the principal cities of Europe. A group of those interested was formed comprising Alitalia, BEA, BOAC, Lufthansa, Swissair, S. G. Warburg & Co., Deutsche Bank and Union Swisse, and a detailed study of the possibilities was commissioned. The concept was to provide 1,000 rooms in London and in Paris, 500 in Rome, 250-500 in Frankfurt and 300-350 in Munich.

BOAC remained responsible through BOAC-AC for the activities of its subsidiary companies only up to August 1972. As from September 1972 BOAC-AC came under the direct control of BAB with a change of name to British Airways Associated Companies Limited (BAAC). In due course, all the shares of the four principal subsidiaries, BAAC Ltd, BOAC Engine Overhaul Ltd, BOAC Restaurants Ltd and IAL were transferred to BAB, together with the College of Air Training (Properties) Ltd, Aeronautical Radio of Thailand Ltd, CITEL (NV) and Citel (SA). Shares in the Airways Housing Trust were transferred to BAAC. Whilst still under BOAC control, BOAC-AC was profitable in both 1969/70and 1971/72 and paid dividends to its parent company. In the last full year, 1971/72 although able to meet BOAC's interest charges on loans, the composite financial results of companies in which BOAC-AC had investments recorded a loss for the year. There was little change in the activities on the airline side. Air Jamaica was reconstituted in 1969 and AC ceased to have any interest. Bahamas Airways went into liquidation in October 1970, Malaysian- Singapore Airlines split into two separate companies, and AC shareholding was disposed of Fiji Airways was renamed Air Pacific with its wider ambitions in that area, while Gulf Aviation became the national airline of the newly independent territories of Abu Dhabi, Bahrain and Qatar, together with Oman, and AC's financial interest was on the way out as a result. AC's developments were mainly on the hotel side of the business. In 1969 the first AC hotel was opened at-Georgetown, Guyana and the generic name 'Pegasus'adopted for hotels where AC had a considerable investment and where it was appropriate. In the same year the Gulf Hotel, Bahrain opened its doors to business. Construction was begun on two new hotels, one near Colombo on a beach site and the other in Kingston, Jamaica. Within Europe, hotel development was in the hands of the European Hotel Corporation in which BOAC held a 16.7% interest.

The decision to form this company had matured in 1969 and agreement reached between the airlines and merchant banks who were financing it. The Corporation's objective was to construct moderately priced hotels in central European centres. Its first two ventures were in London and Paris, the London hotel, named the London Penta, was opened in 1973 at a site in Cromwell Road close to BEA's town terminal. In 1970 the BOAC Board authorised AC to invest in a Hong Kong hotel project being developed by Jardine Matheson, and this eventually became the Excelsior Hotel. A small investment during this period was made in hotels in Mauritius. At the time when BAB became responsible for AC, BOAC's investments in hotels represented by ordinary and preference shares amounted to about £2.8m while loans granted came to £600,000, and some new projects were under study. AC had had a chequered career but for the last nine years or so under the chairmanship of Gilbert Lee with John Linstead as General Manager and later Managing Director, it achieved an annual surplus with only one exception, a far cry from the massive losses of the late fifties and early sixties.

'Speedbird' and the Mail

No book on BOAC would be complete without paying due tribute to the famous emblem that adorned the nose and tails of many of the fleet for over fifty years.

The Speedbird is the stylised emblem of a bird in flight designed in 1932 by Theyre Lee-Elliott as the corporate logo for Imperial Airways. It was initially used on advertising posters and luggage labels. Later, it was applied to the nose section of the company's aircraft.

Theyre Lee-Elliott was a graphic artist and painter working in London in the 1930s, and the Speedbird is among his best-known works. The minimalist, stylish and avant-garde design representing a bird in flight is instantly recognisable and has a timelessly modern appeal.

Lee-Elliott was influenced by the avant-garde work of Edward McKnight Kauffer, and the logo echoes

Kauffer's angular bird forms in his 1918 poster for the *Daily Herald*. Other notable works of his include posters for the London Underground and the Airmail logo. Many of his paintings and original artworks are in the collection of the Victoria & Albert Museum.

With the creation of BOAC in 1939, the logo was retained, continuing to appear on the noses of aircraft throughout World War Two despite the military camouflage that had replaced the airline livery.

From 1950 BOAC gave the Speedbird greater prominence on the aircraft using it on the tail fin, either in navy blue on a white background or vice versa, and also using it widely elsewhere, such as on airport buses. It became a design classic and was used by the airline and its successors – British Overseas Airways Corporation and British Airways – for 52 years.

With the advent of air traffic control and the adoption of call signs to identify aircraft and their operators, BOAC chose the name of their now well-known logo, 'Speedbird', as their call sign.

In the mid-1960s the design of the Speedbird was slightly altered, with a slimmer 'body' and larger 'wing', and on the tailfin coloured gold on a navy blue background. Elsewhere the colours used for it were mostly a combination of cyan and white.

In 1974, BOAC was merged with British European Airways and others to form British Airways. The speedbird logo was retained unaltered but returned to the nose section of the aircraft. A prominent Union Flag design now occupied the fin. The Speedbird survived for another ten years, finally being retired in 1984.

As British Airways prepared for privatisation, a new corporate look was adopted in 1984. Referred to as the Speedwing, the red flash on the lower dark blue part of the fuselage bore a slight resemblance to the original 1930s design. 'Speedbird' continues to be used by British Airways as the ICAO callsign for its main international services - however, on its domestic services, it uses the callsign 'Shuttle'. There still remains evidence of the design 'on the ground' as it were; Hatton Cross Underground station that was opened in 1975 still has 'Speedbird' decorating its platform pillars. Speedbird House was an office block at Heathrow Airport. It was originally the headquarters of BOAC and, until the company moved, of British Airways. There is also a Speedbird Way near the Colnbrook Bypass, close to Heathrow Airport.

Officially the Empire Mail Scheme (EMS) came to an end on 1 April 1947 when the carriage of all-up mail without surcharge was formally classed to have ceased. There were ongoing repercussions of the Scheme in that for some time the Dominions, Colonies and special category states such as Egypt, Anglo-Egyptian Sudan, Jordan and Burma continued to pay preferential mail rates to BOAC.

From then on, mail was classified into three main categories, letters carried by the national airline

Above: Theyre Lee-Elliott poses before projected versions of his 'Speedbird' and 'Air Mail' logos.

Left: his designs no *(author's collection)*t only got used on aircraft; they were also applied to posters and stamps.

originating in their own country, ex-participant mail, that is from those countries which formerly participated in the EMS and international mail such as mail carried by BOAC out of a foreign country. The British Post Office was responsible ultimately for deciding the carriage rate for the first two categories, and the Universal Postal Union (UPU) was responsible for the third category, i.e. mail of foreign origin.

BOAC had been concerned by the somewhat arbitrary attitude by the Post Office in deciding what it would pay for carriage and also at the difficulty of getting the BOAC view across as all the negotiations were conducted with the Post Office by the MCA. However, by 1949 the Ministry had agreed that in future BOAC would negotiate directly with the Post Office. Also in that year rates were approved with the Post Office for second-class and parcel mails at a carriage rate about one-third that for first-class letters.

Mail rates were quoted internationally by so many gold francs per tonne-kilometre, and for convenience, the same measure was used for the other two categories; so far as the Post Office was concerned, it was not prepared to consider any subsidy element in the mail payment nor did BOAC seek it.

As from 1 July 1948 on international first-class mail a rate of 6 gold francs became effective other than for intra-European mails as compared with a mere 2.43 gold francs paid to BOAC by the Post Office for the UK originating letters carried on Commonwealth routes; whilst this low rate was raised in stages to 3.99 gold francs by 1951, and no corresponding increases were made in the rate for ex-participating countries. It was 1953 before the Government gave BOAC permission to raise the rate to these former EMS participants from 2.43 to 4 gold francs.

The reason why four gold francs was chosen arose from the fact that after long and difficult negotiations with the UPU, at a plenary session of the latter in Brussels in July 1952 it had fixed the new rates for international first-class mails at three gold francs intra-Europe and four gold francs for extra-Europe, these rates to apply for five years. Undoubtedly the UPU decision was influenced by top-level discussions with the IATA Executive Committee, for there had been a real risk the UPU might have accepted proposals from certain of its members for a so-called 'unique' rate of three gold francs or even lower, whereas BOAC and the other IATA long-haul carriers would -have hoped for something nearer the former rate of six gold francs.

Early in 1952, the Chairman gave executive responsibility for the formulation of the Corporation's mail rate policy and the handling of all negotiations with the Ministry and the Post Office to the Financial Comptroller, but negotiations with IATA and with the Universal Postal Union were to remain the province of McCrindle.

BOAC was trying to persuade the MCA that rate-fixing should not be left within the narrow confines of the GPO, but that wider consideration was involved requiring high-level Government concern. It was the BOAC view that the principle of 'what the traffic will bear' should be applied. Mail could and should pay more and thus enable lower fares to be offered to passenger traffic which was elastic in its attitude to the cost of travel. The GPO was concerned to get the cheapest possible carriage for mail and were reluctant to put up mail charges to the posting public. But these ideas of BOAC did not produce any concrete results.

Some stability then came into the rating scene in that the UPU having fixed the four gold franc rate for five years, and the Post Office early in 1954 agreed to renew that rate on UK mails for at least a further two years.

Mail revenue was of immense importance to the airlines and particularly to BOAC with its routes

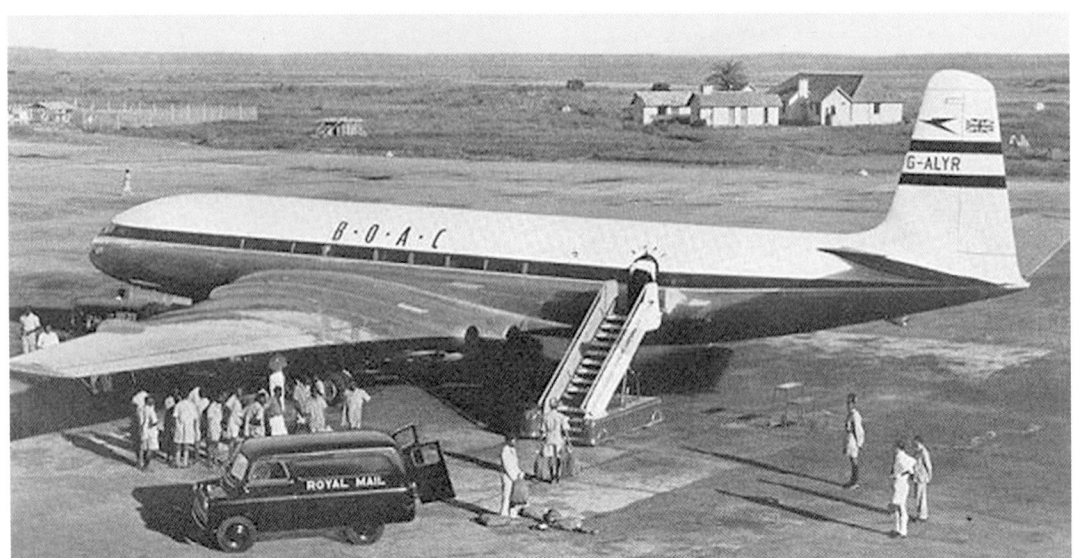

The Royal Mail comes aboard BOAC Comet G-ALYR 'somewhere in the tropics'. *(De Havilland)*

BOAC converted two of their DC-7Cs into dedicated freighters.

touching countries where large numbers of expatriates and persons of British stock were living, particularly North America, Australasia and South Africa. Although the proportion of BOAC's total revenue made up by mail declined over these years, in 1946/47 it was 30%, in 1950/51 25.5% and by 1955/56 down to 22.3%, the gross revenue received from mail carriage went up nearly threefold, from £3.4m in 1946/47 to £9.5m in 1955/56.

Cargo by air was becoming more and more important, as David Wilkinson, a Press Officer for BOAC explained in 1960. *'Fifty years ago, a letter sent from Britain to America took two weeks to arrive; to read an up-to-date London newspaper in Nairobi was impossible; and people laughed if anyone said ' ...and pigs can fly'.*

Today a trans-Atlantic letter takes little more than two days from door-to-door; London newspapers can be bought in almost any big city in the world a day or two after publication, and pigs do fly - regularly! The reason for this revolution: air-cargo.

The very first British international commercial flight, from Hounslow to Paris in 1919, carried one passenger, some grouse, post office mail and some Devonshire cream. Ever since then, mail and cargo have been flown regularly in passenger airliners.

In the early days, sacks of letters and parcels were easier to handle than other types of cargo, so airmail developed faster as a regular business. The post office were quick to see the advantages of this new form of transport - especially in linking together the countries of the Commonwealth - and they lost no time in setting up an airmail service. Today, one of BOAC's most important duties is the carrying of mail on scheduled routes throughout the world. Every jetliner in the Speedbird fleet bears the Royal Mail emblem on its fuselage.

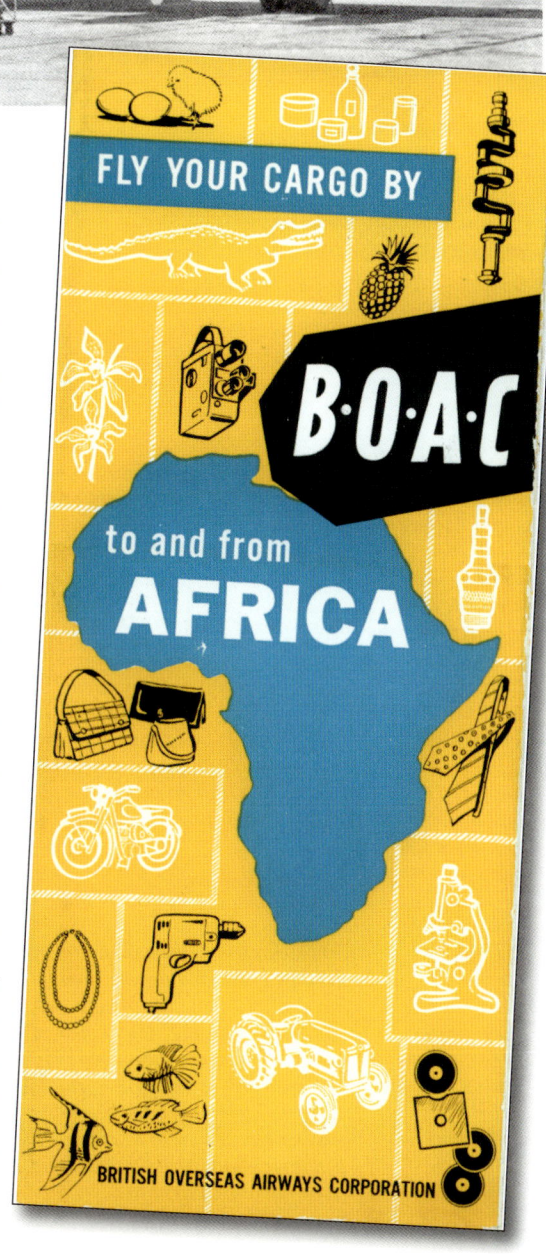

Above: Cargo is loaded aboard a dedicated DC-7C freighter of BOAC at London Heathrow.

Right: One of two hundred and sixty chinchillas - worth about $12,000 in 1960s money - which was flown from Rochester, USA to England.

Each week some 80 tons of letters and parcels are delivered by GPO vans to BOAC's Mail Section at London Airport. In the post office headquarters at Mount Pleasant the mail has already been sorted, according to destination, into bags, each weighing about 44 pounds. The staff of the Mail Section check the bags and make sure that they will be put in the hold in the right order, so that when the aircraft lands at stopping-places along the route, the mail for each area can be ofl-loaded easily and quickly.

When mail for destinations beyond Britain arrives at London Airport on an incoming flight, it is speedily transferred to another service so that no time is lost. In all, about 35 tons of mail comes into London on BOAC flights each week.

By far the busiest season of the year for the Mail Section is Christmas. During the peak period it handles nearly 300 tons of cards, letters and parcels.

There was a time, not so long ago, when aeroplanes were looked upon with scepticism, and people thought that all 'aviators' were mad. So they hesitated about entrusting their valuable goods to the newly-formed airlines. But aeroplanes developed rapidly from 1919 onwards. The public realised that the airline business was here to stay. Adventurous exporters began sending small shipments of goods by air, and were delighted (if a little astonished!) when their goods arrived safely and far more quickly than they had dared to hope.

The huge Empire flying boats of Imperial Airways, with their capacious holds, were ideal for cargo, and the airborne flow of small, valuable or urgently-needed goods between the countries of the Commonwealth began in earnest in the 1930s.

When machines, weapons and supplies had to be shifted speedily in the Second World War it was to the newly-formed BOAC that Britain looked. After the war, this experience was put to good use when Constellations, Argonauts and Stratocruisers began to carry passengers, mail and cargo.

In 1951 BOAC carried 7,000 tons of cargo. In 1960 the figure had risen to 13,000 tons! Today at London

Airport there are specially- built warehouses, trucks, vans, trolleys and fork-lifts to aid the teams of highly-skilled staff who deal with the huge daily flow of freight in and out of London. To make things easier for the shipper, a team of men and women clerks operate a round-the-clock reservations and advice service. The shipper simply dials 'SKYport 5511' and asks for 'Skyload'. Operating from twenty-line switchboards, the cargo reservations unit answers - and within minutes a booking has been made.

The goods are collected by a cargo van and taken to the export warehouse. Here the documents, weights and measurements are checked. The consignments are then placed in the appropriate section, according to destination. A little while later they are loaded on to a lorry and driven out to the aircraft.

Sometimes the items are too big to be loaded normally. A motor car, for instance, requires an expertly handled fork—lift so that it can be loaded in exactly the right spot. Freight of this size, which is too large for the belly-hold of a passenger aircraft, is carried in specially adapted freighter aircraft.

If you visited a BOAC cargo warehouse at London Airport, you would see endless stacks of parcels, packets, boxes, cases, crates, bottles and sacks — destined for the four corners of the earth. Listen for a moment and you might hear the chirpings of several hundred day-old chicks. Warmly housed in light and airy cardboard containers, they are usually bound for Africa or the Middle East.

Pull aside the heavy, insulated wrappings of a 'fish

Left: a press picture of a publicity-minded chimpanzee posing with a BOAC stewardess before setting off to star in a 'Tarzan' movie.

Below: With a shipping label tied carefully to it's mane, this valuable English-bred racehorse embarks on a BOAC DC-7 frieghter bound to Toronto, Canada.

It did not matter if it was a ton of Aberdeen Angus walked up the ramp to be loaded through the rear door...

or an MG Miget loaded by fork lift through the front - all was valuable revenue earning freight for BOAC.

muff' and you would see a world of brightly-coloured fish swimming about in plastic bags filled with oxygenated water.

When fish were first sent by air they usually travelled in glass-tanks or jars, which were heavy, costly and breakable. So plastic bags were tried, and proved to be highly successful. Today several bags at a time can be sealed and wrapped in one fish muff. Packed in this way the fish stay warm and safe for many hours. (Incidentally, BOAC carries more tropical fish than human passengers!)

Fish are the least obtrusive of the creatures which form part of the airline's live cargo: even in their muffs they weigh very little and take up only a small amount of space. There are other animals, however, which weigh as much as a dozen passengers and which need a whole compartment to themselves. During the past few years BOAC has flown bulls, elephants, tigers, lions, zebras and even rhinoceros - all of which are very hefty cargo indeed.

Hundreds of monkeys are flown every year between India and other parts of the world... Snakes, crocodiles and alligators from time to time leave their jungles and swamps behind and take quite happily to the air. Chimpanzees, chinchillas and Cheetahs fly, as well as chiffchaffs. In fact, about the only exotic animal which cannot fly, nor is ever likely to, is the giraffe!

As well as helping to keep the zoos re-stocked, air cargo is proving very valuable to breeders of livestock in different parts of the world. Instead of having to send their costly pedigree animals on long sea journeys, exporters can now despatch them by air and know they will arrive in a matter of hours instead of weeks.

Race-horses are usually very highly-strung animals, and it was once thought that to send them on an air-journey would be asking for trouble - until someone tried it and found that the animal seemed happier in the air than when rattling along the ground in a horse-box. Nowadays British bloodstock is regularly flown to *destinations as far away as Japan.*

As well as livestock and bloodstock, Britain breeds the finest dogs in the world. They are a very valuable export, and every year hundreds of them travel in BOAC airliners on their way to new homes abroad. Each dog (which must be in perfect health and not too young to be parted from its mother) is carefully prepared for its journey. Complete with feeding instructions, collar and lead, it is taken to London Airport shortly before the flight is due to depart. Then it is settled comfortably in a specially tailored wooden kennel. At each stopping-place on the journey, the dog is taken from its kennel and exercised until it is time to go on board once more.

Nobody would take the trouble to send an animal, bird, fish or reptile on a globe-trotting trip unless it was of considerable value. Most of BOAC's live cargo is worth a great deal of money - sometimes many thousands of pounds - and the utmost care is taken to see that it arrives in perfect condition.

This same care is exercised with all types of cargo. It is not just a case of loading a mound of boxes, crates and parcels into the hold of an aircraft and leaving it to rattle about during the journey. Everything must be packed, labelled and stowed away securely.

The head loader must make sure that he maintains an exact balance in the aircraft, and that the full load meets the specified weight. The details of his work are recorded on a 'Load and Trim' sheet, eventually signed by the captain.

Almost every BOAC jetliner when it sets off on a journey has a variety of precious goods stowed in its hold. It may be precious minerals from Africa, or priceless antiques from England. It may be delicate flowers, such as orchids from Malaya; or luxurious food, such as caviare from Iran. It may be a consignment of rich brocade cheong-sams from Hong Kong or silken same: from India. Each needs its own special type of packing.

In some instances, experience has proved that the less packing used, the better. An aircraft's hold is so compact and secure that a delicate piece of machinery - which sent by other methods of transport would need heavy and expensive packaging - can be safely stowed away with only a dust cover to protect it.

A manufacturer of neon tubes was once worried by the number of tubes which broke inside the tough packing cases he provided. Then it was suggested that the tubes should merely be clipped to a light, strong piece of board, and placed on top of other cargo in the hold. The idea worked perfectly, and the maker was saved the expense of packing as well as the cost of replacing the broken tubes.

A familiar sentence on hotel menus throughout the modern world is '...specially flown from...' New Yorkers can breakfast on fresh Scotch haddock or Fleetwood kippers. Visitors to Barbados in the West Indies can feast on fresh Canadian salmon. Canadians can enjoy fresh Dublin Bay prawn. These are just some of the luxury foods which are flown regularly on BOAC and associate services.

Thus, in many ways, every hour of every day, from Manchester to Montreal, from London to Lima, aircraft of the Speedbird fleet are helping to open up the trade-routes of the air, and to bring the world to your very doorstep.

Relations with British Independent Airlines

The Air Corporations Act 1949 reserved the operation of scheduled services between the UK and foreign countries to the Corporations and their Associates. The term 'associate' is defined as being an undertaking which was associated with the Corporation under the conditions of any arrangement for the time being approved by the Minister as being an arrangement calculated to further the efficient discharge of the function of the Corporation.

Government policy to allow Independent airlines greater scope than the 1949 Civil Aviation Act permitted was implemented by invoking Section 12 of that Act, which confirmed a 1947 Order establishing an Air Transport Advisory Council (ATAC): under this the Minister had the power to refer to the ATAC any question *'relating to facilities for air transport by air in any part of the world'*. It was thus possible to use the ATAC as a British licensing authority even though this was not the original purpose of its creation.

The Minister, Lord Pakenham, after discussion in Parliament, issued a directive to the ATAC dated 26 September 1950. This directive, while confirming government policy as being that internal and external scheduled services should generally be undertaken by the Corporations, wished the associate agreements to be continued with some modifications. The Council was asked to consider and to make to the Minister recommendations on applications from Independent operators for scheduled services on particular routes. New facilities were not to overlap nor compete with existing and planned services of the Corporations and were to have a licence not exceeding five years' duration. Inter alia, the Minister, would review any recommendations in the light of existing bilateral arrangements, and the terms and conditions of employment of staff would have to be no less favourable than those negotiated through the NJC procedure for the staff of the Corporation.

The new Conservative administration, however, wished to give even greater scope to the Independents and to follow a Debate in the House of Lords in December 1951 the Minister on 30 July 1952 issued new terms of reference to the ATAC, replacing the 1950 directive. The Corporations' position was not to be undermined, and they would have the exclusive right to operate first and tourist class scheduled services on existing routes. The Corporations and the Independents alike would be free to apply to operate all new types of service, such as Inclusive Tours (ITs), and scheduled services over any new routes: the Independents as before would operate under Associate Agreement with the Corporations. The scheduled service and all-freight routes then operated by BOAC were set out and were reserved to it. Colonial Coach (CC) services between points in the UK and points in the Colonies were permitted even along BOAC's reserved routes, provided they catered for a lower class of service than usual scheduled services: this covered type of aircraft, stage lengths, baggage allowance, and passenger amenities as well as lower fares with the frequencies being related to the new type of traffic. Licences for seven to ten years were provided for. During Parliamentary Debates and in discussion with the Chairman, the Minister made clear that the ATAC would continue to make recommendations to him but that the final decision rested with the Minister.

The 1952 Terms of Reference governed the way in which the ATAC licences were issued over the last four years of Sir Miles' Chairmanship. Effectively, as the Minister stated in a letter to the Chairman in October 1954, they curtailed the Corporations' monopoly position under the 1949 Act. Whatever the likely future effect on the revenues of the Corporation any new licence's use by an Independent, BOAC would have to sign an Associate Agreement or in refusing to do so face a direction from the Minister and, eventually, amending legislation. This might well produce a worse final position so that although they came near to refusing to sign, the Chairman and the Board in the end always accepted that prudence dictated compliance.

The Minister and his successors sought to help the Independents by persuading shipping interests to invest in them, stressing, in particular, the great future in freight carriage by air: by leaning on BOAC to reach 'spontaneous' working arrangements with them; and by ministerial interpretation in their favour of the scope of services open to them and to the Corporation, mainly as to charters, long-term passenger contracts, trooping and the conditions under which Colonial Coach services were to be operated.

The question of trooping was a matter of considerable consequence during the early part of Sir Miles' Chairmanship. In May 1950 the Air Ministry called for tenders for the carriage of large numbers of troops between the UK and the Far East and the UK and the Middle East for a three-year period from early 1951. Such a contract would have provided employment for the Hermes for which no use could be found. A memorandum was sent to the MCA which pointed out that trooping lay within the powers of BOAC under the Act, that BOAC alone had the domestic set-up and route organisation to provide year-round high-frequency services as required and moreover that such an operation would provide BOAC with valuable experience in the field of low-cost operations. No contracts, however, were obtained by BOAC. Airwork obtained the London-Fayid contract and in November 1951 hired four BOAC Hermes with an option to purchase. Indeed a consequence of BOAC's not being awarded any trooping contracts was that the Chairman in September 1953 reported to the Board that no further Hermes could at the time be sold as BOAC would have to drop its price as it had no further use for the aircraft: in fact Hermes services were withdrawn the following month. Eventually in 1954, the then Minister ruled that BOAC could not tender for long-term trooping contracts on the ground that this would entail maintaining aircraft specifically for charters.

In July 1952, the Chairman was pressed by the Minister to give an undertaking that BOAC would not apply to the ATAC for new all-freight licences on any route: as a result, the ATAC granted an all-freight licence to Airwork over the Atlantic for ten years from May 1953. In reply to a query from the Chairman, the Minister said this was not an exclusive licence, but BOAC would have to justify to the ATAC the need for a second British operator on the route.

While reserving its right to seek such a licence itself, BOAC began talks with Airwork and Furness Withy, its shipping backer, for some means to co-operate together: this had the active support of the Minister. A factor in these talks was the grant in May 1953 of a scheduled service licence to Airwork for the London-Aden route. BOAC had been operating over this route for several years by linking its London-Cairo service to an Aden

Competition came from many directions, even within the UK including airlines such as Cambrian with their Viscounts such as G-AOYG seen here...

...and BKS - a name taken from the first letters of the founders surnames; James W. 'Jim' Barnby, T. D. 'Mike' Keegan, and Cyril J. Stevens.

Airways charter between Cairo and Aden: despite this BOAC's ATAC licence was made worthless by virtually restricting services to the on-carriage of traffic travelling from points behind London or going beyond Aden. The Minister when so advised asked the Chairman to reach a compromise with Airwork.

By the end of 1953, a broad understanding had been reached between the Chairman and Myles Wyatt, Airwork's Chairman. When this was reported to the Board in January 1954, Whitney Straight went on record that in his view the proposals represented a very formal surrender of BOAC's rights with possible future financial implications which the Board should only accept under a Ministerial direction. The understanding envisaged BOAC not applying for an Atlantic all-freight licence and co-operating with Airwork on the Atlantic by handing over to them such freight as it could not accommodate on its scheduled services and by undertaking not to limit passenger seats to make way for freight on the Atlantic and elsewhere should it arise. Airwork for its part would not operate to Aden and would not object to BOAC doing so. Airwork insisted that BOAC should cease to enter into long-term passenger contracts since these were vital to Independents: this was entirely unacceptable to BOAC, and so the talks broke off until the Ministerial policy could be established. An agreement was eventually drawn up and signed in June 1954 which endorsed the Atlantic/Aden arrangements but did not cover long-term passenger contracts.

In early 1954 the Board had become much exercised at developments since the 1952 policy had come into being: the timing was related in particular to two current issues — Colonial Coach services and the possibility that the Minister might take heed of Airwork's obsession over long-term passenger contracts. The Board accordingly decided to see the Minister en bloc - an unprecedented approach which underlined their severe misgivings as to the way policy was moving. The Minister agreed to meet the Board on 3 March 1954. In a prepared statement the Board expressed concern on specific issues:-

1. The way in which the Minister had forced talks towards 'spontaneous' working arrangements between BOAC and Airwork by means of pressure on BOAC on the Atlantic and the Aden licence position.
2. The way in which, despite considerable concessions proposed by BOAC, the discussions had broken down on Airwork's insistence that BOAC ceases to undertake long-term passenger contracts:

    BOAC was already earning £8m annually from existing contracts and there appeared to be virtually no type of passenger traffic which could not be made the subject of such arrangements by Independent companies, such as under the travel club or so-called 'affinity club' concepts.

3. Colonial Coach services had to be a lower class of service than those of BOAC: existing licences already permitted similar passenger amenities to those of BOAC and the same baggage allowance: Furthermore, applications were before the ATAC for Colonial Coach services to Singapore and Hong Kong using Hermes/Viscounts and clearly operating over stage lengths at least as long as those of BOAC.
4. The proposed extension of Colonial Coach services would inevitably affect the international fare structure by combination of cheap long-distance and local IATA fares.

In conclusion, the Board accepted that BOAC could maintain aircraft for open market charters, but long-term contracts were a different issue: undertook to negotiate constructively with Airwork:and stated that BOAC could not sign further associate agreements for Colonial Coach services unless the international fare structure was safeguarded and the services made genuinely third class. The result of the meeting was that the Minister ruled that the Corporation could not apply for long-term trooping contracts: that it could enter into long-term passenger contracts provided these could be regarded as being within its own scheduled service pattern: and that on existing Colonial Coach services baggage and passenger amenities standards should stand but would be reduced under any new licences: present type aircraft alone would be permitted.

In the event the ATAC did not grant the

Singapore/Hong Kong applications: thus since the 1950 policy had been introduced services had only been allowed to East, Central and West Africa by Airwork/Huntings and from Central Africa to London via East Africa for CAA.

BOAC had formed close arrangements with Skyways, the only significant independent carrier without shipping backing and for a time leased its aircraft to operate a freighter service to Singapore and for capacity into the Gulf area. In March 1954, proposals were formulated under which BOAC would acquire a 25% interest in Skyways and Skyways would buy eighteen Hermes from BOAC for trooping and charter work, undertaking not to apply for scheduled service licences without BOAC's agreement. This package was sent to the Minister and was approved, only to have it withdrawn an hour before the two airlines were to sign the agreement on the pretext that the sale price of the Hermes was too low: BOAC was able to demonstrate beforehand that the loss on the residual value of the aircraft was small and that no other offers were likely to be so high. The real reason for the Minister's volte-face appears to have been last-minute political pressure on behalf of shipping interests with investments in other independent companies.

In 1954 talks were also held with Hunting Clan, at which BOAC agreed to help its tramper services by handing over surplus freight to the Company on the African routes. Hunting Clan expressed its anxiety at the financial results of the Colonial Coach services, which it wished to convert to tourist services. In November 1955 Airwork advised BOAC that the first year of the Atlantic freight service had resulted in the loss of £0.5m and that a similar figure was expected for the next two years: their shipping partners were unwilling to support such losses and unless a way could be found to reduce them the services would have to be withdrawn. Simultaneously the Minister asked the Chairman to consider quick ways of helping Airwork, suggesting for instance that BOAC might transfer passengers who could not be carried on its scheduled passenger services to Airwork's freighter. He was also asked to consider for the longer term the possibility of sharing traffic on the Corporation's existing routes with selected Independents coupled with financial participation in such companies. The Board replied they could see no way further to help Airwork and that it was opposed in principle to sharing traffic with independent operators and thus reducing its earning powers: furthermore there were no financial resources available to invest in independent companies. A consequence of the withdrawal of Airwork's freight service was that discussions between BOAC and Furness Withy for a possible Airwork service between Bermuda and New York in partnership with BOAC had to be broken off.

Those were difficult years for BOAC which felt itself to be under constant attack by the Independents who had the ear of the Government during the Conservative Administration.

Chapter Seven

Enter the 707

On 22 June 1956 the Chairman's Aircraft Requirements Committee submitted a lengthy paper on the Corporation's future aircraft needs to a special meeting of the Board. This demonstrated that the Corporation had no aircraft on order which could counter the non-stop jet competition which was sure to start on the North Atlantic from the winter of 1959 with the deployment by Pan Am of Boeing 707-320 aircraft they would be receiving by August 1959. The traffic target set BOAC in the paper was 50% of the business into and out of the UK on services to the East and Middle West of the USA and Canada and with expected new routes to the West Coast of the USA. All this amounted to an estimated growth in BOAC's traffic of 28% compound per annum for the first five years while BOAC was making up for lost ground and 11% per annum thereafter, recognised as a formidable target at the time.

The paper to the Board considered the claims of some types, but the only possible British contender as a turbojet was the proposed Comet 5 with a likely delivery date in 1962, at least two years later than the American 707s and DC8s, and with a much higher direct cost. Several turbo-prop developments from the Bristol stable were considered, all of which were on paper only and none of which on information then available was competitive in speed or cost with the turbo-jets. In contrast to these paper aircraft, the Committee had been able to make a detailed assessment of the Boeing 707 and Douglas DC8 on the strength of manufacturers' detailed engineering specifications and performance guarantees, both of which showed speed and cost advantages over all other possible contenders. It was hoped that it would prove possible to equip either aircraft with Rolls-Royce Conway engines. As to numbers, the Committee assessed the requirement as thirteen aircraft in 1960, two in 1961 and two in 1962, a total order for seventeen with an assumption, taken on Government advice at the time, that neither Prestwick nor Manchester airports would be developed to take the large jet types.

The Board supported these proposals and agreed that the Corporation should send a mission to the West Coast, USA, for discussions with Boeings and Douglas to examine the technical and economic aspects of the aircraft in detail; moreover, each manufacturer was to be asked to grant options on the number of aircraft in question. An assessment of the total capital outlay involved was around £48.5m.

The Chairman wrote to the Minister on 22 June 1956 making it clear that the Corporation had already had discussions with De Havillands and Bristols and then set out all the arguments for an American purchase; he sought the Minister's support for the dollar commitment on the grounds that the competitive aircraft would enable the Corporation to increase its dollar earnings which otherwise would be considerably reduced.

Options were obtained as required from both Boeing and Douglas for seventeen aircraft and a BOAC team was despatched in July to visit Boeing at Seattle and Douglas at Santa Monica under the leadership of Campbell Orde, the Development Director.

Alan Colin Campbell-Orde was educated at Sherborne School, Sherborne, Dorset, England. He fought in the First World War, seeing active service in Belgium, where he was wounded. He gained the rank of Flight Sub-Lieutenant in the Royal Naval Air Service and the the rank of Flying Officer in the Royal Air Force. He was decorated with the Air Force Cross in 1919.

He was one of the original commercial pilots on London-Paris route with Aircraft Transport & Travel Ltd. He was Instructor and Advisor to the Chinese Government in 1921 at Peking, China. He became an Instructor and then Chief Test Pilot for W G Armstrong-Whitworth Aircraft Ltd between 1924 and 1936. During his time here he flew the maiden flights of several aircraft including the AW.15 Atalanta on 6 June 1932 and the A.W.38 Whitley Bomber on 17 March 1936 .

He was Operational Manager for British Airways and Imperial Airways between 1936 and 1940 and was Operational Director of BOAC between 1940 and 1943. He was on special duties for BOAC with Transport Command, Royal Air Force between 1943 and 1944. and then became Assistant to the Chairman of BOAC between 1944 and 1946. Between 1946 and 1957 he was the Development Director for BOAC. He became an aviation consultant in 1958.

On the return of the mission - which had found little to choose between the two types - it was decided to come down in favour of the Boeing fitted with Rolls-Royce Conway engines, given the type number 707-420, on the grounds of Boeing's more significant experience in the production of large jet projects, albeit almost all for the military.

It was a matter of immense importance to BOAC's

On taking delivery of their first 707, BOAC made much of the event, sending out a lavish brochure to travel agents. 'Jet Powered by Rolls-Royce. The BOAC Rolls-Royce 707, biggest and fastest of the world's jetliners, is something special. It is powered by Rolls-Royce Conway jet engines. Four of these great by-pass turbo-jets drive the 707 through the sky. And what a performance they give! They sweep you along at ten miles a minute, anything from five to eight miles up and that's high above any tricks the Atlantic weather may play! At take-off, they develop a total thrust of 70,000 pounds. With these superb engines and a fuel capacity of over 19,000 Imperial gallons, BOAC's Rolls-Royce 707 has a range, with full payload, of more than 4,500 miles - over a thousand miles farther than London-New York! Thrust reversers, one installed in the tail pipe of each engine, provide something like 50 thrust in the reverse direction - massive extra braking power from the Rolls-Royce engines when landing! Each engine also has a sound suppressor, so there's no jet roar to disturb you as you speed on your way. Flying in the Rolls-Royce 707 is supremely quiet and vibrationless!

Right: Handing over at Seattle. In front of the first BOAC 707 are, from left to right, Bill Carlyon and Tom Spalding (Boeing); Ivor Lusty (BOAC resident representative); Serge Gorney (Boeing); Charles Abell (BOAC chief engineer); J L Uncles (707 project engineer); Capt. H J 'Dexter' Field (BOAC chief technical pilot); and Tom Gillan (BOAC inspector).
(all DGR Photo Library)

future that Government should accept its arguments in favour of the American 707. The Corporation saw only too clearly that this new generation of aircraft with non-stop North Atlantic range and low operating costs would make possible a lower standard of fares than had previously seemed possible as early as 1960. BOAC thought that no British product could match these American aircraft and without them, the Corporation would be in an impossible competitive situation.

The Minister was informed of these views and discussions were held with the technical experts in the MoS who made their own comparisons between a developed Britannia, type 430, the Comet 5, the DC-8, and the Boeing 707-420, and came out with the surprising conclusion that in terms of payability the Britannia 430 headed the list followed by the DC-8, then the Comet 5 and last in the order the Boeing 707. BOAC could not accept these conclusions, which in any case were based on existing runway lengths - a quite modest runway extension would be enough to raise the 707 payloads to BOAC's estimate and give satisfactory economic results.

Over the years the very favourable terms, contractual provisions and safeguards which were put into place between Boeing and BOAC led to very strong suspicions that something was rotten in the state of Denmark - especially as these were far more so that could be expected from any British manufacturer. These suspicions were further strengthened when it was revealed that Boeing were prepared to 'buy back' fourteen of BOAC's Stratocruisers as part of the deal!

It was not long before a whole raft of new nicknames appeared - all versions of the BOAC theme, but slanted towards the American context: Boeing Only Aircraft Club, Bought On American Credit and Bring Over American Cash being just three examples.

It would seem that the MoS evaluations had paid too little attention to the difference in speed between the Britannia 430 and the Boeing 707, a difference of 100 miles per hour in favour of the Boeing, and this would certainly be reflected in achievable passenger load factors. All these points were set out in a paper to the Minister who was expected to give a decision by October 1956. To ease this decision, BOAC had indicated in its paper that it could need around twenty aircraft to cover future requirements on African and Eastern routes, and it was foreseen that the Minister in permitting the Boeing purchase might well require the Corporation to place an order with British manufacturers to meet that need.

The Government's approval to the purchase of Boeing 707-420 aircraft was announced in the House of Commons by the Minister on 24 October 1956. The permission was limited to fifteen aircraft, not the seventeen asked for by BOAC, and the Minister's statement included the words *'this purchase is therefore regarded, both by the Government and by BOAC, as an exceptional measure to bridge the gap until a new British type is produced'*. It also included reference to discussions proceeding between the Corporation and De Havilland's for a suitable British type for the Eastern hemisphere routes.

Negotiations with Boeing for a contract were then pursued and by November 1956 the main terms were already outlined. Deliveries of the fifteen aircraft were to be between December 1959 and December 1960 and the initial price per aircraft was to be $5,172,684 with no escalation on the airframe but 12½% on the Rolls-Royce Conway engines. The contract based on a Boeing specification was signed on 8 November 1956 and a BOAC team was despatched urgently to Seattle to work out with Boeing's changes to that specification to meet BOAC's particular requirements. By May 1957 detailed estimates of the overall cost of the order were available, including aircraft, spare engines and spares. Along with contingencies, the resultant figure came to £43,428,000 which included a dollar content of $87,081,000. The Board of BOAC gave their approval to this capital expenditure on 9 May 1957.

On 21 December 1957 the first production series 707-120 made a successful maiden flight, and its C of A was granted on 23 September 1958. Early in the following year, BOAC's estimate of performance for the 707-420 model with the Conway engines was well up to expectations, and with the planned runway extension at London Airport, a direct operating cost of about 4.7p per commercial ton-mile looked achievable. By June 1959 BOAC's first 707-420 had been successfully flown at Seattle. Before granting a C of A, the ARB called for modifications to the rudder boost system and fins related to directional control. Boeing had, therefore, to develop modifications to the aircraft consisting of the fitting of a ventral fin, a fully boosted rudder control system and an extension of the existing fin.

Pilots had problems adapting to the slower increase in lift as power was applied during descents resulting in

A company 707 touches down at Heathrow following a delivery flight from Seattle.
(BOAC via author's collection)

some abnormal situations. The popular press was quick to provide coverage of these incidents, including the loss of control during simulated engine failure of both engines on one side of the aircraft causing the aircraft to dutch roll. This often resulted in incidents where engine pods were scraped along the runway on landing - something that could only happen with seven or eight degrees of roll! Rumour has it that many of BOAC's most distinguished old pilots - many who had served during the war - discovered that the lively, fast and lethal enthusiasm of the 707 for Dutch rolling made it only too easy to 'scrape a pod' and that after such an 'incident' they soon finished up in management!

BOAC - and Boeing - would have preferred that the media presented the 707 into service as everything was going smoothly, but that was not the case. That modern transport aircraft had their problems was again emphasized when an American Airlines Boeing 707, taking off on a training flight, began to spin and crashed on Long Island. Newspapers headlined: *'jinx strikes at Boeings again'*. A Boeing representative said: *'The aircraft apparently got away from the pilot. Perhaps he made too tight a turn, or was demonstrating one of the emergency procedures to the others.'* In fact, with a

BOAC produced a myriad of promotional material for the 707 for both public and travel industr alike.

G-APFE, which when operating as Speedbird 911 crashed onto the slopes of Mount Fuji in Japan on 5 March 1966. *(author's collection)*

Glistening in the sunlight, BOAC 707 G-ARRC is seen at Manchester.

heavily swept wing an accidental side-slip, perhaps induced by sudden engine failure, creates roll in the opposite direction to that intended and may temporarily overcome corrective action. Every 707 incident was attracting maximum publicity, and a second appeal was made to Congress to ground these aircraft pending investigation. In Britain the Air Registration Board, in the interests of economy and flight-testing, decided that their own pilots should participate in all future prototype tests, commencing with the Vanguard, and would include the Board's special safety trials such as abortive take-off, engine or systems failures.

Dutch roll was a problem inherent in swept-wing jets and was kept in check by the yaw damper function of the autopilot. The yaw damper could not be used at take-off or on the landing approach. Unlike piston transports, the 707 had to be 'flown' positively onto the runway rather than aimed at the threshold and 'floated' on. Pilots who tried to land the 707 like a piston risked undershooting the runway.

As some incidents and accidents during crew training demonstrated. There were specific disconcerting characteristics that neither Boeing nor the British Air Registration Board was happy about.

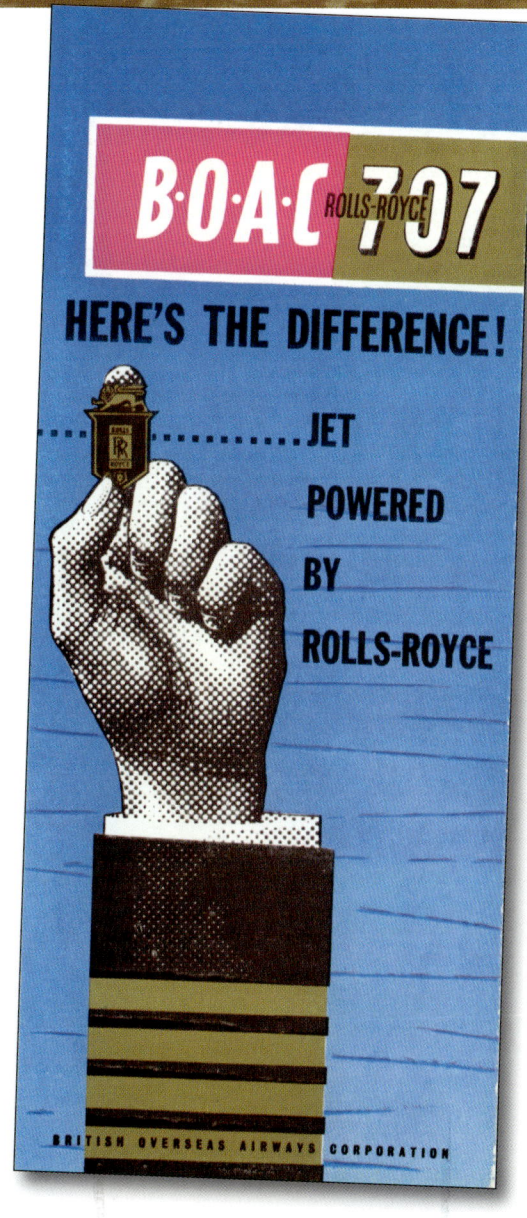

Pilots found the rudder system on the 707 somewhat different to what they had previously been used to. The original rudder system as fitted on the 707-120s had aerodynamic control only, but later variants have had power-boost assistance for more prominent angles of rudder movement. For the first ten degrees deflection, the powered trim tab moved the rudder aerodynamically. Between ten and fifteen degrees the power booster began to take over and above fifteen degrees it was entirely sufficient. However, this meant that there was no trim relief at higher angles and for structural safety reasons, the boost 'gave way' at air loads greater than 180-foot pounds of force. To use this effectively required training and practice as a heavy application of rudder gave a noticeable 'lag' as the booster became effective and then a rapid deflection to the booster's limit. The tendency was to add aileron and spoiler inputs, leading to overcorrection and Dutch roll. A new booster that worked throughout the full range of deflection was designed to meet UK Airworthiness Registration Board (ARB) certification requirements and was later fitted to all 707s, but it was still not enough.

The situation was understood by pilots, and proper corrective action was a matter of training and practice. However, apart altogether from any criticism of general handling, the characteristics of the system were such that, with the higher take-off power available from the Conway and later J75 marks, the unmodified 707 could only be certificated to ARB standards by raising the minimum control speeds and thus unsatisfactorily restricting its take-off and landing performance.

There was a known requirement for increasing the keel area; indeed this was one demand from the British ARB; a modification that became known as the 'ARB fin' which not only increased stability but also acted as a tail bumper to prevent over-rotation on take-off. The ventral fins were of at least three different shapes, with straight or curved leading edges and different areas.

These changes also involved a rationalisation of the entire rudder-control system. The rudder was fully powered through its whole range of movement. The power supply and appropriate systems were duplicated for obvious safety reasons - and, more important, to provide maximum rudder deflection, without delay, in emergency conditions. The pressure was supplied at 3,000 pounds per square inch for low-speed flight and 1,000 pounds per square inch for high-speed flight, thus providing two different power 'ratios' with the changeover controlled by a Q-operated switch. The trimming system operated through the Q-pot mechanism, over the whole range of rudder movement.

'Q' feel is related to the aerodynamics and precise flight conditions that apply at the time of the control demand. As the aircraft speed increases so does the aerodynamic load in a mathematical relationship proportional to the air density and the square of velocity. The air density is relatively unimportant; the squared velocity term has a much greater effect, particularly at high speed. Therefore it is necessary to take account of this aerodynamic equation; that is the purpose of 'Q' feel. A 'Q' feel unit receives air data information from the aircraft pitot-static system. In fact, the signal applied is the difference between pitot and static pressure, and this signal is used to modulate the control mechanism within the 'Q' feel unit and operate a hydraulic load jack which is connected into the flight control run.

In this way, the pilot was given feel which was directly related to the aircraft speed and which greatly increased with increasing airspeed. It was usual to use 'Q' feel in the tailplane or rudder control runs; where this method of feel was used depending upon the aircraft aerodynamics and the desired handling or safety features. The disadvantage of 'Q' feel was that it was more complex and only became of real use at high speed. There was no duplication in the yaw damper system, which remained inoperative during the take-off/climb and approach/landing stages of flight.

Most of BOACs -436s were delivered directly to London with empty cabins, but two were fitted with seats and galleys shipped out from the UK and picked up passengers from Montreal on the way.

Overall, problems with the aircraft itself included damage to skin, flaps and cargo doors by snow and slush thrown up by higher taxi speeds and trouble with the complex electrical wiring. The phenomenon of 'sonic fatigue' caused by engine noise created problems with certain areas of the airframe, in particular, the tailcone. This was cured by replacing some magnesium components with aluminium ones and by lining the tailcone with fibreglass.

Spare parts availability was an early problem, made worse by minor changes on the Boeing line that came about as part of a pre-delivery design improvement programme which saw changes made between the specification given to the airlines and delivery itself, including everything from new doormats to leading-edge flaps. American found that if a part were to be changed on the production line, subcontractors would often stop making it immediately, leaving Seattle's stocks as the only source.

As a result deliveries to BOAC were delayed for this work to be completed. It had been hoped to start services at the start of the summer season in April 1960, but it was mid-May before five aircraft had been delivered and BOAC was able to begin services between London and New York on 27 May 1960. Further summer season but modifications to the 707s to meet ARB requirements delayed the entry into service until 6 June on East Coast routes and until 18 June to Canada. Consequently, it was necessary to retain Comets on the North Atlantic until October 1960 - before they could be released for their

Rear Admiral Sir Mathew Slattery, Chairman of BOAC between 1960 and 1963 and former Chief of Naval Air Equipment explores the flight deck of a company Comet 4 that was undergoing maintenance as the striped label top-right shows.

You'll find life good aboard a B.O.A.C. Rolls-Royce 707

There's a warm welcome for you from the courteous cabin staff – there are seats more comfortable even than your favourite armchair at home – there's food to rival and surpass the best restaurants in the cities over which you are flying – and, result of the four Rolls-Royce Conway Jet engines, there's exhilarating speed. Slung in pods beneath the wings, these great by-pass turbo-jets sweep you along at ten miles a minute – way, way above the weather. At take-off they develop a total thrust of 70,000 lb. With a fuel capacity of over 19,000 Imperial gallons, this plane has a range, with full payload, of more than 4,500 miles. Thrust reversers, one installed in the tail pipe of each engine, provide something like 50 per cent thrust in the reverse direction – massive extra braking power for landing. Each engine also has a sound suppressor, so there's no jet roar to disturb you as you speed on your way.

A warm welcome... luxurious seats... superb food ... exhilarating speed... and silence. That's why life is so good on board the B.O.A.C. Rolls-Royce 707.

Economy or First Class ... B.O.A.C. takes very good care of you

Part of a very comprehensive promotional brochure extolling the virtues of the R-R Conway powered Boeing 707s of BOAC. The style was part Madison Avenue advertising-speak with some 1960s kitch artwork thrown in for good measure.

original task of serving the Eastern Hemisphere routes.

By 15 June the aircraft were operating a daily Monarch service to New York as well as flights to Toronto. Initially, a cabin layout of 34 first class and 97 economy class seats was selected.

So began what was to become a highly successful marriage between the Boeing 707 airframe and the Rolls-Royce Conway engines. They both served BOAC well and were in full operation right up to the time of BOAC's disappearance into British Airways on 1 April 1974.

Following Sir Gerard d'Erlanger's decision to resign for health reasons, the announcement of the new Chairman for BOAC was made by the Minister of Aviation, Duncan Sandys, on 20 June 1960 to take effect on 29 July 1960, the day following Sir Gerard's resignation. The chosen successor was Rear Admiral Sir Matthew Slattery. This came as a surprise to the staff, although Sir Matthew was a sensible choice; he had been intimately connected with the manufacturing side of the aircraft industry, first at Short Bros. and Harland, and then at the Bristol Aeroplane Company, and prior to that had been the Chief of Naval Air Equipment from 1945 to 1948. In his position as a Director of Bristols, it had fallen to him to negotiate the final settlement over the Britannias with Sir Gerard in 1958.

Unlike Sir Gerard's part-time post, Sir Matthew's appointment was as full-time Chairman with Sir Wilfred Neden as part-time Deputy Chairman and Smallpeice carrying on as Managing Director. It was a good team with specialised knowledge in Engineering, Personnel Relations and Finance but adverse circumstances were to prove too strong, and the team lasted only three and a half years. Even so during that relatively short period in command, Sir Matthew had to deal with three different Ministers, briefly with Duncan Sandys, then Peter Thorneycroft and finally Julian Amery.

Two months before he resigned Sir Gerard sent a comprehensive document to Duncan Sandys giving what could be termed 'a broad outline as we now see it' of the Corporation's financial outlook. He was able to tell the Minister that for the financial year just completed, 1959/60, BOAC had made a record, operating profit on its operations of over £4m, giving a result. Just better than break-even after payment of interest in full, an improvement of £2.25m over the previous year, and that Associated Companies' operations were also better than the previous year by about £2m although still with a deficit of just over £1m. As an indication of the expectations within BOAC at that time, the chairman included in his letter forecast results for the next three years up to 1962/63; for each of these forward years the forecast showed profits on BOAC's operating account before interest ranging from £5m to over £6m and with small surpluses in each of the years after allowance for interest. As to the Associated Companies, Sir Gerard's view was that it was never envisaged that these should be allowed to operate at a substantial loss and that there was no reason if properly managed, why they could not be made at least to break even. He supported his predictions by referring to the airline's success in reducing costs, especially within the Engineering Department, and to the low operating costs of the new fleet of Boeing 707 aircraft shortly to enter service as compared with the costs of existing fleets of DC-7Cs, Britannias and Comet 4s.

On the critical subject of growth, he reminded the Minister that over the years 1957 to 1962 BOAC would be nearly trebling its output, a deliberate policy adopted shortly after he took over the Chair and with full Ministry support. He then concluded his letter with a forthright stand at that time at around £13m after deduction of capital reserve - attributing this to Comet 1 disasters, late delivery of Britannias, high costs of introducing British aircraft into service and losses on Associated Companies. He then put it to the Minister that there was a case for granting BOAC relief from paying interest on lost capital, as otherwise the Corporation was being prevented from accumulating reserves or depreciating current equipment more rapidly.

d'Erlanger drew attention to the possible capital losses in the future arising from the likelihood of BOAC's DC-7Cs and Britannias being outmoded by the new large jet aircraft. Even so, his last optimistic words to the Minister expressed the confident belief that BOAC would earn sufficient profits to pay interest not only on capital employed but also, if it had to, on the lost capital.

It was in this atmosphere that Sir Matthew Slattery took up the reins in July 1960. And there was reason to look to the immediate future with confidence. The final outcome for 1959/60 confirmed broadly Sir Gerard's figures to the Minister, and in spite of the large increases in capacity the overall load factors in 1957/58, 1958/59 and 1959/ 60 had all been at a level sufficient to exceed the overall break-even load factors on the operating account before interest with unit operating costs falling in each of these years. The 1960/61 financial year had started well and by mid-summer traffic for BOAC was still buoyant.

There were clouds on the horizon - other international airlines were not proving to be so fortunate, the delay in introducing BOAC's Boeing 707s had been expensive in

Right: Now Sir Basil Smallpeice. Managing Director of BOAC with his wife Lady Kay at Buckingham Palace on the occasion of his investiture into the Royal Victorian Order.

Below: a publicity picture of a company 707 at Ancorage, Alaska - clearly from their stance, the huskies were non-too-happy about being on the ramp!

BOAC 707 G-APFB is seen at Kai Tak, Hong Kong in what appears to be the earlier colour scheme with the letters writ large on the aircraft roof.

lost business. Then there was the Civil Aviation (Licensing) Act which received the Royal Assent in June 1960 and abolished the Corporation's statutory right to be the sole British operator of scheduled services on the worldwide routes with possibly damaging consequences: and finally BOAC was still in the throes of its expansionist policy with planned capacity growth of over 30% for each of the next two years, 1960/61 and 1961/2, as the Boeing fleet came into full operation and was thus exposed to the risks inherent in such a big leap forward over a few years.

In contrast to Sir Gerard's Chairmanship which was heavily involved in the vital considerations surrounding the acquisition and ordering of fleets of jet aircraft in the future years, Sir Matthew took over when these aircraft decisions had been primarily taken, and it fell to him to deploy the new fleet of Boeing 707s and to carry on and complete the programme of expansion set by train by d'Erlanger in 1957. His term of office began on a hopeful note, but circumstances became progressively more difficult as the months went by and traffic flows fell away just as the major airlines, including BOAC, put massive capacity increases on the market with their new fleets of big jets. By 1961 the situation was severe; 1962 saw little improvement, and by September of that year the Minister had put in hand an inquiry into the Corporation's financial affairs.

Delivery flight via The Pole

Such was the level of interest shown in the introduction of the 707 into service by BOAC that numerous reports appeared about the delivery flights from America. This Flight Log is from the Delivery Flight from Seattle to London Heathrow 707 G-APFN, which took place on 18 November 1960.

'*Our take-off weight was 130,500kg, of which 72,000kg represented full tanks. We needed, with all allowances, 66,500kg of fuel for the 4,210 n.m. flight. Estimated flight time 8hr 42min on a minimum-time track, with average +24kt wind component, calculated by BOAC's New York dispatching office and received in Seattle by telephone. The 72,000kg of fuel would give us 11 hr 48 min endurance.*

Track: Carmi, Churchill, Frobisher, 64°N at 50°W, 56°N at 10°W, Bush Mills, London. Flight level 330 to 90°W, 370 to 20°W, then 410. Take-off about 1700hr local, 0100Z (GMT). Take-off clearance: "Climb on runway heading to 3,000ft: left turn to heading 340° to intercept 030° radial of Seattle VOR: climb NE-bound until 15,000ft via direct Carmi: maintain flight level 330. Transition height, 24,500ft because of the mountains. Gets dark during climb, red rotating beacons reflecting off pods; flight deck almost Christmas-like in red and white lights on grey panels: everyone head-down working hard, except pilots peering into night sky. We press on to cruising height, mostly using DR plot and scattered NDBs with occasional VOR. Talking to all sorts of stations on VHF and HF, asking for position reports to be passed to BOAC at Montreal.

Distinctly Canadian accents on radio. Change heading from 025° to 060°. Meet jarring turbulence: navigator's plot shows sharp wind-change: radar shows thunderstorms: temperature drops rapidly; lights dimmed and captain stares into black night, hand on autopilot heading control. This is a jet stream—and rough! Decide to climb straight to 370 to get clear, and notify control. Using both VHF and HF almost constantly. Pass Tippo Lake at 0202Z estimating Churchill at 0300Z. Dull, furtive veils of northern lights snaking above us - ADFs tuned to Frobisher NDB, no astro. Outside air temperature -52°C. Hear SAS over-the-Pole flight asking to climb from 280 to 310 at 0449Z, position 70 °W, 66 °N. KLM flight is there too. Northern lights seem to have gone. We talk to 'Leeway' on VHF.

0445Z: Note from co-pilot Lee, "Leeway is defence radar at Frobisher: we saw their lights on the ground: have now returned to compass steering: will get radar fix on No 2 VHF at about 0455Z: now reporting to Goose on No 1 HF." Our report, read from a form, gives estimate for 64°N, 67°W as 0509Z, the wind found, fuel state and consumption, speed, ETA for London and much besides. Goose asked to repeat to Gander and Montreal for BOAC, to Sondrestrom for ATC. Sondrestrom cannot understand,

B·O·A·C ROLLS-ROYCE 707

You will see the world-famous Rolls-Royce symbol inset in the metal surrounding the engine.

1 Search Radar 2 Flight Deck 3 Forward Passenger Entrance
4 3 Toilets 5 Cloakroom 6 Forward Galley
7 First Class Lounge & Bar 8 First Class Cabin
9 Economy Class Cabin 10 Rear Galley 11 Rear Passenger Entrance
12 Cloakroom 13 3 Toilets 14 Rolls-Royce Conway jet engine

Going back to the days of Imperial Airways, the airline was famous for its airliner cutaways - BOAC continued the tradition with this glorious one of a 707. *(BOAC via author's collection)*

so Goose changes HF frequency to try again. 'Leeway' fixes us by radar at 120 miles. SAS and two other BOAC aircraft talking on HF. Navigator plotting all the time; engineer fills in fuel tables every 5,000kg, about every 40 min. Pressurizing on one turbocompressor and two direct engine bleeds. We call Prestwick on HF, apparently without reply.

0530Z: Northern lights sneak up again. ADF tuned to Kook Island NDB, mid-west of Greenland, and we see its lights below. Whoever lives there? No. 2 ADF getting Christiansund NDB, 320 miles away on southern tip of Greenland. At 37,000ft: TAS 475kt; two minutes up on ETA; winds northerly; engines at 88% r.p.m.; radar tilted down 7° for mapping.

0542Z: Temperature -55°C. Air has been smooth for hours. Captain and navigator still hard at it, co-pilots and engineer relieved. Passenger cabin a dark, empty tunnel— only nine seats fitted. Dead of night, northern lights stealing about.

0625Z: At 35°W and 37,000ft. Hope to climb at 30"W. Three min ahead of plan. No VHF contacts. Iceland cannot hear our HF, so relaying via Sondrestrom. Expect to contact weather ship on VHF at 0645 and get fix. Nearest to Iceland at 0700. Many other aircraft south of us calling Gander. Receive HF weather broadcast from Shannon giving shallow fog for most British airfields; also Canadian maritime weather broadcast from Gander. Frobisher has a 9,000ft runway good for a diversion. Our point of no return relates to Gander. But now we have the feel of the other side and are heading south-east for Britain.

0800Z: Wake with a start from sleep to see a hard yellow, copper and pale green dawn rising over us. Still making 480kt true on 132°. ATC has held us down to 37,000ft; passing 10°W and estimating Bush Mills at 0830Z. Windscreen frames now thickly coated with frost. The sun begins to shine dazzlingly straight in at the windscreen, and shades are down, lights turned low. Outside temperature -48 °C. Captain still in seat. Navigator makes complete table of airways check-point ETAs for Red 1 and Amber 1 via Belfast, Isle of Man, Wallasey, Lichfield, Daventry, Beacon Hill and Watford to LAP's runway 28R.

Descent to begin at 0852 and to last 24 min at mean TAS of 364kt, using 800kg fuel. Engineers plan pressurization management between bleeds and turbos when throttled back on descent. ETA London 0916Z with 19,000kg of fuel remaining at 1,000ft. The tip of Ireland is painting well on radar at 60 miles. Sun is blinding. A leaden sea visible between dollops of cloud thrown almost up to our level in polar maritime cold air. IAS 250kt; M0.82; r.p.m. 88%; o.a.t. − 48 °C; cabin height 6,000ft. Navigator hands time plot to co-pilot and relaxes slightly. HF weather reports in French. Cillard RAF radar (in Scotland?) has us. English voices, clipped and calm in welcoming efficiency.

0820Z: Ireland in sight. Centre and reserve tanks now dry, remaining fuel distributed in wings.

BOAC ROLLS-ROYCE 707

Monarch

CROWNING LUXURY

LONDON – NEW YORK

BRITISH OVERSEAS AIRWAYS CORPORATION

0837Z: Cillard loses us and we switch to Scottish Airways control. Estimate Isle of Man at 0843. Prepare-for-descent checks read out. Landing weight will be 77,500kg — very light — VREF 126kt, target threshold speed 135kt, maximum threshold speed 149kt.

0842Z : Pass Isle of Man, in sight below, together with coasts of Wales, Ireland, England and Scotland, and request descent clearance for 0851. At 0846 cleared down to flight level 210 and call Preston. Throttle outers to 68% and inners to 87% r.p.m. Descending at M.O.68 at 700ft./min. Wallasey at 0853.

0900Z: The Pennines lava-like in valley fog and snowy tops. Joddrell Bank telescope like a deployed parachute far below. Atlantic charts and manuals being cleared away. Pass Lichfield, estimating Daventry at 0907, tuning beacons, change to London control. 250kt i.a.s., jolted in rough air. Daventry at 0907, estimating Beacon Hill at 0913. Cleared to flight level 190. Watford on No 1 ADF, Dunsfold on No 2 ADF. Find Beacon Hill by Flying Dunsfold range leg to a bearing from Watford. Under London radar surveillance from Daventry. Cleared to flight level 080. Don't confuse Beacon Hill with Woburn, check with ADF. 1,000ft/min now at 150. Wheels rumble down for airbrake effect, slow to 200kt at 2,500ft/min; trying to make Watford at 8,000ft. Over Watford at 11,000ft radar takes us straight on to a southerly lead-in for ILS, asks our rate of descent.

Runway visibility 1,500yd. Still on autopilot, in cloud. Flap coming down. See Greenwich through a hole in cloud, then Crystal Palace. A helicopter is reported leaving Battersea. Approach checks read. Autopilot-coupled glide-path and localizer armed. Radar vectors us on to

Above: BOAC's Monarch service from London to New York.

Right: a passengers view of the Rolls-Royce Conways as fitted to BOAC 707s.

centre-line. Speed coming back to 150kt at 3,200ft. QNH set on co-pilot's altimeter, QFE on captain's. Height 2,100ft, glidepath coupler engaged at 152kt, going down at 900ft/min into dull mist. Melted frost dripping fast from window frames. Captain's hand poised on control wheel. Windscreen wipers working hard. Lead-in lights now dimly in view, but no trace of runway. BEA engineering base comes into sight to our left, co-pilot postively identifies runway and tells captain.

We surge in past the lights, the captain cuts the autopilot and holds off. When I think we are still 100ft up, the main wheels touch smoothly, the nose comes down, spoilers are popped out, reverse thrust pulled. Further end of runway still out of sight. The captain takes the nosewheel tiller and starts braking while the co-pilot holds the column forward and calls the decreasing speeds down to 60kt. We turn off with some runway to spare, switch to airfield control frequency. Shutting down checks begin.

We are home. Chock-to-chock time 9hr 15min for 4,210 n.m.: we took off at about 1700 hr Seattle time and it is now 0130 by that reckoning - it's time for bed. But here in London it is 0900hr or so and a new day is just beginning.

Into Service

BOAC was delighted with the financial figures that came back through the introduction of the 707 into passenger service, and by January 1961 was seeking Government permission to purchase three more 707s to cover the forecast capacity gap up to 1967/68 when the VC10s were due to be in full operation. The case for these additional aircraft rested on four factors - the need to provide competitive large-jet services between Manchester, Prestwick and New York and also between London and airports in the Caribbean area, traffic growth on the Atlantic beyond 1962/63: the fact that the Government had limited BOAC's 707 order to fifteen aircraft instead of the seventeen initially asked for: and finally the expected decline in the competitiveness of Comets and Britannias.

Surprisingly permission to proceed with the purchase was received in April 1961 without undue delay, but BOAC was concerned to learn that it would be required to pay import duty on the three aircraft; the chairman protested to the Minister on the grounds that comparable British aircraft were not available in the required timescale and moreover, no duty had been paid on the original order for fifteen.

The Minister was not prepared to change his ruling except to indicate that if these three additional 707s were sold to an overseas purchaser after the entry into service of the VC10, the duty would be refunded.

Treasury approval to the purchase having been obtained, the order for the aircraft was signed on 25 May 1961 for delivery in March 1962, February 1963 and March 1963 at an estimated cost of £8.8m the bulk of which was to be dollar expenditure. This purchase raised the 707 fleet strength to eighteen aircraft, and two more 707s were added in the autumn of 1962 brought in by Cunard Eagle following the formation of BOAC-Cunard.

The division of BOAC's commercial activities into three groups of routes had become well established by the time Sir Matthew took over from Sir Gerard. The General Managers of Western, Eastern, and Southern Routes were

A once common sight on the road from Victoria to Heathrow Airport was BOAC's fleet of fifteen Roe bodied Atleantean coaches. They had large luggage areas downstairs with 16 seats and only another 18 upstairs. Passengers could check in at Victoria and be whisked straight to security at the airport. This picture must have been taken after 1966, for that is the year of the 'age-date-letter' suffix - in this case the letter 'D' on the coaches.

on the Board of Management, responsible for the financial results of their respective Route groups to the Managing Director. The system was functioning well, and staff worldwide were in no doubt as to where they fitted into the overall organisation and to whom they were responsible.

Western Routes, covering the North and South Atlantic and the Pacific areas, was broadly equivalent in size regarding the load carried to the sum of Eastern and Southern Routes - not surprising since the traffic on the North Atlantic route was and still is the densest of any long-haul route. The year 1960 saw the gradual introduction of the new Boeing 707 fleet to the central North Atlantic destinations, reaching San Francisco in October and Hong Kong over Honolulu and Tokyo in December. In March 1961 a new service was inaugurated with 707s to Los Angeles via what was euphemistically called the Polar route. In the Winter of 1960/61 valuable 707 utilisation was obtained when most needed by introducing these aircraft to the winter holiday routes between USA and Bermuda, Nassau and the Caribbean. Similar operations were undertaken on behalf of BWIA with Britannia 312s. October 1960 also witnessed the first Skycoach services with B.312s between the UK and Bermuda, Nassau, Jamaica, Barbados and Trinidad at fares well below economy class and with modest standards of comfort and service and restricted to residents of the UK and of the colonial territories concerned.

Fare reductions were considerable. Economy fares were introduced on the Pacific, and on the North Atlantic, the jet economy fare was reduced. To cater to the particular emigrant movement, very low emigrant fares were brought in from London to the USA and Canada.

BOAC's cargo promotion received a boost in the US and Canadian markets with the DC-7F freighter service began in December 1960 from London via Manchester and Prestwick to Montreal and New York. The DC-7 continued to provide this cargo outlet until superseded in October 1963 by the specialised cargo aircraft; the CL44 leased from Seaboard World Airlines.

In April 1961 BOAC re-opened its service to Lima, Peru, which had been missing from the route pattern since the Constellation 049 days; 707 aircraft were used, extended down from New York via Nassau to Lima. In June of that year, a service to Washington was inaugurated, again with 707s and again via New York. Following the grant of the new traffic rights by the Germans, an experiment was made in April 1962 to boost revenue by extending the Los Angeles-London service to Frankfurt, but this failed to reach expectations, and the service was withdrawn eighteen months later. Much the same fate befell the direct service London-San Francisco which had an equally short life; both routes fell victims to the retrenchment policies of late 1963. These withdrawals were disappointing, so too were the poor results on the Mid-Pacific route from San Francisco via Honolulu and Tokyo to Hong Kong. BOAC had believed for a long time that it must be right to share in the rapid growth of traffic in the Pacific area which had led to the establishment of these routes. A personal survey of all the BOAC stations in the Pacific by the MD in the autumn of 1961 had presented an encouraging picture of the potential and endorsed BOAC's policy of expansion. However, by 1963 the losses led to the abandonment of the Los Angeles service and the direct San Francisco service, while the long London - New York - San Francisco - Honolulu - Tokyo route was becoming increasingly difficult to justify although it formed part of BOAC's round-the-world operation with 707 aircraft; this had been opened in March 1962 when 707s on Eastern routes via India and 707s on Western routes via New York met in Hong Kong.

BOAC was also at this time having problems with its operations into the Middle West of the USA. Traffic available to BOAC between London and Chicago could not support a daily 707 in the face of severe American competition; various plans were tried including routeing via Montreal, Boston and Detroit; there was an absence of mail, and the average revenue rate was low due to poor first class business. This Middle West operation remained difficult for many years until steps were taken on both sides of the Atlantic at Government level to bring capacity into a more reasonable relationship with traffic offering. On the Mid and South Atlantic routes there was little change over the three years or so. The Britannia 312s operated across the central Atlantic to Bermuda and on through the Caribbean to Caracas and Bogota, being superseded by the 707s in April 1963. The Comet service to the East Coast of South America, terminating at Santiago, was struggling for economic survival by the second and third year after the route's re-introduction. Although the first year's operation had been relatively satisfactory due to the lack of jet competition, by later years, competitors had deployed their large American jets, and South American Governments were being restrictive; consequently, BOAC's losses on the route had become insupportable by the close of the 1962/63 financial year. General Manager, Southern Routes - who had assumed responsibility for the East Coast South America route in October 1962 - was convinced by the end of 1963 that the route could not be commercially viable for some years - not surprising when the Comet break-even load factor on the route was 91% and even the coming VC10 would offer only about 70%.

Wash and Brush Up!

No airline liked its aircraft to appear dirty and travel-stained - it also made good economic sense to keep them clean to reduce drag, and therefore fuel consumption - So it was with BOAC, as this item shows: *'A big aircraft like the 707 can pick up quite a lot of dirt on its round-the-world travels: sand from the Middle East, dust from India, soot from Tokyo. By the time the travel-stained jetliner has completed a few hundred thousand miles it is in need of a good wash and brush-up.*

Cleaning an average-sized motor car outside your front gate is one thing; shampooing a jetliner is another. But BOAC's maintenance men now have the problem more than beaten.

Keeping an aeroplane spic-and-span has always been a laborious task - carried out mainly by mopping and hand-polishing. It still needs plenty of hard work, but with modern methods and equipment it is no longer just a wearying chore.

A shift from BOAC's Chemical Cleaning Unit set to work with brooms and hoses giving a company 707 a wash and brush up. Every speck of dust and dirt had to be removed to prevent corrosion.

It was hard work to clean a Boeing 707 - plenty of elbow grease was needed! *(all BOAC Press Office via author's collection)*

If you happen to be passing BOAC Hangar 8, London Airport, at the right time of day, you would see a group of gentlemen dressed as if about to set out on an Antarctic Expedition. These are the men of the Chemical Cleaning Unit, and their job is to keep the silver, white and blue livery of BOAC jetliners gleaming brightly.

Three groups of eight men operate a two-shift service. Using ramps and ladders, or climbing on to the aircraft's massive wings, they spray a cleaning solution on to the bodywork. Then they get to work with their long-handled brooms and hose-pipes.

The chemical used, which is a very strong version of liquid detergent, removes every speck of dust and dirt. It also deals effectively with London smog, and keeps the aircraft shining clean and free from corrosion.

Special solutions are used to clean parts of the wings and engines marked by heat or deposits from the exhaust. If after months of exposure to the elements the aircraft's surface becomes dull again, another chemical is used to bring back its shine. In all it takes eight men four hours, forty gallons of chemical and four hundred gallons of water to clean one 707'

Enter Cunard

In October 1960 Eagle Airways (Bermuda) Ltd acting as a Bermuda based airline obtained the right to operate scheduled services, including its Gulfstream service between London - Bermuda - Nassau - Miami and a Skycoach service between West Atlantic points and London. This led to BOAC and Eagle arranging the Skycoach operation to agree together the type of service and fares, and to pool what was a very limited exercise. By the time the pool agreement came into effect in April 1961, Eagle Airways (Bermuda) had been taken over by the Cunard Steamship company, and re-named Cunard Eagle (Bermuda) Ltd and this first relationship between BOAC and Cunard Eagle developed into something far more comprehensive in the following year.

Eagle Airways was owned by Harold Bamberg, a former wartime pilot who formed the airline on 14 April 1948 with a nominal capital of £100 as Eagle Aviation Ltd at Aldermaston. The initial fleet comprised two wartime bombers converted for carrying fruit and vegetables. Both aircraft saw extensive service along with a further two others during the Berlin Airlift.

The airline acquired Air Freight Ltd with three more Halifaxes later the same year. Eagle acquired three Avro York aircraft in late 1949, followed by eight others, and used these until early 1955 for both passenger and freight charters. Eagle Aviation moved to Luton in 1950. For most of its existence, the company's head office was located at 29 Clarges Street, London W1.

By 1951, Eagle Aviation had won its first regular Government trooping contracts, including the first regular contract awarded by the War Office for trooping flights between the UK and Singapore starting in August 1951. This helped keep its fleet of six Halifaxes and nine Avro Yorks busy and employed 100 people including 12 pilots. Operations moved to Blackbushe Airport in 1952, followed a year later by the launch of secondary scheduled services in association with BEA, from whom Eagle had purchased a large fleet of Vickers Vikings.

During 1953, Eagle Aviation's steadily growing passenger charter operations included for the first time aerial cruises around the Mediterranean. Following Eagle's decision to sell the Yorks to rival UK independent Skyways for £160,000, the airline expanded from charter work into scheduled services from its new base at Blackbushe Airport, using Vickers Vikings. Eagle, which by that time had set up Eagle Airways as a new company to run the scheduled side of the business (leaving Eagle Aviation in charge of all non-scheduled operations, including trooping flights), inaugurated its first scheduled service on 6 June 1953 from London (Blackbushe) to Belgrade (via Munich). It also began operating domestic flights within the UK and additional international services to secondary western European destinations.

In 1954, the Ministry of Aviation granted Eagle permission to operate a limited programme of a new type of low-fare service that combined air travel and overseas holiday accommodation at a cost substantially below the aggregate of each component if purchased separately. This new concept enabled the airline to circumvent regulatory restrictions that prevented private airlines from competing with their state-owned counterparts.

When the Thomas Cook & Son travel agency declined Eagle's offer to take on the role of the airline's tour operator, Eagle acquired the Sir Henry Lunn Ltd travel agency chain. This made the airline one of the pioneers of the British package holiday industry and probably marked the first occasion in the UK in which an airline became vertically integrated with its in-house tour operator - that is where an airline owns or is owned by a tour operator or both are part of an integrated travel group. British Eagle also acquired the Polytechnic Touring Association in the 1950s and formed Lunn Poly from the two agencies in the mid-1960s.

Eagle's first inclusive tour (IT) flights operated to destinations in Italy and Spain (including Majorca). Between 1955 and 1960, many of the airline's aircraft carried the Eagle Airways operating name.

By 1957, the summer IT programme included for the first time 15-day, all-inclusive packages to Spain's Costa Brava - it was also the year Eagle joined IATA.

On 26 July 1957, Eagle formed an overseas subsidiary, named Eagle Airways (Bermuda), in preparation for the launch of transatlantic scheduled services between Bermuda and New York, using Viscount 800 turboprop aircraft. Within a year of launching its first transatlantic scheduled route, the airline's North Atlantic scheduled operation extended to Montreal, Baltimore, Washington and Nassau.

Harold Bamberg, the Head of Eagle Airways with the airline's chief pilot, Harold Watkins.

One of a pair of Boeing 707s operated by Cunard Eagle Airways.

This sequence of events started when Cunard Eagle applied to the Air Transport Licencing Board (ATLB) in January 1961 to operate scheduled services from London to New York and other Eastern seaboard points, to the Middle West and Canada. Before the hearing commenced, the Company announced that it had ordered two Boeing 707 aircraft with an option on a third.

BOAC objected strongly to the application and the hearing took place between 16 and 30 May 1961. A formidable array of Cunard-Eagle and BOAC executives was there to give evidence. Counsels spent hours questioning and wrangling, and there were barrages of statistics in the course of six long sessions. Each party tabled its estimates of future traffic increases on the North Atlantic, Cunard Eagle at 20% annually and BOAC at 17%; however, it was later announced that the actual growth of traffic in the year 1961/62 was 0.6%.

BOAC based its case primarily on the fact that material diversion of traffic would occur from its services and that its financial commitments, particularly for British aircraft, were such as to provide for all traffic estimated over many years ahead.

Despite the objections, the ATLB saw fit to grant Cunard Eagle a licence valid for fifteen years from 1 August 1961 to operate a daily service between London or Manchester or Prestwick and New York; additionally, Cunard Eagle would be permitted to serve Philadelphia, Boston, Baltimore and Washington but each service would have to call at New York. The ATLB dismissed the application for services to the Middle West and Canada but there was a disquieting note in their reason for doing so, namely that the routes were not yet sufficiently developed by BOAC to warrant the introduction of British competition at this stage. The obvious implication being that it would only be a matter of time before new applications would be made for those routes.

'We are surprised and deeply disappointed by the Board's conclusions...' commented a BOAC spokesman. '...We shall appeal in the strongest possible way. The Board's decision appears to accept material diversions of traffic from BOAC if Cunard- Eagle operate successfully. If that is the case BOAC's position will be undermined - yet BOAC had understood that Parliament's intention as embodied in the Civil Aviation Licensing Act 1960, was that this should not occur.'

BOAC decided to appeal against the decision and the appeal took place before Sir Fred Pritchard, the special Commissioner appointed by the Minister. Among BOAC's grounds for appeal was the fact that traffic in the summer had fallen far short of expectations and that even one very poor year would have a significant effect on forward estimates. BOAC therefore submitted new evidence to show that the forecasts made at the original hearing by both itself and Cunard Eagle were likely to turn out to be much too optimistic; secondly that there was no existing or potential need for the Cunard Eagle services which would only result in wasteful duplication of capacity already being provided by the Corporation.

Both sides were represented before the Commissioner by Counsel, the Commissioner eventually upholding

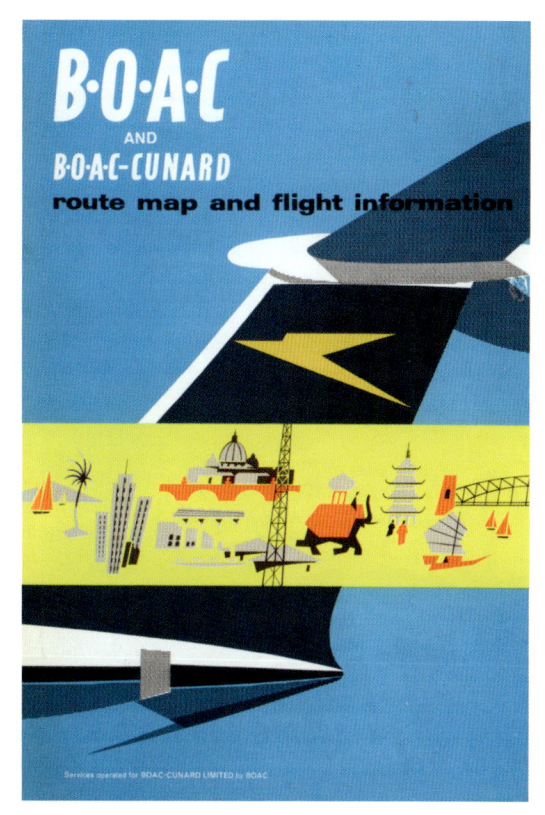

BOAC's appeal on the grounds put forward and in particular that the grant of the licence to Cunard Eagle would result in diversion of traffic from the Corporation's services. On 21 November 1961 the Minister accepted the Commissioner's decision and Cunard Eagle's licence was revoked. The Minister stated he had been influenced by the amount of capacity BOAC had on order, by the weakness of the market and by the extent of over-capacity on the North Atlantic, demonstrated by the industry load-factor in the past summer season.

Behind the scenes - and indeed behind the backs of Harold Bamberg and his airline board - Cunard were getting more and more frustrated with the politial lethargy in the UK and with the bureaucratic regulatory authorities on both sides of the Altantic over the what they saw as the lack of passenger rights. So, early in 1962 Cunard approached BOAC to see if some mutually acceptable deal could be worked out. As a strong card, two 707s - which were soon to be delivered could be offered in the

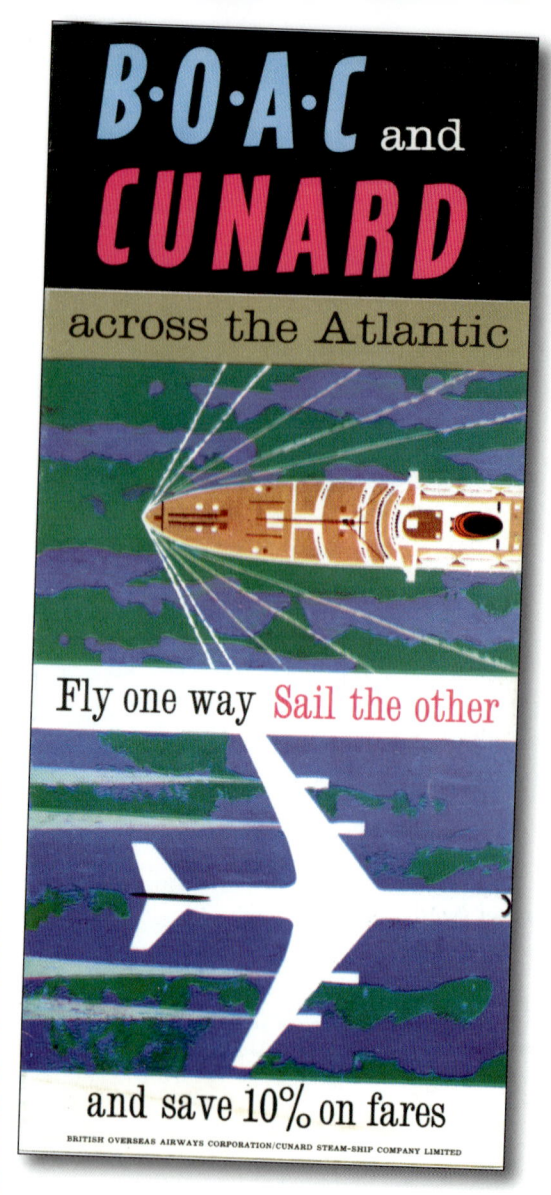

BOAC-Cunard certainly pushed the Gourment and Glamous aspect to North Atlantic Travel. *'You can fly one way, sail the other, across the North Atlantic - booking the entire journey through any BOAC or Cunard office or Appointed Travel agent'*. So said a BOAC Cunard promotional leaflet advertising the North Atlantic services.

If the sea portion of your trip is completed within Cunard Thrift Season you qualify for a 10% Round Trip Discount on both your air and sea tickets! What an opportunity to save money on your transatlantic trip!

The European sea or airport need not be in the same country! You can fly from New York to London, Glasgow or any other point in Europe served by BOAC and sail back by Cunard from Southampton, Liverpool, Cherbourg, Le Havre, Greenock or Cobh.

In the same way passengers travelling from Europe to North America can return to any European point served by BOAC or Cunard and still qualify for the 10% Round Trip Discount in the Thrift Season.

negotiations, but privately, having lost the North Atlantic case Cunards were worried about finding adequate use for the aircraft. To BOAC this seemed a way of closing a potentially dangerous gap in its defences bearing in mind that Cunard Eagle had the support of Bermuda and its aircraft were registered there and of preventing a loss of revenue resulting from Cunard Eagle's Gulfstream service. BOAC also hoped to obtain benefits from feeding two more 707s into the fleet and from a combined selling effort of Cunard's and BOAC's worldwide sales shops.

The concept of a joint company to operate Atlantic air services was put by Sir Matthew to the Board in April 1962 and met with approval; apart from all else it was felt that such an arrangement would be well received politically.

The BOAC view was that in the division of interest in any new company the Cunard share should be less than 25%. However the final draft of the Agreement between the two companies gave the authorised share capital of BOAC-Cunard Ltd with 75% subscribed by BOAC and 25% by Cunard. Surprisingly, the two companies were summoned by the Chancellor of the Exchequer on 5 June 1962 who insisted that Cunard should have a 50% interest. On Cunard's protesting it could not find such a large investment, the Chancellor accepted a 70/30 split.

The final agreement was dated 6 June 1962 and provided for the formation of BOAC-Cunard Ltd as from 24 June 1962 *"to provide air services (passenger, passenger cargo and cargo, both scheduled and non - scheduled) to, from, and within the agreed area, defined as USA east of 100° West, Mexico, Central America, Gulf and Caribbean, Peru, Ecuador, Colombia, Venezuela, and British Guiana"*. It was also agreed that BOAC-Cunard (BCL) should acquire all the share capital of Bahamas Airways and Cunard Eagle (Bahamas) Ltd and Cunard-Eagle (Bermuda) Ltd, also the 10% interest of BOAC in BWIA; on this last there were problems with the Government of Trinidad and Tobago which prevented acquisition of the holding. Part of the arrangements provided that Cunard and its subsidiaries and associates should not operate services where BCL or BOAC were operating other than Europe, the Mediterranean islands and the UK. The fleet of the new Company was to consist initially of eight Boeing 707s put in by BOAC and two by Cunard.

Sir Matthew was appointed Chairman of BOAC-Cunard Limited and Sir John Brocklebank, Chairman of Cunard Steamship Co Ltd, was made Deputy Chairman. Smallpeice was also appointed to the Board and subsequently became Managing Director, BOAC had three further directors, Granville, Lee and Stainton and there were three more from the Cunard side, Harold Bamberg and Ashton-Hill from Cunard Eagle and Taylor from the Cunard Steamship Company. Bamberg and Ashton-Hill resigned about a year later when the former agreed to buy a major interest from Cunard in Cunard Eagle Airways; at that time Cunard obtained a covenant from Cunard Eagle that it would not compete in the agreed area with BCL or with BOAC.

Smallpeice took care to explain to BOAC staff, who looked askance at the new Company, that although aircraft would have BOAC- Cunard on their sides and the new name would eventually appear on sales shops, the Company would in fact be staffed by BOAC personnel and things would go on as they had in the past plus the advantage of joint marketing with Cunard. Sir Basil made clear that: *'For all practical purposes we are still the same people, doing the same work, flying the same routes, but adding the great marketing advantage that the name of Cunard means. As it is the intention that BOAC-Cunard should not, at any rate for the present, have any staff of its own, BOAC staff in the new company's area will continue to be BOAC staff on the same terms and conditions as before. Aircraft will have 'BOAC-Cunard' on their sides, sales offices may have the new name across the front, and notepaper may carry the new heading, but our staff will still be BOAC's.'*

Lord Douglas came into the act: *'The BOAC-Cunard agreement is bound to have profound effect on this country's civil aviation policy. The private airlines led by Cunard-Eagle and BUA have made great play of the advantages of competition between British airlines. Their arguments are rendered quite meaningless by this new move.'*

It got off to a slow start in its publicity and image to the public since it did not receive a US CAB permit until 25 March 1964 and until then its name did not appear on aircraft or in publicity material, and even then BOAC had

Bamberg's proudest moment came when he merged with the Cunard shipping company and put 707s on the transatlantic routes. But it was not to last; in February 1963 the Atlantic venture ended in defeat and, addressing his employees from a maintenance platform at London Heathrow, he announced that he had bought back control of the company.

The 1960s was definately the age of the jet-set film stars, celebrities, royalty, pop stars... all wanted to see and be seen and could often be 'caught' by the camera posing on the aircraft steps.

Below: *'Oh, flew in from Miami Beach BOAC; Did'nt get to sleep that night...'* so went the first two lines of the Beatles song *'Back in the USSR'*. The Beatles were regular passenger, as three of the 'Fab Four' are seen here having something of a laugh with a stewardess and an oversized comb.

Right: it was not just the British Invasion of the USA - Tamla Mowtown and soul music did a very good job of taking things in the reverse direction - and they used BOAC too! Here's a publicity shot of the Supremes in March 1965, when it was still Mary Wilson, Florence Ballard, and Diana Ross.
(both Press Office via PHT Green Collection)

to be clearly shown as the air carrier.

In some political quarters the agreement between BOAC and Cunard was regarded with disfavour and this became evident at the second reading of the Air Corporations Bill in November 1965. Prior to the debate the Ministry checked with BOAC as to whether it was still desirable from BOAC's point of view to continue with the joint company. BOAC confirmed there was no reason to believe otherwise. Subsequently questions were raised in the House about an interest free loan from BOAC to BCL and the Ministry asked for details for placing in the Commons library. The Board of BOAC were opposed to a commercial transaction being disclosed but a note was prepared for the Minister to show interested Members of Parliament. The Opposition pressed the point and eventually BOAC had to assure the Minister it would not make any loans at less than the ruling commercial rates to its subsidiary or associated companies in which there was a private stockholding.

The essence of the Opposition's complaint was that to provide public money free of interest to a partially private company put private banks in an uncompetitive position to provide finance.

By 1966 Cunard were having problems in the shipping field with declining traffic and seamen's strikes; they could foresee the requirement to provide 30% of the capital needed to acquire the new aircraft for BCL, running into many millions. Sir Basil Smallpeice, who by then was Chairman of Cunard decided that the Company's resources were not sufficient to meet such heavy expenditure and, moreover, they were compelled at that time to sell assets to make good the cash loss suffered in the seamen's strike. It was accordingly mutually agreed to dissolve BCL.

The last Board meeting of BCL was in September 1966. The termination agreement provided for BOAC to pay Cunard £11.5m for their shareholding in BCL and the cost of altering insignia of aircraft and buildings was to be shared 70/30 BOAC/Cunard. The Agreement also provided that until June 1972 and for nine years thereafter Cunard were not to provide or operate air transport services themselves or by any subsidiary or associated company.

Many in BOAC considered the whole affair misconceived from the start and that the commercial results were negligible. What it did however, was to eliminate for a time potential competition on the Atlantic routes from Cunard-Eagle, but it is questionable whether such competition would have amounted to much, especially as the Corporation's Atlantic aircraft were by then fully competitive.

Incidents and Accidents

The first 707 accident happened to G-AFPN on 24 December 1960. when the aircraft made a precision approach radar descent to land on runway 23 Left at Heathrow at the conclusion of a scheduled flight from Chicago, IL to London-Heathrow Airport with en route stops at Detroit, Montreal and Prestwick. It touched down nearly halfway along the runway, and as the captain was not able to bring it to a stop on the remaining length of the runway, it ran onto the grass surface beyond the runway end. The main landing gear units collapsed, and the aircraft was extensively damaged.

Then on 5 March 1966 tragedy struck half a world away. 707 G-APFE was operating Flight 911 a scheduled service from San Francisco to Hong Kong via Honolulu and Tokyo. The airliner was expected to arrive at Tokyo Airport at 16:45 on 4 March. However, due to poor meteorological conditions at Tokyo and because the precision approach radar of the GCA was out of service, it diverted to Fukuoka and landed there at 18:00. After staying overnight at Fukuoka, Flight 911 left for Tokyo at 11:25 and landed there at 12:43. The aircraft was prepared for the next leg to Hong Kong, and a flight plan was filed for a flight following the instrument flight rules via Oshima on airway JG6 to Hong Kong at FL310.

At 13:42 hours the crew contacted ATC requesting permission to start the engines and clearance for a VMC climb via Fuji - Rebel - Kushimoto. The aircraft left the ramp at 13:50. It was instructed to make 'a right turn after takeoff', and departed Tokyo Airport at 13:58. After takeoff, the aircraft flew over Gotemba City on a heading of approximately 298 degrees at an altitude of approximately 4900 m and indicated airspeed of 320 to 370 knots. The aircraft, trailing white vapour, then suddenly lost altitude over the Takigahara area, and parts of the aircraft began to break away over Tsuchiyadai and Hiramatsu. Finally over Tarobo at an altitude of approx. 2000 m, the forward fuselage broke away. The mid-aft fuselage together with the wing, making a slow flat spin to the right, crashed into a forest at the foot of Mount Fuji. The forward fuselage crashed into the forest approx. 300 m to the west of the above site and caught fire. One hundred and twenty-four people lost their lives.

Another 707 incident happened to -436 G-APFP on 31 March 1967 which had been away from Heathrow crew training at Stansted. It seems that the aircraft had recently undergone maintenance and the last time the aircraft's nose gear had retracted the wheels were somehow offset by a few degrees to starboard. When the gear was subsequently selected down, the port nose wheel lodged on the edge of the wheel bay and stalled the activating jack and the emergency lowering system. Following the landing, a fire caused by burning magnesium in the Nose Landing Gear fairing was quickly extinguished by the Airport Fire Service.

Speedbird 712 was a BOAC service operated by Boeing 707-465 G-ARWE from London Heathrow Airport bound for Sydney via Zurich and Singapore. On Monday 8 April 1968, it suffered an engine failure on takeoff that quickly led to a significant fire. The engine fell off the aircraft in flight. After the aircraft had made a successful emergency landing, confusion over checklists and distractions from the presence of a check captain contributed to the deaths of five of the 127 on board.

The actions taken by those involved in the accident

Left: 31 March 1967 - the end of a crew training flight finishes with G-APFP on its nose at Heathrow.

Below: 707 G-APFE on finals at Heathrow. The aircraft was later to crash on Mout Fiju, Japan with tragic results.

The remains of 707 G-ARWE operating as Speedbird 712 on the runway at Heathrow

resulted in the award of the George Cross posthumously to stewardess Barbara Jane Harrison. Two other crew members received awards; a BEM and an MBE. As a result of the accident, BOAC changed the checklists for engine severe failures and engine fires, combining them both into one checklist, the 'engine fire or severe failure' checklist.

After an intensive investigation metal fatigue was ultimately blamed for the failure of the number five compressor wheel in the number two Rolls-Royce Conway turbofan engine, a failure that initiated a rapid chain of failures. The crew's omitting to shut off the fuel to the engine was blamed for the rapid growth of the fire and the loss of the aircraft. Check Captain Moss had accidentally cancelled the fire warning bell instead of the undercarriage warning bell. Moss had also issued orders to Captain Taylor, in breach of the standard protocol for his duties. However, the report on the accident also stated that Captain Taylor had briefed Moss to act as an extra set of eyes and ears inside and outside the cockpit. Moss's actions, therefore, could be seen as working within that remit. Although Moss had alerted the crew to the fire, none of them was aware that the number 2 engine had fallen off until after the evacuation on the ground.

By 1969 BOAC's 707-436s were beginning to approach the end of their life as first-line scheduled aircraft. A major programme of renovation was carried out ready for sale as the 747s came into service. BEA Airtours needed for larger aircraft to replace their Comet 4Bs and a satisfactory deal was arranged with BOAC under which two 707-436s went to Airtours in January 1972 and a further five by March 1973. All the increase in BOAC's capacity apart from the one further Super VC10 in May 1969 came from increases in the 707-336 fleet.

By that summer the 336Cs had increased to six aircraft and were proving invaluable both as cargo and passenger carriers. Early in that year, it became clear that if the Corporation were to make any serious inroads in the non-scheduled markets, it would have to acquire and apply specific capacity to that end. With the invaluable ability now open to it to purchase aircraft 'off the shelf' from Boeings at a time-lag of about a year, a case was made for two more 707-336C aircraft provided with kits for conversion to passenger layout when needed. These two aircraft were for delivery in March 1970 at a total cost of £8.05m, with an option on a third for delivery a year later. The intention was to use this additional capacity for the fast-growing movement of so-called VFR (visiting friends and relatives) traffic between the UK and Australia/New Zealand and between the UK and Canada. BOAC also had at the back of its mind that more long-range capability would be advantageous if and when Polar and trans-Siberian routes could be opened up to the Far East. Board of Trade approval for the purchase was received provided overseas borrowing financed the whole dollar cost.

By early summer 1971, the 336Cs had grown into a flexible fleet of nine aircraft and the wisdom of the original decision by the planners in 1960 to go for this type rather than the CL44 had again been justified. It was not long before a decision was taken to purchase more aircraft. In October 1969 the Board approved the case for two 336B airliners in unconvertible passenger configuration, one of which involved taking up the option obtained from Boeing earlier in the year. This purchase was explicitly linked to planned operations on the London-Moscow-Tokyo route expected to begin in summer 1971. The long Moscow - Tokyo sector was demanding on the range and for this reason the 336C aircraft with its strengthened floor and consequent higher empty weight did not give adequate payload, hence the need to go for the lighter 336Bs.

The ten aircraft now in the 336 fleet was increased to eleven when in November 1970 a convertible 336C was ordered from Boeings to replace the capacity loss caused by the destruction of the Super VC10 at Dawsons Field.

Above: at first glance this looks like a beautiful photograph of a BOAC 707 - then you see the registration is G-BOAC, never worn by a 707 - and you realise it is a supurb rendition of artwork used for a company promotional postcard. *(BOAC Press Office via author's collection)*

IATA negotiations.

During the 1960 to 1963 period, significant changes took place in the commercial policies of the major world carriers. Not only were more national carriers coming on the scene, but large jets such as the 707 and DC8 revolutionised the marketing aspects of the airline industry. The volume of both passenger and cargo capacity was increasing, and although the International Air Transport Association (IATA) agreement was reached on the level of fares and rates, there was no control on the amount of capacity a member airline could direct to a specific route. Under some agreements frequencies were laid down, but where the full principles of the Bermuda type agreement applied, there was freedom to operate without restriction subject only to review between Governments, a problematic and ineffective control. BOAC had met this situation, where it became possible to do so, by partnership and pooling agreements which involved sensible capacity planning between the partners.

The problem route was the North Atlantic. There was no shortage of passenger traffic during the holiday season, but this was highly peaked and also directional, creating problems of filling empty seats in the winter months and those periods when the traffic was moving mainly in one direction. The solution was to widen the market, spread its incidence and bring air travel within reach of more people, primarily by giving fare inducements for out-of-season travel. This is what the significant operators set out to achieve through the IATA machinery. During those three years, there were three major fare and rate Conferences, Cannes in October 1960, Chandler in Arizona in October 1962, and Salzburg in September 1963. All were concerned with the problem of setting fares at levels which would not only bring in more business but would also increase total revenues and push up load factors.

BOAC's experience over the four years April 1960 to March 1964 is enlightening. In 1959/60 the traffic revenue per CTM on scheduled services was 15p, and the corresponding revenue per passenger mile was 2.6p. By 1963/64 the former figure had fallen to 10.5p, and the latter to 2.2p brought about by the introduction of economy class and promotional fares, and by a swing away from first class and tourist class travel to the cheaper opportunities. If the load factors had responded adequately to the enticement of more affordable travel as was hoped, the financial results might have been satisfactory, but for BOAC this did not happen. The fall in the rate of revenue coincided with a drop in the passenger load-factor from 61.4% in 1959/60 to 47.6% in 1962/63 with some recovery to 52.7% in 1963/64; the overall load factor followed the same trend.

October 1960 saw three crucial innovations; first the introduction of skycoach fares to British cabotage points in the Caribbean, Africa and the Far East, well below the economy fare and available only to residents of the UK and the Colonial territories served; second, the introduction of economy class fares throughout the Far East and Australia and two months later on the Pacific routes; third a seventeen-day excursion fare applicable during the winter months on the North Atlantic. Low emigrant fares were introduced between the UK and the USA and Canada. A further innovation was the setting up of a tour-basing fare for agents one third below existing tariffs to be used solely for the construction of agents' inclusive tours. In 1961 economy class travel was extended to the South Atlantic route on BOAC's operations to the East Coast of South America.

In March of the following year after much bargaining within IATA, group fares were agreed on the North Atlantic for parties of twenty-five or more travelling together. It was thought such a group fare facility would be more rewarding than the 17-day winter excursion fare which BOAC believed had been counterproductive financially although generating more traffic. In fact, 1963 saw few developments in the fare structure. Traffic in the previous year had been poor, and airlines went to the Traffic Conference at Chandler in a defensive mood seeking status quo, although promotional fares in particular areas were agreed. One step taken to improve revenue was a change in the return fare rebate from 10% to 15% to take effect from the start of the summer season, but subsequently, this was thrown into doubt by the refusal of the US CAB to agree to the change. The European carriers fought this decision with the help of their Governments and succeeded in obtaining agreement to introduce the change in July 1963 but only after consenting to a small reduction in the economy class fare on the North Atlantic. By the summer of 1963, BOAC was preparing its fare policies for the forthcoming IATA Traffic Conference in Salzburg. The problems were still much the same; if fares were to be reduced, significant new traffic had to be secured if there were not to be a reduction in overall revenue.

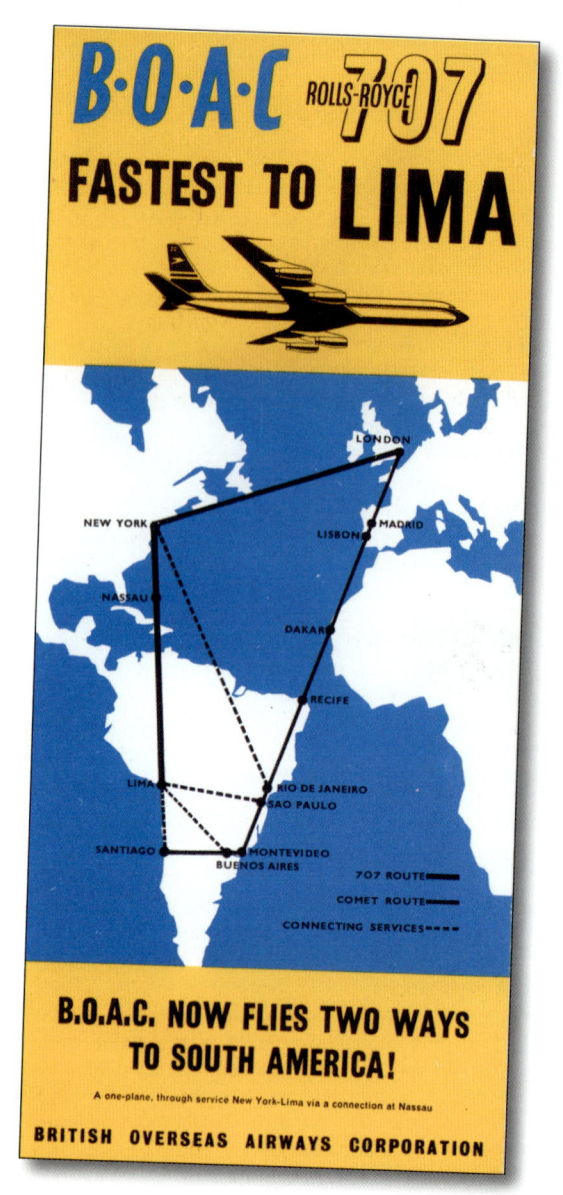

Many of the large airlines, especially on the North Atlantic, were increasingly concerned at the competition from charter operators who were estimated to carry around 18% of the total traffic, so there was some support for even lower economy fares which would be more competitive; certainly the US CAB made no secret of their desire to see a lower fare level and had circularized American carriers to that effect. However, after two sessions, one in September 1963 and a second in October, agreement at Salzburg could not be reached. The CAB Chairman toured round European Governments in November, but the outcome remained undecided at the close of the year. It was becoming more and more challenging to reach unanimity at the IATA Traffic Conference, and this was a situation which would continue and lead eventually to radical changes within the IATA machinery.

With its routes straddling the world, such issues were beginning to cause BOAC many problems in scheduling, especially the flooding of Atlantic routes by US capacity. BOAC's Far East services were also being cut into by the increase in low-fare charters from London by British Independents and sometimes by reciprocal charters from the other end. BOAC had been aware the way the market was developing and was already engaged in wholesale selling through Contract Bulk Inclusive Tours (CBIT) and Special Group Inclusive Tours (SGIT) and had been instrumental in pressing for the acceptance of these facilities by other long-haul IATA carriers.

The Corporation's philosophy was consistent - to provide the basic fares and capacity to carry the low fare business on scheduled services rather than charters. It pioneered the so-called 'early bird' fares to Bermuda, Bahamas and Antigua from March 1970, where the passenger had to buy his ticket at least four months in advance of travel and was not able to change his reservation or to obtain a refund. By these means, BOAC could avoid last-minute cancellations and could plan for high seat factors. Other scheduled airlines were, however, slow to follow. The Corporation was deeply concerned at the extent of the competition from non-scheduled carriers - British, American and Canadian - who were operating large numbers of affinity group charters; unfortunately, the rules were open to extensive abuse, and it was almost impossible to prevent individuals getting in on the affinity flights by claiming bogus qualification.

The seriousness for the scheduled operator was only too apparent on the UK-Canada route whereby season 1970 charter traffic formed 50% of the total. It was not just the affinity group loophole which was worrying; ATLB had granted licences to non-scheduled airlines for inclusive tour charter flights where the price of the tour to an individual could be less than the scheduled fare to the same destination, which offered a passenger the opportunity to travel on the flight without making use of the hotel accommodation provided, thus undercutting the scheduled fare. It seems strange that the other carriers planned were so reluctant to follow BOAC's lead in pressing for low ticket prices with reasonable booking conditions. Several operators set up subsidiary companies, not members of IATA, through which the parent company could match charter competition. In December 1970 the Corporation decided it had no alternative but to follow suit having failed to obtain in IATA a suitable fare package.

BOAC had a good reasons for taking this action. Not only did the UK-North America market offer the most significant potential of low fare holiday traffic but the emergence in the UK of the second force airline with the declared intention of serving both the scheduled and non-scheduled markets meant that the challenge had to be met, otherwise its operations planned would wither away. Accordingly a subsidiary called British Overseas Air Charter Limited (BOAC Ltd) was established: At the same time the Corporation's backup organisation for the marketing of charters was strengthened; results proved the effectiveness of the new uninhibited approach, as the revenue which could be attributed to the charter type of traffic increased from £4.2m in 1970/71 to a record £12.6m in 1972/73.

In June 1971 the Presidents of the European Airlines had a discussion, the subject being the battle between the scheduled airlines and the so-called supplemental carriers known as non-scheduled airlines, for the North Atlantic market. The American CAB was pressing for further relaxation in the charter rules and, while the Europeans wanted to support their scheduled carriers and resist these pressures, they made it clear they could do so only if the airlines themselves could reach agreement on a more competitive fare structure.

A pair of company 707s, with a pair of Comets behind, surrounded by support vehicles.

The observance of the rules controlling affinity group charters had become so loose, especially on the route to Singapore and Bangkok, where the Independents, particularly Caledonian/BUA, were taking loads of passengers ostensibly as charters to those two points who then found their way to Australia, that the DTI stepped in to tighten up the rules, but legitimised these carryings meanwhile so as to create a breathing space for the Government and the industry to work out a solution.

The Minister decided to grant exemption from the CAA rules governing affinity group charters for a total of one hundred and eleven flights to South East Asia from August 1971 to March 1972. To seem to show impartiality, DTI gave BOAC similar permission for sixty-six exempt flights to Kuala Lumpur, Singapore and Bangkok and twenty-five to India, but was most surprised when BOAC at once took up the offer. The DTI insisted that the facade of a charter should be kept up and so BOAC had to set up a subsidiary company, Overseas Air Travel Ltd (OAT), which would charter the capacity for BOAC Ltd. As BOAC's Annual Report for 1971/72 points out this created the farcical situation that *'...OAT, a BOAC Subsidiary, sells seats to the public on a series of flights which it has chartered from BOAC Ltd, another BOAC Subsidiary, which in turn arranged for BOAC to operate the aircraft on its behalf'*.

Exempt flight permissions were extended numerous times by the DTI, but ended in December 1973 to Kuala Lumpur; exempt flights to Singapore had ceased in August 1973, being superseded by new CAA charter arrangements, under which BCAL was able to operate for a time under Singapore Government permission. Exempt charters to India ceased for everyone in July 1973 as new low fares were introduced on scheduled services at little above the charter rate.

APEX or ABC by TGC to XYZ?
BOAC played a leading part in urging Governments and other airlines to bring order into the chaos of the low fare markets. By 1971 charters had become virtually indistinguishable from scheduled services and so posed a genuine threat to scheduled operations. The method urged by BOAC, whose Commercial Director, Richard Hilary, was in the forefront of the campaign, was the concept of advance purchase of tickets well before flight departure, and the creation of Advance Booking Charters (ABC) which could be marketed by the non-scheduled and also the scheduled carriers who would dispose of a portion of their scheduled capacity on a wholesale basis. The British Government put its authority behind this solution which involved the virtual abandonment of the discredited affinity group basis. The European airlines now began to support the advance purchase philosophy and early in 1972 submitted their ideas to CAA: in February 1973 BOAC was granted a licence by the CAA to operate ABCs and OAT was licensed as a travel organiser to make the arrangements.

The fares may have become cheaper for the travelling public, but the rules regarding their use became increasingly complex, forcing BOAC and other airlines to publish explanations as to just what the new ticketing arrangements were. The standard approach seems to have been leaflets that took the form of 'Questions and Answers' and they reveal just how difficult it was to fulfil the requirements of travel:

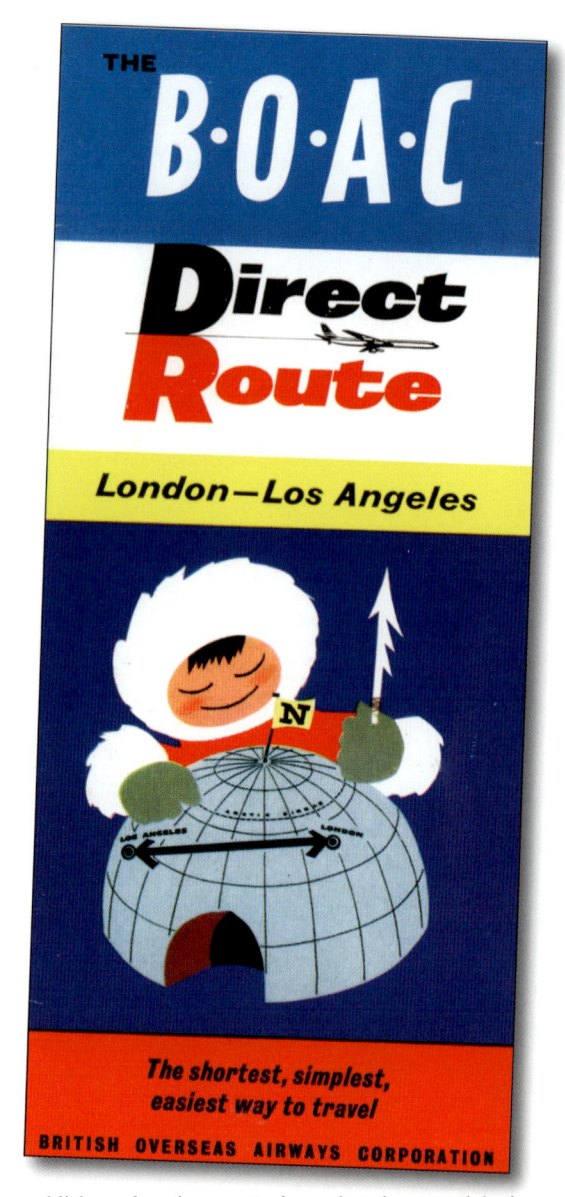

What is an Advance Booking Charter?
An Advance Booking Charter is a round-trip air travel facility which was introduced by the Civil Aviation Authority in 1973 for journeys to and from NorthAmerica and the Caribbean area. It is designed to bring relatively low-cost air travel within reach of the public as a whole, without requiring passengers to be members of any club or organisation. Its main feature is the requirement for the passenger to book in advance. ABCs can be organised in this country only by an Air Travel Organiser licensed by the Civil Aviation Authority. To travel on an ABC, you must book your flight at least 60 days in advance and stay abroad for at least 14 days, including travelling time (or at least ten days if you leave after 31 October and return before 1 April).

Are ABCs reliable?
To operate an ABC, airlines have to hold a special licence, and only Air Travel Organisers licensed by the CAA are entitled to organise them. The CAA has the power to take away the licence of anyone it considers to be unfit to hold one.

What are the drawbacks with ABCs?
Generally, ABCs do not offer the same guarantee concerning the date and timing of flights which is expected of a more expensive scheduled service. ABC fares can be kept low only if aircraft are full or nearly full, and air travel organisers must, therefore, have some flexibility in making their arrangements and may change the date and times of flights. They also reserve the right to cancel under-booked flights, or to combine several flights in one. However, flights may not be combined less than 30 days before the date of the departure, although flight arrangements may be changed for other reasons after this date.

Most ABC passengers do not encounter these difficulties, but you should bear the possibilities in mind when choosing between an ABC and scheduled service, and you should read the conditions of booking carefully before committing yourself.

Can I fly anywhere in the world?
Not yet. At present ABCs are being licensed only to and from North America and, broadly, the Caribbean. Similar arrangements may be extended to other countries, but this will require the agreement of the foreign governments concerned.

What do I have to do to book an ABC?
You can book an ABC with a licensed Air Travel Organiser or through a travel agent acting for an Air-Travel Organiser. You must make your booking more than 60 days before the flight, and give your name and address. You will probably be asked to pay a deposit, although this is not required by CAA rules.

Why more than 60 days?
Because the list of passengers for an ABC flight, together with their addresses, has to be in the hands of the CAA at least 60 days in advance of the date of departure. In fact, you are advised to book at least 70 days ahead, and preferably even longer so that the Air Travel Organiser can process the bookings and send his list to the CAA in time.

What happens if I want to cancel my booking?
If you have to cancel your booking, for example through illness, you may not be entitled to a refund of any money you have paid. Booking conditions usually provide for financial penalties on a sliding scale for cancellation by the passenger, and you would be wise to read such provisions carefully before committing yourself and to ensure against these, and perhaps other risks.

What happens if I cannot get on the ABC flight I want?
If one Air Travel Organiser cannot accommodate you on his flights, you can always try others. If you are unsuccessful in finding a suitable ABC flight, the alternative is to book on a service of some other kind, but this will almost certainly be more expensive.

Do I have to be a member of a club or society?
No. Anyone may book ABC flights.

Will I risk being stranded in a foreign country?
The Air Travel Organiser is required to issue all ABC passengers with a return ABC air ticket, showing clearly the date of the outward and return journeys, and the airline they will be flying with. The same Air Travel Organiser must arrange the outward and return flights. These and other measures are designed to reduce to a minimum the risk that passengers will be stranded.

Are charter flights safe?
The safety requirements for aircraft registered in the UK are the same for all aircraft whether they are operated on charter or scheduled flights.

Can I use an ABC flight for a business trip?
Yes, if you can book more than 60 days in advance and are away for at least 14 days including travelling time (10 days in winter).

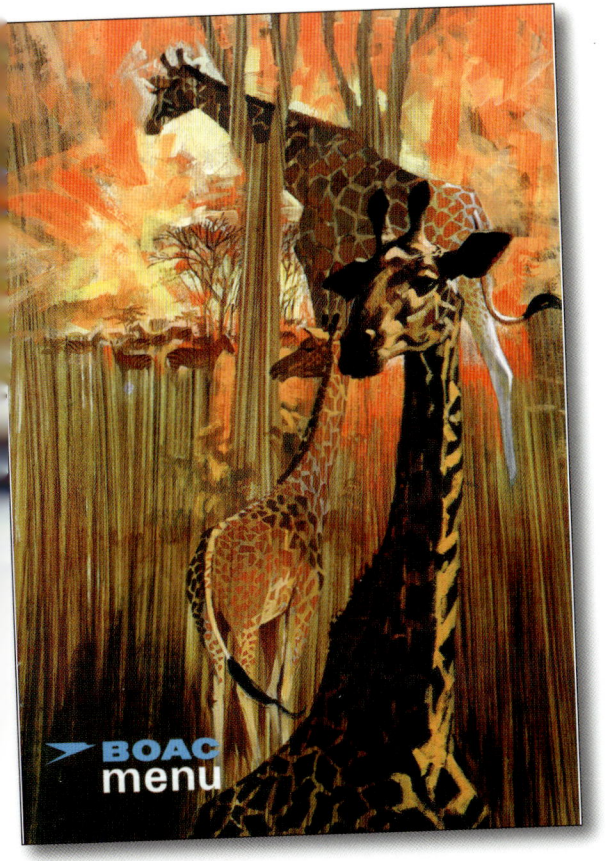

BOAC's Menu Cards progressed from simple typewritten sheets to quite large, full-colour booklets with covers that reflected the global nature of their services...

Can I book a single ticket or a single ticket with a voucher or an open-dated ticket for the return journey?
No. The rules require ABC passengers to be issued with a return ticket showing dates of departure and return, and the name of the airline. Anyone selling you a single ticket for an ABC flight, or a voucher to be exchanged later for a return ticket, is breaking the rules.

What happens if I miss my outward or return flight?
Although in an emergency, and with the CAA's prior permission, the rules allow passengers to be carried on flights other than those on which they are booked, there is no guarantee that you will be carried if you miss your flight. Your contract with the Air Travel Organiser will usually be for specific ABC flights, and carriage on alternative flights is (subject to the CAA's prior permission) at the discretion of the Air Travel Organiser and airline concerned. You should, therefore, ensure that you arrive in good time to catch your flight. If you miss it, you may have to make your own alternative arrangements, finding yourself both seriously inconvenienced and out-of-pocket.

What information can I expect when booking an ABC flight?
The Air Travel Organiser or the travel agent should be able to tell you the name of the airline operating the outward and return flights, the dates of the flights, the type of aircraft and the airports of departure and destination. If the person offering you the flight refuses, or is unable to provide this information, you should be very cautious about entering into any commitment or parting with any money.

How much will my ABC flight cost?
ABC fares are not regulated by the CAA but are left to the commercial judgment of Air Travel Organisers. You may find it worthwhile to compare prices. However, fares may be increased because of rising costs, including fuel costs. You should ask your travel organiser about this before you book.

By the summer of 1973, BOAC was blocking off scheduled services and using the capacity to operate ABCs; a move needed by the failure of the US and the UK Governments to reach agreement on advance purchase excursion (APEX) fares for individual travel. BOAC relying on the support of the CAA had in fact sold over 17,000 APEX seats to the USA and Canada, and on the refusal of the CAB to accept the APEX principle, all the US bookings had to be transferred to BOAC's ABC operations or scheduled services. The two Governments could agree no comprehensive policy - in the UK the CAA approved the plan of ABCs while the CAB in America introduced Travel Group Charters (TGCs). The introduction on the UK-USA route of these new facilities took place on 1 April 1973, with each country applying its own rules. The UK phased out the discredited affinity group travel, although this continued to be available in the US. It was nearly two years later, in January 1975, that

...on the flights into the United Kingdom BOAC often used images that reflected 'all things British'.

agreement was reached between the carriers and approved by the Governments for the introduction of APEX fares on North Atlantic routes, a belated triumph for what was by then the Overseas Division of BA, although BOAC had already been able to bring this type of fare to the South African route from November 1973 by agreement between the UK and South African Governments.

The second threat to BOAC's position on the North Atlantic came from its scheduled airline competitors, Pan Am and TWA. These two carriers had been building up frequencies on the USA/UK route not only from New York but also from Chicago and Washington to such an extent that they were swamping these routes with capacity. Figures taken from the American Airline Guide for May 1969 showed US airlines operating one hundred and seventy seven frequencies per week to the UK and only sixty to France. In that year BOAC began to urge the UK Government to take action under the terms of the Bermuda Agreement, which provided for examination of capacity if this were getting out of line with potential traffic offering and that there should be a fair and equal opportunity. Informal talks were held with the Civil Aeronautics Board at Government level, but no progress was made.

The London-Chicago and London-Washington routes were particularly problematic; traffic density was far less than on London-New York and did not justify the capacity being deployed by both Pan Am, and TWA. Both carriers, but especially Pan Am, were making use of points behind

Right: The prestige of the so-called 'Royal Tours' were always important for BOAC and the Corporation always went all-out to gain publicity. This picture, taken on 21 January 1963, relates to the eight stewardesses chosen to fly with the Queen and Prince Philip on their flight to Fiji and Australia.

Below: taken around the same time, this colour shot shows the severity of the dark navy uniform against the crisp white blouse. A few years later the girls were wearing frivolous paper dresses on board! *(BOAC Press Office via author's collection)*

their bilateral gateways in the USA and points beyond London and BOAC's contention was that they were using London as a point of convenience even where the volume of traffic would support direct flights into Europe.

As part of BOAC's campaign, Granville spoke at the International Aviation Club in Washington in November 1969 about the Bermuda Agreement and North Atlantic services. Shortly afterwards Robert Burns, Deputy Secretary at the Board of Trade, and BOAC's Planning Director had further discussions with US officials in Washington. But restraint was not on the cards in the US and they thought the UK would not press the matter to the extent of calling for a capacity review. More talks in Washington in May 1970 still produced no positive results and, unfortunately, the problem BOAC was having in introducing its 747 reduced the effectiveness of its arguments. However, during that summer the general introduction by the US international and domestic operators of the wide-bodied jets caused even the Americans to take stock; a significant development was an approach to the CAB by some of the domestic carriers advocating capacity limitation over a wide range of domestic routes. Another round of UK/US talks at Government level took place at the UK's insistence in Washington in January 1971 but again met with stonewalling although the British tabled detailed statistics to support their case. By then Pan Am and TWA were having severe financial problems caused by

A company 707 converted into a cargo hauler.

over-ordering the large jets and by weak traffic growth; prospects for some capacity limitation especially on London-Chicago and London-Washington looked more hopeful.

An agreement was reached for the summer of 1971 that Pan Am and TWA would act with restraint on London-Chicago and London-Washington, Pan Am withdrawing its 747s from the Chicago route and TWA from the Washington route. The persistence of the British at last won through, and by the end of 1971 similar talks were being held by BOAC with National Airlines on the London-Miami route; these were unsuccessful and made it necessary for the DTI hold talks in Washington at the end of April 1972. No agreement could be reached: National Airlines were, however, told by the UK authorities that the airline's summer schedules to the UK would be restricted unless capacity was reduced.

In June National had their frequency restricted by the British Government to four 747 flights per week instead of the daily service planned. The US State Department reacted strongly and acrimonious meetings were held in London between the two Governments. In retaliation, the CAB required BOAC to file its schedules in advance. Common sense eventually prevailed, and an agreement was reached with National to operate five 747 services per week during the rest of the summer of 1972, and each airline would operate daily 747s in the 1973 season by which time the market would have grown. Honour was satisfied, and BOAC owed a debt to the DTI representatives for the lively way they had argued the case. It had created a valuable precedent in relationships with the US. So it was proven that the Bermuda Agreement, had teeth.

In addition to the fierce competition from US operators, BOAC in its last years had to face a threat from British carriers licensed to compete with it on its traditional routes. Although the ATLB had turned down Laker Airways application for a Skytrain service from Gatwick to New York in December 1971, it had granted all BCAL's requirements on trans-Atlantic routes in February 1972, including New York and Los Angeles, and BCAL became a North Atlantic scheduled operator in April 1973 when it commenced Gatwick-New York services. Later in 1972 BCAL applied for licences for a wide range of routes, namely to Atlanta or Boston with optional calls at Washington, Houston and Dallas, to Toronto and Vancouver, to Perth and Sydney via the Seychelles, and to Singapore via Bahrain or Karachi. BOAC opposed all these applications. But by August 1973 BCAL had been granted Boston and/or Atlanta, and/or Houston and also Toronto and Singapore, but was turned down on Australia. The outcome was considered so severe to BOAC's interests that it decided to appeal against the North Atlantic licences but without success. The Conservative Government was determined to open up new opportunities for the private sector. The new CAA also overturned the ATLB decision and gave Laker a licence for Skytrain. When the Minister overruled the CAA, Laker took legal action as to the Minister's powers concerning the CAA and won. So he was free to go ahead with his Skytrain plans, but this was after the end of BOAC.

Chapter Eight

Try a Little VC Tenderness!

When Harold Watkinson announced in the House of Commons on 24 October 1956 that the Government had agreed to BOAC purchasing Boeing 707 aircraft, he sugared the political pill by including in his statement a reference to the fact that '*...the Corporation is urgently discussing with the de Havilland Aircraft Company Limited the technical details of an aircraft that will be sufficiently flexible to meet their requirements and command a ready sale in world markets. As soon as a satisfactory specification has been agreed, BOAC will place an order*'. In fact, the Minister's agreement to the Boeing order was contingent on BOAC placing an order for twenty aircraft from British manufacturers for use on Eastern routes able to cope with less developed airfields with which it was assumed at that time the Boeing 707 could not.

In reporting this to the Board, the Chairman said that the Treasury and the MoS would only agree to finance the new British aircraft if it was purchased from De Havillands to preserve the balance and welfare of the manufacturing industry. So initially BOAC's freedom of action was severely limited.

De Havillands showed little enthusiasm and presented BOAC with details of the Comet 5 only the day before a meeting between BOAC and the Ministry of Supply. At that meeting, De Havilland's made it clear they would not proceed with the project unless they received orders for fifty aircraft, and as they could see no possibility of that, BOAC was freed from the restriction on its purchasing policy.

Over the next three months the Chairman's Aircraft Requirements Committee were busy evaluating for the Corporation's route pattern a whole range of aircraft possibilities, eleven types in all including both turboprops and pure jets from four possible contenders, de Havilland, Bristol, Vickers Armstrong and Handley Page, matching their projects against the known performance and economics of the B.312, and the Comet 4, and the assessed performance of the Boeing 707-420. The de Havilland DH120, a swept wing Comet, was also considered but discarded as unsuitable.

In BOAC's view, the argument as between turboprop and pure jets was settled; it rightly considered on the grounds of speed, passenger attraction and above all economics that the suggested Brittania 422 turboprop with Orion engines would be uncompetitive against foreign operators using Boeing or Douglas jets. What the date in service BOAC was looking for was around 1963 or 1964 to take over from the Comet 4.

After a process of elimination, BOAC's first choice was the Handley Page HP.97, a civil version of the Victor bomber, and it so informed the Minister on 25 March 1957, but with significant reservations. The Victor had been developed as a military aircraft, and there would have to be technical examination as to its suitability for civil operations particularly in regard to the fatigue life of the structure; the delivery dates of the HP.97 could well be in 1961, two years too soon for BOAC resulting in early retirement of the Comet 4; this could be very expensive if it happened unless Comet 4s could be sold at a reasonable price. Finally, BOAC expressed doubts about Handley Page's capabilities for such a world-wide project, but as insurance proposed a consortium of manufacturers headed by Vickers, a name now well known in international aviation circles following the success of the Viscount.

On 29 March 1957, the Minister called a meeting with BOAC at which he threw doubt on any possibility of Vickers accepting such a proposal. Three days later the Chairman met Sir George Edwards, Managing Director of Vickers Armstrong (Aircraft) Ltd, who confirmed his opposition to any involvement with a crescent wing aircraft like the HP97, but said he would be prepared now to see if he could design a jet aircraft

Sir Basil Smallpeice with Sir George Edwards of Vickers (centre) and Keith Granville (left) at the signing of the contract for thirty-five VC10s on 14 January 1958.

to meet BOAC's needs. These covered a turbo-jet airliner preferably with four Conway engines, as in BOAC's 707s, able to fly London-Johannesburg with one stop and London-Sydney with three stops, with better airfield performance than the 707 and with a direct operating cost similar to and at least as fast as the 707.

Vickers acted quickly, and by the end of April 1957, BOAC had received details of a four-engined Mark 5 VC10 with Conway engines with a passenger capacity of one hundred and twenty-three in five abreast seating in high-density seat pitch. The four engines were positioned at the rear of the fuselage as this would result in a clean-wing design with aerodynamic advantages, with better cabin access and less cabin noise. Vickers also believed that engines submerged in the wings, similar to the Comet, would be difficult to sell in the United States; moreover, a rear-engined aircraft with an unusual 'look' would make it easier to market as something different from the American products. BOAC had no objection subject to the aircraft's handling characteristics and weight being satisfactory.

After evaluation, the Chairman's Aircraft Requirements Committee concluded that the Mark 5 VC10 was as good as anything BOAC could hope to get in the time scale and on Eastern routes showed up at least as well as the 707.

Discussions then took place between BOAC, Vickers, and MoS and the Ministry of Transport and Civil Aviation (MTCA). BOAC was prepared to order twenty-five aircraft with an option on a further ten at a fixed price of £1.6m each; for their part, Vickers alleged their breakeven point was forty-five aircraft, and they could not start the project without a firm order for thirty-five at a fixed price of £1.7m. So the question of aircraft numbers loomed large even at that early date.

At a meeting of the BOAC Board on 26 April 1957 financial questions were discussed, and it was decided that the Corporation needed thirty-two aircraft for the Eastern hemisphere routes and a further three for South America. Approval was given to order for thirty-five jets with an option on an additional twenty, delivery to commence in 1963 and that a fixed price of £1.6m would be acceptable, although discretion on that was left to the Chairman. When these terms were referred to Vickers, they countered by asking BOAC to place an order for thirty-five aircraft with an option on twenty more at a fixed price of £1.57m per aircraft as at 1 January 1960 but with escalation after that. The Chairman thought that after an allowance for increase the average cost per aircraft would work out at around £1.77m.

Four days later the Chairman wrote to the Minister telling him of Vickers latest proposal and that the BOAC Board was prepared to accept the Vickers offer for thirty-five VC10s, with delivery commencing in 1963 with twelve aircraft and the balance of one per month after that, together with an option for twenty more. The opportunity was to be exercised by August 1962. He estimated the total cost including provision for spares at £77.5m and reminded the Minister that this would mean over-stepping the Corporation's statutory borrowing powers but asked for approval to proceed. Having received the Minister's agreement, announced in the House of Commons on 22 May 1957, BOAC sent Vickers a Letter of Intent on the same date.

From then until the end of the year discussions continued with Vickers with the object of improving the aircraft for its Eastern hemisphere work. As a result, the shape of the fuselage cross-section was altered by substituting the Boeing 707 type for the previously proposed Vanguard type cross-section, thus increasing the potential passenger seating from 123 at 34-inch seat pitch, five abreast, to 135/147 at 34-inch pitch but six abreast. There was also an increase in base operational weight from 117,000 lbs to 125,000 lbs. Coming out of these discussions was a new concept of making the aircraft not only still suitable for Eastern Hemisphere use but also to have Atlantic capability, which would be made feasible by using a new mark of the Rolls-Royce Conway engines, the IIIc. This 'worldwide' version would have a somewhat larger wing area, greater fuel capacity and generally a strengthened structure. On first estimates it looked as though the operating costs might be acceptable although still above that of the 707.

By the end of the year the Board had before it all relevant papers about the world-wide version, including an outline specification, aircraft diagrams and a detailed evaluation of the aircraft as applied to the Corporation's route pattern. On the North Atlantic, the estimated elapsed time London-New York was almost identical to the 707 but direct operating costs were higher although doubts were expressed about the size of the difference due to differing take-off performance assessments between the American and British aerodynamicists. The price of the worldwide version was £1.702m per aircraft as compared with the £1.57m of the original version for which Ministerial approval had been received in the previous May. With the background that there was little or no possibility .that dollars would be made available to the Corporation for any further purchases of American aircraft, the Board decided that the worldwide version offered the best prospects but until Vickers could improve performance figures for the North Atlantic BOAC could not contemplate an option order for twenty over and above the underlying order for thirty-five. As to the latter, now that the aircraft would have considerably higher seating capacity than the original proposal of May 1957, the thirty-five should be reduced to thirty at a price of £1.702m each with an option for five more exercisable within one year from the date of the contract. Smallpeice was therefore instructed to conclude a contract with Vickers accordingly.

Vickers was only prepared to accept a reduced order for thirty aircraft if the price per aircraft were increased by £80,000 so that they were no worse off. In the meantime, the Royal Aircraft Establishment had reassessed the VC10 performance, and had come out with estimated direct operating costs for the North Atlantic which showed it could be on a par with the Boeing 707 - in fact the RAE threw doubt on the 707 meeting its performance guarantees. However, Vickers, the sellers of the aircraft, in their guarantees to BOAC made no such claim, and the Corporation had gone ahead although its estimate of the direct operating costs was higher than for the 707, and although it had initially insisted the VC10 costs should not exceed those for the Boeing. Apparently on the strength of the estimate from the RAE the Board on 9 January 1958 approved an order for the full thirty-five aircraft at the price of £1.702m each and the Minister was so informed and gave his

Right: One of the two pair of Rolls-Royce Conway engines.

Below: VC10 production at Weybridge.
(both Vickers Aircraft Ltd)

Four VC10s come together in the final assembly shop. . (Vickers Aircraft Ltd)

approval, although he did ask for assurances that the economic performance of the VC10 would be comparable to the 707, mainly as to whether Vickers was committed to achieving that. He received a reassuring letter from the Chairman on 25 February 1959 that the Corporation was satisfied on the point, although Vicker's guarantees remained as they were. The contract with Vickers was signed on 14 January 1958.

The specification aircraft had 129 seats at 34" pitch, and its all-up weight was 299,000 Ibs with Rolls-Royce Conway 3C/15 engines. There was still doubt that the aircraft could in fact match the 707 on unit costs and talks between Vickers and BOAC were almost continuous to see what improvement could be made, either by increasing the seating with structural redesign or by the latter together with the fitting of improved Conway 4 engines with increased take-off power, or, and this was the recommendation of the Chief Engineer, that the 129 seat variant as ordered should have Conway 4 engines installed. In March 1959 the Board accepted the Chief Engineer's recommendation provided installation of the Conway 4s did not exceed £35,000 per aircraft. The proposed aircraft had undergone significant changes since the original specification in May 1957 and this together with runway extensions at New York and London Airport had reduced the estimated direct operating cost.

In October 1959 Harold Watkinson was replaced by Duncan Sandys as Minister, but as Minister of Aviation covering both the manufacturing and airline interests, consequently, the health of the aircraft manufacturing side was also his concern. Vickers was having financial problems attributed to having underestimated the cost of producing the Corporation's thirty-five VC10s, and BOAC was expecting an approach from the Minister asking it to give Vickers further VC10 orders either of the existing specification or a developed and larger aircraft. Smallpeice, therefore, instructed his Planning Group in November 1959 to produce a review of the Corporation's capacity needs up to 1966/67 with particular reference to the VC10, together with a unit cost comparison between the Standard VC10, the 707-436, a possible enlarged 707, the type 520, and a Super VC10 as now put forward by Vickers with significantly increased passenger capacity at 191 seats. The 707 types still offered the lowest direct operating costs on the North Atlantic as compared with the Standard VC10 and with Vickers new Super VC10 concept, with much earlier delivery dates by up to four years for the 707-436; moreover, Vickers looked for a minimum order for twenty Super VC10s, and if less, their price would rise significantly.

The conclusion reached in the Planning Group's paper was that no increase should be made in BOAC's commitment to the VC10 beyond the thirty-five already ordered and that the Corporation's right not to have to decide before August 1962 on the option for more aircraft should be maintained. The Board came down firmly in accord with that conclusion.

The matter did not rest there, for less than two months later Vickers were confirming to BOAC and the Minister that they had under-estimated the cost of producing the VC10 and as a result, their financial situation was such that they might not be able to continue with the aircraft's production. However, they went on to say that they could do so if BOAC were to order immediately ten Super VC10s at a price per aircraft of £2.707m. BOAC was deeply concerned at this development since its plans for 1964 onwards wholly depended on the availability of the VC10.

On 7 January 1960, the Board instructed the Chairman to discuss the matter with the Minister, who had already made it plain that he hoped BOAC would be able to place further orders as Government policy for the aircraft manufacturing industry was involved. At a special Board meeting on 15 January 1960 Sir Gerard d'Erlanger reported on a meeting he had had with the Minister who was told that the Corporation was considering placing an order with Vickers for ten Super VC10s. In the Board minute of the time, the following appears: *'the Minister's attitude was that, while he would not press the Corporation to undertake any such commitment, an order for Super VC10s would be helpful to Government policy for the aircraft manufacturing industry, and it was a matter for the Corporation to decide. The Minister had further stated that it was not the intention of the Government to prejudice the planned expansion of the Corporation'*.

The re-assurance regarding expansion had been sought by BOAC in the light of the Government's favourable attitude towards the growth of British independent airlines.

On 19 January the Chairman wrote to the Minister informing him that the Corporation had told Vickers it was prepared to order ten Super VC10s at £2.4m each, but he went on to state that *'although we believe the need for that number of additional trans-Atlantic aircraft will exist in 1965/66 we would not decide to place an order for them now if we were left to our own commercial judgement . . . Commercially speaking we would prefer to delay a decision until more information about these possible developments (i.e. developed 707s and Supersonic aircraft) was available, and until nearer August 1962 which is the option date we have on our current contract with Vickers for ordering more aircraft'*. The letter also contained a request for assurance that any change in the Air Corporations Act and the terms of reference of the new Licensing Board would not prejudice BOAC's expansion.

The Minister replied that there was no doubt BOAC's decision would be most helpful to the aircraft industry, but the only assurance he gave in regard to BOAC's expansion was that such plans *'would be taken fully into account by the Air Transport Licensing Board'*.

After further discussions over the price, BOAC agreed to an order for the ten Super VC10s at the cost of £2.55m per aircraft but did so on the expectation that if Vickers again ran into financial difficulties over the VC10 production, the Government would step in to assist them. The Minister was told on 17 February 1960 that BOAC's negotiations with Vickers were completed and the contract was awaiting Governmental approval before signature. Three weeks later the Minister gave the go-ahead, Treasury clearance having been obtained,

The twin-jet pods in the tail show these aircraft to be Vickers VC10s. The clean wing was the key to high speed, though seen here with the powerful flaps down. G-ARTA was the prototype. *(Vickers Aircraft Ltd)*

and on 23 June 1960 the contract for ten Super VC10s between BOAC and Vickers was signed after Vickers had finally offered satisfactory noise level guarantees.

Neither the Ministry nor BOAC was at that time in any doubt that there would be a cost penalty on the VC10 against the Boeing 707 not only on the North Atlantic but also on Eastern and African routes where it was now accepted that the 707 could operate satisfactorily. All this was made clear in a letter from the Chairman to the Minister on 12 May 1960, although he expressed the hope that better airfield performance from the VC10 might narrow the gap.

This then was the state of play on the whole VC10 issue at the time of Sir Gerard's resignation in July 1960. The Corporation was committed to order for thirty-five regular VC10s plus an order for ten Super VC10s, the latter with passenger seating of around 191, a higher seating capacity than anything BOAC had previously considered. The incoming Chairman would undoubtedly have preferred to have had more room for manoeuvre.

There is no doubt that when Sir Matthew became Chairman in mid-Summer 1960, he was gravely concerned at the VC10 situation. There was not only a question mark about the total numbers on order — forty-five aircraft, thirty—five Standards and ten Supers - but also as to the operating costs of the aircraft compared with the Boeing 707.

Indeed, when giving evidence before the Select Committee on Nationalised Industries in March 1964 Sir Matthew Slattery, who followed Sir Gerard as Chairman, told the Committee that he had asked him shortly before the latter left BOAC in July 1960 not to commit himself on the ten Super VC10s until he, Sir Matthew, had had a chance to consider the matter. Sir Matthew quoted Sir Gerard as saying *'I am sorry, but it is too late. I have been under very strong pressure to do this. It is necessary for the Minister's plans in connection with the formation of the British Aircraft Corporation'* - a reference to the Government's plans to restructure the aircraft manufacturing industry.

After discussions within BOAC and mainly taking account of the Chief Engineer's views in favour of a somewhat smaller Super VC10, Slattery and Abell visited Vickers in January 1961 to sound them out. After further study, it was concluded that both financial and operational advantages would result if the Corporation were to abandon the 187/191 seater and opt for a smaller aircraft accommodating 159 passengers. Vickers was able to introduce a Super VC10 of that size into the delivery programme from the sixteenth aircraft. The Corporation, therefore, decided to modify the entire VC10 order from thirty-five Standards and ten Supers to fifteen Standards and thirty Supers, the Standards with 135 and the Supers with 159 economy seats. Such a change involved an increase in the potential overall seating capacity of the fleet equivalent to over one aircraft extra, but in response to a Board Member who suggested it would, therefore, be sensible to reduce the Supers order from thirty to twenty-eight aircraft, the MD was in favour of maintaining the full number, and this was agreed.

Negotiations then proceeded between the Chairman and Sir George Edwards of Vickers on the question of the overall price for the revised order, and a figure of £101.4m was finally agreed, being an increase of £6.3m above the amended price for the original order. The Treasury were not willing to give their agreement to any increase above the amount in the original contract; BOAC, therefore, had no alternative but to negotiate with Vickers for the cancellation of three of the Standard VC10s to remain within the Treasury mandate. To compensate Vickers for the cancellation, the Corporation had to pay in effect a sum of £600,000: the £510,852 already paid as progress payments on the three aircraft was then put against the cancellation fee.

With the April 1961 contract hardly dry, the Treasury demanded cuts to the Super VC10 order. The Minister of Aviation, now Julian Amery, rejected this out of hand. In October he wrote to the Treasury: *'I would remind you that the VC10 project was closely considered by Ministers in 1959 when Vickers were saying they would have to go out of the civil aircraft business altogether if this programme was not further supported.*

this is the finest airliner in the world

BOAC VC10

The BOAC VC10 is a superlatively comfortable jetliner to fly in. Almost everyone who has ever been aboard it will agree.

The seats give you plenty of room. Air conditioning and pressurisation are ultra-comfortable.

The engines have exceptionally high thrust for the weight of the aircraft. This means you take off and land almost before you know it. (For instance, you are airborne half a minute from brake release.)

The powerful fanjets are by Rolls-Royce – an indication of the little extra luxury you'll find at every point of a VC10 flight.

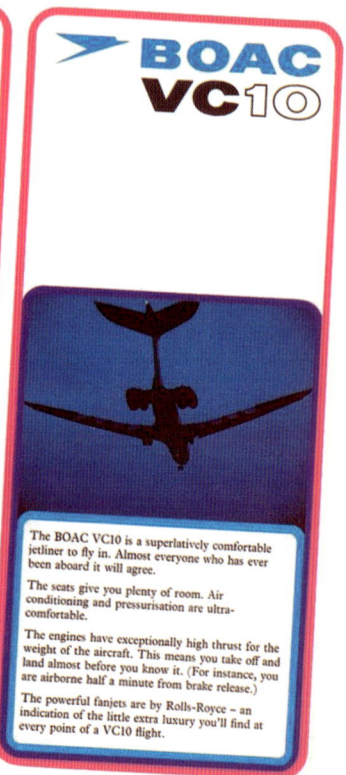

Above: An early public relations photograph of the VC10 in flight.

Right: Jack Bryce, who piloted the VC10 G-ARTA on its first flight, leaves the aircraft. Behind him is assistance pilot Brian Trubshaw, who later tested Concorde. *(both Vickers Aircraft Ltd)*

It was out of this situation that the Government decided must continue with the further support represented by the BOAC order for 10 Super VC10s, and additional support for development, and this additional support was one of the considerations which led to the formation of BAC.

To go back on all of this at the present stage, would be quite unthinkable. To continue the uncertainty at this stage would cause enormous difficulties with Vickers, with the inescapable implication that the VC10 order might be substantially reduced or even cancelled.

A fundamental reassessment at this stage, with all the delay it would incur, the doubts that it would cast on the only big British jet in prospect and the uncertainty it would create in BOAC, would be a major calamity".

At the time in April 1961 when Sir Matthew was in touch with the Minister over the question of the import duty on the three extra Boeings, he also raised the subject of the comparison between VC10 and the Boeing 707. He told the Minister that on his tours abroad he had made a particular point of stressing the merits of the VC10 in reply to Press questions as to how the VC10 could be expected to be superior to the 707. It was in BOAC's interest, quite apart from the national interest, to promote the new British aircraft. In saying all this, the Chairman was concerned that the Minister should not get the wrong impression. Now that the 707 was in service, its costs were a known factor, and BOAC was estimating that in a full year's operation of VC10s the same seating capacity could be provided by 707s at a lower annual cost of between £7m and £11m. To eliminate this difference BOAC estimated it would have to obtain at least six points of load factor higher on the VC10s than on the 707s, and while there might be some passenger attraction to the newer aircraft, it would be a daunting task to make good such a vast difference.

The Chairman's reasons for raising the matter were set out in these words *'there will come a time when it will be my duty to bring these facts to light in my public report to the Minister. I am very loath to do so this year . . . , but I have to consider the morale of all those who*

Above: another view of the VC10, showing the 'clean' wing and rear-mounted Rolls-Royce Conway engines.

Left: A BOAC stewardess poses in the First Class area of a VC-10

The maiden flight of the first VC10 took place at Weybridge on 29 June 1962, and from 8 November the main test programme began. Delivery to BOAC was due to start in January 1964.

In October 1962 the VC10 flight management appointments were made with Captain Rendall as Flight Manager, Captain Frank Walton as Deputy, and Captain Jack Nicholl in charge of training.

The drastic and unexpected decline in airline traffic growth in 1961/62 threw into disarray the forward estimates for later years. Although there was some improvement in 1962/63, it was not sufflcient to make up for the 'lost' year of 1961/62, and by November 1962 when BOAC could take a balanced view of the year's likely outcome, it was explicit that forward estimates of required capacity would need to be revised downwards. On 26 November General Manager Planning put a paper to the Board of Management which indicated that in the year of full VC10 deployment, i.e. 1967/68, BOAC would have ten or eleven spare large jet aircraft. The Board accepted at its February 1963 meeting that a surplus of the order of ten aircraft would exist and considered that BOAC should be relieved of the commitment by reference to Government or by negotiating with Vickers. The Chairman told the Minister of the conclusion reached, putting the surplus VC10s at ten to thirteen, and he also informed the Chairman of Vickers, Lord Portal.

Vickers requested that no announcement should be made at that time and gave an undertaking that postponement should be made at their request of detailed discussions would not involve the Corporation

work so devotedly for the Corporation's financial results, criticism that might be tempered if the facts of what the Corporation is doing to sustain the British aircraft industry and our interests abroad were more widely known'.

More attempts were made to forecast the cost differences between the two aircraft but so much depended on the assumptions taken.

Eventually, BOAC called in Stephen Wheatcroft in March 1962 to do an assessment; he arrived at an annual difference of £10.7m by 1966/67 which might reduce down to £7.5m in later years. So much continued to hinge on the passenger appeal of the VC10, an unknown factor at that time.

in additional cancellation charges. The Minister then asked BOAC to complete an exercise to assess the economic effect of replacing thirteen of its 707s by Super VC10s. The financial assessment turned out to be overwhelmingly in favour of retaining the full fleet of 707s. At a meeting with the Minister and Vickers on 6 May 1963 the Chairman made clear he was awaiting a proposition from Vickers about the excess aircraft; not surprisingly the Minister was anxious to see if some solution other than cancellation could be found. By the end of May 1963 the Corbett Report was already in the hands of the Minister, and until he had had time to decide his reaction to it, the issue of VC10s numbers (and other items) was put into cold storage.

Meanwhile, the production of the Standard VC10 was well advanced. The first aircraft for BOAC had flown on 8 November 1962 and performance data were becoming known facts rather than estimates. By March 1963 test flights had shown that the aircraft's performance was marginally above minimum guarantee, but Vickers was still confident that eventually, the aircraft would be considerably better than that. The problem was primarily one of reducing the drag, and Vickers undertook to carry out the necessary work on the Super VC10; on the Standard VC10 some but not all of the modifications to reduce drag were undertaken by Vickers ensuring that at least minimum guarantees of performance would be met and Sir Matthew agreed to accept delivery of the twelve aircraft on that understanding.

By May. 1963 the Minister and the Chairman had flown in the VC10 and were impressed by its passenger appeal. In an article, the MD wrote in the *'BOAC News'* to all staff he listed some of the good points about the aircraft: a lower noise level in the cabin; a good pressurisation system; a landing speed lower than the 707 and greater airfield flexibility. An extensive advertising campaign was undertaken by the Corporation to promote the VC10 world-wide once any doubts about its C of A date had been removed. *'The Times'* came out in November 1963 with a full supplement on the aircraft well publicised with advertising material by BOAC and in its advertising campaign it made the most of the new shape of the VC10, its quietness in the passenger cabin under the slogan of 'swift, silent, serene' and its slower landing speed as compared with US jets.

There was a proposal to modify up to eight of the Super VC10s to a convertible passenger/ freight version as the result of a recommendation by the Corporation's air cargo study group. These aircraft would be provided with a large freight loading door, strengthened floor, movable dividing bulkheads and removable cabin furnishings. However, this proposal became caught up in later events and never materialised.

A programme of crew training was undertaken in readiness for the start of passenger services early in 1964, and in October 1963 a 1,000 hours development programme began with intensive flying by BOAC crews

Left: the economy section of the VC10.

Below: G-ARTA poses before a suitably dramatic cloudscape for this publicity picture.

Captain Athrlstan Sigrid Mellersh 'Flaps' Rendall BSc, Senior Captain (First Class) BOAC , Flight Manager VC10s.

Rendall had a distinguished flying career; born on 3 January 1914, he was educated at Gresham's, Akworth and Leeds University. Joined Imperial Airways in 1936, he transferred to BOAC; Flight Superintendant Hermes and Yorks 1950-1953; Flight Superintendent Britannias 1953-1957; Flight Manager Briannia 102 Flight 1957-1960. He was awarded the Queen's Commendation for Valuable Services in the Air in 1954.

mainly on routes to the Middle East and to West Africa but also across the Atlantic to Canada. Engineering staff were gathered at Headquarters from overseas to undergo courses to familiarise themselves with VC10 maintenance needs and to acquire the necessary licences. Commercial staff were educated on the aircraft's main selling points and given visual presentations on the massive advertising programme planned for the new year. As BOAC said in its annual report *'every precaution that could be conceived by forethought and planning was taken to ensure the smooth introduction of the VC10 scheduled service'*. The subsequent success of the original programme on the routes confirmed how thorough had been the preparations.

More changes...
With the appointment of Sir Giles Guthrie as the new BOAC Chairman and the replacement of a number of board members, the Minister had told the Commons on 21 November 1963 that he had invited Sir Giles to prepare a plan during 1964 to make the Corporation financially sound, and on 1 January 1964, his first day at BOAC Headquarters, he set about the task. He summoned the General Manager Planning, Winston Bray, and instructed him to critically examine the whole of BOAC's route structure with the object of identifying the immediately profitable routes, those likely to be profitable shortly and those whose profitability even in the long term was problematical. This critical study also to referred to areas of the world where the Corporation was not operating but was missing out by not doing so. From this study, it would then be possible to calculate numbers and types of aircraft required and from those decisions would emerge the number of staff needed.

This was the three-stage plan to put BOAC on the road to recovery. The fourth vital step - a reconstruction of the Corporation's finances - would follow later after detailed negotiations with Government.

At the end of February 1964, Bray handed the Chairman a composite plan known as G1 covering all the years up to March 1968 and including estimated financial results year by year. This plan met the Chairman's target of an operating surplus of at least 10% of revenue and required a fleet of twenty Boeing 707s, twelve Standard VC10s and eleven Super VC10s plus two large-jet freighters, preferably the Boeing 320C cargo version. However, Sir Giles felt it would be prudent to increase the forecast surplus and accordingly the plan was slightly reduced in scope, calling for the same number of 707s and Standard VC10s but a reduction in Super VC10s from eleven to seven. The revised G2 plan concentrated effort on routes where BOAC as the national carrier had an inherent commercial advantage, particularly to Commonwealth countries, to countries where substantial expatriate population existed, such as South Africa, and to English speaking North America; it simplified the route pattern by eliminating unprofitable intermediate stops and speeded up services to the Middle East. It excluded the loss-making New York-Washington sector, cut out the New York - Lima extension (although this was subsequently replaced in the plan) and concentrated European calls at three points, Rome, Zurich and Frankfurt.

The only significant change recommended was the closing of the route to the East Coast of South America because there were no foreseeable prospects of making the route profitable. As to new routes, it was envisaged that BOAC would extend its operations from the West Coast of the United States to New Zealand and Australia provided operating rights could be obtained from the Americans and at that time continuation of the unrewarding mid—Pacific route to Tokyo, and Hong

Kong would be reviewed. If operating rights became available from the Russians, it was recommended that Moscow should be used as a transit point to the East on some services in place of Cairo or Beirut.

The Chairman and the Board accepted the recommendations in this overall G2 plan at their May 1964 meeting although it was well understood that it had serious implications, in particular, the closure of the South American route and the requirement for only seven Super VC10s by 1967/8 out of the thirty on order. As to closing the route to the East Coast of South America, the General Manager Southern Routes, Trevor Glover, had tabled an exhaustive study at Management and the Board in February 1964, covering the history and prospects. The route had had a chequered career.

Opened in 1946 by BSAA, a loss of nearly £3m had been incurred by 1949 when BOAC took over operations on the merger with BSAA. Losses continued, and by 1954 when BOAC temporarily closed the route following the Comet disasters, a further loss of £2.8m had been suffered. Pressures then built up, particularly from the Foreign Office, for BOAC to reopen the route and so the Corporation carried out an on-the-spot market study in depth in 1957.

The forecast results from this study still showed losses. However, the Corporation resumed operations when the Comet 4 became available in 1960, but by the end of March 1964, it was established that a further £2.75m would have been lost, making a total of well over £8m since 1945. Even were the VC10s to be applied to the route, losses were forecast to continue, which would have been inevitable since the VC10 would have required an overall load factor of 75% to break-even, not an obtainable figure considering the low frequency imposed by the South American governments and the high route costs. Also changed routings, for example via Accra in West Africa, made little difference to the forecast result. The solution would be for the Government to provide a subvention because

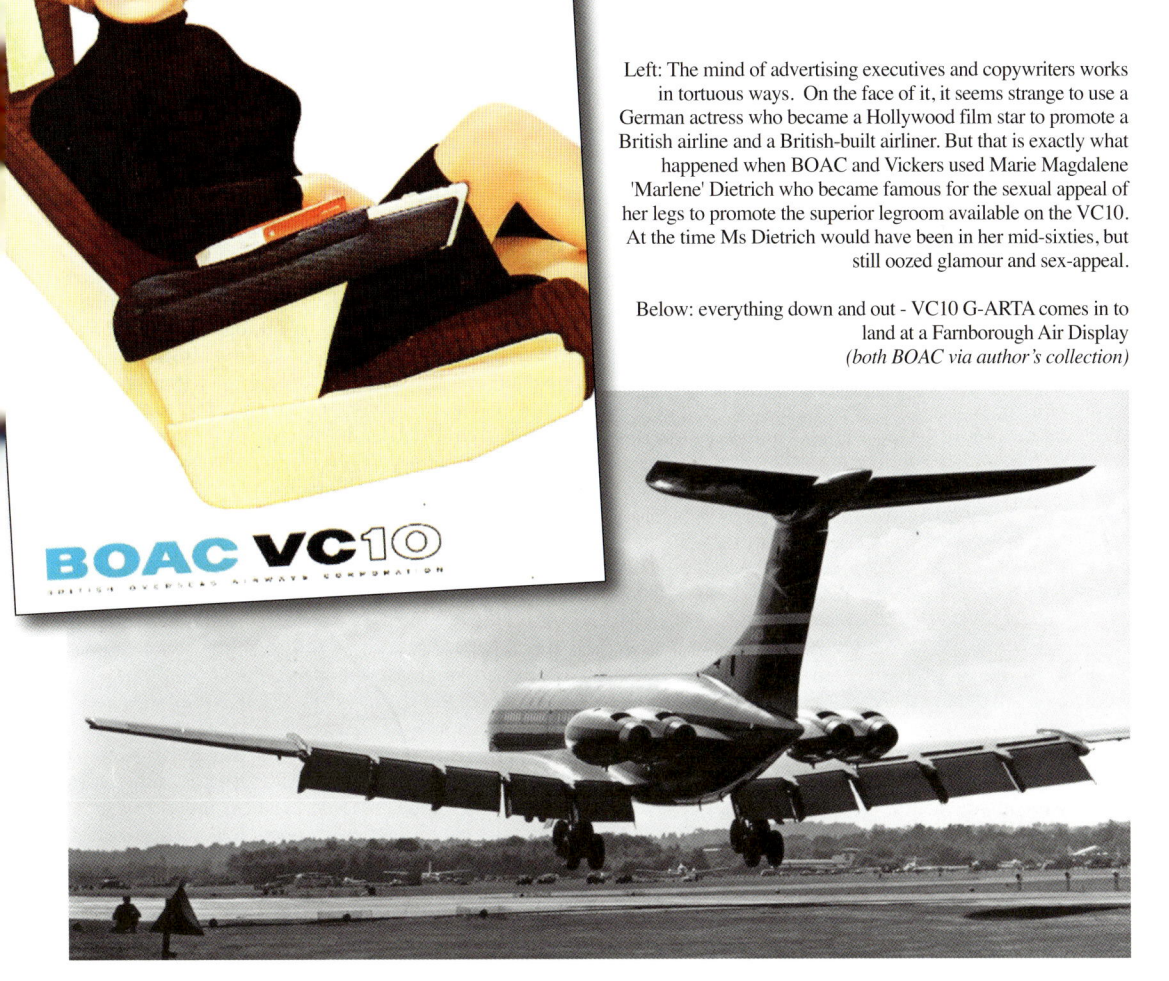

Left: The mind of advertising executives and copywriters works in tortuous ways. On the face of it, it seems strange to use a German actress who became a Hollywood film star to promote a British airline and a British-built airliner. But that is exactly what happened when BOAC and Vickers used Marie Magdalene 'Marlene' Dietrich who became famous for the sexual appeal of her legs to promote the superior legroom available on the VC10. At the time Ms Dietrich would have been in her mid-sixties, but still oozed glamour and sex-appeal.

Below: everything down and out - VC10 G-ARTA comes in to land at a Farnborough Air Display
(both BOAC via author's collection)

G-ARVC on a very slushy apron following a snowstorm.

the route was of national importance, but any subsidy would need to be around £1.3m per annum. These conclusions were put to Government in March with a request that a quick decision be given. It was not until September that the reply came from the Minister that the Government was not prepared to pay the subvention and BOAC, therefore, closed the route in October. Subsequently, the route was handed over by the MoA to British United Airways (BUA) who were prepared to operate without direct Government financial assistance. An even more startling outcome of the plan was the proposed cancellation of twenty-three VC10s, leaving only seven in the fleet. The Chairman quickly concluded that a small fleet of only seven Super VC10s would be less advantageous to BOAC than the alternative of increasing the Boeing fleet and cancelling all thirty Supers. This the Board in May decided to do. The planners were required to amend the G plan and substitute 707s for Super VC10s; in doing so, it became possible to carry out the proposed operations with only six 707s compared with seven Super VC10s; and it was this plan which was handed to the Minister in June 1964. BOAC's fleet would then consist of twenty-six Boeing 707s (twenty 707-436s and six 707-320s) and twelve Standard VC10s, plus possibly two 707-320s for cargo work. Not unnaturally there was a grave concern in the Government at any prospect of cancellation, of all thirty Super VC10s and protracted discussions took place up to Cabinet level and between the Minister and Sir Giles and his team. In the course of these BOAC indicated it might well need a further eight subsonic aircraft beyond 1968.

A number of alternatives were considered including the possibility of BOAC taking more than seven Super VC10s and selling 707s to make room for them. Here the disadvantage would be that the Corporation would have to pay the full price for the Supers and receive only the second-hand price for the 707s, resulting in a massive difference in cost; moreover, there would be full amortisation to consider on the Supers whereas the 707s were already half-amortised. Many differing figures were quoted, but Sir Giles reckoned it would cost BOAC about £15m more to operate seven Supers than six 707s over the life of the aircraft.

The Chairman had an urgent meeting with the Minister on 7 July when he was informed that a small Cabinet Committee chaired by the Prime Minister had been considering BOAC's request to cancel all thirty Super VC10s and order a small number of Boeing 707-320s. The Minister later told BOAC that its case was accepted by the Committee but that as public money would be used in paying cancellation charges to Vickers, or in reimbursing BOAC for the extra costs of using Super VC10s instead of 707s, the Government felt it had the right to state how it should be spent. This being so, BOAC would be asked to take at least twenty Super VC10s out of order for thirty. Permission to purchase six 707-320s would not be given, and although the first seven Supers would need to be taken on the current delivery programme, the balance could be put back to 1968 with the suggestion that at that time 707-436 aircraft could be sold to accommodate further Supers. However, favourable consideration would be given to BOAC's requirement for two 707-320Cs for cargo use.

The Minister added that it had been additionally agreed that the Supers above seven would be provided at the book value of any 707-436s displaced. Moreover, the Government would meet the cost of care and maintenance of any Supers delivered ahead of BOAC's commercial needs and would cover the excess operating costs of the Standard VC10s. All these arrangements would be written off by the Government so that no continuing burden would fall on BOAC.

The intention was to put BOAC in the same financial position as if it had the all-Boeing fleet it had asked for, and the Chairman felt in these circumstances all he and the Board could do would be to agree, mainly as the Minister's remarks made it clear that BOAC was not voluntarily taking delivery of the Supers but only at the requirement of the government.

These matters would be aired in Parliament. Meanwhile the Chairman received a letter from Sir Richard Way, the Permanent Secretary to the Ministry of Aviation in July 1964 which made it clear that the Government was likely to overrule BOAC's commercial interests in this matter but as Sir Richard wrote that *'...steps will be taken to put the Corporation in a position to operate such a fleet on a fully commercial basis.'*

Sir Giles then issued a newsletter to all his staff setting out the BOAC case, and the reasons why he and the Board were pressing Government to agree that the Corporation should take only seven Supers, although he must by then have known that the Government answer would officially be negative. On 19 July 1964, he had a further meeting with the Minister, who subsequently discussed the issue with the Prime Minister at Chequers.

All these comings and goings were before a statement made by the Minister to the Commons on Monday 20 July which finally put to rest all the speculation. He dealt at length with the background to the issues and then announced Government policy. BOAC would be required to take seventeen Super VC10s out of the thirty ordered, seven to meet the planned needs up to 1967/8 and a further ten for growth beyond that date. Of the balance of thirteen aircraft, three would go the RAF, and as work on the remaining ten had progressed very little, they would be put in suspense for later consideration. The following day Sir Giles circulated to all staff the Minister's statement together with a message from himself. He welcomed the decision as confirming Government's acceptance of BOAC's monetary judgment as to the route structure and as to the number of aircraft required. He made it clear that BOAC was not committed to taking the last ten aircraft and that Government would ensure the Corporation would not be worse off financially as a result of taking the Super VC10s than it would have been under BOAC's proposals. But what was also of great importance to BOAC was the Minister's statement on the subject of restructuring the Corporation's finances, when he told the House *'I have assured Sir Giles Guthrie that it is the Government's intention to take whatever action may be necessary to reorganise the Corporation's capital and financial structure so as to enable it to operate as a full commercial undertaking with the fleet of aircraft now planned and with those which may be ultimately selected.*

The detailed implementation of this assurance will be worked out between my Department, the Treasury, and the Corporation in the context of any steps necessary to put BOAC on its feet financially'. It was thus only a few short months after the receipt by BOAC of the Minister's guidance letter of 1 January 1964 telling it to act commercially that the Corporation was being required against its judgment to behave differently. Sir Giles, therefore, insisted that all concerned should understand the circumstances. This was made inevitable by the inclusion in BOAC's 1963/4 Annual Report of

...to G-ASGD 'somewhere in the tropics', BOAC VC10s roamed the world. *(BOAC via author's collection)*

copies of the correspondence between the Minister and the Chairman, namely the Minister's letter of 31 July 1964 and Sir Giles' reply of 12 August in which he confirmed that discussions would take place with the British Aircraft Corporation on the phased deliveries of the seventeen Super VC10s and also as to whether the last ten aircraft put in suspense could be in any way made of commercial interest to BOAC in the 1970s. With the number of VC10s settled for the next few years, the emphasis switched to the third part of the plan, the staff numbers required (although there remained much to do before all the repercussions of the reduced VC10 order had been sorted out, as will appear subsequently). Sir Giles warned the staff in July 1964 that there would have to be a considerable reduction in the overall numbers although it was hoped that normal wastage would take care of a significant part of any such modification and he set up a Manpower Deployment Group to produce a detailed plan.

In December 1964, the Minister, Roy Jenkins, set up a committee *'...to examine the future place of the aircraft industry in the British economy',* under the chairmanship of Lord Plowden. The Labour Government was looking for firm recommendations from this Committee in favour of nationalisation; in the event the Committee came down on the side of some degree of nationalisation of the two main airframe companies and included one crucial recommendation in respect of civil aircraft production, namely, that the manufacturers should aim their products at the world market and not precisely at the British Air Corporations.

The VC10 had great passenger appeal, and the image was used on a whole range of marketing tools, like this map of LONDON Heathrow AIRPORT.

G-ASGN rests between worldwide flights.
(both BOAC via author's collection)

'Going up like a homesick angel' - a BOAC VC10 demonstrates its superior takeoff performance. *(Vickers Ltd)*

In the course of the Plowden inquiry, BOAC was asked specific questions about its future ordering policy and its forward aircraft needs. This gave the Corporation a welcome chance to put forward views based on its own ordering experience over the years. It made many points of constructive criticism of the British aircraft manufacturing industry; on long-haul airliners, the time lag between the contract and first delivery was too long, and delivery dates were unreliable; although contract performance guarantees were usually stated with wide tolerances, aircraft were frequently offered for delivery with performance deficiencies. Progress payments in the course of production formed too high a proportion of the aircraft price; spares costs were too high: long-haul aircraft had to be ordered in excessive numbers to get production started: manufacturers had encountered vicissitudes financially necessitating extra-contractual measures on the part of BOAC. British long-haul airliners were not competitive in performance with American counterparts: and preference by Commonwealth and other partner airlines for American jet aircraft prevented BOAC from achieving the economies to be derived from the operation of similar airliners on partnership routes. To back up each of these points, BOAC submitted to the Committee factual examples.

With the problems of VC10 numbers having been clarified by the Minister's statement in the House on 20 July 1964, it now remained for BOAC and the British Aircraft Corporation to sort out between them an extended delivery line and the financial implications. There also remained the unanswered question of what should be done about the last ten Super VC10s of the original order for thirty. Lengthy discussions took place between BOAC and BAC, and in March 1965 Sir Giles wrote to Sir George Edwards reviewing the outstanding points. On the Standards, he reminded Sir George that although BOAC had accepted all twelve aircraft, BAC had still to show compliance with the noise guarantees. As to the Supers, of the first nine aircraft, all in passenger configuration, five should be delivered by end April 1965, but as to the remaining four BOAC would require deliveries to be stretched out to match traffic growth but agreement on delivery dates had still to be reached and there remained uncertainty as to whether the aircraft would comply with the minimum performance guarantees. It was understood that three Supers would go to the RAF leaving eight jets which BOAC had initially been asked to be in the convertible configuration for either passengers or cargo or a mixture of both, but which BOAC would now almost certainly require in standard passenger layout. As to the last ten aircraft, production was in suspense, and BOAC could see no requirement for them. In his letter, Sir Giles, in fact, proposed that seven Supers out of the first twenty aircraft should be cancelled in return for which BOAC would waive compliance by the standard VC10s with noise guarantees, and a similar concession would be made on the first ten Supers (the Supers never did meet the noise guarantee). BOAC would also forego damages for late delivery on the first five Supers. Finally, included in the proposed package was the cancellation of the last ten Supers. The assumption was taken that the three Supers would still go the RAF and that BOAC would order some BAC-111s for its Associated Companies. This package deal lapsed, but negotiations continued and, meanwhile, in June 1965 the Chairman informed the Minister that BOAC had positively no requirement for the last ten Supers, that these aircraft should now be formally cancelled and any cancellation charges defrayed by the Government. He also told the Minister that BOAC was likely to have a firm requirement for five new 250 seater aircraft with improved economics, for delivery commencing 1968 or

1969. If the Super VC10 could be developed into such an aircraft in time to be competitive, then BOAC would take five of them in substitution for seven Supers.

By November 1965 the air had cleared and subjected to Government approval agreement had been reached between BOAC and BAC. By then BOAC had received six Supers and had agreed with BAC a stretched delivery programme for the remaining eleven aircraft. With the last three coming in 1969 there would be a cost to BOAC of £3.6m. On the previous ten aircraft, BOAC's position remained unchanged, and BAC had accepted their cancellation provided BOAC paid a cancellation charge of £750,000 per aircraft. BOAC agreed to consider this subject to being satisfied that the figure was reasonable and that the Minister gave his consent. Before putting the case to the Minister, a detailed and final assessment of the pros and cons of taking the last ten Supers was carried out by the Director of Planning. The conclusion reached that it would be financially better for BOAC to pay the £7.5m cancellation charge and retain its flexibility to order a new generation of aircraft for the 1970s rather than to be committed to the last tranche of VC10s which could not all be taken usefully into the fleet until around 1973, thirteen years after the VC10 was finally designed. The Chairman wrote to the Minister on 27 January 1966 confirming the Corporation's decision to cancel and on 8 February agreeing to pay the £7.5m to BAC. No definite reply came from the Minister who hoped to delay the whole matter until the impending General Election. But the issue was too urgent both for BOAC which wanted to clear the compensation payment in the Accounts for 1965/6, and for the parent company, Vickers, who wanted a quick financial settlement to a loss-making project. Accordingly Sir Charles Dunphie, Chairman of Vickers, and Keith Granville called on the Minister on 8 March, who reluctantly agreed to refer the matter to the Prime Minister at the same time making it clear that if the cancellation went ahead BOAC would not receive permission in the future to buy any more passenger aircraft of comparable size to the Super VC10. BOAC accepted this condition and a letter from Granville to the Minister on the following day confirmed the understanding but included the words "we shall require some aircraft with substantially increased passenger carrying capability with seating more than say 250, for delivery from about 1969 onwards. It is our understanding that against such a requirement it will be open to us to seek your authority for the purchase of such aircraft, if necessary from foreign sources when the time comes". A Commons statement finally settled the issue on 9 March by the Minister, Fred Mulley explaining the circumstances and confirming the cancellation of the ten aircraft in suspense since 1963. While all these high policy issues were being debated, progress on the Supers proceeded unchecked. The aircraft received a full C of A on 19 March 1965 and were put into passenger service on 1 April on the London - New York route. During the discussion with the Minister on 8 March 1965, the subject of a stretched version of the Super VC10 was raised, but the Minister made clear then that there was no prospect of any such development going ahead and Sir Charles Dunphie for Vickers agreed.

However, this possibility came up again three years later in May 1968 when BAC put up a proposal to BOAC for a 191 seater stretched Super VC10 to take over from the existing VC10 fleet in due course. There was no doubt about the response from BOAC; by then the Corporation was committed to a fleet of Boeing 747s and had enough capacity up to 1973 with further 747 options in hand. Moreover, by that time new aircraft from Lockheed and Douglas in the 300 seater range with wide-bodied fuselages were guaranteed to be available and this would make the proposed narrow-bodied stretched Super VC10 uncompetitive. BAC was told in June 1968 that for these

A tug tows G-ARVG past the BEA hangars at the easterly end of Heathrow.

A pleasing study of G-ARTA in BOAC colours about to land at Vickers' Wisley airfield (Vickers Ltd)

reasons BOAC would not be interested.

That said, in the event, the VC10 proved to be a very popular aircraft with passengers, attracting 10% higher load factors than the Boeing 707. It would go on to give decades of stalwart service as a tanker/transport for the RAF. But the commercial data are unyielding: a total of 12 Type 1101 VC10 were purchased in 1964-65, followed by 17 Type 1151 Super VC10s in 1965-69. After the last aircraft was delivered in February 1970, the production line closed, with just fifty-four airframes having been built. On the other hand, 1010 Boeing 707s and 556 DC-8s were sold internationally. Even the loss-making Convair 880 had 65 sales.

The VC10 as initially conceived was built to a narrowly drawn specification: later attempts to broaden its appeal and to improve its economics had a marginal effect. Even the advantages of short field performance from high altitude airports were negated mainly as runways were lengthened to take the American jets (often helped by U.S. aid programmes).

At the beginning of 1969, BOAC's fleet consisted of sixteen Super VC10s, twelve Standard VC10s, eighteen 707-436s and five 707-336Cs. One Super VC10 had still to be delivered to complete the order and this, the seventeenth aircraft, was received in 1969. The original fleet of twelve Standard VC10s was reduced by one when in December 1969 WAAC (Nigeria) Ltd purchased G-ARVA. At the time of the merger into BA in April 1974, the fleet had been further reduced to nine by the sale of two aircraft to Gulf Aviation Ltd, but that is getting ahead of the story somewhat.

Then came the problems of terrorism. In September 1970, four jet airliners bound for New York City and one for London were hijacked by members of the Popular Front for the Liberation of Palestine (PFLP), who had already gained notoriety for several previous hijackings in support of a free Palestine. Three aircraft were forced to land at Dawson's Field, a remote desert airstrip near Zarqa, Jordan, formerly Royal Air Force Station Zerqa, a desert airstrip, one that quickly became the PFLP's 'Revolutionary Airport'. By the end of the incident, one hijacker had been killed and one injury reported.

Hijacking is the term applied to the unlawful seizure of an aircraft by a person or persons. While not new - the first recorded event is from 1931 -the end of the 1960s saw a significant increase in the number of hijackings worldwide.

The episode evolved over some days, in the full glare of the world's media. On 6 September, TWA Flight 741 from Frankfurt, a Boeing 707, and Swissair Flight 100 from Zürich, a Douglas DC-8, were forced to land at Dawson's Field.

On the same day, the hijacking of El Al Flight 219 from Amsterdam (another 707) was foiled; hijacker Patrick Argüello was shot and killed, and his partner Leila Khaled was subdued and turned over to British authorities in London. Two PFLP hijackers who were prevented from boarding the El Al flight, hijacked instead Pan Am Flight 93, a Boeing 747, diverting the large plane first to Beirut and then to Cairo, rather than to the small Jordanian airstrip.

After three days, on 9 September, the British government confirmed that there were no British citizens among the hostages taken. They spoke too soon, for shortly after lunchtime the news broke that three Palestinians had taken control of Speedbird 775 between Bahrein and Beirut while underway from Bombay to London. These terrorists were not part of the original plan but acted in support of the PFLP to pressure the British government in releasing Leila Khaled, who was taken a prisoner in the failed hijack attempt three days earlier. After refuelling in Beirut, Super VC10 G-ASGN is the third aeroplane to end up at the Jordanian desert airstrip.

The one hundred and five passengers and nine crewmembers joined the rest of the hostages and could only await their fate. Over the next days some hostages were moved to a hotel in Amman while others were released. The hijackers demanded the release of several prisoners from different countries and to put pressure on the different governments they kept fifty-six hostages amongst which the flight crews and Jewish passengers. In anticipation of a strike they left Dawson's field on 12 September and destroyed the three airliners with

Above: BOAC VC10 G-ASGN burns after being blown up by hand grenades at Dawsons Field in Jordan.

Right: Captain Cyril Goulborn returns to the UK with his wife following his ordeal in the desert.
(both BOAC via author's collection)

explosives. The BOAC VC10 was the first to be blown up with the 707 and DC-8 following. Images of these explosions and their aftermath were shown on television screens around the world.

The PFLP kept the flight crew and passengers, amongst them Captain Cyril Goulborn, as prisoners in locations around Amman which were under control by the Palestinians. In a counter strike by the Jordanian forces some of the prisoners were released over the next few days while the last prisoners were let go in exchange for Leila Khaled and three prisoners from Switzerland at the end of the month.

As the VC10 burnt down the tail dropped to the ground but was left relatively unscathed. A team was sent out to retrieve it and this horizontal stabiliser later served as a spare part for the rest of the BOAC fleet, thus enabling a fatigue modification on the stabiliser to be carried out without major downtime of any aircraft.

While the majority of the 310 hostages were transferred to Amman and freed on 11 September the PFLP segregated the flight crews and Jewish passengers, keeping the Jewish hostages in custody, while releasing the non-Jews.

In 1971 a short-lived coup had taken place in Sudan. The two leaders of the coup who had been out of the country were heading back to Khartoum on BOAC flight BA045 when the aircraft was asked by Libyan air traffic control at Benghazi to land at that airport *'for the safety of the souls on board'*. According to BOAC the captain, Ray Bowyer, asked for a clearance back to Rome from the Maltese controller who was nominally directing the flight at that point. Having turned the aircraft around and just minutes away from leaving the Libyan FIR the Maltese controller revoked the clearance and repeated the earlier request from the Libyan authorities, ordering them to land at Benghazi.

The aircraft had to circle the airport for a while to burn off fuel and the Captain used this time to speak to the two Sudanese men on board. They insisted that the Captain should take no action which would endanger the other passengers. After the landing at Benghazi the two Sudanese were removed from the aircraft by the Libyan authorities. The other passengers and crew remained on board for almost 90 minutes after which they flew back to London.

There were rumours of threats by the Libyans against the VC10. Although no passengers or crew saw anything, and the state of the Libyan Air Force fighters made it unlikely, there could have been an interception or the threat of one. Because of this the decision by the Captain was sensible. The outcome however was tragic. The two men, Majour Farouk Hamadallah and Lt Col Babakr El-nur Osman, were handed over to the restored regime in Sudan and were executed within a day.

Flight International called it 'an unprecedented act of government-sponsored piracy' and ICAO expressed 'grave concern'. While hijackings were not uncommon in those days this was the first time that a government not only condoned an act but actually openly contrived it. Questions were also asked about Maltese involvement but the article in *Flight International* did not have the answers.

Chapter Nine

Management Crisis As The Debts Mount Up

It has been questioned by many as to just how did a situation develop in BOAC and between BOAC and the MoA over the years 1960 to 1963 that led to the resignations of the Chairman, the Deputy Chairman and the MD in November 1963 plus a shake-up of the Board as a whole, little more than three years after the new Chairman had taken over in July 1960. Was it a crisis created by the losses in 1961/62 and 1962/63 or was it a more deep-seated malaise with its origins in earlier years?

These were not easy questions to answer from the avelanche of words written and spoken at the time. Perhaps it is best to record the events as they occurred which had a direct bearing on the final outcome and to attempt to bring to some conclusion the various views expressed in the Swash and Corbett Reports, the resulting White Paper of November 1963 and the statements of those directly involved to the Select Committee on Nationalised Industries in 1964. Indeed the crisis itself was a watershed in BOAC's existence and brought out into the open matters which ought to have been thoroughly discussed and given proper recognition much earlier.

The Swash Report came about when following the poor financial results relating to BOACs involvement with Middle Eastern Airlines, and Mideast Aircraft Services Company (MASCO) came under the Minister of Aviation who requested an examination of the situation by S. V. Swash, a former chairman of Woolworths in the UK. In his report to the Minister early in 1962 he was critical of the Corporation's record that while the policy of the investments was sound - namely, to secure feeder traffic and protect BOAC's rights and other interests in the Middle East - the benefits and disadvantages of the investments in his view were not adequately assessed from time to time and too much time was allowed to pass before withdrawal

The management of BOAC in 1960. From left: Ken Bergin (Director of Personnel and Medical Services), Tommy Farnsworth (Deputy Chief of Flight Operations), Charles Abell (Chief Engineer), Ken Bevan (Financial Comptroller), Keith Granville (Deputy Managing Director), J B Scott (Chief Economics Officer), Basil Smallpeice, Ken Staple (Secretary and Legal Adviser), Gilbert Lee (Commercial Director), Ross Stainton (GM Western Routes), Derek Glover (GM Southern Routes) and Basil Bampfylde (GM Eastern Routes). *(BOAC Press Office via author's collection)*

The public tunnel at Heathrow in the late 1950s with the central island in the middle. The Airport Control Tower is very noticeable, as is the Police Control Box on the central roundabout in the foreground. *(author's collection)*

took place. The Ministry was later to admit that the findings of the Swash Report were a significant factor in its decision to call for a much more extensive inquiry into BOAC, resulting in the Corbett Report in 1963. This report was comissioned by the Government a year earlier, was received by Julian Amery, the Minister of Aviation who reported to Parliament later that year in a White Paper, although much remained confidential and not for publication. What was revealed on 10 December 1963 were the costs: these, including that of the three management consultant firms which were associated with Mr. Corbett, amounted to £22,258. These costs were borne by the Ministry of Aviation.

When Gerard d'Erlanger became Chairman in 1956, there was a surplus on the Balance Sheet, and by the end of his first year in office, this had been marginally increased to £440,000. In 1957/58, there was a small loss on BOAC's operating account, but after bringing in Associated Companies and interest on capital the result was a loss of £2.8m for the year. This with adjustments became a loss of £3.49m, creating a deficit of £3.045m as at 31 March 1958 - high, still a manageable figure.

It is noticeable that the losses on Associated Companies were already over £0.5m in that year alone, four times the operating loss on BOAC's mainline activities. The year 1958/59 saw a sharp turn for the worse in the deficit; although BOAC on its operations made a surplus of nearly £0.9m, interest charges were mounting up, which turned this surplus into a loss of £2.1m. However, by now losses on Associated Companies after allowing for interest payable had reached the enormous sum of £3.1m, thus increasing the total deficit for the year to £5.2m. A direct consequence of the Comet 1 disaster was the hurried purchase of second-hand propeller aircraft on the market,

Batting them in! Britannia G-ANBE is welcomed home after another service. *(BOAC via author's collection)*

A large part of BOAC's financial problems came about by the activities of the Associated Companies and the enforced writing down of the value of it;s propeller fleet - here both are combined in one picture; Britannia G-ANBH in Nigeria Airways markings.
(BOAC via author's collection)

but when these were retired, they were sold at below book values, so that an amount of £1.12m had to be provided for such losses.

Additionally, there was a write-off of £15m for development expenses both on the Britannias and Comets, and all this increased the total deficit for the year to £11.6m after minor adjustments. Thus, by the end of the financial year 1958/59, the accumulated deficit stood already at the alarming figure of £14.7m. The year 1959/60 was better, with BOAC's operations achieving a surplus of £4.3m before interest: but after taking account of Associated Companies losses and payment of interest, the accumulated deficit just topped the £15m mark.

The Chairman had already made public in the Annual Report for1958/59 his view that the responsibility on the Corporation to pay interest on all its capital whether or not profits had been made was a heavy burden which many of its competitors did not suffer to the same extent, having a proportion of their capital in the form of equity. With an accumulated deficit of over £15m, the Corporation was already in a dangerous situation, especially as it did not have to pay interest on interest.

Operations for 1960/61 produced an almost identical result, with a surplus of £4.26m, but by the time interest charges and increased losses on Associated Companies had been taken into account there was a deficit of nearly £2m with the accumulated deficit exceeding £17m. However, 1961/62 was to prove a disastrous year. On BOAC's operations there was a deficit after the interest of £10.9m; on top of this there was a further £2m or so losses on Associated Companies, plus the cost to BOAC of disposing of its investments in MEA and BWIA, which added a further £1.47m. The Chairman decided it was necessary to substantially increase the obsolescence charges for all propeller aircraft and to write down more quickly the Comet 4s in the face of the widespread introduction of the

Heathrow North side, with a pair of BOAC Britannias, a company DC-7C and a DC-6B of Pan American World Airways.
(BOAC via author's collection)

large jets. For these purposes an amount of £31.7m was provided, covering DC-7Cs, Britannias and Comet 4s. The provision was also made for a possible loss of £1m on disposal of aircraft employed within Associated Companies. All these sums together with certain adjustments included the penalty of £674,790 from reducing from fifteen to twelve the number of standard VC10s from Vickers, produced a staggering total deficit of £50.1m accountable in that year alone. As a result, the accumulated deficit soared to £67.3m.

In 1962/63, the last full year of Sir Matthew's chairmanship, there were further losses both on BOAC and Associated Companies, increasing the accumulated deficit once again, up to £80.1m. Interest charges alone reached £7.7m, and clearly, the situation was out of hand. The Board and Management were deeply concerned at the way things were going and by the autumn were seeking changes which would require Ministry approval.

A paper sent to the Ministry laid out the Corporation's views on how its capital should be reorganised. The airline was not earning enough profit to meet the fixed interest on its borrowings and therefore had had to borrow more money to finance the deficit on which again it had to pay interest so that a vicious circle was building up. The solution proposed by the Corporation was that the accumulated deficit (at that time only £13.25m) should be written off as lost capital and that in future part of its capital needs should be in the form of income bonds.

The Permanent Secretary called a top-level meeting on 22 December 1960 at which the Chairman, Deputy Chairman, Managing Director, Chairman BOAC-AC, Legal Advisor and Financial Comptroller were present. It was to enable the Ministry to explain its ideas on the White Paper shortly to be issued by the Government on the *'Financial and Economic Obligations of the Nationalised Industries'*. This involved setting a financial target which could be a return of 8% on working capital. BOAC pointed out the difficulties in the international air transport industry on previous experience of achieving anything like an 8% return, and still considered that lost capital should be written off. It then suggested a compromise under which the Corporation would be relieved of interest and progressive redemption of lost capital.

These ideas were set out in a paper to the Permanent Secretary at the Ministry on 24 January 1961 and included a separate target for capital employed in the Associated Companies and to the Corporation receiving a directive from the Minister if it had to undertake uncommercial activities for political or other reasons. More controversial was the proposal that, if the Corporation was encouraged to buy aircraft for national or social reasons which were not the best available, it should be compensated. The reply from the Permanent Secretary, Sir Henry Hardman, was not entirely discouraging but said: *'there would be the greatest difficulty in any proposal for writing off BOAC capital and you must not take it that the Department is committed to a policy of this character'*.

In February 1961, the Ministry asked BOAC to provide an estimate of its capital investment and borrowing requirements up to 1965/66, a five year period as part of the Government's target exercise; the Corporation complied with the request but added reservations about making long-term forecasts. In April 1961 BOAC produced its budget for 1961/62 showing a surplus on BOAC operations of £267,000 and a deficit on Associated Companies of £1.22m with an overall result as a deficit of £955,000. When the budget was under examination Management paid attention to routes forecast to be big losers; in every case, there was a substantial contribution to the Corporation's overall financial result which would merely be worsened if the services were discontinued, yet the Ministry later claimed that BOAC had been unwilling to consider any route changes.

However, as the year progressed, it was evident that all airlines were in a period of reduced traffic growth, not foreseen in budgets. The position was unusually severe on the transatlantic routes where traffic into and out of the UK was showing scarcely any growth whereas capacity by all carriers had been massively increased. BOAC, in this its last and fifth year of exceptional expansion, had increased its capacity by nearly one third above 1960/61 and so was peculiarly vulnerable to the absence of forecast traffic growth. This situation continued throughout the summer months, and by the end of July there was a deficit of

Left: George Edward Peter Thorneycroft, Baron Thorneycroft, CH, PC (*b*.26 July 1909 *d*. 4 June 1994) was a Conservative Party politician. He served as Chancellor of the Exchequer between 1957 and 1958, when he resigned, along with two junior Treasury Ministers, Enoch Powell and Nigel Birch, because of increased government expenditure. Thorneycroft returned to the Cabinet in 1960, when he was appointed Minister of Aviation by Macmillan.

Right: Harold Julian Amery, Baron Amery of Lustleigh, PC (*b*.27 March 1919 *d*. 3 September 1996), was a politician of the Conservative Party, who served as a Member of Parliament (MP) He served as Secretary of State for Air (1960–62), followed by Minister of Aviation (1962–64).

From a BOAC publicity sheet comes this image of what it took to put a Comet 4 in the air and keep it there Group 1: Flight Manager, Works Manager (Aircraft Maintenance). Group 2: Flight Captain (West); Flight Captain (East); oflicer i/c Training; Deputy Flight Manager; Works Superintendent; Comet Project Engineer; Fleet Supplies officer; Fleet Planning officer; Fleet Inspector. Group 3: Flight Planning Officer; Manager's Secretary; Cabin Services Offieer; Flying Staff Administration;Communication Officer; Flight Operations Officer. Group 4: Captain; Co- Pilot; Flight Navigation Oflicer; Flight Engineering Offieer; Flight Steward; Second Steward; Flight Stewardess ; Stewardess. Group 5: Instrument Mechanic; Foreman; Assistant Foreman; Chargehand; Section Inspector; Inspector; Secretary; Upholsterer; Aircraft Scheduler. Group 6: Loader Driver (Freight); Loader Driver (Aircraft Catering); Driver (Passenger Coach); Sgt. (Security); Air Ministry Met. Forecaster; M. of A. Marshaller. Group 7: Station Duty Officer; Service Control; Receptionist; Station Operations. Group 8: Rigger; Electrician. Group 9: Driver; Cleaner; Leading Hand Fitter; Leading Hand Metal Worker. Group 10: Shell-B.P. Airfield Fuel Operators. *(BOAC via author's collection)*

£4.25m on BOAC's operations alone, representing a worsening against the budget of £4.75m, and there was also a loss to date of £818,000 on Associated Companies.

During this period an Engineering strike occurred at London Airport, but this failed to account for the total revenue shortfall. It was decided to reduce frequencies in the winter months and to plan Atlantic capacity for 1962/63 at about the same level as 1961/62 and to dispose of as many DC-7Cs and Britannias as possible. In the light of subsequent events, it is noteworthy that the Minister, Peter Thorneycroft, in September 1961 re-appointed Sir Basil Smallpeice as a Member of the Board for a further five-year term from 1 October 1961. On 9 October the Minister wrote to the Chairman asking to be informed as to the likely out-turn for the full year 1961/62; in his reply, Sir Matthew took the opportunity of raising once again the subject of lost capital. He told the Minister that the year could show an operating loss including Associated Companies of around £12m after interest with some capital losses which might amount to £1.5m and added *'the millstone of lost capital gets heavier as the years go by and this, I suggest, is something we shall have to discuss when we get down to talk with your secretariat on the White Paper on the Financial and Economic Obligations of the Nationalized Industries'.*

Sir Matthew then referred the thirty-one Britannias and ten DC-7Cs in the fleet which would sell for considerably less than their written down book value if put on the market; the book value was £31m by March 1962 and £27m of this he felt ought to be regarded as lost capital. He gave his view that it would be sensible to write off all the lost capital which could amount to as much as £58m. The whole issue was now out in the open. In his report to the Board in October 1961, the MD also recommended that the propeller fleet should be written down in the books to their estimated market value as at 31 March 1962, that in 1962/63 interest charges on the accumulated deficit should be added to the deficit and not charged against current operations, and that a more stringent plan should be put in operation for 1962/63.

However, the development services - the 'Polar' route to Los Angeles, the trans-Pacific service and the South American West Coast service - were to be retained despite the incurred losses. The Board supported all these proposals.

The financial results of the twenty-eight weeks to mid-

October 1961 showed a total combined BOAC and BOAC-AC loss after the interest of £9.6m, but the Chairman held out some hope that the officials at the MoA *'appeared to be receptive to the suggestion that means should be found to write off the Corporation's lost capital'*. At that time his hopes were disappointed.

Discussions with the Ministry of the five-year target lapsed while the Cunard-Eagle Atlantic case was being argued before the ATLB but once completed, talks resumed and new BOAC figures put to the Ministry in January 1962. The exchanges were unsatisfactory and by April the Chairman felt the issues were somewhat fogged as the result of requests from the Ministry for a mass of detail and, as he put it, to the Permanent Secretary *'the number and variety of assumptions that can be made are such as to make the final answer quite worthless'*. One such asked for by the Ministry which surprised BOAC was that it should *'operate with the equivalent of twelve VC10s less than you have been planned'*. This would mean the cancellation of twelve VC10s, but as the Corporation had already asked the Ministry officially for authority to cancel nine VC10s and replace them with six long-range 707s, any suggestion of cancelling a further twelve seemed unexpected to say the least.

The exercise was running into difficulties. The Ministry was pressing for firm figures while BOAC was struggling with the problems being created by the fall in 1961 traffic which had upset all forecasting; what is more, the Corporation was still making a case for the write off of lost capital and for a reduction in the number of VC10s. In May both sides concluded that further discussions towards setting a five-year target were a waste of time until the critical policy issues had been sorted out. This was a relief to BOAC, all of whose energies were devoted to trying to cope with the decline in business.

1961/62 finished with a substantial loss, £11m on BOAC's operations and £3.4m on Associated Companies, both including interest charges. For 1962/63 BOAC budgeted for almost a no-growth year in capacity in the expectation that the modest traffic growth forecast could be met and raise load-factors to reasonable levels. Interest on lost capital was not charged in the budget against the year's operations. A surplus of £1.37m was forecast for BOAC but after expected losses on Associated Companies were taken into account, there would only be a marginal surplus. But by summer 1962 results were again falling below budget, so the MD instructed the Chief Economics Officer, Jim Scott, and General Manager Planning, Winston Bray, to prepare a confidential study to ascertain how a profit could be assured in 1963/64 and what steps, however severe, would have to be taken to achieve it.

Plans were produced which included only those routes which could be expected to produce profits; London - Hong Kong or Sydney, London - Johannesburg via East Africa including Dar-es-Salaam and Aden, London - Lagos, London - New York and some extensions, London - Montreal and Toronto. The operation could be produced from a fleet of twenty Boeing 707 aircraft which would be available in 1963, eighteen of BOAC's plus a further two brought in by BOAC-Cunard. It was estimated a plan of this kind could earn a profit of £10m, reduced to £5.5m after interest on capital but with no interest charged on the accumulated deficit.

The consequences of contraction were extreme. Twenty countries served by BOAC would disappear from the network, and thirty-eight foreign cities would have no British service; that meant a loss of £21m foreign currency annually, and there would be redundancies in the airline of up to 10,000 staff with obvious implications in the way of Union difficulties and compensation payments. The Treforest engine overhaul base would have to close; most if not all of the VC10 order would be abandoned with a high cancellation penalty. It was evident that such a drastic remedy was not a practical proposition.

A further alternative was merger with BEA if the Government were willing, and so perhaps improve the results of both Corporations; however for the present both Management and the Board felt the right policy was to carry on with existing plans and fleets, seek Government help to permit the Corporation to undertake contract trooping and to restrict poaching of British traffic by the fifth freedom carriers such as KLM, Sabena and SAS, whilst BOAC itself would give even greater attention to the marginally profitable routes.

Another Associated Company was East African Airways Corporation, who changed the markings of Britannia G-AOVD by replacing the BO of BOAC with EA, and then putting their lion logo on the fin. *(BOAC via author's collection)*

Inside the maintenance hangar at Heathrow, with four spare props on their handling trollies parked in front of a Britannia in for scheduled checks. *(BOAC via author's collection)*

At the close of the first five months of the 1962/63, financial year results were worse than the budget, due to a lower rate of traffic growth. Thus the outlook was still gloomy. A complete draft of the 1961/62 Annual Report went to the Board in August 1962 and subsequently to the Ministry. In this draft, there were two important chapters. Chapter III dealt at length with the circumstances which had contributed to the asset values of the Britannia, the DC-7C and the Comet 4 being overstated in the Corporation's books to the extent of £37m, later revised to £32m. It set out in detail BOAC's experience over the sixteen years since 1946 during which it had operated seventeen different types of aircraft; it gave the history of the Brabazon Committee, of the Britannia and Comet projects with all the vicissitudes which occurred and referred to the price the Corporation had had to pay for pioneering advanced technologies, in particular of the technical troubles that compressed the useful operational lives of the Comet Is, Britannia 102s and 312s into uneconomically short spans, curtailed the competitiveness of the Comet 4s and made necessary emergency purchases of Constellations, Stratocruisers and DC-7Cs.

It talked about the role BOAC had played over the years 1946 to 1961 during which it was regarded to a large extent as an instrument of national policy, which had influenced in so many ways, not least in its aircraft purchasing policy; but went on to say that a new phase was now to begin as a result of the Government's White Paper of 1961 under which the Corporation would be directed to make profitability its first consideration and be expected to achieve the profitability target imposed by the Government. Chapter IV then dealt with the need for a fundamental re-adjustment in the Corporation's capital structure, referred to the accumulated deficit as lost capital on which there was no prospect of earning any return, yet the Corporation had to pay interest on it to the Government. It set out the Corporation's view that the value in the books of the Britannias, DC-7Cs and Comets needed to be heavily written down, adding that sum to the total accumulated deficit and then stated that the Corporation had recommended to the Minister that this sum should be written off.

When the Ministry received the draft its officials took exception to the wording in these two chapters, firstly expressing concern that the Ministry might be held to have agreed to the accumulated deficit being written off, and secondly that the historical account of the Corporation's post-war fleets was likely to be damaging to the British aircraft manufacturing industry. The Board had no alternative but to rewrite the two chapters to meet the Ministry's requirements but felt so strongly that they instructed that the original drafts of chapters III and IV should be filed with the Board papers.

Immediately upon his appointment in July 1962 the new Minister of Aviation, Julian Amery, was being advised by his Department that a dangerous situation had arisen in BOAC so he decided to set up a private inquiry by outside consultants. At the same time, he asked Sir Matthew to give him a paper on the future of BOAC which he provided in August 1962. Although the views expressed in the papers were his own, it was subsequently tabled at a BOAC Board meeting in October. The paper listed seven significant issues which the Chairman felt the Minister ought to consider and, since every single issue was ultimately sorted out much in the way recommended by Sir Matthew, and having regard to the Minister's subsequent action in respect of Sir Matthew's own position, these seven points are quoted in full. The Select Committee on Nationalised

Publication in September of the Annual Report of 1961/62 showed a deficit of £14.4m including Associated Companies, but this was by no means the final figure. The Chairman's determination to bring the capital value of the fleet to a realistic figure by writing down the values of the Britannias, DC-7Cs and Comet 4s was reflected in an additional provision for obsolescence amounting to the colossal figure of £31.7m. Certain other adjustments were made including the amount of the penalty in cancellation charges on the three VC10s, and all these increased the accumulated deficit - now standing at £17.2m - to a total of £67.3m. 1961/62 had been a disastrous year, but now the Corporation's Accounts represented a correct position.

At the press conference on the Annual Report in October 1962, the Chairman was explained how certain of the losses on Associated Companies had occurred and referred to the history of Kuwait Airways, at that time one of the big losers, where BOAC had undertaken financial responsibility for the airline at the request of the Minister. Sir Matthew stated that although losses had been foreseen, and the Government had agreed to the inclusion in the accounts of a note explaining that BOAC was involved at the Government's behest, what had apparently been overlooked was the fact that such losses would have to bear interest and so increase the amount permanently on loan on which again interest would have to be paid. It was on this issue that the Chairman made his much-publicised statement that it was *'bloody crazy finance'*. The Press seized on his words and made much of it.

Sir Matthew later said that at least he obtained a great deal of publicity for his view which he had been pressing on the Minister that BOAC ought to have a proportion of capital on which interest would be paid only when results justified it. Subsequently, Julian Amery said in the House of Commons in November 1962 *'to dismiss as 'bloody crazy' the idea that one should pay interest charges on borrowed capital was not something which in itself commended itself to me'*. So Sir Matthew's words became quoted completely out of context.

On 25 September, one week after the date of the Corporation's 1961/62 Annual Report the Minister asked John Corbett, a partner in the firm of Chartered Accountants, Peat, Marwick, Mitchell & Co., to undertake an inquiry into the financial affairs of BOAC. This commenced immediately and was well under way by the following month and the Chairman was telling his Board that the Minister had apparently decided to make it a private inquiry solely for his own information. Needless to say, the Chairman pressed the Minister to make the report when it appeared available to the Corporation.

Traffic remained depressed, and the outlook was still gloomy. This was not helped by a debate on the Second Reading of the Air Corporations Bill on 6 November 1962. The Minister reviewed the relationship of his Ministry with the Corporation, his remarks leaving the impression that the initiative had always been with the Corporation on such matters as aircraft procurement, the Associated Companies, and the obsolescence policy, and that except in the case of BOAC's commitment in Kuwait Airways, the Ministry had

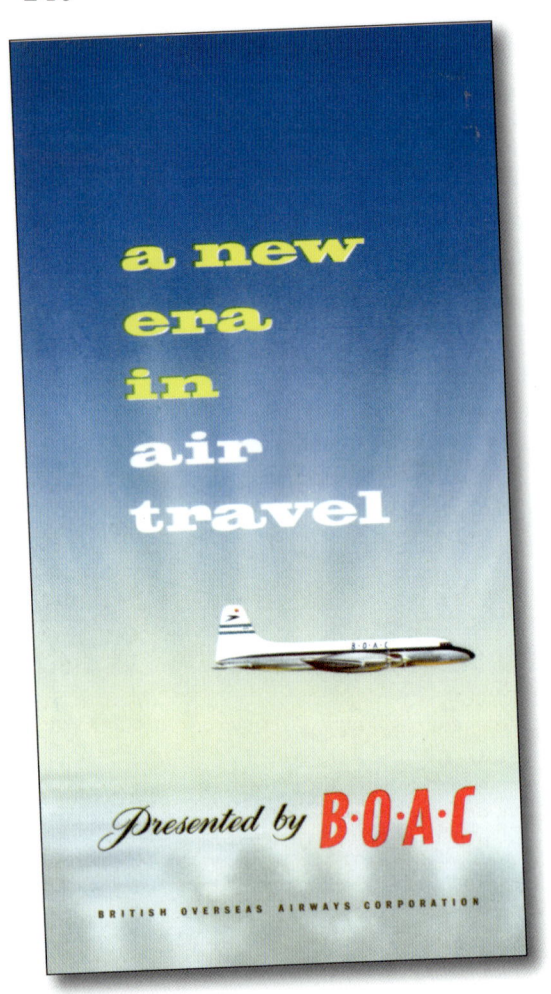

Industries in 1964 evidently took the same view since the complete text of the Chairman's paper was included as an appendix in their report. The seven issues were:-
1. A major policy decision was required regarding the aims of the Corporation. It has to be wholly commercial in outlook or should it take account of other national and Commonwealth interests and, if so, how is the Corporation to be reimbursed in its accounts for any uncommercial activity?
2. Some £70m of lost capital should be written off.
3. The remaining capital should be reorganised to provide for some 50% being in the form of equity capital.
4. Greater continuity of Management was desirable.
5. A sufficient increase in the scale of salaries fixed by Her Majesties Government (i.e. Chairman, Deputy Chairman, Managing Director and others) is desirable to enable the Corporation to recruit competent and experienced men to fill higher management posts.
6. The Corporation should continue at its existing scale of effort as planned in recent years and for which aircraft have been ordered.
7. Some more informal integration with BEA is desirable to provide for a more united British effort.

left to the Corporation's commercial sense decisions on these matters. He included in the list of issues BOAC's commercial agreement with Cunard in the previous summer, yet it was a matter of record that the division of capital for the new BOAC-Cunard company had been altered, contrary to the Corporation's wishes, by the intervention of the Government; moreover the Corporation found it hard to accept that in the fields of aircraft procurement and the Associated Companies there had been no Ministerial pressures.

One of the issues which figured much in the censures on BOAC in the White Paper was the matter of its obsolescence policy. This went back to 1956 when there was criticism from the Minister that not enough allowance was made for the depreciation of the fleet, so BOAC and BEA agreed on a common policy acceptable to him. This involved writing off aircraft by equal annual amounts over their working life to an estimated realisable value on disposal. For the Britannias and DC-7Cs introduced at that time, a seven-year life was taken with a residual value of 25% of the cost. However, technological advance affected the lives of the propeller aircraft including the DC-7Cs, and the Corporation decided in the 1958/59 accounts to write down the DC-7C fleet by some £6.5m. This was agreed with the auditors who recommended that provision for exceptional capital losses should be shown as a deduction from the total capital liabilities on the Corporation's balance sheet. The matter was eventually discussed with the Minister, Harold Watkinson, who was not prepared to defend the Accounts as approved by the Board because such losses had not yet been incurred, nor did he favour treating the item as lost capital. In consequence, a special note for the 1958/59 Accounts was agreed with the Ministry pointing out that the book value of certain aircraft due for retirement in 1960 might exceed their then realisable value. For the years 1959/60 and 1960/61 the same procedure was followed in the Accounts, the contents of which were, as usual, discussed at length between the Corporation and the Ministry before publication.

When it was time to prepare the Accounts for 1961/62, Sir Matthew took the view that the excess book values on the Britannias and the DC-7Cs must be written down, especially as the whole Britannia 102 fleet was being withdrawn during 1962. Accordingly, the Corporation had discussions with the Ministry aimed at having the Minister's old ruling altered and the necessary amounts for additional obsolescence shown in the Accounts. There was confusion as to why provision had not been made in the earlier years; the Corporation took the view that the Minister had the power to dictate to it and in fact had done so in 1959; hence the absence of proper provision, whereas the Ministry claimed that whatever had been said at the time was no more than advice to the Corporation.

In the Commons Debate on 6 November 1962, the Minister had this to say on the subject *'what I cannot accept so easily is that the depreciation in the value of the Corporation's aircraft fleet has only been fully revealed this year. This is not a situation which has come upon us in the night. It has been building up for two or three years past, and in any business efficient management must call for a clear understanding and a clear presentation of the balance sheet'.*

The Minister was questioned in the House again on 19 November and seized the opportunity to seek to absolve his Ministry of any responsibility for the obsolescence situation by stating *'We have no powers to direct the Corporation in this matter. Advice had been given from time to time when our advice has been sought'*. So in response to a further question, he said *'it is not our duty to intervene in the Accounts or the figures. We are responsible for the presentation of the Accounts, but not for the totals included in those Accounts. When our advice has been sought, we have given advice. Sometimes that advice has been taken and sometimes not, but our advice is on the form and not on the quantities'*. This became one of the issues examined in the Corbett Report. The Corporation felt it had

From a maintenance point of view, the Britannias caused something of a problem, as they would not fit inside some of the aircraft servicing facilities. The solution was to cut holes in the doors so that part could be 'left outside' and the doors closed around the rear fuselage. *(BOAC via author's collection)*

A bit of a mixture - BOAC Britannia G-AOVG outside the now-Grade II listed maintenance hangar at Heathrow, wearing additional BEA stickers. *(BOAC via PHT Green Collection)*

Opposite page: A contemporary inflight magazine page showing passenger facilities available for anyone travelling on a company VC10. The explanation is in eight languages.

been unjustly treated by its Minister in the Commons debates not only over the obsolescence issue but also the impression given to the House that there had been no occasions when the Corporation had been pressed to act in the national interest rather than in its commercial interests. The Chairman let off steam on these matters in a long letter to the Minister on 12 December which he closed by welcoming the appointment of John Corbett to inquire into all these issues.

By March 1963 the results for the year 1962/63 were almost complete with four weeks to go, but continued to show heavy losses; it was expected that the increase in the accumulated deficit would be about £12.7m increasing the total to around £80m. The budget for 1963/64 was therefore pitched at a conservative level, but even so, there seemed a prospect of barely achieving a break-even result. On 24 May 1963 Corbett submitted his Report to the Minister, and only three weeks later BOAC's Secretary was reported to his Board that four members of the Board had had their appointments renewed by the Minister for only one year ahead, Sir Matthew Slattery until 28 July 1964 and Sir Sir Wilfred Neden, CB, CBE, until 19 June 1964. Sir Wilfred was born in 1893. He studied economics at London University and joined the Army soon after the outbreak of World War 1, serving overseas with the Yeomanry and Royal Field Artillery. He was granted a regular commission in 1917, retiring in 1922 to join the Ministry of Labour. Following a spell in the Aliens Branch, he became responsible for recruitment under the Military Training and National Service Acts 1939-45; later he was appointed Undersecretary of the Military Recruiting and Demobilisation Department. In 1948 he became Director of Organisation and Establishments, a position he held until 1954 when he was appointed Chief Industrial Commissioner. Sir Wilfred retired from the public service in 1958, joining the board of BOAC in December of the that year.

Two of the part-time board members were placed on short-term appointment renewal until 19 June 1964. Sir Walter Worboys was born in Western Australia in 1900. He was educated at Scotch College, Western Australia, and Oxford University. He was chairman of British Tyre and Rubber Industries, a director of Westminster Bank and British Portland Cement Manufacturers; he was formerly commercial director of ICI and had held many important posts in the chemical industry. Lionel Poole was born in 1894. From 1919 to 1943 he was a national organiser and full-time officer at his branch of the National Union of Boot and Shoe Operatives, and from 1943 to 1949 was assistant general secretary of the Union; he was the general secretary from 1949 to 1959. During 1957-58 he was a member of the TUC General Council.

From the time of the Minister's decision to ask for the Corbett Inquiry in autumn 1962, there was no meaningful contact on Policy matters affecting the Corporation's future between the Minister and the Chairman and Sir Matthew's papers to the Minister of August and November 1962 remained unanswered. Once the Government had the Report in May 1963, it concentrated its attention on how the Report should affect its attitude to BOAC and on the production of a White Paper setting out its conclusions and the steps proposed to be taken.

The Corbett Report sought to cover a vast amount of ground in the remarkably short time of eight months; to

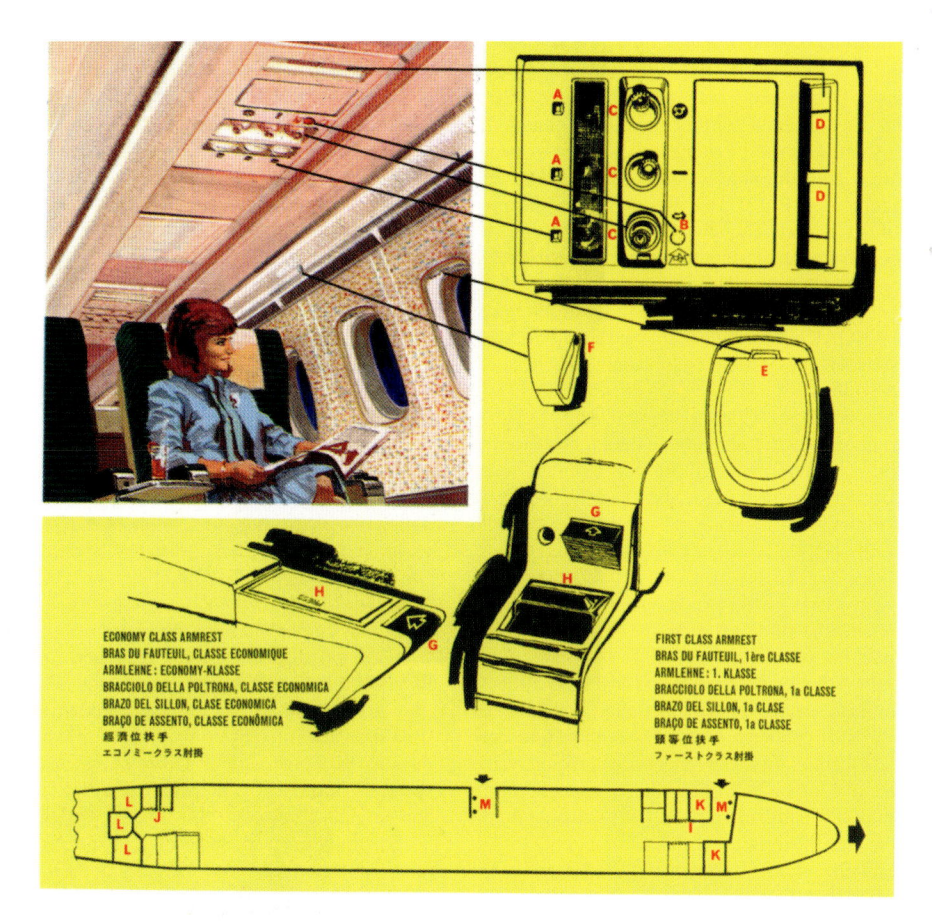

B·O·A·C VC 10

A Reading light switch **B** Call button **C** Cold-air ventilator **D** No smoking and fasten seatbelt sign **E** Window blind **F** Coathooks **G** Seat recline button **H** Ashtray **I** First Class drinking water fountain **J** Economy Class drinking water fountain **K** First Class toilets **L** Economy Class toilets **M** Extra points for electric razors

A Lampe individuelle **B** Bouton d'appel **C** Bouche d'aération **D** Signal "Défense de fumer—Attachez vos ceintures" **E** Rideau **F** Patère **G** Bouton de réglage du siège **H** Cendrier **I** Fontaine d'eau fraîche, Première Classe **J** Fontaine d'eau fraîche, Classe Economique **K** Toilettes, Première Classe **L** Toilettes, Classe Economique **M** Prises supplémentaires pour rasoir électrique

A Leselampen-Schalter **B** Klingel **C** Frischluft-Ventilator **D** Lichtzeichen mit Hinweis: Bitte nicht rauchen, bitte anschnallen **E** Fenstervorhang **F** Garderobenhaken **G** Verstellvorrichtung des Sitzes **H** Aschenbecher **I** Trinkwasser-Spender, Erste Klasse **J** Trinkwasser-Spender, Economy-Klasse **K** Waschräume der Ersten Klasse **L** Waschräume der Economy-Klasse **M** Zusätzliche Steckdose für elektrische Rasierapparat

A Interruttore della lampadina da lettura **B** Pulsante per la chiamata del personale di bordo **C** Congegno per l'immissione d'aria fresca **D** Segnale di "vietato fumare" e di "allacciare le cinture di sicurezza" **E** Tendina **F** Gancio appendiabiti **G** Regolatore del sedile **H** Portacenere **I** Acqua potabile (nella Prima Classe) **J** Acqua potabile (nella Classe Economica) **K** Tolette di Prima Classe **L** Tolette di Classe Economica **M** Prese supplementari per rasoi elettrici

A Botón para luz individual de lectura **B** Botón de llamada **C** Ventilador **D** Luz de aviso "No fume —abróchese el cinturón" **E** Cortina **F** Percha individual **G** Botón para graduar el asiento **H** Cenicero **I** Surtidor para agua fresca, Primera Clase **J** Surtidor para agua fresca, Clase Económica **K** Toilettes, Primera Clase **L** Toilettes, Clase Económica **M** Enchufes adicionales para máquinas de afeitar eléctricas

A Interruptor de luz para leitura **B** Botão de chamada **C** Ventilador de ar frio **D** Luminoso "Não fume e aperte o cinto" **E** Persiana **F** Cabide **G** Botão para ajustar o assento **H** Cinzeiro **I** Bebedouro de água, Primeira Classe **J** Bebedouro de água, Classe Econômica **K** Toalete, Primeira Classe **L** Toalete, Classe Econômica **M** Tomadas adicionais para barbeadores elétricos

A 閱讀燈鈕 **B** 喚人按鈕 **C** 冷氣通風器 **D** 請勿吸煙及綁緊安全帶燈號 **E** 百葉簾 **F** 衣架 **G** 座椅調整鈕 **H** 煙灰碟 **I** 頭等位沙濾水泉 **J** 經濟位沙濾水泉 **K** 頭等位洗手間 **L** 經濟位洗手間 **M** 電鬚刨插掣

A 読書灯点滅スイッチ **B** 呼びボタン **C** 冷風換気孔 **D** 掲示板—"禁煙・座席ベルトを締めてください" **E** 日よけ **F** コートハンガー **G** 座席調節用ボタン **H** 灰皿 **I** ファーストクラス水飲み場 **J** エコノミークラス水飲み場 **K** ファーストクラス洗面所 **L** エコノミークラス洗面所 **M** 電気剃刃専用コンセント

assist him in his work John Corbett employed two firms of consultants, one to examine BOAC's Engineering and Maintenance Department and the other the Sales Department. He took as the starting point of his inquiry the year 1956 when Gerard d'Erlanger took over from Sir Miles Thomas. In addition to a study of the organisation and Management of BOAC, the Report looked into the extent of BOAC's losses and how they had arisen, into the capital losses on aircraft, development costs, the Associated Companies, the factors both political and economic which had affected and were still affecting the Corporation's results, the BOAC-Cunard link-up, and, in particular, the VC10 story so far as it had developed up to the time the Report was written.

In his conclusions Corbett started with BOAC's situation as he saw it: an accumulated deficit of £81m, an extravagantly staffed organisation with thirteen more aircraft on order than it needed, and the VC10 coming along with direct operating costs substantially higher than those of the Boeing 707 which its competitors were flying. As to the losses, he estimated over £40m could be put down to the shortened lives of the DC-7Cs, the Britannias and the Comet 4s and the introduction of five new aircraft types in the previous six years.

There were other matters which had adversely affected the Corporation's finances, and he thought it bad luck that all these - loss of cabotage routes faster than foreseen, increasing competition from the Independent operators, the declining relative importance of London as a traffic centre and the unexpected pause in traffic in 1961 - all combining to create 'the perfect storm'.

As to the number of obsolescence charges, Corbett came to a conclusion that *'by 1959 if not earlier, it should have been obvious that the rate of obsolescence which was being charged in the BOAC Accounts was entirely inadequate'*.

He did not ask why the Ministry officials did not discuss the issue with the Corporation's Financial Comptroller at the time of the draft Accounts for the years 1959/60 and 1960/61; and as to the Corporation's attitude he commented *'I can find no evidence that an increase in the obsolescence charge was considered - there is certainly no evidence that it was ever considered with the Ministry'*.

Regarding the accumulated deficit the Corbett recommendation was that it should be written off as there was no prospect that BOAC could ever earn enough to pay interest on it. He also drew attention to the apparent contradiction in the Government's White Paper of April 1961 on the *'Financial and Economic Obligations of the Nationalized Industries'*, where there was a reference to the obligations on such industries "of a national and non-commercial kind". He felt that the right course was an obligation on the Chairman and Board of BOAC to operate commercially, leaving the Minister to have regard to the national interest, and that where the Corporation was required to perform in the national rather than in its commercial interest, it should be given a directive by the Minister and the Corporation's Accounts should contain a note of the circumstances. The absence of clear guidelines from the Government on this issue had caused doubt in BOAC's mind as to its right course of action in the past.

On organisation, Corbett considered that the Chairman should be supported by appointment full time of a 'proven business administrator' as Deputy Chairman who would spearhead an economy and efficiency drive and succeed in due time as Chairman. Sir Wilfred Neden as part-time Chairman should carry on because of his expertise in labour relations. He agreed with Sir Matthew's views that a Technical Director should be appointed to the Board and that the part-time Board members should be required to devote more time to the Corporation's affairs.

Opposite page: passenger facilities abaord a Corporation 707.

Below: The winter of 1962–1963 - known as the Big Freeze of '63 - was one of the coldest winters on record in the United Kingdom. Temperatures plummeted and lakes and rivers began to freeze over and traffic chaos set in to add to BOAC's woes. Here a Comet 4 of BOAC sits on the tarmac of a very cold London Heathrow. *(BOAC via author's collection)*

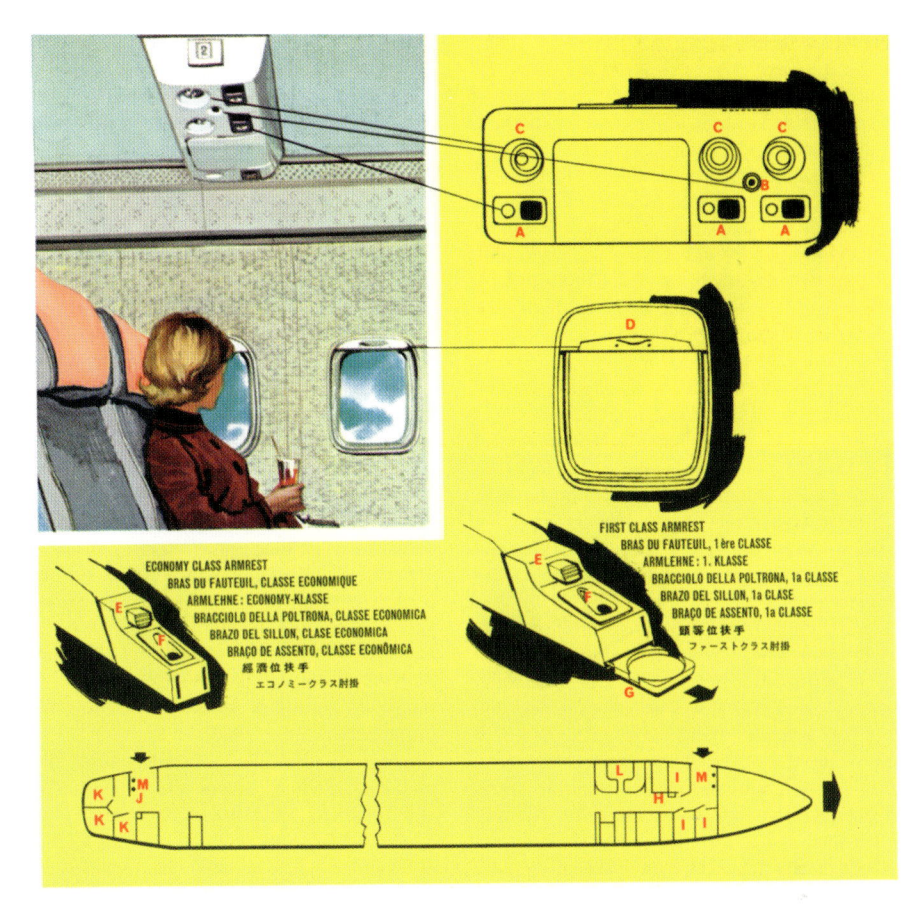

B·O·A·C ROLLS-ROYCE 707
羅斯勞斯707機　ロールス ロイス707

A Reading light switches **B** Call button **C** Cold air ventilator **D** Window blind **E** Seat recline button **F** Ashtray **G** Glass holder **H** Drinking water fountain, First Class **I** First Class toilets **J** Drinking water fountain, Economy Class **K** Economy Class toilets **L** First Class lounge **M** Extra points for electric razors

A Lampe individuelle **B** Bouton d'appel **C** Bouche d'aération **D** Rideau **E** Bouton de réglage du siège **F** Cendrier **G** Emplacement pour poser un verre **H** Fontaine d'eau fraîche, Première classe **I** Toilettes (Première classe) **J** Fontaine d'eau fraîche, Classe Economique **K** Toilettes (Classe Economique) **L** Salon bar Première Classe **M** Prises supplémentaires pour rasoir électrique

A Leselampen-Schalter **B** Klingel **C** Frischluft-Ventilator **D** Fenstervorhang **E** Verstellvorrichtung des Sitzes **F** Aschenbecher **G** Glashalter **H** Trinkwasser-Spender, Erste Klasse **I** Waschräume der Ersten Klasse **J** Trinkwasser-Spender, Economy-Klasse **K** Waschräume der Economy-Klasse **L** Lounge der Ersten Klasse **M** Zusätzliche Steckdose für elektrische Rasierapparat

A Interruttore della lampadina da lettura **B** Pulsante per la chiamata del personale di bordo **C** Congegno per l'immissione d'aria fresca **D** Tendina **E** Regolatore del sedile **F** Portacenere **G** Portabicchiere **H** Acqua potabile (nella Prima Classe) **I** Toilette di Prima Classe **J** Acqua potabile (nella Classe Economica) **K** Tolette di Classe Economica **L** Salotto-bar di Prima Classe **M** Prese supplementari per rasoi elettrici

A Botones para luz individual de lectura **B** Botón de llamada **C** Ventilador **D** Cortina **E** Botón para graduar el asiento **F** Cenicero **G** Portavasos **H** Surtidor para agua fresca, Primera Clase **I** Toilettes, Primera Clase **J** Surtidor para agua fresca, Clase Económica **K** Toilettes, Clase Económica **L** Salita de estar de Primera Clase **M** Enchufes adicionales para máquinas de afeitar eléctricas

A Interruptores de luz para leitura **B** Botão de chamada **C** Ventilador de ar frio **D** Persiana **E** Botão para ajustar o assento **F** Cinzeiro **G** Suporte para o copo **H** Bebedouro de água, Primeira Classe **I** Toalete, Primeira Classe **J** Bebedouro de água, Classe Econômica **K** Toalete, Classe Econômica **L** Sala de estar de Primeira Classe **M** Tomadas adicionais para barbeadores elétricos

A 閱讀燈鈕 **B** 喚人按鈕 **C** 冷氣通風器 **D** 百葉簾 **E** 座椅調整器 **F** 煙灰碟 **G** 杯架 **H** 頭等位沙濾水泉 **I** 頭等位洗手間 **J** 經濟位沙濾水泉 **K** 經濟位洗手間 **L** 頭等位休息室 **M** 電鬚刨插掣

A 読書灯点滅スイッチ **B** 呼びボタン **C** 冷風換気孔 **D** 日よけ **E** 座席調節用ボタン **F** 灰皿 **G** コップ受け **H** ファーストクラス水飲場 **I** ファーストクラス洗面所 **J** エコノミークラス水飲場 **K** エコノミークラス洗面所 **L** ファーストクラス談話室 **M** 電気剃刃専用コンセント

As to Sir Matthew himself, the implication of the wording in the Report was a recommendation that he should be confirmed in his position and it described him as *'a firm, fair and efficient leader'*. The Report had nothing to say as to the Managing Director. The consultants who examined the Engineering Department claimed that several million pounds annually could be saved subject to solutions being found in the field of industrial relations, whilst those who looked into the sales and advertising organisation in the UK and Europe concluded that 'the Corporation is getting very nearly the best possible value for both their sales and advertising expenditure'. On the workings of the Financial Comptroller's Department there was some criticism in the Report on the control of capital and that capital expenditure forecasting was not satisfactory; there was also said to be a lack of management development and a falling behind other major airlines in the use of electronic data-processing equipment; and, finally, that considerable savings could be achieved in the Department's overall cost. Corbett himself took the view that the Financial Comptroller had not in the past had a firm enough control over expenditure.

In a general comment on the attitude of the BOAC Managers, the Report said, *'loyalty, amounting almost to devotion, of the senior executives of the Corporation is quite outstanding'*. However, it was not all sweetness and light: some of the strongest criticism concerned the losses BOAC had incurred on the Associated Companies, amounting to nearly £14m over the six years to 1962. Corbett felt that the Board may have thought it was acting in the national interest but maintained that these losses would have been greatly reduced if there had been a proper financial assessment of the benefits as compared with the cost incurred in each case and if expenditure had been properly controlled.

Finally, the Report dealt at length with two VC10 problems - BOAC's conviction that it had thirteen too many VC10s on order, and their high operating costs over the Boeing 707s. On the first issue Corbett believed it had been wrong of the Board in January 1960 when they ordered a further ten VC10s over and above the order for thirty-five as not being in the real interests of the Corporation, and that the Corporation should cancel the thirteen aircraft forthwith although compensation to Vickers would have to be paid unless the RAF could absorb them; on the second issue he recommended that the Government should find some method, difficult though it might be, to recoup BOAC for the amount of the annual penalty arising from operating VC10s.

In its conclusion, the Report made the obvious statement that all BOAC needed were two things - efficient and robust Management and competitive aircraft but added: *'the enormous losses to date chiefly arose because there was neither'*. It recommended the Government to write off the accumulated deficit including further losses on Comets and Britannias which should be written down as at March 1963; to reconstruct the Board and take a sympathetic approach over the problem of excess VC10 operating costs. BOAC should also be allowed to build up a reserve before the Government required payment of any profit which hopefully might arise. On the whole, the Report was balanced in its tone and no sense extreme. It is hard to understand in hindsight why so much secrecy surrounded it because of the contents. The Chairman was eventually permitted to see it and to make available certain extracts to Departmental Heads in BOAC, but it was not available to Parliament nor was the Select Committee of Nationalised Industries permitted access to it during its work which began in November 1963.

In July 1963 the Minister announced he would be producing a White Paper to deal with the recommendations of the Corbett Report; before that became available, BOAC issued on 15 October 1963 its Annual Report for 1962/63 in which the Chairman was able to announce that current financial results were showing marked improvement over the previous year and there was a real prospect of ending 1963/64 with a break-even before interest payments. On 20 November 1963 the White Paper with the title *'The Financial Problems of the British Overseas Airways Corporation'* was issued by the MoA. It was a strange document of only eighteen pages with little concrete in the way of facts to justify the change of BOAC's management which followed. In fact, a series of Machiavellian machinations worthy of Sir Humphrey Appleby only

Left: Maurice Harold Macmillan, 1st Earl of Stockton, OM, PC, FRS (*b*.10 February 1894 – *d*. 29 December 1986) was a British statesman of the Conservative Party who served as Prime Minister of the United Kingdom from 1957 to 1963.

Right: Sir Wilfred John Neden (*b*. 24 August 1893, *d*. 1978) part time board member of BOAC)

Opposite page: passenger facilities aboard a Corporation De Havilland Comet

253

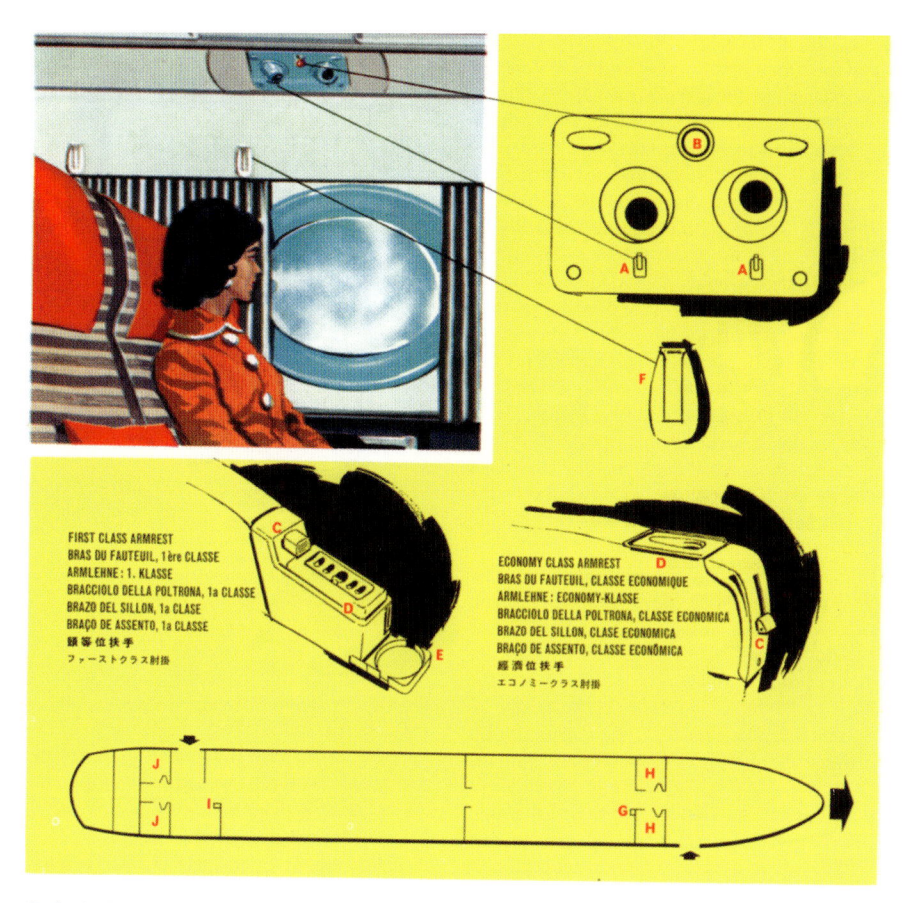

B·O·A·C COMET JETLINER
慧星機 コメット

A Reading light switches **B** Call button **C** Seat recline button **D** Ashtray **E** Glass holder **F** Coat hook **G** Drinking water fountain, First Class **H** First Class toilets **I** Drinking water fountain, Economy Class **J** Economy Class toilets

A Lampe individuelle **B** Bouton d'appel **C** Bouton de réglage du siège **D** Cendrier **E** Emplacement pour poser un verre **F** Patère **G** Fontaine d'eau fraîche, Première classe **H** Toilettes Première classe **I** Fontaine d'eau fraîche, Classe Economique **J** Toilettes Classe Economique

A Leselampen-Schalter **B** Klingel **C** Verstellvorrichtung des Sitzes **D** Aschenbecher **E** Glashalter **F** Garderobenhaken **G** Trinkwasser-Spender, Erste Klasse **H** Waschräume der Ersten Klasse **I** Trinkwasser-Spender, Economy-Klasse **J** Waschräume der Economy-Klasse

A Interruttore della lampadina da lettura **B** Pulsante per la chiamata del personale di bordo **C** Regolatore del sedile **D** Portacenere **E** Portabicchiere **F** Gancio appendiabiti **G** Acqua potabile (nella Prima Classe) **H** Tolette di Prima Classe **I** Acqua potabile (nella Classe Economica) **J** Tolette di Classe Economica

A Botones para luz individual de lectura **B** Botón de llamada **C** Botón para graduar el asiento **D** Cenicero **E** Portavasos **F** Percha individual **G** Surtidor para agua fresca, Primera Clase **H** Toilettes, Primera Clase **I** Surtidor para agua fresca, Clase Económica **J** Toilettes, Clase Económica

A Interruptores de luz para leitura **B** Botão de chamada **C** Adjustador do assento **D** Cinzeiro **E** Suporte para o copo **F** Cabide **G** Bebedouro de água, Primeira Classe **H** Toalete, Primeira Classe **I** Bebedouro de água, Classe Económica **J** Toalete, Classe Económica

A 閱讀燈鈕 **B** 喚人按鈕 **C** 座椅調整器 **D** 煙灰碟 **E** 杯架 **F** 掛衣鈎 **G** 頭等位沙濾水泉 **H** 頭等位洗手間 **I** 沙濾水泉 **J** 經濟位洗手間

A 読書灯点滅スイッチ **B** 呼びボタン **C** 座席調節用ボタン **D** 灰皿 **E** コップ受け **F** コート掛け **G** ファーストクラス水飲場 **H** ファーストクラス洗面所 **I** エコノミークラス水飲場 **J** エコノミークラス洗面所

discovered via the thirty-year rule demonstrates the possibility that the Government were asking the current BOAC senior management team of Sir Matthew Slattery Sir Wilfred Neden and Sir Basil Smallpiece to resign by the end of the year. But more of that later.

It was ironic that as late as 8 November Sir Matthew in a message to all staff said that *'the momentum of BOAC's financial recovery is now so great that it could carry us to a £3m profit at the end of the financial year - enough to pay the interest on our active capital'*. In announcing the changes in the House of Commons on 21 November, 1963 the Parliamentary Secretary to the MoA, Neil Marten, ran over all the issues brought out in the Corbett Report and the White Paper as grounds for change, saying it would be premature to write off BOAC's accumulated deficit, that no decisive case had been made for merger with BEA, and that the Minister had decided that the Managing Director's duties should in the immediate future be undertaken by the Chairman. As Sir Matthew Slattery was not prepared to undertake the double task - he was resigning at the close of the year and that Sir Basil Smallpiece was taking the same course of action after seven years as Managing Director; in consequence the Minister had invited Sir Giles Guthrie to take on the double task from the New Year and to produce a plan for profit within twelve months. When the Minister himself spoke during the same debate, he denied there had been undue interference in the Corporation's affairs either in aircraft ordering or matters concerning the Associated Companies. He quoted the matter of obsolescence charges and the findings of the Swash report on BOAC's handling of MEA and MASCO and said that by July 1962 things were so bad that either a public or private inquiry into BOAC was essential.

He announced that the new Chairman of BEA, Anthony Milward (later Sir Anthony) would be appointed to the BOAC Board and that Sir Giles Guthrie would retain his seat on the BEA Board. The new chairman was asked to produce, *'within a year, a plan for putting B.O.A.C. on its feet financially. This will involve a review of the organisation of the Corporation, of its route structure and the composition of its aircraft fleet.'*

Three other Board members of BOAC were asked to resign to facilitate the changes - Lord Rennell, Lord Tweedsmuir and Lionel Poole - but all were unwilling to resign until their periods of appointment lapsed in June 1964. The White Paper had started off by referring to BOAC's reputation for safety and service as being second to none and that it was the second largest long-haul airline in the world; that in the previous ten years it had nearly trebled its traffic and revenue, but that like some other airlines it was not paying its way. The Paper then referred to the accumulated deficit standing in the 1962/63 Accounts at £80m. Before analysing this sum, the Paper gave a brief historical review of BOAC, drawing attention to the problems on aircraft acquisition over the years and the short lives of the Comets and Britannias and mentioned the loss of cabotage points and the emergence of many new long-haul airlines which had coincided with a decline in the rate of traffic growth and an increase in restrictive practices by foreign governments. It made no mention of the increased opportunities given to the British Independent operators by the Government at BOAC's expense. A breakdown of the £80m deficit followed to indicate where the losses had built up - £35.4m on aircraft withdrawals and extra amortisation, abnormal costs in aircraft development (£5m), the Associated Companies' loss (£15.4m), the penalty for the reduction of the VC10 order from forty-five to forty-two (£0.7m), interest on capital (£34.2m), operating results after taking credit for Exchequer grants (£14.3m surplus), fleet insurance reserve (522m) and miscellaneous items (£1.8m).

Original caption: *'Following the resignation of the chairman and the No. 2 managing director of BOAC over matter of government policy, it has been announced that Sir Giles Guthrie (right) has been appointed to be the new chairman of B.O.A.C. It has also been announced that Mr. Anthony Milward, Chief Executive British European Airways will take over as chairman of B.E.A. when the present chairman Lord Douglas Of Kirtleside retires next March, in addition Mr. Milward will have a seat on the board of B.O.A.C'.*

Criticism of BOAC was made in four areas - amortisation of aircraft, initial costs of new aircraft, investment in Associated and Subsidiary companies, and operating results. On depreciation, the Paper charged that BOAC had made no change in its rate until the 1961/62 Accounts when as a result of a revaluation of aircraft an additional significant amount of amortisation was brought in. On the introductory costs of new aircraft, the Ministry case was that an airline of BOAC's importance must be first in the air with new aircraft of better performance rather than acquiring proven aircraft pioneered by other operators, and that there was a considerable advantage to be had by buying aircraft designed specifically for its requirement. On Associated and Subsidiary companies, the White Paper, while admitting that the Corporation's commitment to Kuwait Airways had been undertaken at the wish of the Government, went on to express the view that even if BOAC's policy of investment in all these companies were right, the implementation of the policy was unsatisfactory, although it mentioned that BOAC's commitments in two of the biggest losers, MEA and BWIA, had been almost eliminated. When it came to operating results, the White Paper blamed BOAC because its rate of expansion and service frequencies from 1960 was unduly optimistic. In addition to these four grounds for criticism, the White Paper then attacked the efficiency of the Corporation's internal

BOAC DC-7C G-AOII of BOAC Cargo Service. (BOAC via author's collection)

organisation and Management.

In referring to the Corbett Report, without actually naming it, the White Paper dismissed the Corporation's sales, publicity and advertising activities as 'well directed', then criticised the cost levels of the Engineering and Maintenance Department and the effectiveness of the financial control which it maintained had 'not been afforded sufficient importance within the Corporation'.

In the White Paper's conclusions there was an essential statement of Government policy as to what the aim of the Corporation should be - it was to operate as a commercial undertaking and 'if national interest should appear, whether to the Corporation or to the Government, to require some departure from commercial practice, this should only be done with the agreement or at the instance of the Ministry of Aviation' Neither the Chairman nor the Managing Director made any public attempt to rebut the criticisms in the White Paper on the praiseworthy grounds that the task for all in BOAC was to look forward rather than debate the past. But a great deal was said in the Press following the Debate in the House of Commons on 2 December on the second reading of the Air Corporations Bill. Subsequently, and to the satisfaction of the staff in BOAC, all these matters were dealt with at length during evidence before the Select Committee on Nationalized Industries early in 1964. BOAC was not at all unhappy at that outcome.

There was never an adequate public discussion on the White Paper, partly because the Corbett Report was kept secret and somewhat because events had coincided with the assassination of President Kennedy which diverted the attention of the Press and the public. In the last BOAC staff news of 1963, Sir Basil said his farewells to the staff from his regular column expressing *'deep regret that I shall no longer be a member of the very fine team who at home and overseas on the ground and in the air have built up a first-class world-wide airline'*. It must have been some satisfaction, if galling, for Sir Basil to be able to put alongside his farewells in the BOAC news, a heavy type heading reading '£6m better off and heading for a profit'. The Corporation's results for the first thirty-two weeks of the 1963/64 Financial Year showed an operating profit of £3m before interest and that the upturn in revenue was continuing. A year later, on 18 November 1964 it was again being discussed in Parliament:

Mr Lubbock asked the Minister of Aviation if he will now publish the Corbett Report on the British Overseas Airways Corporation.

Mr Roy Jenkins: *Having read the Corbett Report for myself, I can assure the House that there is little of substance in it with which it is not already familiar from the thorough investigation of the Select Committee on Nationalised Industries. It is nevertheless a fact that when*

Comet G-APDL outside the maintenance hangars at Heathrow. (BOAC via author's collection)

Mr Corbett was asked to undertake this inquiry, he was told that it would '…be regarded as private and confidential'. This was the basis on which he took evidence from the various people with whom he discussed the problems of the Corporations. Moreover, the Report as it stands contains a certain amount of information which would, in any case, have to continue to be treated as confidential for commercial reasons.

In view of all this, and bearing in mind that the Report is now nearly 18 months old and that what we want most of all is an end to the controversies and recriminations which have gone on between British Overseas Airways Corporation and the Government during the past year, I am convinced that the balance of argument is against publication. In any case, I should not consider the cost of publication a justifiable public expenditure at the present stage.

I have, however, told Sir Giles Guthrie that he should now feel free to show his copy of the Report to senior members of British Overseas Airways Corporation, whose range of duties lies within the scope of the Report.

Mr Lubbock: *Does the Minister appreciate that many of the criticisms in the Corbett Report are shown by the Select Committee to be completely unfounded? Will he now congratulate Sir Matthew Slattery on the excellent results achieved under his chairmanship and, in particular, on the wonderful profit and loss account that has just been produced?*

Mr Jenkins: *I think that Sir Matthew Slattery will already have found, in many of the comments published on recent results, a substantial note of congratulation, which I certainly share.*

Mr Maude: *Is the right hon. gentleman aware that we welcome what we regard as his very well justified decision, as there are obviously certain things in this Report that would be commercially valuable to competitors of BOAC? Is he aware that one of the purposes in setting up this inquiry was, in his own words, to end 'recriminations and controversies' that had previously taken place, and that this is surely the main achievement of this Report?*

Mr Jenkins: *Yes, but the hon. Gentleman will note that I did not underline the judgment that it was right to set up this secret inquiry in the first place. I do not think, whatever the intention was, that, during the time it took place, the existence of the inquiry helped to do away with the atmosphere of recrimination.*

In hindsight, Sir Matthew Slattery was an unlucky Chairman being someone who was in the wrong place at the wrong time. He took over at a time when the top Management and Board had become too attentive to the views of its sponsoring Ministry, an attitude which had became marked during the time that Harold Watkinson had been Minister and Sir Gerard had been Chairman; as a result, uncertainty increased within BOAC as to its obligation to the national rather than commercial interest.

He inherited the mistakes of the immediate past, three of which came home to roost with severe consequences in the course of his three and a half years at the helm. The first was the decision by the Board in January 1969 to increase the firm order for VC10s from thirty-five to forty-five, resulting in Sir Matthew having the unhappy duty of informing the Minister that BOAC would have too many aircraft; the second was the lack of firm financial control over the results of the Associated Companies, partly due to the way in which these Companies had been removed from the direct financial control of BOAC proper in 1956, although Sir Matthew was quick to rectify that situation and put in train a new initiative under Granville which enabled BOAC to disengage itself from the massive losers, and brought the results of these Companies into profit for the Corporation within the space of four years; and third the failure to adequately provide for the amortisation of the fleets, although here there was considerable doubt as to how far the responsibility for this rested with the Ministry and how far with BOAC.

From boardroom critisms to the glamour of the airways - two girls who used to do the rounds in the wards of English hospitals were now doing 'rounds' eight miles above the earth. Miss Mavis Street (right), from Bournemouth, and Miss Olive Casey, (left) from Galway, both former nurses qualified as BOAC stewardesses.

Mavis, 25, spent four-and-a-half years at Guys Hospital, London, where she became a staff nurse. She was at the West End Neurology Hospital, London, before she joined BOAC Olive, 23, did her nursing training at Northampton General Hospital, studied midwifery for six months at St. Thomas's Hospital, London. Before joining BOAC she spent several months in France as English teacher to a French family.

With BOACs advertising material in the early 1960s often all is not what it seems. Many of the colour photographs, although undoubtedly 'real' have a strange almost 'arty' quality about them when compared to todays images. Also, many of the aircraft have their registrations altered to what could be called fake ones. A good example of this is this picture taken in Rome during the refuelling procedure thought to have been around 1963.
(BOAC via author's collection)

Indeed, Sir Matthew was determined to see that proper provision was shown in the Accounts and he overruled the then Financial Comptroller, Ken Bevan, who left the airline. There appears to be no satisfactory answers from the Ministry officials at hearings before the Select Committee on Nationalized Industries in 1964 as to why they had not pressed BOAC earlier to make proposals to revise the rate of amortisation.

Sir Matthew could not be held responsible for the policy of expansion. This commenced with full Ministry approval under the d'Erlanger regime, and commitments to aircraft had been made accordingly. This policy began in 1957/58 and worked out satisfactorily over the first three years with an increase in capacity of 121% compared with a growth in traffic of 90%, but the sudden and unexpected pause in traffic growth in 1961/62 was not foreseen by any of the intercontinental operators. There was little BOAC could do but ride out the storm since all routes, even the worst hit, made some contribution to overheads and the results would have been worse if aircraft had been left on the ground.

Another charge brought by the Corbett Report and the Ministry concerned the slowness in introducing measures in BOAC to reduce engineering and maintenance costs, and the charge was endorsed by the Select Committee in 1964. But the radical changes in work procedures which were required were challenging to hurry through in the face of the industrial relations problem they created in a susceptible area. In fact, a significant reorganisation in the Engineering and Maintenance Department had mainly been completed by the time of the Corbett inquiry and BOAC's costs had reached levels at least as good as those of its main international competitors, as Abell pointed out to his staff in a riposte to the November 1963 White Paper.

Sir Basil Smallpeice was in a different position from Sir Matthew, since he had been in the post of Chief Executive for seven years at the time of the Corbett Report and so had been directly concerned with the unsatisfactory period under Sir Gerard's chairmanship. Sir Basil was not responsible for the results of Associated Companies during the d'Erlanger period - that responsibility lay with Sir George Cribbett - but the losses on these Companies went back even earlier into the Miles Thomas' era when Sir Basil was Financial Comptroller and then Deputy Chief Executive, and he had played an important role in the Corporation's original involvement with MEA.

The confused issue of the amounts provided for amortisation over the years 1957/58 to 1960/61 also fell in the period when Sir Basil was Managing Director, although this was a matter for the Financial Comptroller to deal with in the first instance. But the gains to BOAC during Sir Basil's time as MD were impressive, and not brought out as they should have been in the November 1963 White Paper. Admittedly the introduction of the large jets had done much to produce these reductions, but that does not detract from the very real improvement in efficiency BOAC achieved over these years.

Day-to-day events press hard on any airline management in such a volatile industry, and no detailed public case was ever made by BOAC to answer the criticisms directly: however, the Corporation in its 1963/64 Annual Report, written under the new Chairman's direction, including extracts from the 1964 Report of the Select Committee which by no means gave full support to the views expressed in the White Paper. The Committee felt that the uncertainty of Corporation's role caused some of BOAC's major difficulties; on the subject of the overall deficit they said *'Had the Board of BOAC understood as they now do that their responsibility was to operate on commercial lines, events would have followed a different course, and the deficit would have been less serious'*. On the subject of aircraft, the Report had this to say *'Your Committee are of the opinion that the Board of BOAC is responsible for BOAC's commercial interests and no one else's and that they should not have accepted responsibility for any part of the manufacturing industry's problems. These were for the aircraft industry and the Government to look after, and far from pressing BOAC to accommodate them against their commercial interests, the Government should have been insisting that the Corporation take the clear-cut view of their commercial interest that has since been given to the new Chairman, Sir Giles Guthrie, as his term of reference'.*

The question must be asked why such drastic action was taken by the Minister. After all, Sir Matthew in his paper on the future of BOAC in August 1962 sent to the Minister at his request, set out what needed to be done and in fact mainly was done after Sir Matthew's departure. Clearly the core of the problem lay in the attitude of the permanent officials at the MoA. It must be recalled that Julian Amery came in as the new Minister only in July 1962, two months before the Corbett inquiry was officially set in train, and would himself scarcely have had time to become thoroughly acquainted with all the background.

Nevertheless, he was faced with the unpleasant task of having to present BOAC's 1961/62 accounts to the House of Commons where both the massive loss and the issue of the inadequate amortisation would be publicly revealed. It must be expected that he had had harsh words to say to his permanent officials as to why the situation had been allowed to develop; the officials in turn no doubt put the blame on the Corporation and particularly on the Managing Director. The issue coming on top of the adverse Swash Report on BOAC's involvement with MEA and MASCO, and the frustration felt by the Ministry officials over the failure to agree a five year target called for by the White Paper all added up to a lack of confidence on the part of the permanent officials in the Corporation's financial direction and in the Managing Director. And so the Corbett study was put in motion.

The Corbett Report had nothing but good to say about Sir Matthew, and yet he resigned before his term of office had expired. Looking back many years later at the events of 1963, Sir Matthew gave his account of events leading to his resignation. He had been initially appointed in July 1960 for three years and in the Spring of 1963 inquired of the Minister as to whether his term of office would be extended. The reply he got was that until the Corbett report had been received, it would be embarrassing to reappoint him for a further three years but the Minister asked him to stay for an additional year to July 1964 without a commitment beyond that date.

It was only during the IATA meeting in Rome in September 1963 that some unofficial inkling of what was in the wind leaked out - it seems the BEA team there, Lord Douglas, Anthony Milward and Sir Giles Guthrie knew what was coming. When Sir Matthew returned to London, he was summoned by the Minister and told that Sir Basil would have to go. This was a surprising turn of events since only the Board could appoint the Managing Director of the Corporation although the Minister could remove him from the Board. Sir Matthew stoutly defended Sir Basil's record and firmly stated that if the Minister removed him from the Board, he, Slattery, would recommend to his Board that Smallpeice was reappointed as General Manager.

The Minister's response was to ask Sir Matthew to work out his time to July 1964 combining both jobs of Chairman and Managing Director, but that after that date he would be relieved by Sir Giles Guthrie. As the Minister would not alter his mind about Smallpeice, Sir Matthew asked permission to resign on 31 December 1963 which was agreed, at the same time expressing his view that it was not a good idea to combine the two jobs.

It was ironic that Sir Matthew Slattery and Sir Basil Smallpeice heralded a period of healthy traffic growth and BOAC finished that financial year with a record operating surplus. The new Chairman, Sir Giles Guthrie, wrote in the 1963/64 Annual Report *'if BOAC were as inefficient as the White Paper implied, it could not have achieved the dramatic financial recovery which the figures for 1963/64 disclose. The new Board takes no credit for this result. The foundation had been laid by its predecessor. The operating plan had been worked out in detail, capacity allocated, sales targets fixed. The arrangements made were sufficiently flexible to enable BOAC to take full advantage of the upsurge in traffic which occurred'.*

Once again the Corporation had a new leader, the ninth in a matter of twenty-four years. Sir Giles Guthrie was an unknown quantity to the staff of BOAC, whose morale had been badly shaken by events. The new Chairman faced a stiff test of leadership when he took up his post on 1 January 1964. He had one inestimable advantage, Sir Matthew had already pointed the way ahead in his memoranda to the Minister of August, and November 1962 and the official minds were now ready for the change.

Bricks, mortar and technology.
The years 1964 to 1968 saw rapid growth in BOAC's property developments both at London Airport and at John F Kennedy Airport, New York, the name having been changed from Idlewild Airport in December 1963 following the assassination of the president a month earlier. At London Airport there were some projects. The new headquarters building for BOAC, Speedbird House, was finally ready in May 1966 for occupation by the Board, Management, the Commercial Divisions and the Planning Department, thus freeing the main hangar building to become the home of the Engineering and Flight Operations Divisions as initially intended. A specialised building was constructed to house the new complex of computers, known as Boadicea House after the initial letters of the equipment - British Overseas Airways Digital Information Computer for Electronic Automation. The building required standby power supplies, air conditioning and arrangements for radar screening and the total cost was over £1.25m.

Training facilities at Cranebank were extended to cater for an engineering apprentice school. In June 1966 land was leased in the south-west of the airport to build and equip a modern air cargo terminal alongside a similar structure for BEA. The buildings were to house up-to-date handling equipment for computer control by at an estimated cost of £4.75m with a target date for use in 1970. In 1967 the Board approved an extensive building programme for the construction of a new hangar ready for the 747 aircraft, for a new workshop block with multi-storey car-park above, and an air-conditioned office block at a total cost approaching £9m. Work on this massive programme began in 1968.

In the West End of London the arrivals side of the extension to Airways Terminal, Victoria, came into use in March 1968, so completing the second phase of the building's development. On a site opposite Airways Terminal arrangements were made with London Coastal Coaches whereby the latter would erect a building and BOAC would lease it for additional office accommodation and car parking in the basement.

On the technical side, there were some critical innovations towards greater efficiency both in the air and on the ground. The Air Navigation (4th Amendment) Order 1963 made mandatory the provision of flight recorders on aircraft. BOAC had already used flight recording for its technical purposes, so decided in 1964 to equip its aircraft with advanced 'black boxes' to provide not only the data called for by the Order to be of value in accident investigation, but additional data to give more information about aircraft performance, fuel consumption, cruise speed and so on. In 1966 BOAC's Air Safety Committee recommended that the fullest use should be made of the data and by 1968 the potential of the flight recorders was being used to the full to give information or help in maintaining flight operational and technical safety standards. Navigation procedures were revolutionised by the development of the inertial navigation system, a fully self-contained system on the aircraft independent of ground aids, whereby the aircraft flight path could be continuously established and displayed to the pilot by means of a digital computer. After evaluation of the system, Board of Trade approval was received for installation in the Corporation's aircraft.

Progress was also made in developing automatic landing devices on the Super VC10 using the Elliott system and Bendix automatic pilots. This would allow landings at 100 feet decisions height and 1,300 feet visibility (category 2) and subsequently at 700 feet visibility (category 3). Early in January 1967, the Chairman in a Super VC10 made several landings with the Elliott Automatic equipment in control at the RAE's airfield, Bedford. A significant problem with the system was its dependence on ground" facilities which were at that time available at few airfields. In due course, automatic landings at London Airport on Super VC10s became standard practice when conditions made it desirable. One of the difficulties engineers were faced with was the maintenance of the complicated electronic system in the new aircraft types. In 1964 BOAC pioneered new automatic test equipment called TRACE. This was computer-based equipment which could be used to analyse the complex circuits in the repair shops and track down the origin of any fault, thus reducing the time taken in rectification and cutting down the number of spare units held as replacements.

Electronic Computers - as opposed to the human variety - began to assume more and more significance in BOAC's internal affairs. A start had been made in Sir Matthew's time but the vast capital programme BOAC was to undertake had to await the arrival of Peter Hermon who joined BOAC in 1965 and three years later was made a Member of Management.

Some progress had been made before his arrival on the scene. By 1964 the Corporation had installed IBM 1401/1410 computers to supply a data processing system at London Airport for accounting and analytical tasks and had concluded arrangements with Collins Radio whereby they would carry out message-switching employing an electronic data processing installation. This was a significant advance in speed and accuracy of message distribution and also saved outlays on staff.

In July 1965 Hermon put a detailed paper to the Board

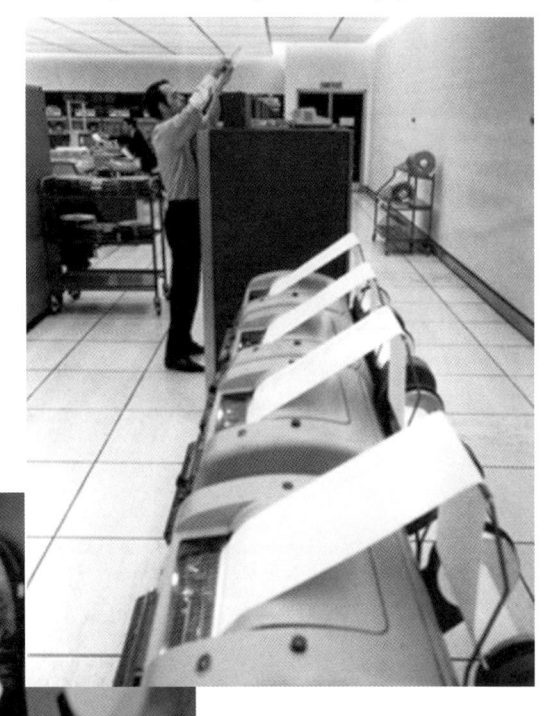

Inside Boadicea House - **B**ritish **O**verseas **A**irways **D**igital **I**nformation **C**omputer for **E**lectronic **A**utomation. Today it looks like dinosaur technology, but in 1970 it was very much state of the art. *(BOAC via Hugh Jampton Collection)*

recommending the setting up of an integrated computer system at London Airport at a cost of around £3.3m. The objective was to establish a system which would provide a range of information to serve Management and staff over a full field. The proposal covered the purchase of two IBM 360 Model 50 computers and associated equipment, to be housed in a new building to be built at London Airport as part of the BOAC complex. The Board gave their agreement, and subsequently, the expenditure was approved by the MOA after satisfying itself that no British alternative equipment could match IBM capabilities.

The new computers were to be applied first to the task of passenger reservations at London and subsequently at New York; they would be linked to satellite computers at main sales points which would provide instantaneous booking information on small screens to the reservations staff. Reservations were thus the priority followed by passenger check-in and aircraft weight and balance. On reservations work alone it was confidently expected that quicker service to customers, more accurate information, but the use of last minute cancellations, and slower staff increases would all produce positive financial benefits.

In October 1965 approval was given to an estimate of £1.25m for a building to house the computers and work on the urgent task of modernising reservation began. Meanwhile, the Chief of Information Services, as Hermon was described, completed a survey of how computers could benefit BOAC more widely. On 22 April 1966 he put his case to the Board far more extensively than before, since it included aspects in all areas of BOAC, not only reservations and passenger check-in but also planning, flight operations, engineering, finance, and personnel up to the year 1979/80.

The July 1965 proposals would now be superseded, and under this new comprehensive scheme three IBM Series 360 model 65 would be acquired. In all the total cost of the whole Boadicea project, as it was known,- up to 1979/80 was reckoned at £33.5m, whereas over the same period savings were estimated at £48m by staff and equipment which would be needed if the computer plans did not go ahead. This was a massive expenditure but the Board to their credit approved it in principle but asked for quarterly reports on progress and spending committed.

Methods of management in BOAC would be revolutionized, and in consequence, there was an urgent need that staff from top management down should know what computers could and could not do. To ensure that all concerned were better informed, Peter Hermon gave a series of informative talks on how the system would work in practice and what its scope would be. Good communications worldwide were a fundamental part of the Boadicea project, and immediate steps were taken to put more urgency into the work of the team concerned with this aspect. By April 1967, it was clear that reservations in London and New York would not be switched over to Boadicea on 1 April 1968 as initially planned mainly due to the IBM programming, but as Hermon pointed out 'the work is right on the frontiers of present experience'. By July 1967 the first two IBM 360 processors model 65 were installed in the new Boadicea House and the third at the close of the year. In September 1967 a plan was approved to add to the Boadicea complex a Ferranti computer installation for use on the export side of BOAC's new cargo terminal at London Airport at an additional cost of £1.5m. By now enthusiasm in the Corporation for the computer use had built up to a point where the 1400 series was overloaded, and the waiting list for tasks to go on the new 360 processors was mounting. It was already clear that reservations work would consume more of the 360s time than had been foreseen, leaving only reduced night hours for other applications.

In January 1968 it was decided to extend the Boadicea departure control system at Prestwick, Montreal, Toronto, to the East Side Terminal New York and the gate lounges in BOAC's new terminal at New York. One of the nightmares associated with a centralised computer system for reservations was the unlikely event of the tapes being destroyed by fire or other means, and so throwing worldwide bookings into chaos. To ensure against this, copy tapes were held at a location separate from Boadicea House. Inevitable problems arose with such advanced equipment, and these caused short shutdowns early in 1968, but by mid-summer, these difficulties had been overcome. In July Hermon listed the main applications under development or partly operational already; this list came to thirty-six areas covering practically every aspect of the Corporation's work, from revenue accounting to payrolls, from aircraft utilisation analysis to message switching, from tariffs to output budget. In autumn 1968 passenger reservations were changed into Boadicea with details of all BOAC flights for a year ahead; all sales and reservations centres worldwide were then linked to Boadicea over the teletype network. In November 1968, the new small television type screens were installed in New York and Airways Terminal, London and were extended to other stations thus putting BOAC ahead of all airlines in the extent and sophistication of its reservations set-up.

Hermon was able to justify the purchase of a fourth IBM 360 computer, partly to take over from the older 1401 and 1410 models which had become unreliable but also to be able to undertake even more experimental work on Boadicea.

On 7 November 1968, William Rodgers, Minister of State, Board of Trade, formally opened the new Boadicea complex at London Airport and exchanged messages over the system with Sir Patrick Dean, HM Ambassador to the United States, in BOAC's New York offices.

BOAC was thus well and truly launched into the new era of digital management, no mean achievement in the face of advanced equipment, lack of trained staff, and the development of software programmes never before attempted. The advance had been so rapid and so evident that other airlines began to come to BOAC to purchase its know-how.

Eventually Boadicea evolved into BABS - the British Airways Booking System - which became an integral part in the skullduggery practiced by British Airways against a number of other airlines with the 'Dirty Tricks' scandal of the early1990s - but that's another story!

Chapter Ten

A New Chairman - And The 747

When taking up his post with BOAC, Sir Giles Guthrie was forty-seven years old with enthusiasm for aviation that begun while a schoolboy. He had learnt to fly at an early age and in the thirties took part in national and international air races. In 1938 Sir Giles joined British Airways as a junior traffic officer serving on the Continent; indeed he was in charge of the Warsaw station when war broke out. During the war, he served as a fighter pilot in the Fleet Air Arm, but at the close of hostilities entered the City in the insurance and banking world where his family interests lay. Sir Giles resumed an active interest in aviation in 1959 when he was appointed a part-time member of the Board of BEA, a position he still held on taking over as Chairman and Chief Executive of BOAC on 1 January 1964, on a five-year appointment from the Minister, Julian Amery. Sir Giles was a man of great personal charm and open in his dealings with staff and quickly won over the support of his management team.

After all the criticism of BOAC from both Government and the Press, morale in the airline was low and it required the new chairman to have the gift of inspired leadership. He hit the right note in his message to staff on his first day in office when he said *'We shall have to fight for the Corporation. In a fight, leadership can engender an atmosphere that breeds enthusiasm. I promise to do my best, and I count on you for your support'.*

Even before he became Chairman, Sir Giles wrote to every member of BOAC's top management, telling them he would be bringing in one senior official from BEA, David Craig, to assist him, but apart from that would be making no immediate changes in the management team itself until some time had passed to enable him to assess individual capabilities. He then called an informal meeting of Management at 55 Park Lane to make personal contact and possibly to alleviate some of their concern felt at the appointment of a new and unknown entity from BEA.

Before he took up his new post, Sir Giles had a bright idea of what needed to be done to put BOAC's house in order; something that was not surprising since he had the benefit of Sir Matthew Slattery's earlier suggestions to the Minister on which no action was taken. Furthermore, he insisted on receiving from the Minister what Sir Matthew failed to obtain, namely, clear guidance as to the aims of the Corporation, whether they were to be solely commercial or whether or to what extent considerations of national interest and prestige should be taken into account. This guidance was set out in a letter from the Minister to Sir Giles dated 1 January 1964:

I have sometimes been told that there is uncertainty on the part of British Overseas Airways Corporation as to how far its role is commercial and how far it should take account of considerations such as public service and national prestige which might conflict with a purely commercial interest. The Government do not consider that there is any solid justification for such uncertainty. Nevertheless, it may help to remove any misunderstanding if I give you more specific guidance on this issue.

It has all along been understood that it was the duty of the Air Corporations to make themselves self-supporting as soon as possible. The Chairman of BOAC in 1952 made this clear in a direction to his staff with which I find myself in complete agreement. He said, in the annual report of the Corporation for 1952, that it should be imprinted on the minds of all staff that the Corporation was essentially a commercial undertaking; that the financial aspect of every single activity mattered; and that the ultimate test of the Corporation's success was not only the standard of public service provided but also the normal business criterion of whether it could be made to pay its way. The White Paper on 'The Financial and Economic Obligations of the Nationalised Industries' (cmd. 1337) explains what is meant by a nationalised Corporation paying its way. The minimum requirement is that the revenue should be enough to meet all the items properly charged to revenue, including interest, depreciation and also to provide for reserves. It is accordingly the Government's aim to agree with the nationalised Corporations on financial targets with the object of achieving as soon as possible yields approximating more closely to industrial yields, allowance being made for obligations which are undertaken by nationalised Boards for reasons other than their strict commercial interest. I hope it will be possible eventually to agree on some such financial target for BOAC as I have been able to do with BEA.

The immediate task, however, is for BOAC to achieve the break-even point after meeting interest and depreciation. How this can be done is a matter for you, but for my part, I am also much concerned in view of my responsibility for financing the accumulated and continuing deficit. After consulting you, I have informed Parliament in the course of the second reading Debate on the Air Corporations Bill 1963 that a plan for achieving break even will be produced by the Corporation within a year. This plan may well raise major questions of policy, including questions about the size of the fleet and the route network operated by the Corporation, and you will no doubt wish to discuss its implications with me.

There are many points where national policy and the aims and activities of the Corporation interact. It is highly desirable that a close relationship between the Ministry and the Corporation should continue and, with it, the constant interchange of ideas on matters of common interest.

There is already established machinery under which the Corporations seek the Ministry's approval for the ordering of new aircraft. There is also the Transport Aircraft Requirements Committee where the points of view of all the interests concerned are brought together. But the choice of aircraft is a matter for the Corporation's judgement. It has been the aim of the Corporations to buy their aircraft as far as possible from British sources, and I trust that this policy will continue. There may, of course, sometimes be occasions (as when the Boeing 707s were purchased in 1956) when the choice of foreign aircraft is unavoidable.

It is important that the interchange of views between the Government and the Corporations should not blur the fundamental responsibility of the Corporations to act in accordance with their commercial judgement. If the national interest should appear, whether to the Corporations or to the Government, to require some departure from the strict commercial interests of the Corporations, this should be done only with the express agreement or at the express request of the Minister. How losses, if any, resulting from such a political decision should be presented in the accounts will depend on the circumstances in each case.

Julian Amery.

The letter had immense significance in the remaining years of BOAC's existence. It left no doubt in the minds of the Board and Management as to where their first duty lay and what was equally important stated categorically that the choice of aircraft to be acquired by the Corporation was a matter for the Corporation's judgment. This was music in the ears of top executives who had lived through years of coping with ministerial pressures over aircraft purchases. A second important step in the plans to revive BOAC's fortunes was a re-organisation of the Board. The Minister had told the House of Commons on 21 November 1963 that the Chairman of BEA, Anthony Milward (later Sir Anthony) would be appointed to the BOAC Board and similarly Sir Giles would remain on the BEA Board, both as part-time members. On 1 January 1964 the three new members, Sir Giles, Anthony Milward and Ron Smith (from the Trade Union side), took up their duties. Shortly afterwards on the retirement of Kenneth Staple, the Secretary and Legal Advisor after 31 years service, Charles Hardie (later Sir Charles) was appointed to the Board and given a particular interest in BOAC's financial affairs in the light of his extensive experience as a director of many commercial companies. On 2 December 1963, the Minister had confirmed the continued part-time appointments of Sir Walter Worboys and John Booth, and of Gilbert Lee as a full-time member and re-appointed Keith Granville whose term was shortly to expire. At the same time he announced the names of three more part-time members who would join as Board vacancies arose. Sir Duncan Anderson, a distinguished civil engineer known for his work on the construction of the Kariba Dam on the Zambesi river, Lord Normanbrook, a former Secretary to the Cabinet with immense knowledge of the Whitehall scene, and Arthur Norman (later Sir Arthur) who had been running the De La Rue Company for some years. Sir Walter Worboys resigned on 31 March 1964, due to the pressure of other commitments, and Sir Duncan Anderson took his place.

When the terms of office of Lord Tweedsmuir and Lionel Poole expired in June 1964, their places were filled by Lord Normanbrook and Arthur Norman, thus completing the Board. Lord Rennell continued until the expiry of his term on 20 November 1964, having served on the Board for over ten years.

Like so many others, Kieth Granville was an 'Imperial Airways man' having joined the company in 1929 as a ten-shillings a week trainee. *'It was small, although it seemed big at the time. The Head Office was in Charles Street between the Haymarket and Lower Regent Street. There were a few flights a day to France, Switzerland, Belgium and Germany, and a certain amount of charter work including humping gold shipments from capital to capital as the value of gold and currency changed. There was an air of enthusiasm with every man and woman prepared to do their own and any other job 24 hours a day if need be. Most interesting of all, a wall map of the world showed the few existing routes in Europe and a series of dotted lines to the Middle East, India, the Far East, Australia and Africa. All were marked in stages of 200 to 300 miles, which was then the limit of an aircraft's operating range. Consequently, there were no dotted lines on the Atlantic.*

The staff were few. The Chairman, Sir Eric Geddes - part time - was housed at his Dunlop office in St James's. The Managing Director was George Woods Humphery, his deputy Harold Burchall, Traffic Manager Dennis Handover, Secretary S. A.Dismore, Doris Moore, Bertie Young, Freddie Groves and the only two baggage porters,

Keith Granville, Chairman of BOAC in his office.
(BOAC via author's collection)

Gearing and Butcher. There was a small Boardroom and a few spare offices on the first floor of the building.

The ground floor was a miniature departure area with the office of the Traffic Manager alongside. The commercial hub of the business was in the basement which housed a reservations office with about half a dozen telephones and one direct to Thomas Cook's. The booking sheets were very simple— one stage jobs on one piece of paper. It all seemed bigger at Croydon Airport where John Brancker, the other trainee, and I did a stint. We possessed a fleet of 21 aircraft with a total value of less than half a million pounds. The operational organisation consisted of Chief Engineer Hall, Air Superintendent Brackley, Chief Pilot Wolley Dod, Chief Accountant Quin Harkin and many other well-known names.

The atmosphere was of great expectation because of the plans for opening the air route to India were to materialise early in the year, and Africa was obviously not going to be very far behind.

We knew we were in an exciting and expanding business, but we also knew profits would be small, and that there was going to be a lot of hard work for modest pay. Imperial Airways, in fact, sometimes had great difficulty in meeting its weekly wage bill. I was told that cash for paying the staff at Croydon was sometimes obtained by cashing a Croydon Branch cheque at a West End branch on Friday, knowing it would not be cleared through the Croydon bank until Monday, by which time with any luck the weekend business, particularly for Paris and Le Touquet, would provide enough money to be rushed down to Croydon by car first thing on Monday to meet the cheque.

We were a public company, but received subsidies from the Government through various means, including a payment on the basis of horse power miles which probably had much to do with the decision to operate four-engined aircraft rather than twins and trimotors.

Many other great personalities were around at Croydon at the time, particularly on the flying and engineering side Captains Horsey, Wilcockson, Olley, Perry, 0. P. Jones, Bailey, Powell, Strawson, Malabbott, McMeeking, to name only a few.

The mid-day London- Paris service was called the Silver Wing, and even in that year of 1929, there seemed to be a lot of American tourists. The travel system was that each passenger was allowed 100 kilos free, including his own weight. On average women did better than men and one male passenger, I believe a Daily Mail journalist, had to pay very heavy excesses on his personal weight even before taking any baggage into account - and that incidentally would apply to me today if the rules were still the same.

Croydon Aerodrome with its steeply sloping green grass - was probably the busiest civil airport in Europe. All the well-known companies, Air Union - the forerunner of Air France - Lufthansa and KLM and many others were operating from it. The competition was as keen then as it is today.

The load factor on European services was 71%. This

BOAC used at least one swing-tail Canadair CL44 for cargo work during the 1960s. (BOAC via author's collection)

Technical data
(747 compared with 707)

264 The 747, built by Boeing of Seattle, is a sub-sonic jet aircraft following the same basic design as the well-proven 707 but offering roughly two and half times the 707's passenger capacity.

	747	707		747	707
OVERALL LENGTH	231' 10"	152' 11"	ENGINES	4 PRATT & WHITNEY JT9D-3A	4 ROLLS-ROYCE CONWAY R.CO.12
HEIGHT GROUND TO TOP TAIL FIN	63' 5"	42' 6"	THRUST	4 x 43,500 LB	4 x 17,500 LB
WING SPAN	195' 8"	142' 5"	CRUISING SPEED	568 — 600 MPH	540 MPH
MAXIMUM ALL-UP WEIGHT	316 TONS	140 TONS	RANGE*	5,000 STATUTE MILES	4,000 STATUTE MILES
MAXIMUM CABIN WIDTH	21' 4"	12' 4"	CREW	18 OR 19	10 OR 11
CABIN CEILING HEIGHT	8' 4"	7' 7"	TYPICAL PASSENGER CAPACITY	27F/335Y	16F/130Y

with full fuel capacity, using maximum take-off weight

At London (Heathrow), New York (JFK) and most other airports served by the BOAC 747, boarding and disembarkation is by means of telescopic weatherproof "fingers", involving no steps.

Even before the 747 was put into service BOAC produced a large format booklet designed to help sales agents answer any enquiries from the travelling public. *'It has been produced for reference purposes and should not be distributed to the public, for whom seperate promotional material is available'*. The booklet went on to warn *'We have tried to make sure that the information contained in the booklet is accurate, but it is possible that modifications may be made in the light of operational experience'*. In the event, BOAC were over a year late in intoducing the 747 into service. *(BOAC via author's collection)*

fell to 57% as soon as the Indian route was opened. Capacity, limited though it was, was still more than demand - all rather like today and even the dividend of 5% paid to the shareholders for 1929/30 was much the same as the BOAC dividend last year.

Expansion - and planning for the future

Two new aircraft designs were ordered during Sir Giles' period, the Boeing 707-336C cargo version and the Boeing 747. Meanwhile the Standard and the Super VC10s were put into passenger service, as well as the 707-336C initially on cargo work, and the Conway-engined 707-436 received a significant overhaul to extend its life to 60,000 hours.

Not long after Sir Matthew became Chairman in July 1960, he had shown interest in the development of BOAC's cargo business and especially in the build-up of a fleet of cargo aircraft. He set up a cargo study group; at that time BOAC was operating DC-7Cs on the North Atlantic, and the study group concluded that a more specialised and economical aircraft was required. After exhaustive analysis, it was decided to dry-lease a Canadair CL44 swing-tail freighter from Seaboard World Airlines for the Atlantic operations and this new venture began in October 1963.

The Study Group continued its work on the long-term requirement, and Sir Matthew brought in an outside consultant to lead the group's investigations. The consultant's view was that BOAC should acquire a fleet of seven CL44s for its longer-term cargo operations. BOAC's Planning Department disagreed, being convinced that the solution was to buy two Boeing 707-320C all cargo machines and build up a larger fleet as demand arose.

In the end the planners won the day, and the Board supported the purchase of two Boeing aircraft. It was then a matter of persuading the Government that this was the right course to take, not an easy task because the Minister had stipulated at the time of BOAC's purchase of fifteen Boeing 707-436 aircraft that there should be no further purchases of American aircraft. BOAC based its case on the lack of any suitable British aircraft for Atlantic cargo workable to cope with huge sizes and weights. A successful case was made, and BOAC was authorised by the Ministry to proceed with the purchase of two Boeing 707-320Cs (designated for BOAC as 707-336C) at an overall cost including spare power units and other spares of £6.5m.

The Seaboard World Airlines lease ran out at the end of October by which time it had been hoped to have delivery of the first 336C, but delays arose in getting the aircraft ready for service, and BOAC had to make urgent arrangements to fill the gap; fortunately it was able to charter another CL44 - this time from Flying Tigers - until mid-January 1966. Labour troubles at Boeings caused the delays to the two 707-336Cs, the need to make the flight deck of the aircraft compatible with BOAC's existing 707-436s so that crew could be interchangeable along the routes between passenger and cargo aircraft, and by the need to fit the new aircraft with flight recorders. The two jets were finally ready by mid-January 1966 to take over from the leased CL44 on the North Atlantic route. A new era had dawned for BOAC in the cargo business with the introduction of these fast modern aircraft fully palletised and able to carry containerised loads. Increased investment in cargo terminals at London and New York followed, and sophisticated ground equipment was essential to handle the full range of loads which could now be accommodated. Already by May 1966, the Board had authorised the purchase of a third 707-336C for delivery in September 1968 although if needed delivery could be brought forward to September 1967. BOAC was now enjoying the benefits of being able to order aircraft 'off the peg' in small numbers making ordering less of a gamble than in the past. In it subsequently became necessary to take advantage of previous delivery and this third aircraft was received at London Airport on 1 December 1967. The reason for urgency arose from the accident to a 707-436 in Japan in March 1966. Fortunately, this third 707-336C could be converted to a passenger layout and was used to fill the gap.

There was now a new problem - the necessary re-work of the 707-436 fleet to prolong their life to 60,000 hours. This would mean in effect one passenger aircraft being out of service from the end of 1968 to the end of 1969. So it was decided to purchase a fourth 707-336C for delivery in August 1968 and thus enable the third aircraft to continue in passenger service for a further year to December 1969

when the 707-436 programme would be completed, and - so avoid any loss in momentum in the Corporation's cargo plans. Government approval to the purchase of the fourth aircraft was formally received on 12 May 1967. Already by force of circumstances, BOAC was beginning to build up a highly flexible fleet of 707-336Cs. This trend continued since before the fourth aircraft could be delivered in August 1968 BOAC was faced with another unexpected capacity shortage. On 8 April 1968 a 707-436, G-ARWE crash-landed at Heathrow and was a write-off. Payment was immediate by the insurers, and BOAC sought urgently for a replacement. Fortunately, Saturn Airways had an aircraft shortly coming off the Boeing 320C delivery line which they were willing to assign to BOAC. The aircraft was in passenger configuration but needed considerable modification to make it standard with BOAC's existing aircraft. As it was urgently needed only essential modifications were carried out to enable it to be put into service in August 1968; remaining modifications were completed later during the winter months.

BOAC was finding its growing 336C fleet immensely useful especially as some of them were being used in a dual purpose role not envisaged initially when the MoA permitted the purchase. Ready convertibility from cargo to passenger layout brought dividends since there was a distinct tendency for the peak periods of cargo and passenger traffic to be complementary. By the summer of 1968 - there were five 336Cs with BOAC. In November the Board approved the purchase of a sixth aircraft owned

Signs of the future - Keith Granville holds a model of the new Boeing 747 in front of the skeletal stucture of the first building to house it. *(BOAC via author's collection)*

by Kleinwort Benson and lying at Frankfurt, at the cost of £4.7m which was payable in sterling. There were the usual standardisation modifications to be carried out, and the aircraft did not come into service until May 1969. It was to prove a useful addition in that it enabled BOAC to meet its commitments with QANTAS over the UK - Australia migrant traffic and as a support aircraft on BOAC's forthcoming Polar route to Japan. The Corporation was enjoying the fruits of freedom to make its own decisions on aircraft purchasing under the terms of the Minister's letter of 1 January 1964 to Sir Giles and the machinery withinBOAC for arriving at decisions about new aircraft types to buy was working well. The Commercial Division put their estimates of future traffic flows into the Planning Committee which had advice from the Aircraft Evaluation Committee as to new aircraft proposals coming along from the manufacturers with evaluations of their capabilities. A case would then be made to the Chairman's Future Aircraft Group where all aspects, commercial, financial and technical were taken into account. It was through this procedure that the momentous decision was reached to keep aircraft accommodating 150-160 passengers such as the 707-436s and the Super VC10s to aircraft capable of seating 400 or more passengers.

BOAC then considered a proposal a new flight catering centre at Heathrow, made necessary by the rapid growth in passenger traffic and the need for greatly improved working conditions in the preparation of aircraft meals. The original design had been approved by the Board in 1970, but the scheme was deferred for reasons of economy, completion being put back to 1975. When the project was revived in 1973, escalation of building costs was so high that a simpler layout was approved at the cost of around £4.5m, on land leased from BAA on the south side of Heathrow. When the project was put to DTI for capital expenditure approval, the Minister required the Corporation to get outside tenders from private firms for doing the actual catering task. BOAC felt strongly it should do its catering as it had done over the years; even so, outside quotations were obtained, and BOAC's costs were proved to be lower than any submitted. All this caused more delay. The Minister was still not satisfied and posed a series of questions all pointing to a desire for BOAC to subcontract the catering to private firms; consequently, all construction work was frozen, and the Corporation had to employ an outside consultancy firm to try once more to prove to the Ministry that the BOAC case was sound. It was not until November 1973 after an unnecessary delay for six months that permission to proceed on the BOAC plan was received from the Minister, Michael Heseltine. The delay cost the airline £400,000 in escalated building expenditure.

The way forward
No one could doubt Sir Giles' philosophy: he spelt this out clearly on his first day in office when he said to the staff *'Are we a prestige or a commercial airline? A lot of nonsense has been talked! There is no prestige in owning a Rolls Royce if it takes you to the bankruptcy court. There is no prestige in being the national carrier if it costs the nation millions. But we shall enjoy tremendous prestige throughout the world as soon as we consistently take good care of passengers at no cost to the taxpayer'*. His objective was to put BOAC's interests first and to make the airline profitable. His positive direction to the staff to that effect swept away much of the uncertainty of the past.

In October 1964 the result of the Group's ten months of deliberations was explained to the staff. The objective was to achieve a run-down of 3,700 in two and a quarter years solely by voluntary methods. This would represent an overall cut of 18% with the Engineering Department taking the most massive cut of 1,830, followed by the Commercial Department with 820, Flight Operations with 610 and a further 440 from the Service Departments such as Finance and Personnel. Those staff who were eligible and chose to leave would receive a leaving payment under the scheme based on length of service, age and current salary. There was, in any case, an 'age problem' in BOAC where nearly 44% of the staff fell in the 47 to 55 age bracket, and the scales of payment would be up to two and a half years salary plus cash in place of general notice. In administering the scheme, the Corporation was anxious not

A well known picture, but one that shows very well the size difference between the 707 and the 747. *(author's collection)*

Flight Attendants from the launch airlines pose in front of the prototype 747 at Seattle. (BOAC via author's collection)

to get rid of people with promising futures but on the other hand personnel managers at home and overseas were required to make sure that any whose best interests would be served by accepting the scheme should not be left in any doubt about the limited prospects if they stayed.

It appeared that the scheme might run into severe difficulties. Details had leaked out to other nationalised industries, particularly the Coal Board and British Rail which had expressed concern to the Government that the arrangements were too generous. The two Deputy Chairmen of BOAC then had a meeting with the Minister and with the Chancellor of the Exchequer but the Board decided to stick to its guns and go ahead.

This staff severance scheme finally finished in December 1966, and at that date, nearly 1,300 staff had accepted the particular terms and left the Corporation. As it happened, the hoped-for staff reductions were not achieved, partly because Sir Giles had initially been given an exaggerated idea of the extent of the overmanning and partially because business picked up so well that more staff were needed than at first thought. Even so, the numbers came down quite sharply and fell below 20,000 in January 1965 for the first time for five years. By 31 March the numbers were down to 19,641 and a year later to 18,918.

Sir Giles had dealt with three of the four stages in his plan for recovery. He had decided the route pattern, had agreed with Government the numbers and types of aircraft for the next few years and had started work on thinning down staff numbers. He was, therefore, able to send a summary of his total plan to the new Labour Minister, Roy Jenkins, on 19 February 1965 to meet his original obligation to the Minister at the time he was appointed. In doing so, he assured the Minister that BOAC could remain profitable provided it had the support of Government on three main issues. The first was the approval and implementation by Government of a financial reconstruction for BOAC in the manner already discussed with the Minister. The second was the continued recognition by Government that BOAC operated in a competitive international market and would have Government support in the matter of traffic rights; and third the continued acceptance by Government of the fundamental responsibility of BOAC to act in accordance with its commercial judgment in accordance with the Minister's letter of 1 January 1964. The question of restructuring the Corporation's capital and writing-off the accumulated deficit still had to be settled. In his letter to the Minister on 10 July 1964, a copy of which went to the Chancellor of the Exchequer, Sir Giles proposed that the opportunity should be taken at the time of sorting out the Super VC10 numbers to write off the accumulated deficit as at 31 March 1964. This included the full provision to cover the disposal of resources no longer needed given the new operating plan. He received a reply from Sir Richard Way which left no doubt that the Government agreed with the concept of reorganising the Corporation's capital and financial structure. This would take time, and in any case, the Corporation wanted to be sure that a realistic figure would be used for the once-only capital write-off.

Charles Hardie conducted the complicated negotiations with Government on this delicate issue, then Deputy Chairman of BOAC, together with Derek Glover, the Financial Director, and in February 1965 Hardie was able to report a satisfactory outcome to the Board. The Minister made a statement in the Commons on 1 March. On 31 March 1965 BOAC had £176m on loan capital outstanding, made up of £52m Airways stock and £124m of Exchequer advances. The Minister was able to tell the House that the Government agreed to cancel £110m of Exchequer advances as at 31 March 1965 thus reducing BOAC's liability for interest by about 526m. Thus, after writing off the accumulated deficit of £81m as at 31 March 1964, there was left some £29m as an initial reserve. Against this reserve would have to be put the cost of the staff severance scheme and the stretched delivery programme for the Super VC10s under the new plan. It was expected that those items would take up about £10m, leaving £19m for credit to the profit and loss account for use as need might arise. This settlement satisfactory as it was did not meet all BOAC's requirements. It did not cover the transfer of three Super VC10s to the RAF for which BOAC wanted reimbursement of its progress payments. The Government were not prepared to accept BOAC's claim for higher operating costs of the Standard VC10 as against the comparable Boeing 720s, nor would they take BOAC's view that some consideration should be given it in respect of the developmental costs of new aircraft. The higher price of the Super VC10s and their higher operating costs compared with the Boeing 707-320s would in effect be met by the £19m balance.

These arrangements were finally set out in the Air Corporations Act 1966 which provided that all outstanding liabilities of BOAC as to advances made before 1 April 1965 were extinguished. It also ensured that all obligations for repayment of the balances of BOAC stock not redeemed as at 1 April 1965 were transferred from BOAC to the Treasury, that £31m was considered as borrowed by BOAC from the Minister on 1 April 1965, and finally that £35m was deemed to have been paid by the Minister to BOAC on 1 April 1965 as an Exchequer investment. So BOAC then had capital liabilities of £66m of which loan capital amounted to £31m on which it had to pay a fixed rate of interest every year and Exchequer Dividend Capital of £35m which would be remunerated out of profits only if the financial results justified it. These new capital arrangements were taken into BOAC's Accounts as from 1 April 1965 and marked the final plank in the structure of the Chairman's recovery plan. Undoubtedly the Corporation owed much for this successful outcome to the negotiating skill and financial wisdom of Charles Hardie ably supported by Derek Glover. There was one other important issue which necessarily formed part of the whole plan; this was the need to settle the new administrative organisation. On 3 April 1964, the Minister appointed Keith Granville as a full-time Deputy Chairman and Charles Hardie as a part-time Deputy Chairman and consequent upon this Granville relinquished his chairmanship of BOAC-AC which was assumed by Gilbert Lee. Changes following the resignations of Sir Matthew and Sir Basil had already been made in the direction of BOAC-Cunard, Granville being appointed Chairman and Stainton Managing Director but early in 1964, Sir Giles took over the Chair.

On 14 May 1964, Sir Giles announced to staff his revised BOAC's top management structure that was to come into force on 1 June 1964. The Board policies would be put into effect by the Board of Management made up now of eleven members headed by the Chairman in his capacity as Chief Executive. Commercial, Operations and Engineering remained, assisted by a range of Service Departments. The Commercial Division retained its method of decentralisation through four geographical regions, Western, Eastern, Southern and European, each under a Regional General Manager responsible for the financial results of his Region. The European Region was a new development, which was formed in June 1964 covering stations in the UK, Ireland and Continental Europe. The Board of Management was to meet weekly and once a month there would be a composite meeting between the Board of Management and the thirty-four Members of Management - later increased to forty and irreverently known among staff as the forty thieves!

The Chairman as Chief Executive, therefore, had fourteen top officials reporting to him direct. His load was lightened, however, because Granville was given interest in international relations and commercial affairs, while David Craig with the title of Senior General Manager without Departmental responsibility was concerned with the general efficiency of the organisation, becoming a trouble-shooter to deal with severe internal problems or deficiencies in the airline. Directly reporting to the Chairman was a post of Chief of Information Handling to provide an up-to-date service of information and communications, a position subsequently filled with distinction by Peter Hermon, brought in from the Dunlop rubber company in 1966. The Board of Management had ten members plus the Chairman - namely Keith Granville, full-time Deputy Chairman, Gilbert Lee, Chairman BOAC-AC, David Craig, Senior General Manager, Derek Glover, Financial Director, Ross Stainton, Commercial Director, Captain Denis Peacock, Flight Operations Director, Charles Abell, Chief Engineer, John Gorman, Personnel Director, Winston Bray, Director of Planning and Robert Forrest, Secretary and Solicitor. John Gorman took over the post of Personnel Director in July 1964 when Dr Bergin ceased to be concerned with personnel affairs.

Apart from the weekly Board of Management meeting, the Chairman held a brief daily meeting in his office with the heads of the three operating divisions plus the full-time Deputy Chairman and the Senior General Manager, joined later by the Director of Planning.

The Corbett Report contained two specific recommendations as to Board membership, one concerning the desirability of appointing a qualified Technical Director to oversee both Flight Operations and Engineering and that the Chairman should have a full-time Deputy Chairman; a 'proved business administrator' from industry to support him and in due course to succeed him. The first of these recommendations was implemented by the Minister when with the Chairman's willing agreement he appointed Beverley Shenstone as a full-time Member of the Board; for many years he had been Chief Engineer of BEA and was released by BEA to take up his post. The appointment was for two years, and on his retirement in December 1966, the position of Technical Director lapsed. The second recommendation perhaps, fortunately, was never implemented and was made inappropriate in any event by the promotion of Granville to be Deputy Chairman shortly after Sir Giles took over and by his subsequent appointment also as Deputy Chief Executive.

On 3 April 1965, John Booth retired. He was the longest-serving Member, having joined the Board on 1 August 1949 from BSAA where he had been Chairman. On January 1966, George Chetwynd was made a Board

Member and in June 1967 following the death of Lord Normanbrook, Sir Richard Way, formerly Permanent Secretary at the MoA, came into the vacant position on 1 November 1967. Finally, on 16 September 1968 Ross Stainton, Commercial Director, joined the Board prior to this promotion in the following year to be Deputy Managing Director. There were two significant changes to the Board of Management. In 1968 Charles Abell became Engineering Director, concerned with engineering policy and particularly with the choice of new aircraft, while current engineering matters were taken over by John Romeril. The second change was the appointment to the Board of Management of Peter Hermon as Information Handling Director, an indication of the importance given to the computerisation of BOAC.

All these steps to recovery taken by the new Chairman brought increasing benefits as the months went by, but this position was significantly strengthened by the early financial results achieved, primarily due to the continued upswing in traffic which had begun in the summer of 1963. The fiscal year 1963/4 had finished with a record operating surplus of £8.7m; in 1964/5 the operating surplus was even higher and again a record at £16.8m, and after payment of all interest charges, including those on the accumulated deficit, there was a clear profit of £8.8m. The reversal of fortune was swift, morale in BOAC responded, and the voices of Press and public criticism began to be muted.

Considering another wide-body.
There was one other new type that BOAC was concerned with in this period; this was the Lockheed L-1011 TriStar. Late in 1969 the Corporation was giving thought to its aircraft needs through the 1970s, taking into account not only growthbut also the likely retirement of all or most of its 707-436s and the Standard VC10. It had three firm but alternative projects for study; first more 747s, second a mixture of 747s and one or other of the coming trijet aircraft the Lockheed 1011-84a version and the Douglas DC10-30, and third more 747s, plus a possible new smaller four-engined type from the Boeing stable. After detailed analysis and taking account of Flight Operation's preference for four-engined aircraft, the decision at that time came down in favour of building up the 747 fleet and waiting to see whether a smaller four-engined aircraft was likely to appear on the scene more suitable than the 747 for taking over the thinner traffic routes from the 436s and the Standard VC10s. The decision was made known to the Board of Trade and to the Lockheed Corporation. The British Government's interest in the L-1011 TriStar was centred on the fact that the aircraft was to be fitted with the Rolls-Royce RB211 engine.

In March 1970 the Corporation re-examined the L1011-8 long range version at the request of the Ministry of Technology and the Board of Trade, which were anxious to see Rolls-Royce develop a bigger engine of over 50,000 lbs thrust for which the launching of the long range L-1011 would provide an outlet. But BOAC came to the same decision as before. With the purchase by BEA in the summer of 1972 of six short range L1011s and an option on a further six and with the approaching merger of BEA and BOAC, the issue of a BOAC purchase of what was by then the L1011-2 long range version was raised again. BOAC's answer remained the same, but because BEA would have spare capacity in its L-1011 fleet, the possibility was examined of using BEAs L-1011-1 on shorter BOAC routes to the Middle East. By October of the following year the pressures on BOAC to commit itself to the L-1011-2 had increased, especially from the British Airways Board whose Group Planning Director felt thata case for the aircraft could not be ruled out. In October 1973, however, Sir David Nicolson wrote to the Minister finally confirming that no immediate order would be placed for the longer range type. The economic case for the -2 was unconvincing and to acquire a second new major type in 1976 (Concorde being the other) would have put a great strain on Overseas Division. It was left to BA in later years to build up a fleet of 1011 TriStars, including both short and longer range versions.

British Airways were to acquire a sizable fleet of Lockheed L-1011 TriStars - BOAC on the other hand considered the type, even came up with this model and colour scheme, but never ordered one.
(author's collection)

The BOAC Passenger Terminal at JFK Airport. (BOAC via author's collection)

Developments at BOAC Bases

The largest construction project approved and started during Sir Giles' time was the construction of BOAC's own unit terminal at JFK airport, New York.

When the decision was taken in June 1964, a total cost of $25m was envisaged. Work began in January 1967, but by then the estimated cost had escalated, even though Air Canada would pay a percentage since it intended to make use of the terminal for its local services into and out of New York. The whole project was challenging, due to the restrictions placed on the design by the Port of New York Authority and to the nature of the ground which was only just above sea level, requiring elaborate 'de-watering' during construction. The terminal finally came into partial use in June 1970.

Those were very good reasons for the delays, but there was strong belief that much of the 'problems' were caused by the Mafia-controlled Teamsters union who spent spent years seeking bribes, embezzling money, and engaging in extensive extortion and labor rackets as well as beatings, vandalism and even bombings in an attempt to control the construction and trucking industries under its General President James 'Jimmy' Hoffa. Investigators and prosecutors have long called Kennedy International Airport 'the mob's private candy store.' With cargo valued at more than $80 billion moving in and out of the airport every year, law-enforcement officials say that the Gambino and Lucchese crime families have been silent and violent partners in thefts and hijackings for over many years.

The passenger terminal was commissioned on 30 June 1970, but notably had suffered from a series of fires - in one week alone seven occurred, which became a police matter as evidence showed that some were likely to have been arson. The extent of the big fire in January 1970 can be gauged from the fact that insurers agreed on a settlement of over $2m. In spite of a further severe fire on 26 August 1970, the formal opening ceremony went ahead successfully on 24 September, performed by HRH Princess Alexandra in the presence of the Chairmen of BOAC and of Air Canada; full completion of the terminal was however put back to March 1971. The final cost of the construction was $59.5m, over double the figure in mind when the Board approved the project in 1964; the period of building to completion was just over four years.

New York was not the only place where difficulties were being experienced in getting new facilities into service. The modern air cargo terminal at Heathrow had its problems. Actual construction commenced in July 1967, and import work was transferred to the new accommodation in November 1969. It had been hoped to complete the move with the export section becoming operational in February 1970, but problems with the mechanical handling equipment delayed the progress until August 1970. The complicated machinery installed to select and move loads had become increasingly unreliable, and difficulties were still being experienced when on 22 May 1970 the Duke of Edinburgh formally opened the whole Heathrow Cargo Centre, including the BOAC area.

After repeated attempts to rectify the faults in mechanical handling devices, it was finally decided to re-organise the export section entirely and remove all the cranes and conveyors making up the automatic rack storage equipment. By January 1972, the Corporation had to allocate around £1m for this work and to make the facilities adequate to take care of the significantly increased throughput due to the arrival of the 747s and their cargo containers. Uninitiated visitors who had gazed at the original machinery were heard to say that it was worthy of Heath Robinson himself. So it was not a happy story, as the failure of the equipment to work had caused severe problems, staff discontent and customer dissatisfaction.

BOAC had also been unhappy with its performance on the ground at Heathrow Airport and became convinced that the only way to improve matters was to place all activities connected with the arrivals and departures of aircraft and their turn-round, with the maintenance of all ground equipment including vehicles and with the maintenance and management of all premises in London, under the control of one department, in place of the existing split control. The Deputy Chief Engineer, E. R. Major, was put in charge and after three months' delay in implementing the decision caused by staff anxieties, changed procedures and loyalties, the new department, Support Services, finally got underway on 1 February 1972.

But there was no easy solution to the problems the Corporation faced at Heathrow, which showed themselves in the poor punctuality of aircraft departures and delays in getting their baggage through to the waiting arrival passengers. The volume of traffic passing through the airport had immensely increased and from 1970 the appearance of the 747s with three times the passenger capacity of previous long-haul aircraft put a great strain on all the airport facilities. But there were fundamental problems of expansion linked to the unusual layout of Heathrow itself, with its passenger terminals and control buildings placed in the centre of the airport joined to the outside world by a single tunnel approach for passengers. The British Airports Authority (BAA) was operating under constraint in its capital expenditure which affected the efficiency of the services it provided to the airlines; there were severe delays in setting up loading piers for the new large jets, which meant a continuation of the cumbersome bussing of passengers between aircraft and the terminal buildings, and arrangements for the delivery of baggage from aircraft to the arrival terminals were frequently breaking down under the strain.

BOAC felt strongly that as the national carrier it should be given at least some priority over foreign operators in the use of aircraft stands and loading piers, and that the BAA was adopting too neutral an approach unlike the situation enjoyed by BOAC's competitors at their home airports. To improve matters, BOAC had long been pressing for its exclusive passenger terminal at Heathrow and had put up proposals to the BAA without success. It was ironic that whereas BOAC had by then its passenger terminal at New York Kennedy Airport, at London, its home base, it had to take its share with foreign operators.

There were areas of difficulty mainly out of BOAC's control which caused delays and had repercussive effects on aircraft punctuality. The burden of security, immensely increased following the appearance of hijacking on the international airline scene, and was affecting all operators, BOAC in particular with its high frequencies through the Middle East countries joining passenger loads now had to pass through security checks, and delays could quickly build up when 300 or more passengers had to be vetted for one 747 departure. There were repeated delays due to industrial action by air traffic controllers overseas, particularly in France and all crew and passengers could do was to await clearance before departure was possible.

Delays were also caused by industrial action by the workforce assigned to load and unload aircraft. These difficulties were partly the the result of arbitrary across-the-board staff limitations imposed during economy drives which led to massive overtime being worked with a ready-made weapon for the ground staff by going slow or working to rule.

BOAC's Board of Managements spent over much time in discussing ways of overcoming or mitigating these problems. The establishment of the Support Services Department helped after 1972, but the causes of delays persisted through into the era of British Airways.

Enter the 747

BOAC had been giving thought to what should follow the Super VC10; it believed it would require one more subsonic type before the introduction of the Anglo-French and American supersonic aircraft around the early or mid-seventies. Although there was talk of the Superb VC10 (a proposed 265 seater aircraft developed from the Super VC10) and the Press had much to say about the subject, neither the Government nor Vickers were prepared to go ahead with such an expensive and financially risky project, as had already been made evident in March 1965 during the negotiations over cancellation payments on the Super VC10s. At that time the Corporation was thinking along the lines of an aircraft type to come into service around 1969 with a seating capacity of 250/260 in the economy class layout. In May 1965 the Boeing Company had sent over a team to London to make a presentation of their 707-820 series - a double-deck aircraft - and Douglas were beginning to promote a stretched DC-8 with around 220 economy seats. There followed some months when the airlines were waiting to see what the manufacturers would do. The idea of a huge aircraft was much to the fore as a result of the Lockheed C-5A project for the US military with anything up to 900 seats of a kind, altogether far too large at that time. Size, however, had significant advantages to offer - cheaper seat mile costs, fewer crews, easing the 'slot' problem on arrivals and departures, and potential for increased passenger attraction.

By the late autumn of 1965, the news was coming from Boeings of a 450/500 seat aircraft with good economics - the 707-820 idea having been abandoned. By February 1966 Boeings had made their proposals widely known to airlines in the United States and were using the argument that, even if a successful American supersonic aircraft appeared in 1974 or 1975, the design of their new subsonic type 747 was such that the aircraft could be switched to cargo operations if necessary. BOAC's planners were now giving urgent attention to the prospects of a huge subsonic jet, although were naturally concerned at the vast increase in size from the 707 to the 747 in one leap forward. The commercial planners in Western Region were confident that by 1969 or 1970 they could justify the use of a small fleet of 747s on the high-density routes such as London-New York and London-Montreal although no definite conclusion was reached in that year. Figures in January 1966 put -by the planners to the Chairman's Future Aircraft Committee showed a requirement for twelve 747s by 1973/74 but at that time detailed assessment of the aircraft was lacking. By April 1966, the airline industry had been startled by an order for twenty-five 747s from Pan Am, a substantial potential output of capacity, and, incidentally, a decision which subsequently involved the airline in financial problems. This order sparked off essential studies by other long-haul airlines, including BOAC, all of whom were alarmed at the possibility of being left behind in the competitive race.

Then, on 20 April 1966, experts from Boeings made a detailed presentation of the 747 to BOAC; following this the Corporation up-dated its forward plans which showed that existing subsonic fleet commitments, nineteen 707-436s, twelve Standard VC10s and seventeen Supers would see the airline through to 1969/70 on its passenger operations.

As to the years beyond, on a 10% per annum passenger growth there was a definite need for considerable additional capacity. The Director of Planning put to the Future Aircraft Committee the possibilities for the ten year period' 1969/70 to 1979/80 — more 707-320s, the DC8-63 series with about 225 seats and lastly the 747 with about 440 seats at ten abreast and 410 seats at nine abreast. What the conclusion come to by the Committee was that the 747 offered the best economic case, required fewer departure slots, enabled more people to select the fashionable times of departure, and needed fewer crew. The reverse side of the coin was that it would need a completely new approach in the field of marketing to the travel trade and public.

At the time of the Boeing visit to BOAC in April 1966, the Corporation had asked whether Boeings would offer the 747 with the Rolls-Royce RB.178 engines installed. The reply came that this was not feasible on the grounds of cost and of the delay in deliveries which it would cause.

The Board met on 9 May 1966 to make a decision. At that meeting, it had before it a recommendation from the Future Aircraft Committee for the purchase of six 747s, three for delivery in February, April and May 1970 and three in January, February, and March 1971, a commitment including spares of around £53m nearly all in dollars, and this did not include the full cost of ground facilities, such as hangarage and ground handling equipment. Also included was a recommendation for an option on a further ten aircraft. In his submission at this meeting, the Chief Engineer was confident that agreement could be reached with Boeing over performance and weight guarantees but there was some anxiety about the noise level at take-off. Before any decision was made the Board were able to examine data put before them for the 707-436, the 707-320B, the Superb VC10, the DC8-63 and the 747. The Pratt and Whitney JT9D-1 engines for the 747 had a trust of 41,000 lbs compared with the 17,500 lbs of the Rolls-Royce Conway engines in the 707-436, and a take-off weight of 680,000 lbs, over double that of the 707-436. The 747 takeoff and landing distances were comparable to the 707 series and with its massive sets of undercarriage wheels the 747s runway loading was no higher than for the 707.

With its wide fuselage, the 747 could offer an improved economy class seats, having a cross-section width at cabin level of 19 feet compared with 10 feet 7½ inches on the 707, permitting in the 747 a double aisle in the cabin. At nine abreast seating, the 747 would accommodate at least 400 passengers and still have 4,000 cubic feet for cargo below deck as compared with 1,540 cubic feet for the 707. The Commercial Director, Ross Stainton, was in no doubt that the 747 was the right aircraft for the future. With these facts before it, the Board accepted the recommendation of the Future Aircraft Committee and authorised the Chairman to obtain the Minister's agreement and then to complete negotiations with Boeing.

On the whole, Press comment was favourable to BOAC on this issue. *The Guardian's* comments were typical: *"No one in the British industry any longer seriously believed that this aircraft (the Superb VC10) will be built or indeed that it would make sense for the Government to spend £40m on a vain pursuit of the Americans. The Government, unless it believed we should drop out of the North Atlantic business altogether, must trust BOAC's monetary judgment. This, in the end, it was forced to do over the cancellation of the VC10s, wisely as it now turns out since the final batch of these aircraft would have been delivered to BOAC just at the time when the advent of the new Boeings would have made them obsolete"*.

On a lighter note, the *Daily Express* had a cartoon of a

BOAC's Boeing 747 G-AWNC awaits delivery at Payne Field, just outside Seattle. *(BOAC via author's collection)*

What was to become an iconic advertising image for the 747 - BOAC's G-AWNA gets airborne from Seattle.
(BOAC via author's collection)

stewardess announcing on a 747 aircraft *'jumbo Airways welcomes you aboard. Flying time to London six hours, waiting-for-baggage time London Airport ten hours'*.

Sir Giles saw the Minister on 12 May 1966 and told him the Corporation's decision but found that he was not ready to give his approval, the reason being that the Douglas Aircraft Corporation had informed him they were about to offer BOAC the Douglas DC8-83 fitted with Rolls-Royce engines.

After a false start, Douglas finally put details to BOAC on 31 May 1966 of a type DC10-84, single-aisle with 233 economy seats and gross take-off weight of 380,000 lbs. This called for further detailed analysis by BOAC's experts who again came down in favour of the 747 and a letter to that effect was prepared to go to the Minister. However, at this point, Douglas withdrew its offer leaving the 747 with a clear field. In emphasizing its case for the large Boeing BOAC was able to tell the Minister that already PAA, Air France, Lufthansa and JAL were ahead of BOAC in their decisions to buy the 747. BOAC urged a quick reply as Boeings insisted on a Letter of Conditional Intent by 22 June 1966. Permission from the Minister was received carrying a requirement that as much British equipment be included as possible. On 2 September 1966 a contract with the Boeing Company for the purchase of six 747s was signed, and it remained to work out how to finance the huge amount of dollars necessary. In August the Treasury asked BOAC to borrow in the United States at least 80% of the dollars needed for the purchase, a figure of around $160m for the six aircraft. The Export Import Bank of Washington undertook to authorise an an immediate loan of $35m at 6% with a British Government guarantee and would hope to be able to offer further loans of up to $80.2m; this covered 72% of the requirement, with the manufacturers loaning 8% and BOAC finding the balance of 20%. By February 1967 the current estimate in BOAC of the total cost of the order was £71.025m which included interest on the dollar loan to date, hangar and workshops, tools for Treforest for engine overhauls, a a flight simulator and its housing, traffic handling equipment etc. The flight simulator and housing at Cranebank alone were to cost £1.25m. By May 1967, the Corporation was seeking additional 747 capacity for delivery from September 1971 to April 1972 and a case was made for a further six aircraft at an estimated cost of £56m for the aircraft and £15.5m for necessary equipment including a second 747 hangar giving a total cost of approximately the same as for the first six aircraft. Involved in the case for these aircraft was the expected retirement of at least some of the 707-436 aircraft by early 1972; so these second six 747s would cover both replacement capacity and expansion.

The forward plans for these additional aircraft provided not only for thickening up operations on Western Region but also for 747 introductions in 1972 on routes to India and beyond to Singapore. But when the case was put to the Minister for approval, consent was given for the purchase of only five aircraft, and he stipulated that all the money required for the purchase must be borrowed abroad even though the Corporation had large sums invested in the United Kingdom. Before putting its name to the contract with Boeings, BOAC wanted to be assured that the effect of the devaluation of the pound and the increase in all-up-weight of the aircraft had not seriously upset the economic case which had been initially based on a rate of return of 12%. However, the Financial Director, Derek Glover, was able to reassure the Chairman that the 12% rate could be maintained or even slightly increased. The Board, therefore, gave authority for the conclusion of the deal.

A massive funding operation then began covering all eleven 747s made more difficult because of the Export-Import Bank's insistence that it would supply not more than 40% of the total amount. So BOAC had to turn to the commercial banks to make up the difference at a higher rate of interest. It is an odd reflection that the Treasury did not apply the same stringent ruling requiring overseas borrowing to the UK Independent operators. Eventually, the purchase agreement with Boeings for the five aircraft was executed on 28 June 1968. Shortly before Sir Giles' retirement on 31 December, a case was made for the purchase of a twelfth 747 and the Board at its meeting in Johannesburg gave their approval. At the same time, the total cost of all twelve 747s including associated equipment and facilities was recalculated, and the amount came to the enormous figure of £165,473m which included a dollar commitment of $320m. Included in the total cost was an amount of over £27m for specialised traffic handling equipment made necessary by the height of the 747 passenger door and the need for a useful materials handling system to move the large aircraft containers to and from the terminal buildings and the aircraft hold. Supplies of such equipment had to be available at en route stations although some sharing with other 747 operators had been assumed.

The whole pre-delivery programme for the 747s was a complicated exercise affecting all Departments in BOAC and calling for massive training exercises for flying and ground handling staff. That the plan went smoothly and kept well to the schedule was the result of a particular co-ordinating group set up under the chairmanship of David Craig, where all Departments were represented. Progress was monitored, difficulties aired, and any bottle-necks given urgent treatment. The system worked so well it was followed later for the Concorde pre-introductory programme.

Left, original caption: *'The Monarch Lounge (First Class). This is on the upper deck reached by spiral staircase from the first class cabins. The lounge is fitted with 4 high-wide wing, club-style swivel chairs and settees accomodating 16 passengers. Seats in the Monarch Lounge are non-saleable and will not be used during take-off and landing'.*

Below: Boeing 747 G-AWNA is inspcted by staff and the media after the first arrival at London Heathrow. *(both BOAC via author's collection)*

Delays were caused by engine problems, and by early 1970 a new 747 was coming off the production line once every three days with no upgraded engines for it. Instead, some thirty 747 airframes were parked on the Everett ramp with 9,600 pound concrete blocks hanging off the wings where the engines should have been until enough JT9Ds could be delivered. Behind the 'gliders', or the 'aluminium avalanche' as Boeing engineers called this phenomenon, is the main assembly building, built at a cost of S200 million. Close study of the original photograph show at least one BOAC aircraft without engines in the foreground. (Boeing)

The first of the vast 747 hangars, was completed early in 1970 in good time before the arrival of the first 747 and a year later the new building comprising workshops, multi-storey car-park and offices above was finished, being the most significant building at the airport. Staff moved into the new offices in late 1971. The whole headquarters area then formed an impressive collection of buildings, linked where appropriate by covered walkways at a second-storey level so as not to impede ground movements. As a measure of the size of all this development, the total cost was over £12m. No sooner was Hangar One completed than a case was made for Hangar Two, which it was reckoned would be needed once the 747 fleet exceeded twelve aircraft, expected to be at latest by December 1972. The proposal was accepted and a total cost allowed of £7.5m. However, due to the delay in putting the 747s into service and the need to restrict capital expenditure, the project was deferred by one year, the building being completed in 1974.

When Charles Hardie took over from Sir Giles in January 1969, BOAC had twelve 747s on order all to be powered by Pratt and Whitney engines. In the following month, the large new aircraft was flown by Boeings for the first time on test with satisfactory results.

Advance provisioning for the initial fleet of twelve aircraft was an immense task. Entirely new ground handling methods had to be devised. Not least was the problem of deciding on designs of traffic handling equipment along the routes where the 747 was to operate and of estimating the amount which would be needed taking account of other operators' equipment at common stations. In the initial financial forecasts of December 1968, it was expected that £7m would be required on this item alone to cover about a dozen different types of equipment, including mobile passenger steps, container elevators, transfer and highway transporters, container weighing devices; the list was almost endless. By simplification and equipment sharing this £7m estimate had been reduced by December 1970 to £3.5m, and the total cost of the twelve aircraft and associated facilities had come down to £160.137m.

The Corporation had an option with Boeings for four more 747s and in September 1970 decided to exercise the option with delivery dates in March and April 1973. The purpose was twofold, as replacement capacity for part of the ageing 707-436 fleet and as expansion. The traffic forecasts on which the need was based were optimistic, especially on the North Atlantic, and included in the case was a new feature, the introduction of four times weekly 747 services London - New York - Australia as BOAC's share of a joint daily frequency with QANTAS. With sixteen aircraft in the fleet, it was thought that an increase of 13% in the rate of individual aircraft utilisation could be achieved. The estimated cost including spares was £47.1m, but this did not cover the cost - about £1.3m - of a second 747 flight simulator which would be needed. The Government's approval to the purchase was announced in Parliament on 24 January 1971, but in the event, the option was not taken up with Boeings at that time. By the following June, the situation had radically altered. Several factors occurring at the time called for a re-appraisal of the Corporation's capacity needs.

There was also a matter of pilot pay under the proposed bid-line system. BALPA was insisting that a settlement should be reached by 31 March 1969. There was a wide gap between the two sides. The pilots were seeking a minimum guarantee of £7,500 per annum for top pilots with the ability to go up to £12,000 whereas BOAC was offering £7,700 as a maximum. As BOAC was unable to accept BALPA's claim, BALPA formally notified Granville as MD that its members would strike from 31 March. The strike took place and lasted until 5 April: thus once again BOAC suffered a revenue loss, this time of some £5m. BOAC then sought the help of the NJC in conciliation and through its efforts and with a somewhat improved offer on the table, BALPA called off the strike. Still outstanding were bid-line procedures, higher pay for flying the 747 and a reduction of the crew on the flight-deck, but in the meantime, the Government reluctantly gave its consent to the pay deal.

A difference arose between the Corporation and BALPA over the rules of the bid-line system. At the same time, BOAC was concerned that pay negotiations in preparation for initial training on the 747 had not even started. BALPA being unwilling to discuss this until all other issues had been settled to their satisfaction. Arbitration was completed by August 29 and advised to both sides on September 15 and accepted, with the bid-line system to go into force on 4 January 1970.

Talks with BALPA about the 747 and crewing got underway on 9 September 1969; the agreement was urgent as time was running out for training flying if the aircraft were to be introduced as planned in April 1970, but negotiations soon ran into difficulties. The BOAC negotiators were finding it hard to make progress in the face of BALPA's refusal to make a specific claim for payment. The Board of Trade was kept fully informed, and the timescale was tight as the latest training course to get pilots ready had to start at Boeing's Seattle facility on 24

BOAC's famous 'golden trolley service was to be provided.

BOAC planned their 747 interiors into distinctive cabins - each furnished to look and feel like a separate entity. This was achieved by means of different colours to fabrics, fittings, sidewalls seats, headrest covers, carpets and curtains. As the BOAC advance brochure stated 'This way each projects its own 'personality' as a distinctive, luxuriouse sitting room.

The interior layout artists came produced these design samples: left to right: First Class, Monarch Lounge, upper deck. Main deck, First Class, then three different Economy Class cabins.

November 1969. This date would have to be met if scheduled 747 operations were to achieve the planned starting date. BALPA had made it clear that it had banned its members undergoing training for the 747 until pay and conditions for the aircraft had been agreed.

Early in November, the Prices and Incomes Board issued its report on pilots' pay and BOAC was required by Government to limit its negotiations with BALPA to the basis set out in the report, even though this was well below pilots' aspirations and below what BOAC had already offered. So no progress could be made, and the NJC was informed in December that almost certainly service introduction would be delayed; it expressed its concern both to BOAC and to BALPA. The trade union side of the NJC then offered to meet BALPA, but the offer was not taken up.

The next step was a proposal from BOAC that the issues should go to arbitration but BALPA considered the normal negotiating machinery had not been exhausted. After an agreement with Government, BOAC made a new proposal in March 1970, and by the following month a breakthrough in the negotiations had occurred and progress made under several headings, but by May BALPA had put in a general claim covering pay and conditions on all aircraft including the 747s which was unacceptable to Management. In consequence, a public statement had to be made that the introduction of the 747 would be indefinitely delayed, the first occasion in BOAC's history that the introduction of a new aircraft type had had to be postponed because of industrial action.

With delay now inevitable and with the existing pilots' pay agreement for all aircraft types expiring in December 1970, the Board decided to take time in seeking a comprehensive negotiation which would cover all types, not only the 747. Progress was painfully slow, partly due to changes in personnel in the BOAC pilots' local council and thus in its negotiating team. BOAC had made a new offer to BALPA on 24 June 1970 and advised all pilots of the contents but the summer months slipped away waiting for the new local council to be elected. By September 1970

Today overhead storage bins are very much the norm - but back in the early 1970s they were very much a new innovation.

First Class seats (above) in BOACs 747s were specially designed for the Corporation and were installed in pairs at forty-two inch seat pitch, each pair with a centre console acting as a drinks tray and storage unit. Economy seats (right) were at 33 inch seat pitch.

Both First Class and Economy seats contain armrest controls for reading light, Stewardess call button, recline control, in-flight entertainment headphone socket and controls, along with an ashtray.

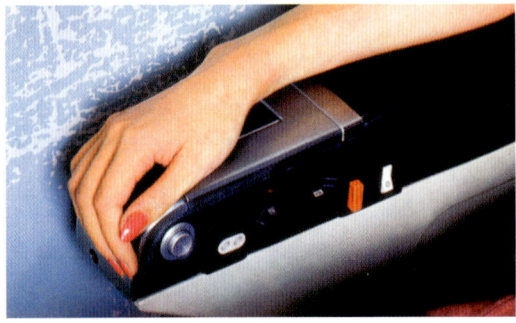

BALPA was ready to resume negotiations and although initial discussions seemed hopeful, BOAC's attempts to have the training ban on the 747 lifted were unsuccessful. An added difficulty in reaching agreement was the insistence of the Board of Trade that they should be kept fully informed and that no improved offers should be made without reference to them; this resulted in a loss of impetus in the negotiations which lapsed for a time and did not recommence until December 1970. Good progress was then made at last, and on 15 January 1971, all outstanding items on BOAC's proposal were agreed under which a basic pay flat rate method was accepted, giving up to a maximum of £9,000 per annum.

Management heaved a sigh of relief since from 15 January 747 training effectively commenced and plans for 747 introduction could go ahead; agreement in good time was also reached on cabin crew claims in respect of 747 operations.

BOAC had failed to persuade the other carriers at the IATA traffic conference in Honolulu to accept its fare package: the loss of work to the second force airline: but above all, the falling passenger demand especially in the UK. All these taken together meant that the 747 fleet could be stabilised at twelve aircraft in 1972 and possibly 1973. So the Board decided not to proceed for the time being with the purchase of aircraft numbers thirteen to sixteen.

Talks between the UK and the US Governments led to Pan Am and TWA cutting back their planned 1971 summer capacity so that only Pan Am operated Washington - London and only TWA Chicago - London; it was hoped at later discussions that capacity between London and Miami might also be rationalised so that BOAC and National provided the same amount. In the face of these considerations, and especially the need to give a proper UK share of the Washington service, there was a case for a thirteenth 747 for delivery in May 1973. If this were not to be provided, there was a risk that TWA would return to the Washington route again. It was therefore agreed to recommend to the BAB that a thirteenth aircraft should be purchased and approval to this was subsequently received from the BAB and from DTI. A case for two more aircraft, numbers fourteen and fifteen, was put to the BAB in September 1972 and approved. These two 747s were for delivery by 1973 for use in the 1974 season; the cost to BOAC was £23.3m. However, with this additional order in hand, Boeings were prepared to contribute £3.3m towards the cost of engine conversion to the-7. This capacity was needed for expansion, especially on the Hong Kong and Australia routes. The Corporation's plan now required the 747s to be in 10-abreast seating in the economy section from 1974 onwards, and this would be feasible with the increase of all-up weight to 733,000 lbs with the -7 'engines. A fleet by 1974 of fifteen aircraft could still be accommodated in the two hangars, and no further construction was necessary. It is woth noting that at that time an inflation rate in the cost of fuel of only 3% per annum was being allowed in the costing calculations.

747 into Service - Eventually!

The technical problems encountered throughout the 747 construction programme were exacerbated by managerial and political infighting and other unnecessary tribulations involving everyone from the US government to Pratt & Whitney, Pan Am, Boeing and other airlines. This conspired to put the brakes on the entire 747 project, and it was apparent that the hoped-for first flight date of 17 December 1968 for N7470, the first aircraft, would not be met. The wings and fuselage were mated in mid-June, but neither the landing gear nor the JT9D engine was ready. The JT9D had been test-run for the first time at East Hartford in December 1966, and in the late spring of 1968 a B-52 was converted to a flying test-bed for further testing, and a single JT9D engine mounted where two J57s had been. Static testing of the JT9D was completed by April 1969. The omens, however, were not good.

The JT9D was showing alarming signs of being prone to power surges and stalling. Also, early testing showed that the forty-six high-pressure compressor fan blades - capable of pumping more than 1,000lb of air per second at full power at sea level - would rub against their casings at the bottom of the engine. While this was anticipated, Pratt & Whitney were utterly unprepared for a further problem with the engine which now occurred during the flight certification test stage, one which could not have been induced during static testing: the ordinarily circular turbine casings were being bent out of shape - in effect they were 'ovalized' - and this was, in turn, causing the high-pressure turbine blades to rub against the sides of the casings. Needless to say, this greatly affected the engine's efficiency and therefore its overall performance.

Pratt & Whitney suggested that this excessive bend resulted from Boeing's decision to mate the JT9D to the wing by a single-thrust link on the turbine case and not – as on previous Boeing jetliners - by a thrust link fitted to the engine compressor casing. At Boeing Everette L.

Left: Meals were to be dispensed from six galleys on board BOAC's 747s - one forward, two amidships and three aft. They contained a total of two single and five double-capacity ovens. In addition there were a pair of microwave ovens, and a completely new range of metal cutlery, crockery and glassware was designed.

First Class passengers were offered a choice of main courses. A special catering unit situated forward of the spiral staircase had hot plates and a micro-wave oven, enabling special dishes to be cooked to order. Prepared chilled meats could be individually cooked in 35-40 seconds.

In Economy Class, 288 meals could be heated simultaneously. Special dietary needs could be met on all BOAC services, provided they were requested at the time of booking. Vegetarian, fat-free, salt-free, Kosher and otherspecial meals could be served.

Above: As the advance brochure stated: *'The thirteen passenger toilets are fitted with make-up table, large mirror, razor sockets and iced drinking water. Top quality cosmetic preparations are freely available'.*

Webb, the head of the technical staff, drew on his experience of stresses gained during the 727 programme and, using engine drawings supplied by P&W, ran a computer programme to prove that it was not the design of the mounting that was at fault, but that the problems were being caused by the engine. Pratt & Whitney carried out a series of modifications and changes to the internal workings of the engine itself, but the bending problems refused to go away.

The JT9D was the first commercial large-fan engine, and it, therefore, introduced many new, innovative features, including an annular, ring-shaped combustion chamber which helped to make the engine shorter than the previous turbine engines. Also, the dimensions of the rear end of the JT9D were markedly different to the size of the 8 feet-diameter fan assembly at the front. Boeing showed that under certain flight conditions, including the high angle of attack during take-off, the different size of the enormous diameter front fan and the turbines behind it were inducing bending stresses in the turbine casing, which soon became distorted. Finally, P&W agreed to look at the outside of the engine. Strengthening the exhaust casing and adding a stiffening ring at the high-pressure turbine case cut the ovalizing tendency by 12% at the exhaust and 15% at the high-pressure turbine, but P&W could not completely eradicate the bending. In fact this problem was only finally solved by a series of measures: first, two 45-degree thrust-transfer members were added to two other points on the exhaust casing: these changes reduced bending by 30% and ovality by 102%. Then a 45-degree 'Y-frame' was fitted to transfer the thrust loads to the intermediate compressor casing; and finally, an inverted 60-degree 'Y-frame' prevented all tendency of the turbine casing to ovalize and reduced bending by as much as 80%.

All this took time, however, particularly given the fact that before they reached Everett, the JT9Ds had still to be flown to Chula Vista, California, where the nacelles were manufactured and installed by the Rohr Corporation. As a consequence, the JT9D was still not ready by the end of 1969. Furthermore, the delays had a knock-on effect and created an embarrassing situation early in 1970 when 747s were coming off the production lines at the rate of one every three days, and there were no upgraded engines for them. Instead, thirty 747 airframes were parked on the Everett ramp with 9,600lb concrete blocks hanging off the wings where the engines should have been, waiting until enough JT9Ds could be delivered. (Boeing engineers called them 'gliders' and the phenomenon of so many 747s without engines went down in folklore as the 'aluminium avalanche'.) Eventually, P&W managed to get the engine to produce 43,500lb of thrust and finally 45,000lb, but the cost to the company was enormous.

Elsewhere, news of the emergent Lockheed TriStar and Douglas DC-10 jets heightened the tension for everyone in the pressure-cooker atmosphere at Everett, while cost-overruns (the 747 programme was the largest drain on cash Boeing had ever experienced) were reaching crisis point. Ultimately, 48-year-old Thornton A. Wilson, who had been handed the reins by 68-year-old Bill Allen on 29 April 1968, would have to initiate a massive cost-cutting regime to pull the company back from the brink of bankruptcy in December 1969. Within eighteen months Wilson made 60,000 employees redundant, half of them production workers. The American banks needed to see something tangible for their money, especially since the hope that the first 747 would be ready to fly on 17 December was a forlorn one.

The roll-out day was set for Monday 30 September and at that point in the 747 still had no landing gear; moreover although the engines were mounted, their only real purpose, apart from making the aircraft appear outwardly almost ready to fly, was as a counterweight to prevent N7470 from sitting on its tail. The pre-production problems also had a knock-on effect on the other 747s on the line. Five 747s were earmarked for the test programme, but all of these aircraft were behind schedule. For instance N747PA, the second 747 and also destined as the first aircraft for Pan Am (*Clipper Young America*), was scheduled to fly on 29 January 1969, but in the event, this aircraft would not be rolled out until a month later.

Finally, 747 chief pilot Jack Waddell, Jess Wallick, flight engineer and Brien Wygle, co-pilot took the prototype 747 N7470 into the air for its first flight on 9 February 1969 and the test flying programme settled down. There had been the usual problems associated with testing a new type of aircraft, and several minor incidents had occurred, but nothing untoward had befallen any of the test aircraft. On 13 December 1969, the first serious accident in the test programme happened when test aircraft number three, N732PA flown by Ralph Cokely left Everett for Renton where all its test equipment was to be removed and airline seating installed before delivery to Pan Am (as *Clipper Storm King*). No 747 had landed on the 5,280 feet runway at Renton before, but N732PA weighed 390,000lb, and at this weight, it was calculated that the distance from a 50 feet altitude to a complete stop was 3,150 feet, without reverse thrust. At Renton, a strong crosswind, gusting at times and intermittent rain did little to help matters, but everyone was still confident that even on a wet runway the 747 would again stop safely in the possible distance.

Cokely brought the 747 in over Lake Washington where the runway runs to the edge of the water, but the aircraft was not high enough to miss an earth bank on the shoreline 20 feet short of the runway and two feet six inches below the runway threshold and the right-hand body gear and wing gear were buckled in the impact. Hitting the runway, the 747 careered along it, showering sparks from its numbers three and four engines as the wing settled. Cokely managed to keep to the runway centreline, and the 747 came to a halt 3,500 feet down the tarmac. Cokely was sacked for this incident.

Behind the scenes, there was increasing drama. Formal FAA certification for the 747 had still not been issued, although the FAA had determined that the engine problems would not compromise safety and had passed these, it was not entirely satisfied with emergency passenger evacuation from the upper deck and wanted these doors redesigned. Pan Am, aggrieved at the lower level of performance on the 747 to that promised initially (it was still 7% short on range), now threatened to withhold $4 million from the final payment of each aircraft until Boeing had corrected several problems. As December dawned, Boeing met brinkmanship with brinkmanship, even threatening to sell the first 747s to TWA instead of to Pan Am. By this time Boeing was in severe financial hardship - final 747 development costs were later put at $750 million - and the recession, which had begun towards the end of 1968, was starting to bite.

A compromise ensued whereby Pan Am would withhold $2 million from the final payment on each 747 delivered until the problems, mostly with the engines, were rectified, paying the remainder in instalments. Finally, on 13 December 1969 N733PA *Clipper Young America* became the first 747 to be delivered to Pan Am when it was flown from Everett non-stop to Nassau in the Bahamas with a cargo of freight and then on to the Pan Am complex at Kennedy Airport in New York. A week later, on 19 December, N734PA *Clipper Flying Cloud* followed.

On 15 January 1970 First Lady of the United States Pat Nixon christened N733PA *Clipper Young America* at Dulles International Airport in the presence of Pan Am chairman Najeeb Halaby. Instead of champagne, red, white and blue water was sprayed on the aircraft. Pan Am chose this aircraft to make the first flight across the Atlantic with paying passengers six days later. On a bleak, bitterly cold evening at Dulles on 21 January, 336 passengers - who had reserved their seats two years earlier at a cost of $375 for a first-class one-way ticket - three flight crew (headed by Captain Robert M. 'Bob' Weeks, New York chief pilot) and eighteen cabin attendants boarded *Clipper Young America* for the first commercial flight of the 747, to London-Heathrow. Delayed by problems with the doors and the cargo hold, there were further problems when Weeks tried to start N733PA's engines. Gusting winds, which were blowing off the mudflats of Jamaica Bay, blew into the JT9Ds' tailpipes, restricting the flow of compressed air and exhaust gas and causing the engines to surge and produce temperature rises. The crew finally managed to stabilise the engines for take-off, and at 7.29 pm, N733PA taxied out for take-off; but almost immediately it had to return to the terminal when the number four engine exhaust temperature ran too high. Later the fault was traced to an insensitive barometric fuel-control system which finally failed due to the high crosswind. All the passengers had to be disembarked and were bussed to the terminal for meals in the restaurants while Pan Am worked out what it could do to prevent the occasion turning into a public relations disaster.

Finally, there was only one solution, and that was to replace the ailing *Clipper Young America* with a substitute aircraft, N736PA *Clipper Victor,* which had been delivered to the airline only the day before and was to be used for training. Delayed by an air traffic controllers' dispute, *Clipper Victor* finally taxied out at around 1.30 in the morning, and 1.50 took off for the flight to London. Much to the relief of Pan Am and everyone else, no further serious problems were encountered, and the trip was made without incident - and without the in-flight movies, since the reserve 747 was not equipped for showing them. *Clipper Victor* covered the route in just six hours sixteen minutes and landed at London-Heathrow to a great reception.

The vast cost of developing the 747 and building the Everett plant resulted in Boeing having to borrow heavily from a banking syndicate. During the final months before delivery of the first aircraft, Boeing had to request additional funding to complete the project repeatedly. Had this been refused, the Company's survival would have been threatened. Ultimately, the gamble succeeded and Boeing held a monopoly in very large passenger aircraft production for many years. However, the 747's introduction into

Up, Up and Away! an unidentifiable BOAC 74 takes to the skies. (author's collection)

A busy day at Heathrow - a company 747, VC10 and just visible at the rear, a 707. *(author's collection)*

service coincided with a sharp downturn in international passenger air traffic. Originally, Juan Trippe at Pan Am had based his airline's 747 requirements on the prediction that this would grow by 15% a year, but by 1970, due mainly to a worldwide recession, it was only increasing by 1.5%. Each and every one of these disasters all made the press and every bad public-relations event for Pan-Am and TWA was in effect passed on to other prospective 747 operations - BOAC included.

Ladies and Gentlemen we apologise for the delay...
The first BOAC 747 passenger service left Heathrow Airport on 14 April 1971 to New York and Bermuda - a year late; by 9 May the service frequency was daily and by August twice daily. By July a daily 747 service was opened to Montreal and Toronto, in November 747s were introduced to Australia over Hong Kong; and in December they took over the daily departure to South Africa from the VC10s.

By November the Financial Director was able to announce that in the first six months of operation the 747s had earned over £12m revenue, were above their budgeted results and had achieved an operating surplus twice that forecast. The anxieties about the size of the aircraft could now be laid to rest. In these early days of operation, difficulties were experienced with the Pratt and Whitney JT9D-3A engines which had fallen short in economy of fuel consumption and durability. By December 1971 the Corporation decided to carry out a modification programme on the engines and by the following March to go the whole way and convert the JT9D-3A engines in the initial batch of 747s to the uprated engines at a cost of about £2.5m. These would improve the aircraft's performance especially out of hot and high airports and would permit an increase in the take-off weight from 710,000 pounds to 733,00 pounds when this was commercially desirable.

Gradually the 747s were introduced to Australia via Hong Kong, and in December they took over the daily departure to South Africa from VC10s. The 747s slowly were deployed on all the good traffic routes including, in 1972, operations to Miami, Chicago, Boston and Detroit. Doubts about the size of the aircraft and its passenger appeal were soon dispelled for passenger load factors for the first four weeks of operation, and New York was in the 70s and 80s. Naturally, there were penalties from being a full year late in introducing the 747; capacity was tight, growth restricted and a hefty cut taken in non-scheduled operations to release hours for scheduled services, but there were some benefits. BOAC gained from the experience of other 747 users and was able to use this experience to better introduce services into the aircraft

When the case for these numbers fourteen and fifteen was put to the Minister by incorporation in the BAB's five-year plan, it fell by the wayside, on the Minister's unwillingness to approve the overall strategy and Boeings had to extend the option to purchase to mid-February 1973. The Minister, Michael Heseltine, was on a visit to Boeings at the time the option expired. This was perhaps fortunate, for on 16 February 1973 he approved the go ahead. The delay caused by the Ministry meant that this additional capacity would go into service several months later than planned and would miss the start of the 1974 year. Before the end of BOAC's life, a case was made to the BAB for a sixteenth and seventeenth aircraft for delivery in November 1974 and April 1975. Some impressive figures were quoted in support of the application. The Corporation was achieving utilisation of nearly fourteen hours per aircraft per day out of its active fleet of thirteen jets in the summer of 1973, and the 747 results were above budget. Data showed that in 1972/3 with a fleet then of twelve aircraft this had provided an operating surplus of over £19m, with an operating cost per CTM of only 3.62p as compared with 5.25p for the Super VC10 and 5.52p for the 707-436. The 747 was proving to be a highly successful commercial aircraft.

Chapter Eleven

The Edwards Report - and Concorde

In 1967 the Labour Government took a decision which over the next few years was to revolutionise the British airline industry. In 1966, civil aviation functions had become the responsibility of the Board of Trade (BoT) in place of the MoA, and in due course Douglas Jay became President. There was considerable unease at the time over the state of the industry. In June 1967 there had been two serious accidents, and the Minister had set up a particular review into safety standards. There was disarray in the Independent sector; the British Independent Air Transport Association (BIATA) had been dissolved, and the Cunard Company's attempt to enter the big league on the North Atlantic failed. In February 1965, the Minister, Roy Jenkins, had laid down a Government policy that in general there should not be more than one British carrier operating on the same route. The ATLB chafed under this restriction for in its seventh Report the Board made it clear it would be prepared to grant licences for parallel services when certain conditions were met.

As a result, Douglas Jay announced in the House of Commons on 26 July 1967 that *'the Government have decided to institute a broadly based inquiry into the civil air transport industry'*.

He proposed a Committee charged with reporting as quickly as possible, hopefully by Spring 1968. Their purpose was '...*to inquire into the economic and financial situation and prospects of the British civil air transport industry and into methods of regulating competition and of licensing currently employed and to propose with due attention to other forms of transport in this country what changes may be desirable to enable the industry to make its full contribution to the development of the economy and to the service and safety of the travelling public'*.

On 11 September 1967, Jay announced the names of the Chairman and Deputy Chairman of the new committee - Professor Sir Ronald Edwards and Sir Hugh Tett. Sir Ronald Edwards was Chairman of the Electricity Council and had occupied the Chair of Economics in the University of London for many years. Sir Hugh Tett was Chairman of Esso Petroleum Company Limited.

On 1 November 1967, the Minister gave the names of further members of the Committee, Alan Fisher, General Secretary of the National Union of Public Employees, Dr M. G. Kendell - Chairman of Scientific Control System Ltd, Philip Shelburne - a partner in N. M. Rothschild & Sons and Captain F. A. Taylor, a former BOAC Captain. A further member was added later, Sir Reginald Wilson, Chairman of the National Freight Corporation, of the Transport Holding Company and Thomas Cook and Sons Ltd. The Board of Trade loaned one of its staff, Gordon Manzie, to act as Secretary to the Committee which then appointed Stephen Wheatcroft to serve as Assessor, a particularly appropriate appointment in the light of his high standing internationally as an airline consultant who had spent some years in BEA.

The Committee made their base in office space rented from the Electricity Council at 30 Millbank, and then proceeded to interview or receive written representations from an immensely broad range of interested parties in the UK, including Government Departments, official bodies,

Below left: Roy Jenkins. centre Sir Ronald Edwards, right; Douglas Jay.

Chambers of Commerce, airport authorities, aircraft manufacturers, and airlines and individuals - a list running into seven pages in their Report. They also paid visits to other countries, notably Australia, Canada and the USA as being especially relevant to their inquiry. There they had discussions with the civil aviation authorities and with senior executives from the airlines - all of whom welcomed them with open arms in the hope that their final report might provide answers to some of the universal problems in the airline industry.

The BOAC Board gave critical thought as to its initial submission to the Committee deciding upon a factual report containing a brief account of the early development of British civil aviation, and then describing BOAC's existing organisation, its place in world aviation, and its plans for the future. The BOAC submission took the opportunity to make four good points - in its view the inquiry had not been set up to bail out BOAC; the investigation was not needed to protect BOAC from competition, there was plenty of that from foreign operators; BOAC was on a firm foundation and could be readily built on further and finally, that BOAC's condition was now healthy and should not be weakened by artificial constraints. The Secretary of BOAC, Bob Forrest, co-ordinated the production of a document which was submitted to the Edwards Committee in November 1967.

On 23 November 1967, the Edwards Committee spent the day at BOAC Headquarters where they were given presentations on the airline's current organisation and its five year plan and subsequently toured the engineering, computer, cabin services and training centres. In January 1968, the Committee said they wished to take up BOAC's offer to submit views in writing on the future organisation of British civil aviation and that it would be helpful to have these before the Corporation would give oral evidence before the Committee in early February. It was decided by BOAC to meet the Committee's request but that the submission should take the form of a set of informal notes rather than a formal statement. In it, BOAC advocated the continuance of the existing split between BOAC in the long-haul field and BEA for short-haul operations. It considered that all international services both long-haul and short-haul and charters should be concentrated in the hands of the Corporations and that the only field for independent exploitation was in air taxi work, crop spraying or other aerial work and that those operations involving passenger carrying should be limited to aircraft of not exceeding twenty seats. BOAC saw no difficulty in the absorption of aircraft, staff and other assets from the Independents into the Corporations. At the same time, a feeler was thrown out about the possible injection of private capital into both BOAC and BEA. On 5 February 1968 BOAC gave its first verbal evidence before the Committee, the discussion centring mainly around the Corporation's aide-memoire. In answer to a question from Sir Ronald Edwards, Sir Giles said that if he had a choice in determining the structure of airline industry in the United Kingdom, he would, if entirely free, opt for a single airline as there would be enormous savings in the aircraft types to be operated.

In considering both the aide-memoire and BOAC's verbal evidence, the Committee must have felt it wanted a final view especially on the case for a nationalised monopoly, so Sir Ronald asked the Corporation to express a broader perspective in the national interest, and said that the Committee could discuss issues raised with BOAC at the second meeting on 6 August 1968.

Accordingly, a lengthy paper was compiled within BOAC entitled *'The Future of the British Air Transport Industry'*. This set out the objectives of and the constraints on the British air transport industry, then asked whether there should be one airline or more than one. It considered alternative industry structures and limitations, emphasised the need for consistent Governmental policy to be clearly defined and finally came to a conclusion. The outcome was a preference for a nationalised monopoly *'almost solely on the grounds that in air transport specifically there is significant opportunity with monopoly to achieve the lowest level of costs and a nationalized monopoly because there are worthwhile objectives from a national point of view that are more likely to be reached if not left to the chance outcome of private enterprise'*.

Finally, the paper referred to the constraints of the present situation which might well limit the prospects of attaining the ideal structure. In addition to the main article which was put to the Edwards Committee as a working document, BOAC also sent the Committee a copy of a paper produced as a brief for a forthcoming IATA Traffic Conference on how the Corporation and other scheduled carriers could meet the competition from charter operators successfully.

Before the meeting of 6 August 1968, Manzie sent BOAC a long list of questions for

Ross Stainton (*b*.27.5.14, *d*. 5.12.11). Having begun his career as a station manager for Imperial Airways in North Africa in the 1930s, he rose to join the board of BOAC in 1968 becoming BOAC's managing director in 1971 shortly before it was brought together with BEA under the newly created British Airways Board, of which he was a founder member. He then served as chairman and chief executive of BOAC from 1972 until 1974, when the airlines merged operationally as British Airways.

discussion. It was evident that the subject most exercising the minds of the Committee was the economies of scale linked to BOAC's policy paper advocating the advantages of monopoly and fusion of long-haul and short-haul routes.

For the meeting on 6 August, Sir Giles took with him the Commercial Director, Ross Stainton, the Financial Director, Derek Glover, Director of Planning, Winston Bray, and Secretary/Solicitor, Bob Forrest.

Sir Giles' opening remarks to the Committee gave a clue to the attitude towards the BOAC case expressed in the Committee's Report. He explained that a small group of senior management had compiled the BOAC policy paper and put it to the July Board. Sir Giles made it clear that his first reaction and also that of some Board members was adverse and that a paper recommending monopoly must in principle be wrong, but that on further consideration it was hard to see what else could logically be suggested. All endorsed it in principle except Sir Anthony Milward who as Chairman of BEA would be making his submission to the Committee.

There were two other points of importance in Sir Giles' opening statement. On the subject of merger, he suggested that apart from any question of cost savings, the area where further examination might prove fruitful lay in the opportunities for increased revenue - a point which was later taken up by the Board of Trade and played a considerable part in the subsequent decisions of the Labour Government; as to a suggestion that a Holding Board should be set up to coordinate the activities of BOAC and BEA and possibly the independent sector, he thought this unworkable.

At BOAC's next meeting with the Committee on 3 September 1968, the subjects were not so explosive. They concerned aircraft procurement, the regulatory structure including questions of safety, airports, air traffic control and route licensing. Shortly afterwards at the Committee's request, BOAC sent a list of innovations introduced over the years, both technical and commercial; they had previously posted another paper on the airline's contribution to the balance of payments, and undoubtedly there was a full flow of information from BOAC.

BOAC's final appearance before the Committee took place on 17 December 1968 when the questions related mainly to finance, traffic 'forecasting, the economics of inclusive tour operations and freight operations. BOAC's policy was to develop its all-cargo fleet sensibly concerning the available capacity on its passenger aircraft having in mind the appearance before long of the huge jets with their greatly increased cargo carrying capability below decks. Regarding inclusive tours, it was BOAC's view and also that of the European scheduled carriers represented by their Governments in the European Civil Aviation Conference (ECAC) that the Independents both American and European should be authorised to operate inclusive tour charters across the Atlantic only to the extent that those operations were desirable to supplement scheduled operations. Sir Giles retired at the close of 1968, and Charles Hardie took over the BOAC chairmanship from 1 January 1969. One of his first acts was to write to Sir Ronald Edwards summarising the points already made to his Committee and confirming the BOAC views.

This was a key document since it re-states the BOAC official attitude, which could have become obscured by the various papers submitted over the previous year. In the long-haul sphere, the strong recommendation now was for no change in the organisation and structure of British scheduled operations. As to short-haul, BEA had put forward their case and while BOAC had presented a paper advocating one publicly-owned airline, '...*we, of course, recognise that in the existing climate this solution would be impracticable in the immediate future'*. It also referred to the possibility of a holding company as a recommendation of the Committee and stated that if such a company were to be superimposed on the two Corporations. BOAC's view was that it would achieve little that could not be achieved under the existing organisation - unless it was to be a first step in effecting integration of all the activities of BOAC and BEA.

Sir Ronald Edwards sent his completed Report to Anthony Crosland, President of the Board of Trade on 1 April, and the Report was published as Cmnd. 4018 on 2 May 1969. It ran to 267 pages plus 23 tables and 38 appendices, with an impressive coverage of all the main outstanding issues in the civil aviation world in the United Kingdom. The Report contained twelve critical proposals concerning the structure of United Kingdom airlines and the means for regulating the industry; it could be said that there was something for everyone. As to high-level strategy, a recommendation called for Government '...*to promulgate by a statutory instrument from time to time as necessary clear statements of civil aviation policy'*, a point which would have full support from the industry.

As to what form British civil aviation should take in the '70s, the Report stated that it should include public, mixed and private sectors. The proposals directly concerning BOAC and BEA were that the two Corporations should be confirmed in their role as the major operators of scheduled air services and should also engage in inclusive tour and charter operations, but the proposal went on to include the setting up of a National Air Holdings Board having financial and policy control over BOAC and BEA. It was stipulated that the two Corporations should retain their identities and that a majority of the Board members of the Holdings Board should also be on the Boards of one of the Corporations or of British Air Services, which was to be a group of mixed ownership regional airlines mainly for domestic routes.

As to the private sector, it was recommended that it should be encouraged to create a what they termed a 'second force' airline, licensed to operate a network that included both scheduled and inclusive tour/charter traffic, both long-haul and short-haul and that where it was decided to license a second British operator on a route, it should be this second force airline. To make this airline financially sound and viable the proposals were to hand over to it some of the Corporation's territory, in return for which the National Air Holdings Board would be entitled to take a financial stake and appoint one or more directors to the second force. Linked to this proposal was the requirement that the private sector must be given a fair opportunity. It was proposed to establish a statutory Civil Aviation Authority (CAA) responsible for economic and safety regulatory functions, for civil air traffic control services, operational research, long-term airport planning and for the main work of traffic rights negotiations. It would be up to this new authority to monitor the financial and managerial resources of the operators on the grounds of promoting stability and safety. The Authority should be encouraged to use its influence in favour' of flexibility and experiment when considering the roles of scheduled and non-scheduled operations.

In enlarging on the concept of a Holdings Board, it was the Committee's view that such a Board would create the combined strategy for the two Corporations, would be concerned with the quality of management, financing and investment and the monitoring of results. Moreover, the Holdings Board would alone deal with the Ministry and would *'...scrutinise, approve and monitor the capital programmes of the individual undertakings'*. It would have overall financial responsibility for all the decisions of the public sector. It was then proposed that the Chairman of the Holdings Board should be part-time and that *'the highest executive post should be that of Secretary-General, who would not be a member of the Board but would be responsible under the Board for corporate planning and such common services as the Board agreed to undertake'*.

The Holdings Board would be the public-accountable body. At the same time, the Report again put emphasis on the need to retain the separate identities of BOAC and BEA, and each would keep its name and a Board of directors, and in stating this, the Report rejected the idea of a merger. Hopefully, the Committee considered that the solution they now proposed *'...will settle the arguments for and against merger which had bedevilled the Corporations and periodically upset the morale of their personnel over the past twenty years'*.

As to the BEA submission, the Report was much more forthright. BEA *'argued strongly, vigorously and cogently the case for retaining its independence'* and the Committee said *'we do not want to see BEA's identity destroyed. We like its fighting spirit, and we have been impressed with the way in which it has presented its case to us'*.

BOAC's response to the Report's appearance was swift. On 23 May 1969, the Chairman wrote to Anthony Crosland setting out the common views of the Board on the Committee's recommendations. It was with two of these recommendations that BOAC strongly disagreed, the proposal to create a National Air Holdings Board and the motion to form by artificial means a second force private sector airline at the expense of the two Corporations.

The letter pointed out that although the Committee agreed that the two Corporations should not be amalgamated this conclusion, the message goes on to point out, was inconsistent with the concept of a National Air Holdings Board as recommended by the Committee.

The letter also damned the Edwards recommendation for a 'second force' private airline if it were to be got off the ground by the transfer from the two Corporations of routes pioneered and developed over the years. The Report estimated that the second force airline would need operations of at least 4,000 million seat miles by 1975; the letter claimed this was equivalent of eight 707s and would lead to a cutback in BOAC's fleet requirement and staff; what is more, the routes transferred would have to be profitable ones. These objections were unchanged by the proposal that the Holdings Board should hold a financial stake in the second force airline in return for the lucrative business hived off from BOAC; it was wrong in principle for the Holdings Board to have a financial interest in an airline with which BOAC would be in whole competition. As to the question of double designation on long-haul routes, the letter expressed the firm opinion that this would not benefit the national interest, for those routes where dual designation might be possible were those where foreign competition was most fierce. On other routes governed by pre-determination of capacity, the second designation would involve cutting back the capacity of the existing British carrier.

There could be no doubt in the Government's mind as to BOAC's feelings on the recommendations. There was opposition, too, from all the Trade Unions in the National Joint Council.

A Second Force...

The concept of an airline to rival the state-owned BOAC was nothing new - there had been a group of independent airlines since before the Second World War. The origins can be traced back to a number of shipping companies that have already been mentioned, but one lineage deserves looking at in greater detail.

Airwork Limited expanded into flying in the 1930s, especially overseas, with two of its ventures metamorphosing from feeder routes into Indian Airlines and United Arab Airlines, although another venture, an airline in Iraq serving oil companies, was less successful. In the Second World War Airwork trained RAF pilots, supplied technical assistance and became a significant aircraft construction subcontractor.

After the war, using a fleet of Vickers Vikings, which were similar in concept to the legendary Dakota, Airwork built up a respectable charter business, much of it for governments. Between 1947 and 1950 it flew 10,000 passengers, including 394 babies, on a twice-a-week service between London and Wadi Halfa and Khartoum for the Sudanese Government. It operated an inclusive-tour business for the Civil Service, as well as getting in on the seasonal activity of flying Muslim pilgrims to Mecca for the Hadj religious festivals and becoming the first trooping airline.

In the middle of 1952, Airwork had begun a new scheduled business between London and Nairobi in partnership with the Hunting group. Twin-engined twenty-seven-seat Vikings left London once a week, taking three days compared with twenty-four hours on some regular BOAC flights. But at £98 single the fares were £42 cheaper. Operators on routes like this to a country's colonies escaped minimum fare control under the rules of IATA.

When Airwork and Hunting introduced their 'colonial coach service' they were fully booked five months ahead only a fortnight after starting flights. Their plans to convert their East African service into a more normal third-class service, sharing the traffic with BOAC on a 30-70% split, caused a major political row.

Airwork had also suffered a significant financial setback with the collapse of an ambitious transatlantic freight service after only nine months for a loss of £650,000. This was an attempt by the airline to grasp a share of the booming business with North America, but when it began its service it found that the British Government was not even prepared to give Airwork a percentage of its mail, refusing either to allow the Post Office to use it or to negotiate on Airwork's behalf with the US for the right to fly passengers on charter. When it came to the crunch, the financial health of the national airline was more critical than enlarging opportunities for predatory private airlines.

Myles Wyatt, chairman and managing director of Airwork, remarked ruefully that, *'We discovered that playing in the first league was a very different affair to schoolboy football.'* For Myles Wyatt, though, it was a crisis that established him in effective control of the company. He had been a sceptic from the start about the scheme. Airwork lacked the minor routes to feed cargo to the main Atlantic run, and the competition was formidable. Wyatt made his reputation by selling three Viscounts and Airwork's place in the queue for two DC6s. The planes were no longer needed and the demand for new aircraft was so great that Wyatt made back enough to recoup Airwork's losses with a handsome capital profit.

Although Myles Wyatt was a personal winner from the affair, the major shareholders in Airwork were not very happy with their investment. They were very substantial men: Lord Cowdray, the polo-playing peer who was reckoned to be the wealthiest man in the UK; Blue Star Shipping, owned by the Vesteys, who rivalled Cowdray for wealth; Furness Withy, another shipping barony; and Thomas Loel Evelyn Bulkeley Guinness, once MP for Bath, a member of White's, Buck's, the Turf, the Beefsteak and the Royal Yacht Squadron. He had a personal fortune inherited from his father Benjamin, who had lost one in the San Francisco earthquake and made another in New York, put at £30 million.

The prominent British shipping groups had diversified into aviation as a defensive reflex. They had enough imagination to see that this modern means of transport might become a threat to sea traffic. The big shipping lines all ran virtual monopolies. Cunard dominated the North Atlantic sea lanes. P&O had the Far East buttoned up. Blue Star ruled the waves to South America. Their management thought vaguely that air traffic must be the same. All they had to do, they reasoned, was get big enough to exert muscle. So when the Government started talking about slices of cake, their reaction was simple - buy.

The first purchase was a good one. It was Gerry Freeman's Transair. Freeman had begun as an air-taxi operator at Croydon airport in 1947 and after a shaky start has gone on to create a very profitable airline. He had made a speciality out of distributing newspapers, by 1952 running over three thousand newspaper flights a year and setting standards of performance the envy of many larger airlines. His annual dinners for Fleet Street newspaper circulation managers were riotously successful.

In 1953 Transair started flying holidaymakers for Vladimir Raitz's Horizon Holidays. Gerry Freeman's aircraft were a fleet of ten Dakotas, all of which were immaculately maintained. Towards the end of 1956, Freeman was planning to move from Croydon to Gatwick and to swap some of his Dakotas for three new Viscounts from Vickers. He had, £5,500,000 in retained profits, but he needed another £1 million to pay for the new aircraft and build a £300,000 hangar at Gatwick. At that time aircraft were considered doubtful security and Freeman was searching for finance. When Airwork offered to buy him out, leave him in control of his airline and make him an executive director of the holding company, he agreed.

He was already part of Airwork when its Board held a meeting at the Savoy at which it decided to buy up more independents. Present were Gerry Freeman and Myles Wyatt, Cowdray's man, Lord Poole, Loel Guinness and Geoffrey Murrant, deputy chairman of Furness Withy of which his father was chairman. The choice in front of them was limited. Apart from Laker, the apparent targets were Eagle Aviation, Morton Air Services, Bristow Helicopters, Silver City and Hunting.

Two of the personalities involved with British United Airways one very much in the public eye, the other prefering to stay very much behind the scenes.

Left: Sir Frederick Alfred 'Freddie' Laker (*b*.6 August 1922 *d*.9 February 2006)

Right: Lord Cayzer; William Nicholas Cayzer (*b*. 21 January 1910; *d*. 16 April 1999)

Freeman was delegated to look at them all. First, he called on Harold Bamberg at Eagle Aviation. Bamberg was interested enough to let Freeman look at his books. Freeman did not much care for what he saw and began looking elsewhere.

Morton Air Services were more modest. Captain 'Sammy' Morton, who had flown scheduled services to Paris with Amy Johnson in the 1930s, was another ex-RAF pilot. He had built up a fleet of DH Doves on charter work, much of it to racecourse meetings on such a regular basis that it almost amounted to scheduled service. Morton had also won the right to scheduled flights from Croydon to the Channel Islands, Deauville, Le Touquet and Rotterdam. Morton succumbed to Airwork's offer. A substantial stake in Bristow Helicopters was bought a little later. What made Bristow particularly attractive to the growing group was that Airwork had a loss-making helicopter subsidiary, half owned by Fisons, involved in crop spraying.

Hunting was a different kettle of fish altogether. It was in many ways another Airwork. The primary Hunting interests had split in two, with one side an aircraft manufacturer and the other, Hunting Air Transport, becoming the subsidiary of Clan Line, another shipping group, although some of the Hunting family still sat on the board. Hunting-Clan were partners with Airwork in the colonial coach services into Africa and shared the 30% slice of freight and tourist business that they finally won from BOAC in the middle of 1957 thanks to Harold Watkinson. Hunting Clan was part of the giant British and Commonwealth Shipping group, controlled by the Cayzer family headed by Sir Nicholas Cayzer. The Cayzers, like Airwork shareholders, were steeped in shipping and saw aviation as a threatening competitor in which it would be wise to have a stake. Moreover, their airline, Hunting-Clan, had also suddenly run into some nasty problems with one of the biggest sides of its flying: trooping. After underquoting for the latest contract Hunting was losing money, while at the same time Freddie Laker was picking up its operational pieces by subcontracting his Britannias to Hunting Clan.

Trooping was still vitally important to the independents, but over the years the Government had squeezed prices until the profit margins had become almost invisible. In 1957 a committee of inquiry headed by Sir Miles Thomas reported that *'So meagre is the return from the principal source of the independents' revenue that the committee finds it difficult to believe that the majority of these companies can continue to operate indefinitely on their present basis of earnings.'*

Ironically, the Tories' initiative in creating a second-force airline embodying most of the independents had the effect of bringing the War Office's 'divide and rule' method of allocating trooping contracts to an end by reducing the number of competitors.

Prime Minister Harold Macmillan brought Duncan Sandys into the Ministry of Supply in autumn 1959 with the express intention of having him break it up and create a separate Ministry of Aviation that could concentrate on the aircraft manufacturers' problems.

While the lesser airlines were chasing each other around the pond, within Airwork it was widely believed that Sandys had put it directly to Wyatt that his airline should take the lead in rationalising the industry. With its powerful and ambitious shareholders, Airwork certainly lacked neither will nor resources. Nor did Wyatt need any encouragement.

The Cayzers were not interested in a quick sale. They wanted far more than that. If there was going to be a significant new airline, they were not going to let it belong to their shipping rivals. This was no takeover: this was going to be a full-scale merger.

By March 1960 the two airlines had confirmed the merger. Some 72% of the shares in the new group was held by shipping companies: Blue Star, Furness Withy and British and Commonwealth. The Hunting Group owned 8%, Loel Guinness and Whitehall Securities, the Cowdray company, 10% each. The balance of power had shifted, with the Cayzers becoming the dominant shareholder.

One of the main problems was the name. The Cayzers were certainly not going to let Hunting Clan operate under the name Airwork. Gerry Freeman solved this. He asked Duncan Sandys to use his influence with the board of Trade to allow the group to be called British United Airways on the understanding that it would not contract the name to BUA and cause confusion with BEA.

By the beginning of 1960, nearly all the pieces of BUA had been brought together. On the main board were Sir Nicholas Cayzer and his nephew the Hon. Anthony, plus Clive Hunting, a jovial, helpful businessman with no interest in becoming involved in day-to-day management of BUA. Neither had the major Airwork shareholders.

But if the shareholders wanted to sit back and collect their dividends, the same was not true of the freewheeling, hard-driving entrepreneurs that had just been bought up. In the big arena of BUA, there was a pack of hard-bitten free-enterprise gladiators, all survivors of the fifteen-year elimination contest of independent aviation. Newly rich, they were no longer working just for money. Freddie Laker for example, did not need his salary of £5,000 a year. What they were used to, though, was power.

BUA was a group in name only. Each of the executive directors ran his own company as an independent fief, jealous of his authority and resentful of any interference from the main board. If any of Freddie's employees wanted to do anything, they went to him. He raised it at board meetings and then went back with the decision.

When he could, he gave his approval first and then presented the Board with a *fait acompli*. It was apparent that someone had to have overall authority if BUA was to make any progress. Although no one said so, the answer was trial by combat. The contenders had to have an excuse for fighting, but it did not take long before the battleground emerged. It was clear from the start that only Wyatt, Freeman and Laker were serious contenders for the job of running BUA.

On paper, Gerry Freeman had the best claim. In 1959 Transair had made a profit of £400,000, making it the most profitable part of Airwork. Freeman's business was run with efficiency that reflected good systems and attention to detail. Freddie Laker, twelve years younger at thirty-seven, was overweight and ebullient, bursting with ideas and enthusiasm and contemptuous of red tape and paperwork. Overshadowing them both was Myles Wyatt. Wyatt was huge, well over six foot and exceptionally heavily-built- eighteen stone of solid muscle and fat. He was twenty years older than Freddie and an influential member of the establishment and Admiral of the Royal Ocean RacingClub to which he had recently presented what was to become known as the Admiral's Cup.

Wyatt ran Airwork the way he sailed his yacht *Bloodhound* - autocratically. His position was solid. With none of the BUA shareholders in control of a majority of the equity, Wyatt could divide and rule as he wished.

Wyatt had joined Airwork in 1934 and was a past master at corporate politics. It was the first time Laker had come up against someone who was his match. Wyatt was older, more significant, more robust, better educated, better connected, better bred and something of a bully. Not that that stopped Freddie Laker, but for once he was up against the head of the herd. To begin with, though, neither he nor Freeman realised quite what a barrier Wyatt was.

The three met in Freeman's Eaton Square flat to thrash out the problem. Freeman, it was decided, should be in charge of all the group's short-haul operations in the UK and Europe. Transair was the apparent vehicle in which to concentrate all BUA's local activities. Laker was to take charge of the long-haul business, the trooping flights and long-distance charter.

Then the question of subordinate roles came up. Freeman wanted Bill Richardson as group chief engineer and Kensington Davison, he stated categorically, was not quite the man to be general manager.

Wyatt told them he could not continue in an executive role if he were seen to have failed to support his men. Surprisingly the other two, he suggested that the simple solution was for him to give up his job as chief executive and take a back seat, leaving the other two to dispose of the top positions and him free to disclaim any responsibility.

Caledonian Airways was the second major player in the 'Second Force' concept. It was a wholly private, independent Scottish charter airline formed in April 1961. Caledonian Airways was the brainchild of Adam Thomson, a former BEA Viscount pilot and ex-Britavia captain, and John de la Haye, a former BEA flight steward and Cunard Eagle's erstwhile New York office manager. It began with a single 104-seat Douglas DC-7C leased from Sabena.

Caledonian proliferated over the coming years to become the leading transatlantic "affinity group" charter operator by the end of the decade. During that period, passenger numbers grew from just 8,000 in 1961 to 800,000 in 1970. The latter represented 22.7% of all British non-scheduled passengers. It also became Britain's most consistently profitable and financially most secure independent airline of its era, never failing to make a profit in all its ten years of existence. By the end of 1970, Caledonian operated an all-jet fleet consisting of eleven aircraft and employed over 1,000 workers. At that time, its principal activities included group charters between North America, Europe and the Far East using Boeing 707s, and general charter and inclusive tour (IT) activities in Europe utilising One-Elevens. This was at the time Caledonian merged with BUA, the most significant contemporary independent airline and leading private sector scheduled carrier in the United Kingdom, and formed British Caledonian.

Waiting - Then Hurry Up!
BEA and BOAC waited for the Government's response to the Edwards Report and took the opportunity of jointly emphasizing to the President of the Board of Trade the extent of the existing liaison between them which in their view made unnecessary the insertion of a new layer - the proposed National Air Holdings Board - between the Corporations and the Ministry. Almost immediately after the appearance of the Report, the Board of Trade instigated an internal study by their Economic Services Division to consider its findings. By June this Division had produced their study and Keith Granville and Henry Marking, the Chairman of BEA, was invited to a meeting at which they were told of the study and asked to produce information as to aircraft operating costs, utilisation, scheduling and traffic movements. It was clear then that the Board of Trade were considering three possible alternatives; closer working between BOAC and BEA through a weak Holdings Board as per Edwards; a strong Holdings Board with complete integration of BOAC and BEA with a single image: and finally any other co-operative arrangement between the Corporations.

Given the two Corporations' doubts as to the validity of the conclusions reached by the study group, the President agreed that a combined Board of Trade - BOAC - BEA study group should seek to resolve the controversial issues. This was especially important as the benefits claimed by

Vickers VC10 G-ARTA of British United Airways, one of the airlines that viewed the concept of the 'second force' airline proposed by the Edwards Report as a potential opportunity. *(author's collection)*

the Group in additional revenue resulting from running long-haul services from the USA via London into Europe as a one-carrier operation which the group put as £27m per annum. This enlarged group began its work, but the President was pressing for a final report by early September 1969 and was not prepared to wait until the group had covered all aspects, although both BOAC and BEA felt more time was needed to arrive at the right conclusions. The outcome was the Government's White Paper published on 12 November 1969. But before the White Paper appeared and no doubt with its imminence in mind, BUA applied to the ATLB for licences to operate a wide range of services between the UK and Africa and also between Gatwick and New York, and at the same time sought to have revoked all BOAC's licences to East, Central and South Africa. The whole airline industry in the UK was anxiously awaiting disclosure of the Government's decisions. These were contained in the White Paper which started off by setting out the Government's view as to what the principal objective of civil aviation policy should be, namely *'...to encourage the provision of air services by British carriers, in satisfaction of all substantial categories of public demand, at the lowest level of charges consistent with a high standard of safety, an economic return on investment and stability and development of the industry. This objective must be set in the context of the need to help strengthen the balance of payments and contribute to overall growth of the economy'*.

The Government, like the Edwards Committee, was reluctant to see the separate operating identities of the two Corporations disappear and so were against the complete merger but opted instead for a single Airways Board. It would be a more powerful Board than that proposed by Edwards and would have full authority and take all decisions *'to secure that the two airlines' fleets and routes were planned and marketed to the best overall advantage'*.

The White Paper added that the Government favoured the licensing of a second British carrier on a scheduled service route where it could be shown that such competition would be in the public interest. So it hoped that a new private sector airline would appear from the amalgamation of existing independent operators and that although there must be some transfer of Corporations' routes to it, the Government would require such territorial concessions to be limited. It was against any financial participation in any such private carrier by the proposed Board controlling the public sector.

BOAC expected to be consulted on the provisions of a draft Bill, required to give effect to these Government proposals and that there would no doubt be discussions with the civil aviation committees of the Parliamentary parties. This happened in December 1969 at an informal meeting between the Conservative civil aviation committee headed by Frederick Corfield and BOAC's Chairman, when it became clear what the Conservative policy would be if and when they returned to power. This was further reinforced by Corfield shortly afterwards in a letter to the Chairman in which he said that the Conservatives were determined to make sure that the independent operators were given greater scope within the industry.

Then, a strange interlude occurred. It seems that BUA was disillusioned at the Labour Government's White Paper and at the Government's refusal to agree to their taking away specific routes from BOAC. BUA then approached BOAC as did British & Commonwealth Shipping (BCS) which controlled BUA to inquire whether the Corporation would be interested in buying out BUA. The asking price was around £10-12m although Granville, conducting the negotiations for BOAC, thought the likely final figure would be more like £8-10m. The President, Roy Mason, was kept informed and raised no policy objection. BOAC was keen to proceed since the purchase would solve many difficulties in Africa currently occurring due to BUA's operations cutting across BOAC's partnership arrangement with African operators. If the sale went through, it would re-establish BOAC on the East Coast of South America route which the airline was now better able to afford and give it a base at Gatwick airport.

The Board of Trade approved the purchase, and everything seemed set for a deal. However, when the negotiations became public early in March 1970, there was an outcry from the Conservatives, but the Minister of State, Geronwy Roberts, told the House of Commons on 9 March 1970 that as BUA had first approached BOAC, it was a commercial and acceptable deal. Caledonian in the meantime in an attempt to stall the issue had applied to the ATLB for revocation in its favour of all the BUA licences.

At that point, the Chairman of British & Commonwealth Shipping claimed that there was no proper authority for BUA to proceed with the BOAC deal as negotiations were in hand with Caledonian. Roy Mason then had to say that he had been misled, and he held up BOAC's purchase to enable Caledonian to bid. In reply to a question from Frederick Corfield in the House on 18 March, Roy Mason said *'In late January this year I was approached by BOAC who sought approval for the purchase of BUA. The initiative came from the British & Commonwealth Shipping Company which is the principal shareholder in BUA. I invited the Chairman of the Company to see me early in February. The clear impression that he gave me was that there was no real possibility that a merger between BUA and Caledonian would take place. On 5 March I authorised an announcement confirming that this transaction was acceptable in principle, subject to further discussion of the financial arrangements. Very quickly after that two important developments took place. First, it was brought to my knowledge that despite the impression that I had been given to the contrary, BCS had been in real and close discussion with Caledonian right up to the time of the announcement by my Department. I had been seriously misled in supposing that there was no prospect that a merger could take place. Secondly, Caledonian made certain applications to the ATLB which materially affected the appraisal of any investment by BOAC and announced it was preparing a rival offer to purchase BUA. In these circumstances, it is clear that my proper course is to withhold approval from the investment proposal put to me by BOAC until the situation has been clarified'.*

The Minister felt, as did BOAC that within the timescale laid down by BCS Caledonian it would not be able to conclude a purchase at the asking price. But all was back in the melting pot when the Labour Party lost the June 1970 General Election, and Roy Mason had lost BOAC the chance to buy BUA. The final chapter in the story came in July 1970. The new Minister of State, Frederick Corfield, accompanied by the Second Permanent Secretary, Sir Max Brown, and the Deputy Secretary, Robert Burns, had a meeting with Granville, by now Deputy Chairman and Managing Director of BOAC.

The Minister opened the meeting with the statement that it was Government policy to encourage Caledonian to purchase BUA and so form a second force airline as proposed by the Edwards Report. He claimed that the transfer of BOAC routes to the new airline would be kept to a minimum and that there would be no public investment in the private sector. The BoT concept was that once the routes to be transferred had been agreed BOAC would not oppose any such application. Granville's reaction was, as the Minister said, predictable. He reminded the Minister that BOAC policy was in favour of purchasing BUA and that *'at no time had BOAC expressed a willingness to give up any of its routes to any other carrier'*. He saw no point in BOAC discussing the matter with Caledonian save that when the Government had decided on routes, BOAC would then need to negotiate for compensation. Finally Keith Granville made it clear, that if anything were done to deprive BOAC of any of its existing routes it would have to be done by Government act and not with the willing consent of BOAC and that any statement made by the Minister in the House should make that point very clear; the Minister accepted that.

A special Board meeting was called on 20 July 1970 and recorded complete opposition to the relinquishment of any BOAC routes to a second force airline as being *'against the commercial interests of BOAC, which is the Board's sole concern'*. It was agreed that if the transfer had to be and the Government had over-riding statutory power to achieve it, it would be in BOAC's interest to discuss matters with the Board of Trade rather than for the Government to reach its own decision without consultation. These were the views BOAC put officially to the Minister on the same day.

The next evening Hardie and Granville were summoned to the Minister's office together with the Deputy Managing Director, Ross Stainton. The Minister, now Michael Noble, explained that the purpose of the meeting was to discuss the Government's intention to action the Edwards Committee recommendation for a second force airline. BOAC's response was to repeat that its policy was to buy BUA and that it would not willingly surrender any routes. The Minister was not surprised at BOAC's reaction but expressed confidence that a merger between BUA and Caledonian could be arranged as the basis of the second force but was concerned as to how the transfer of routes could be effected quickly. The pressure was brought to bear on BOAC to accept without opposition an application to the ATLB by the second force for the route transfers, but Hardie maintained that BOAC would not connive in the route transfers and the Government would have to use its powers to do what was necessary through lawful means. On 3 August 1970, the Government announced that in the national interest the independent sector was to be strengthened and should include a stable company in addition to the two main flag carriers, BOAC and BEA. To

G-AZJM, a Boeing 707-324C of British Caledonian Airways, - formed from British United and Caledonian - prepares to board another load of passengers for a trans-Atlantic flight from Gatwick Airport. BCal as it was known became the second force airline and competed against BOAC for some years.

bring this about there would be some transfer of routes from the Corporations, but that the extent of the transfer would amount to only 2½% to 3% of the annual revenue of BOAC, say a figure of around £6m and that the Government had no intention of making any further transfers.

On 16 October 1970 Michael Nobel, now Minister of State at the Department of Trade and Industry, told Sir Charles Hardie - who was knighted in the Queen's birthday honours list in June that year - that the merger between Caledonian and BUA was likely to be finalised that day. The announcement about routes to be handed over to the new airline would be made shortly after Parliament reassembled on 27 October. It was already evident that the Government was likely to decide on the routes between the United Kingdom and Nigeria/Ghana as the first to be re-allocated. BOAC again emphasised their opposition and told the Minister he would have to be sure he had legal powers to compel BOAC to give up the routes. BOAC would not be a willing victim. On 21 October 1970 Caledonian Airways and BCS announced the sale of BUA to Caledonian effective from 30 November 1970, but there was still no detailed statement from Government about route transfers.

The *Daily Mail* ran a story that BOAC and BEA were to merge, but this was speculation as BOAC pointed out to its staff at home and overseas. On 24 November 1970 Noble told the Commons of the Government's decision to establish an Airways Board which would exercise strategic control over the public sector, and to set up a CAA responsible for economic and safety regulations of the industry. Both innovations were expected to be in existence by 1972 after the necessary legislation. The Minister also said that the necessary new appointments would be announced shortly and that there would be implications for the Boards of BOAC and BEA, but that any question of amalgamation would be left until it could be studied domestically by the new Airways Board.

At the same time, there was a statement to the effect that neither Corporation would lose it identity without reference to Parliament. There was a reference to route transfer which would have to be sufficient to get the new airline off to a good start, but no details were given. BOAC was already aware that the west African routes were at risk but were not told officially until 14 December 1970, twenty-four hours prior to a statement in the commons, yet on 2 December Ray le Goy of the Department of Trade and Industry was already in West Africa to tell the Nigerian and Ghanaian Governments of the impending change of British operator. To carry out its plan, the Government had to introduce a Civil Aviation (Declaration Provisions) Bill. The third reading was on 15 December 1970 and Noble as Minister of Trade under John Davies, Minister for Trade and Industry announced that the routes to be taken away from BOAC would amount to around 2½% to 3% of BOAC's current annual revenue. He then went on to say that BOAC after 31 March 1971 would cease to serve Lagos, Kano and Accra *'with the intention that Caledonian/BUA should serve these directly from Gatwick thereafter'*, and that he would announce later what additional routes were to be taken away.

Regarding compensation, the Minister said: *'these opportunities are not something which has been paid for by the present holders of the licences and are not, therefore,*

Adam Thomson, the Chairman of British Caledonian Airways. From a single Douglas DC-7 making charter flights in 1961, BCal grew to be the ninth largest European airline, with a fleet of 27 jets serving almost 50 international destinations. Passengers warmed to BCal's proud Scottish image, with tartan-clad cabin crew and the lion rampant on its tailplanes. At the peak of its success in the early Eighties, it was voted by travel writers the best airline in the world. However, in the fiercely competitive market - particularly on North Atlantic routes - BCal lacked the critical mass to hold its own, especially as the Thatcher administration favoured British Airways in the run-up to privatisation. After losing a crucial row with BA over allocation of routes in 1984, and suffering a series of other commercial misfortunes, BCal fell, to Thomson's deep regret, to a takeover bid from BA in 1988.
(BCAL)

something for which they should be given compensation when the licences are taken away'. BOAC felt particularly aggrieved at this for the Corporation and its predecessor, Imperial Airways, had been serving West Africa for over thirty-four years, had expended considerable capital sums in building up the business and a fund of goodwill, and had provided aircraft and crews to cope with it. Accordingly, the Corporation was quick to reserve its position on the compensation issue, but it was a lost cause. It had an unsympathetic government.

Immediately after the announcement in the House, the Chairman called a press conference to express BOAC's sadness at the decision. As he pointed out, one of the arguments made much of by Government in favour of the second-force airline was that it would provide the public with the second choice of British carrier, but by substituting Caledonian/ BUA for BOAC that still did not give the public an option.

On 29 March 1971, the Minister announced the remainder of the penalties to be imposed on BOAC to get the second force off the ground. As from 1 July 1971 BOAC was required to cease its activities between the United Kingdom and Libya and in addition Caledonian/BUA were in future to be permitted to carry first-class traffic between London and East Africa, and, although the Minister failed to include it in his statement, between London and Central Africa, all this in spite of contractual arrangements between BOAC and BUA which had many years to run. Once again the Government was compelled to make an Order to effect the Libya transfer *'to put beyond doubt'* as the Minister wrote to Granville *'that in withdrawing you are acting under compulsion'*.

Derek Glover assessed the compensation in a paper to the May 1971 Board; he thought BOAC was due around £10-£11m for the arbitrary loss of business, but not a penny was received. In its 1971/72 Annual Report the Corporation said *'Following the loss of its West African routes, BOAC is obliged to consider whether and to what extent it can continue to operate some routes which were not previously making a proper contribution to the overall result. A first casualty is our mid-Pacific route. We have reluctantly decided temporarily to suspend operations on the San Francisco - Honolulu - Tokyo sectors which have shown consistent losses over a period of years with no immediate prospect of improvement'*.

To put these new policies into effect, the Conservative Government introduced the Civil Aviation Bill 1971 into Parliament with its second reading and debate on 29 March 1971. The Bill became law as the Civil Aviation Act 1971 on 5 August 1971. Its purpose was to establish the Civil Aviation Authority (CAA) and the British Airways Board (BAB). The general functions and objectives laid down for the CAA followed the recommendations of the Edwards Report. In particular, the new Authority was required *'to secure that at least one major British airline which is not controlled by the British Airways Board has opportunities to participate in providing on charter and other terms the air transport services mentioned in the preceding paragraph'*.

This was an objective which could be met both as to scheduled and unscheduled operations under the route-licensing powers of the CAA.

Part III of the Act dealt with the creation of the BAB. The Act laid down the powers of the Board, including the provision of air transport services on charter terms or otherwise in any part of the world. All the activities of BOAC and BEA would be controlled by the BAB which would appoint the chairman and other members of the Corporations. It is not the purpose of this record to follow the story of the new BAB except insofar as it concerns the final years of BOAC. It was explicitly required in the Act that the BAB should undertake a review of the Group's affairs to decide the most efficient way of organising them and to make reports to the Minister accordingly who would then lay such statements before Parliament. Thus it was clear from the Act that the new BAB would have complete control over BOAC and BEA which were to continue with separate identities, each Corporation continuing to have its Board appointed by BAB. However, under Section 57 of the Act, the Minister had the power subject to Parliament to dissolve the Corporations if that were to be the BAB recommendation.

Going Supersonic.

Throughout the five years of Sir Giles' chairmanship, there was never enough concrete information available to BOAC for it to make well-founded projections as to the economics of the Concorde on BOAC's routes. This is hardly surprising since the first flight of any Concorde was not until 2 March 1969 at Blagnac, a year later than the target date of 28 February 1968, publicly displayed since 1965 at the Toulouse works of Sud Aviation. Moreover, the first flight when it took place was by the prototype 001, a smaller aircraft and very different from the eventual production model.

In these circumstances, BOAC was right to be cautious and to maintain its attitude that it would have to be convinced the aircraft would be economic and competitive as set out in the Press Statement in 1962. No other reaction would have made sense, especially in the light of the Minister's letter to Sir Giles of 1 January 1964, and of the stipulations in the Government's White Paper on the *'Financial Obligations of the Nationalised Industries'*. However, this is not to imply that BOAC in any way held aloof from the project; in fact over the five years, 1964 to 1968 the airline co-operated fully with the manufacturers, with other interested airlines and above all with Government Departments, making its long experience in the airline industry available as required. On the technical side, liaison with BAC went on all the time over a wide range of design and maintenance matters.

At the turn of the year when Sir Giles took over BOAC, the Corporation had six Concorde delivery positions on the BAC line, numbers 1 and 2, 4 and 5, 7 and 8, Pan Am having already started the ball rolling in June 1963 by taking so-called options on six aircraft, always subject to BOAC and Air France taking the aeroplane. At Sir Giles' first Board meeting on 9 January 1964 it was decided to reserve six positions for BOAC on the US supersonic aircraft production line; after discussion with the Minister, a BOAC team, David Craig, Derek Glover and Ross Stainton, was dispatched to Washington to negotiate with the FAA. The team was victorious in reserving the required six positions at an initial payment of $100,000 per position.

Air France also acquired six US SST delivery positions, and an agreed statement was made public by the British and the French announcing this new development, making the point that the Concorde and the American SST would complement each other. The statement also referred to the fact that BOAC had decided to place a contract relating to eight (not six) Concordes to soften the impact of the fact that the British were looking to the Americans for supersonic capacity in addition to the Concorde. Moreover, BOAC and Air France were now to put down deposits for Concorde positions which they had not previously been required to do, thus in the eyes of the French demonstrating serious intent by the British to acquire Concorde.

Boeing's attempts to design a satisfactory sweep-wing aircraft had run into insuperable problems, and the Company decided to abandon that design and concentrate on a more conventional delta wing layout. But the sums of public money involved were mounting and so was a criticism of the whole project, both for financial and environmental reasons. It became increasingly difficult to get Congress to approve the annual funding, and the end came in March 1971 when Congress finally refused to continue any financial support, and the whole project was cancelled. The advance payments made by airlines for positions on the production line were refunded. The field was left wide open for Concorde, for the Russian SST - the TU144 - was never thought to be a serious competitor. The disastrous accident to their aircraft at the 1973 Paris Air Show, which was never formally explained as no detailed accident report was ever issued, put paid to any worldwide promotion of the aircraft.

In 1964 the manufacturers were talking about total Concorde orders for 300/400, taking a significant share of the long-haul passenger market. No official price for Concorde was available to BOAC at that time, making realistic evaluation studies impracticable, but there was the talk of the aircraft entering commercial service in mid-1970. By March 1964, BOAC was receiving advance information of proposed Concorde design changes, including an increase in the wing area and revised engine design. Passenger seating was to be increased, and a North Atlantic payload of around 22,000 to 24,000 lbs was hoped for. All this meant a delay of about a year in the date of service entry, but BOAC agreed to make a start on contractual discussions with BAC.

In October 1964, the Labour Party came back to power, and Roy Jenkins was appointed Minister. Almost immediately he was required to handle a delicate situation with the French. On 26 October a White Paper was presented to the Commons on the crisis in the balance of payments, and it included proposals for a re-examination of so-called 'prestige' projects in which Concorde would undoubtedly be involved.

Discussion led by Roy Jenkins followed immediately with the French Ministers. In the middle of November, the French Government gave its response - it saw no reason to review Concorde and reminded the British that the original agreement had no provision for either party to withdraw.

Keith Granville holds a model of the Concorde, whilst standing in front of a poster showing the routes BOAC had planned for the supersonic airliner. (BOAC via author's collection)

The Anglo-French crisis continued until the middle of January 1965 when the Minister announced in the House that the British Government would 'stand by the treaty obligations into which the last Government decided to enter', and the political crisis was over.

These political troubles tended to make the project seem more remote to the airlines, and the atmosphere in BOAC was one of waiting for hard information to enable adequate technical and economic evaluations to be carried out. As Geoffrey Knight wrote in his book *'Concorde, the Inside Story'*: *'...considerable pressure was being exerted on BOAC to place an order or to take out options on Concorde. However, they remained adamant that they were not going to do either until they had a much clearer idea of the capabilities of the aeroplane and its economics and performance. Here was yet another example of BOAC maintaining its financial independence. But who co-ordinates and directs the national interest when one arm of Government decided on massive capital investment in a new aircraft development while another arm of government (sic) displays no great enthusiasm for it?'*

Meanwhile, Sir Giles had written to Roy Jenkins confirming once again BOAC's faith in the stipulations in the Memorandum of Understanding and expressing concern at the preliminary information from the manufacturers to the range, payload, and noise level. An event occurred early in 1965 which had a significant bearing on the prospects for Concorde and the attitude of the major world airlines to it; this was the announcement in April 1965 of the Boeing decision to build a very large subsonic jet, the 747, with lower operating costs than any type yet produced. Among the airlines, there had been discussions at the Technical Committee of IATA where the general opinion was expressed that the seating capacity of the Concorde would need to be raised to improve its economics. As a result in May 1965 the manufacturers announced an increase in Concorde's all-up weight from around 326,000 lbs to approximately 365,000 lbs, and a seating increase which on BOAC's type layout would give 124/128 all economy seats as compared with 114/118 prior to the design changes; these changes were to be put into effect on the two pre-production aircraft. In contrast, the supersonic paper designs of both Lockheed and Boeing were around 500,000 lbs all-up weight with economy seating of 250. But the American companies had their design problems, and BOAC received a letter in July 1965 from the FAA advising that any further royalty payments on positions held on the US SST delivery line, which were due before 1 November 1965, would not now become due before 30 June 1966.

By November 1965 BAC was hoping to complete a production specification by early 1967 which would be acceptable to the airline operators. The Anglo-French Working Party set up by the inter-Governmental Coordinating Committee had assessed Concorde's total operating costs per seat mile as 22% higher over a 2,500 mile sector and 19% higher over a 3,750 mile sector compared with the Boeing 320B, without any allowance for the 320B's cargo potential; no comparisons were yet available against the very large 747. The Chairmen of BOAC and Air France continued their regular contacts to exchange views on Concorde progress, and BOAC's own informal operators committee had further meetings; this committee now included Air Canada, Air France, Air India, FAA, TWA and QANTAS, but little useful was coming out of the American scene on the American SST.

Late in 1965 the British and French civil aviation Ministers each appointed consultants to undertake route application studies and BOAC provided a route pattern where the characteristics of Concorde might be deployed to best advantage, including London—Sydney 'eastabout and westabout, London-Tokyo via Hong Kong, London-New York and London—Montreal.

Throughout 1966 BOAC continued to work on route analyses as up—to—date information became available from BAC and on market research studies in an attempt to forecast potential Concorde passenger traffic at varying fare levels, while separately the MOA officials were busy on their own technical and economic estimates. It was apparent the studies would have to come together at some time, but BOAC was always pressing the point on the Ministry and BAC that there were seventeen or eighteen other operators with tentative places on the production list which should also be drawn on for their estimates. On 1 January 1967, President Johnson announced that the Boeing variable-sweep-wing design with General Electric engines would be selected but future progress was all somewhat vague with further design developments called for, this after operators including BOAC had at the request of the FAA reported their views on the competing designs. As to the Concorde, it was hoped that some considerable degree of standardisation might be agreed between the interested airlines mainly as to flight deck layout, a possibility increased by the so-called Mentzer Group set up by US operators under the chairmanship of the Senior Vice-President Engineering and Maintenance, United Airlines. What would have been useful to BOAC would have been a discussion with Pan Am and TWA on their planners' evaluation and economic studies though all along it seemed that BOAC was far advanced in this field and little of value could be extracted from the potential American users.

There was pressure on BOAC to commit itself to a firm Concorde order even though no aircraft had yet flown and no firm performance figures were available, nor could anyone foretell the reaction of Governments and people on the ground to the sonic boom.

The media had a field day on this issue, and in February 1967 Sir Giles made a public disclaimer that BOAC was in any way anti-Concorde. Asked the question *'Is BOAC anti- Concorde?'* he replied *'Absolutely not. This is rubbish. BOAC will need Concorde - and we will also need the American SST; I see the two aircraft as complementary. BOAC wants both, but I insist they must be commercial propositions'.*

In the following month, six of BOAC's top Management including Granville visited the Sud Aviation works at Toulouse to see the progress of prototype 001 - Sir Giles was in Africa and unable to attend. As was pointed out at the time, it was almost ten years since BOAC was first involved with the aircraft industry in looking at the commercial possibilities of supersonic passenger flying; over that period BOAC had spent countless manhours in studying the issues and in providing operational advice.

In 1967 the great decision was taken by all concerned to spell Concorde with a terminal 'e' much to the relief of BOAC publicity staff who at least now knew what to use in publicity material. In July 1967 BOAC agreed to supply the Anglo-French Working Group with an indication of 747 unit operating costs so that a realistic comparison with Concorde's costs could, at last, be made.

More delays and uncertainties surrounded the American SST programme, with the design objectives still being mulled over by Boeings; however, BOAC decided that it would pay FAA the second £100,000 advance royalty payment on each of its six delivery positions when these became due at the end of October 1967.

By October 1967, BAC/SUD had issued a technical specification but this was a tentative document with a large number of technical queries, making it unlikely that a firm specification could be issued for many months yet. But a real sign of progress was demonstrated on 11 December 1967 when French prototype 001 was rolled out of its hangar at the Sud Aviation works at Toulouse. By now BOAC's planning team was heavily involved in private studies on the possible effect of supersonic operations on subsonic services and how the respective fare levels might develop, and was co-operating with officials in the Ministry of Technology (MOT) and in the BoT who were endeavouring to estimate Concorde's sales worldwide. In all these planning exercises broad assumptions had to be taken about any restrictions on overflying populated areas because of the sonic boom.

At last, the estimated operating costs of the Concorde were agreed within reasonably narrow limits by Government Departments, BAC, BOAC and Air France, but forecasting the impact of the aircraft on the passenger market was another matter.

By July 1968, the specific date of 28 February 1968 had not only been passed without the first flight of 001, but expectations were now towards an overall delay of about a year. BOAC's planners were now assuming entry into service in 1973 and possibly a C of A by October 1971. Several composite route evaluations had been completed for those routes offering the best passenger markets for Concorde, but all showed a considerable worsening of BOAC's overall results and a total cost per CTM for Concorde, about twice that of a mix of BOAC's subsonic aircraft.

At the close of 1968 and the end of Sir Giles' chairmanship, the first flight of Concorde 001 was still uncertain but expected to be early in the new year with the C of A more likely in 1972. The MOT had commissioned consultants to compare route results of a fleet incorporating Concorde with one without Concorde and BOAC was asked to cooperate by supplying traffic flows on particular routes. As to the American SST, information on Boeing's progress was scarce, but the firm was due to put its latest design proposals to the FAA in January 1969; it was becoming clear that a huge investment by the US Government would be called for if the project were really to go ahead. Thus, much remained unresolved on Sir Giles' departure from BOAC and BOAC's final decision on the purchase of Concorde was still over three years away.

When 1969 opened, BAC/SUD was estimating that type certification for Concorde could be completed by September 1972 and first delivery to BOAC in late 1972. The Ministry of Technology forecast certification a few months later than BAC/SUD, but was still indicating that nine production aircraft would be available by December 1972. But in BOAC's G plan the Corporation was cautiously including Concorde in 1974/5 as the full introductory year.

The BOAC planners had discussions in June 1969 with officials at the Ministry of Technology and had kept them informed of the disappointing economic results estimated for Concorde on a limited BOAC route pattern. The Ministry had commissioned Floyd Associates as consultants on Concorde route application from the financial aspect. BOAC had expected to be allowed to see the eventual Floyd Report and to discuss the conclusions, but there was reluctance to enable the report to be examined although a copy was finally made available in August 1969 and BOAC's views on it passed to the Ministry. In general, BOAC thought the Report over-optimistic.

A milestone was passed on 2 March 1969 when the first prototype 001 made its maiden flight with encouraging results. According to Sud Aviation, the aircraft had behaved as the flight simulator had predicted. By April 1969 BAC/SUD had allocated eighteen positions on any future production line, six to BOAC, four to Air France and eight to Pan Am, but only Pan Am had secured its positions by initial payments. In December 1969 the Planning Director put to BOAC's Board a summary of economic studies already completed on the available information. The price of the aircraft at 1974 levels was then taken as £10m, or £12/13m with spares, with an assumed life of 8½ years. Operating cost per seat mile came out as just about double the rate for the 747 and a route application showed that about £1m would worsen BOAC's overall results to £2m per annum for each Concorde aircraft introduced into the plan. All information was under discussion by the planning team with the Ministry of Technology, its consultant and the BOT. Three months later a comprehensive economic study was produced covering the potential Concorde routes, London - New York, London - Sydney and London - Tokyo. As expected, the route offering the most hopeful result was London - New York, with an elapsed time westbound of 3 hours 40 minutes against 7 hours 35 minutes for the coming 747. The final forecast results still showed BOAC as being considerably worse off with Concorde in the pattern, even on London-New York, although that route could produce a small profit on Concorde in isolation.

In all these and later studies, BOAC naturally assumed its main competitors would also deploy Concorde. Comparative figures used at that time between Concorde and the 747 are impressive, Concorde figures being given first: maximum takeoff weight 385,000 lbs versus 710,000 lbs, operational empty weight 173,000 lbs versus 365,000

Above: a previously unseen artwork of a supersonic airliner in BOAC colours and markings. A close study reveals that it is not the Concorde, so can only be one of the US projects.

Left: a rare BOAC Concorde menu cover used in the early 1970s to publicise the introduction of the aircraft.

TIME/CONCORDE

BOAC

lbs typical mixed-class seat layout 20 first and 96 economy compared to 52 first and 292 economy, usable cargo volume 643 cubic feet. versus 5,160 cubic feet with a Concorde range of, marginally, London - New York. Average block - to - block speeds on a typical route pattern were around 910 mph and 500 mph. This detailed study was fully displayed to the Ministry of Technology and Board of Trade senior officials, including a detailed breakdown of Concorde, estimated operating costs. In spite of the discouraging results coming from the studies, BAC was still forecasting total eventual sales of the aircraft at 250 or so.

The Corporation, with the history of aircraft orders from British manufacturers in mind, was meticulous in keeping Government and the manufacturers informed of its studies and conclusions about Concorde and again in May 1970 full briefing meetings were held with top officials of government Departments concerned.

In September 1970 BOAC's planners had had several sessions with BAC to compare studies on a five-aircraft fleet of Concordes injected into BOAC's route pattern to New York, to Johannesburg, to Tokyo, and to Singapore before going on to Sydney. The BOAC study showed a worsening of BOAC's operating results by about £16m whereas the BAC case showed an improvement of £5.8m, both figures relating to the year 1975/6, a gap between the two assessments of nearly £22m. After an exhaustive analysis of the BAC case, the BOAC planners saw no reason to alter their estimates and all the papers for both forecasts were tabled at the September 1970 meeting of the BOAC Board. Following the meeting Sir Charles Hardie wrote on 1 October 1970 to Michael Noble at the Board of Trade summarizing the point reached within BOAC on the issue of Concorde, reminding the Minister that the latest BAC and BOAC studies were with his officials and stressing the fact that in its studies BOAC had gone so far as to assume that all outstanding technical questions, including sonic boom and noise level, would be resolved in Concorde's favour. The Corporation once again stated that it was anxious to operate Concorde in order to participate in the pioneering of supersonic flight and that it was a willing customer for Concorde, but as seen at that time it would be unable to make a case to Government for its purchase that would comply with the Government's own criteria for testing BOAC's investments.

Sir Charles' letter also expressed anxiety that the Government should understand how seriously BOAC had considered the whole question of Concorde operations and *'the reluctance with which we have been driven to the conclusion reached in our economic analyses'*. The letter concluded by asking whether Government would consider the idea of an airline consortium to operate the aircraft to establish what the extent of passenger attraction to SST travel might be. The Minister made no response to this idea, and it was put to him again by Sir Charles early in November 1970. After the lapse of a month, a reply came requesting BOAC not to pursue such a line with foreign operators, since if any arrangement of that kind were to be entered into "the implications for sales would be serious". During 1970 the French Government had consolidated the aerospace industry, and Sud Aviation had disappeared into a larger grouping under the name of Aerospatiale with Henry Ziegler as its dynamic president. It was with this organisation that BAC worked from then on. BAC was telling BOAC by January 1971 that in any future studies a price for the aircraft of $27m should be assumed.

In February 1971 the BOAC Planning Department produced an up-to-date and comprehensive study of Concorde application to BOAC's routes and the estimated financial results with and without Concorde; this was known as the "Blue Book". Presentations based on this were given by the Planning Director to the BOAC Board and to senior officials of the DTI and of the Ministry of Aviation supply. The data in the Blue Book largely confirmed what earlier studies had disclosed, so Granville, now BOAC Chairman, told Sir Max Brown at the DTI on 19 February 1971 that *'on the basis of current knowledge the financial return on the operation of these aircraft would be inadequate and their effect on our profitability would be adverse. Nor indeed could an application for capital expenditure approval measure up to the standards required of us'*.

These comments were made in response to a request from Sir Max Brown for a statement of intent by BOAC as to the number of Concordes it would be prepared to order. Rather to its surprise, the Corporation then received a request from Lord Rothschild of the Central Policy Review staff of the Cabinet Office to answer a wide range of questions about Concorde; many of these should have been addressed directly to Government or to the manufacturers, but answers were given to a number of economic questions. The Corporation's Blue Book was made available to Lord Rothschild, and this contained the answers to several of his questions. Central Policy Review staff visited BOAC and BAC to go over the Concorde economic figures and were followed by consultants, Cooper Lybrand, acting for Lord Rothschild. No one could say there was any secrecy about BOAC's thinking. The Corporation was not shown the results of the Rothschild exercise but in his book on the Concorde Andrew Wilson of *The Observer* wrote that Lord Rothschild told Heath that Concorde would never provide an adequate financial return, but it should be persevered with because of the international political considerations, and that the cost of continuing the project was worth it for the sake of entry into the Common Market. So HMG now agreed with a French Government plan to authorise production of Concordes numbers seven to ten and to initial spending linked to a further six aircraft beyond ten. BOAC's views on this were not asked. Within BOAC a Project Controller, Concorde had been appointed to maintain liaison with BAC at Filton on all aspects of Concorde's development and Captain Jimmy Andrew, BOAC's Flight Development Manager, was to be the first BOAC captain to pilot the aircraft.

With summer leave out of the way, Sir Max Brown called a meeting at the DTI to review the Concorde situation. BOAC was represented by Granville, Abell and Bray. There seemed no doubt now that Government had decided to go ahead and the DTI was giving serious thought to what financial arrangements could be made in the event BOAC took the aircraft. None of the still unanswered questions as to aircraft price, performance and acceptability at airports was resolved; the possible purchase was thus proceeding in a fog of uncertainty as to the practical issues of operating the aircraft along world routes.

However, BOAC was told to continue to assume a Certificate of Airworthiness at the end of 1974 with 1975 as the in-service date. Sir Max Brown put forward the Government's preferred view on financing the purchase; this could be by allocating public dividend capital to cover Concorde's first cost, to be remunerated only if profits were made. At the close of the meeting Granville referred once again to the idea of a consortium of airlines to operate specific selected routes, such as London - New York and London - Australia, but received no encouragement from the Ministry.

Further indication that matters were approaching decision on Concorde ordering was given by a request from Lord Rothschild in October 1971 for additional economic information. BOAC itself had updated the Blue Book where necessary, and forecasts still showed that BOAC's operating results would be considerably worsened for each Concorde operated. Even so, the Board confirmed the policy of doing all possible to introduce Concorde into service subject to adequate financial protection. More information on performance was coming along following the trip of the French prototype 001 to South America, which had demonstrated that the aircraft could be operated satisfactorily in poor weather conditions under air traffic control. There was still anxiety at the noise level at or near airports, especially in the face of growing antagonism from the public to aircraft noise. On all these matters BOAC maintained liaison with Air France at the top level.

On 10 December 1971, the Secretary of State for Trade and Industry flew in the Concorde together with the Minister for Aviation Supply and on their return made it plain that Government was backing the aircraft and there was no question of any last minute cancellation. In BOAC the idea of being given public dividend capital to cover the purchase of the aircraft was being given serious thought. Strangely there was then a lull in activity and BOAC was under no pressure either from Government or from BAC to complete contractual negotiations. In this quiet, the Board approved the policy outlined by the Chairman, that if the Government required BOAC to acquire the aircraft even in the face of the adverse financial effects that this would have on its finances, then there must be discussions about how to arrange finance, otherwise BOAC could not make the usual type of financial justification to the Treasury; moreover, BOAC would need firm assurances on a number of highly critical practical matters respecting noise acceptance, foreign Government traffic rights and clearance for supersonic overflights and also runway improvements at certain vital airports. On finance, BOAC was prepared to proceed if adequate public dividend capital were made available to it under certain safeguards. On top of all this, BOAC was still awaiting guarantees especially on landing distance, engine development, and subsonic performance.

On all these issues the BAB were kept informed, and on 1 February 1972 Granville told Sir Max Brown of the assurances BOAC required. An extraordinary meeting took place on 4 February 1972 at the DTI. Sir Max Brown and his officials again met Granville, Abell and Bray, and for the first time, BOAC was told formally that it was now firm Government policy to go ahead with Concorde and that BOAC and Air France would be expected to place orders. The Minister would issue no directive to that effect since that could well prejudice foreign orders. On finance, the Government had made up its mind to adopt the method of providing public dividend capital, but this would be included within the BAB's complete requirements so that the capital applicable to Concorde would not appear as a separate item. On 27 April 1972 the Minister for Aerospace, Michael Heseltine, wrote to Nicolson explaining his ideas about the financing of BAB's future capital expenditure including that for Concorde. His conception involved the issue of public dividend capital and loan capital of amounts which would bring BAB's debt/equity ratio from its current high gearing of 66:34 to around 40:60 within two or three years and then maintain it at around that level. He assumed for this purpose that BAB's total capital requirement for the five years 1972/73 to 1976/77 would be about £1,000m of which 40/45% would be financed from the BAB's resources, with approximately £200m of public dividend capital and £350m of fixed interest-bearing loan capital issued to the Board.

On the same day, Heseltine joined the BAB at dinner, and discussion centred around BAB's capital structure as a whole along the lines of the letter of 27 April. The Minister thought it should then be feasible for the BAB to produce an average return on net assets of 6%, which he indicated would be satisfactory to Government, as compared with BOAC's target of 12½% and BEA's of 8%.

Concorde production at Filton. Even-numbered aircraft were assembled at Filton while odd-numbered machines were assembled at Aerspatiale's Tolouse site. (BAC)

The Minister added it was not his intention to try to lay down a firm target at that time.

The BAB members were unhappy at the Minister's approach since Sir Ronald Edwards said at the end of the day BAB would be held responsible for results and that it was BAB's duty to take a purely commercial view. The Minister stuck to his view that a 6% return coupled with the provision of £200m equity capital must satisfy BAB.

The hazard faced by a nationalised Corporation in dealing with ever-changing Ministries and Ministers was once again demonstrated when on 3 May 1972 Heseltine made a tour of BOAC's organisation at Heathrow and joined Granville and other BOAC executives at lunch. During the meal and concerning Concorde, the Minister said it was BAB's and BOAC's responsibility to act in the national interest. This caused great surprise to the BOAC officials and Granville reminded the Minister that BOAC's policy was based on the well-known letter from Julian Amery to Sir Giles Guthrie of 1 January 1964 which laid a duty on BOAC to act in its commercial interest and for Government to impose when necessary any national interest decisions.

The Minister's response was to say that that was an out-of-date concept. BOAC was so concerned at this reply that the Chairman wrote to Nicolson at BAB telling him what had occurred. Because of significant differences between the BAC and BOAC assessments on Concorde, the Minister's next move was to call for a confrontation between them on the economics of Concorde as applied to the BOAC route network. This took place on 5 May 1972 in the Boardroom at Airways Terminal, London. There was a large gathering, including the Minister with top officials from DTI, Nicolson as Chairman of BAB with five of his Board members including Sir Ronald Edwards, Granville as Chairman of BOAC with a BOAC team including Bray as Planning Director who was to make the BOAC case and finally Sir George Edwards and a top team from BAC.

BOAC and BAC then made detailed presentations and answered questions from the Minister and others present. Even compared with what BOAC considered to be its optimistic assumptions on performance and route clearances from foreign Governments, the BAC answers were still far more favourable to Concorde. Subsequently, the Minister asked his officials to prepare a comparison of the two cases, and on 9 May BOAC amplified some items and set out the latest estimates giving a total investment cost of £18.05m per aircraft including escalation to December 1973 at 6% and spares and spare engines. Asked by the Department of Trade and Industry to assume a basic aircraft price of £13m in July 1971, BOAC then gave the revised total investment cost per aircraft of £21.52 million. All the BOAC presentation material was supplied to DTI so that no possible misunderstandings could arise. A flurry of activity followed in BAB, and a Concorde subcommittee met on the same afternoon as the presentation to decide policy, in particular, what approach to make to Government to recoup BAB for the loss of around £159.5 million forecast by BOAC. The discussion was inconclusive other than that the Chairman of BAB and BOAC should see the Minister and press for a somewhat higher figure than the £159.5m. It was evident that matters were moving towards a final plan for BOAC's Concorde purchase. On 12 May 1972, the meeting with the Minister took place. The airline fielded a strong team including the two Chairmen, Sir Kenneth Keith with three other BAB Members and BOAC's Planning Director. Granville pressed hard for an amount of around £4 million per annum per Concorde to cover the adverse effects, but the Minister was only prepared to offer a lower financial target, the provision of public dividend capital and the acceptance by Government of a debt/equity ratio of 40/60%. BAB/BOAC on the other hand were seeking not only public dividend capital but a debt/equity ratio, say 35/65, and if necessary a fund of up to £75m to cover worsening results. On this last point, the Minister called in Geoffrey Knight of BAC who said he thought that Government assistance to operate the aircraft would not have a serious effect on foreign sales if it were done sensibly. The Minister then tabled a draft of the statement he would be making shortly in the Commons, asking that BAB/BOAC should study this in readiness for a further meeting on 15 May 1972.

This meeting took place in the House of Commons with much the same attendance as three days earlier. Little came out of the discussions; the Minister stuck to his belief that he could not agree on a subsidy but accepted Sir Ronald Edwards' suggestion that on the basis of a range of assumptions there could be a comparison between what was then seen as necessary to make Concorde profitable as against what might eventually transpire, but the Minister linked this to adjustment in the target rather than, as Sir Ronald had proposed, to adjustment in the interest payments. The Minister said he was prepared to include in his statement to the House a reference to BOAC acquiring Concorde, but there would be no quantification of Concorde effect to avoid denigrating the aircraft. Any such calculation could, of course, be done in the future. In the course of the meeting Abell for BOAC pointed out that the problem of noise was still outstanding and also certain aspects of the Rolls-Royce Olympus engines so that in effect no progress had been made, nor was it likely in the near future, in obtaining clearances for the aircraft along the routes. BAB/BOAC left the meeting with the task of working out a range of results for Concorde based on a series of parameters, these results to be used as a yardstick. So the outcome of the meeting left BAB/BOAC as far away as ever from securing any special Government development grant or subsidy. The dilemma that BAB/BOAC were in was expressed in a paper agreed by BAB on 15 May, the day on which the Minister would make his Commons announcement. As the paper stated, *'on the one hand it can be said that the principle of reducing our performance to cover up an uneconomic national investment is unacceptable, on the other hand it can also be argued that Air France will have the Concorde anyway, so that BOAC must have it too and that BOAC should share in the development risk of going supersonic in this event'.*

Exchanges then took place between the Minister and his officials and BOAC over the wording of the Minister's statement to be made in the House on 25 May, which would announce the purchase by BOAC of five Concordes. The Corporation was anxious to ensure that the wording offered protection to BOAC in the event that the Concorde results severely worsened its overall financial results.

On 18 May Granville had another meeting with the Minister and at the BOAC Board meeting on the following day tabled an essential piece of paper which recommended that he be authorised to proceed with the ordering of five Concordes subject to five provisos: one, that certain amendments be agreed to the draft of the Minister's statement; two, that the text of a draft letter from the BAB to the Minister concerning the parameters to provide a base point from which future financial negotiations with Government would commence must be acceptable to both sides; three, that technical assurance were given verbally by Geoffrey Knight of BAC to him were confirmed in writing; four, that certain technical matters regarding noise level and contract as put to the DTI in Abell's letter of 19 May were satisfactorily settled by Governments; and five, that a satisfactory outcome was reached in negotiations with Rolls-Royce over the spare engine contract. BAB endorsed Granville's authority to proceed as both BAB and BOAC felt there could be no possibility of misunderstanding on the part of Government as to BOAC's conclusions on the likely effect of operating Concorde on BOAC's profitability, '...particularly', as Granville said to the BOAC Board *'through the medium of the Blue Book dated February 1971 as amended in October 1971 and of the BOAC document dated 10 March 1972 formally sent to the Minister by the Chairman of BAB on 28 April 1972'*.

By 24 May, the day before the Minister was to speak in the Commons, Nicolson wrote to him sending the assumptions which BAB were making about Concorde operations and attaching a tabulation of varying parameters and a range of likely effects of the financial results over the life of Concorde. This range varied through a break-even point to a marked worsening of results. Nicolson then said that BAB interpreted the coming statement by the Minister to mean that if the actual results fell in the range of columns showing a worsening, there would be a case for BAB to ask Government, using the words in the coming Ministerial statement, *'to take steps to ensure that the Board maintain a sound financial performance'*; on that understanding BOAC would proceed to place an order for five Concordes.

The Minister merely acknowledged this letter on 26 May 1972, after he had made his statement to Parliament the previous day. The portion of his speech relevant to BOAC and Concorde is quoted here in full since it brings to a close the long chapter of events from as far back as 1959 when Basil Smallpeice, appointed a supersonic transport committee in BOAC to study possibilities - a stretch of thirteen years to BOAC's commitment to purchase: *'For BOAC there is the need to expand its subsonic fleet, and, of course, the great challenge of Concorde. The Government, having weighed all the relevant investment considerations, believe that an investment in Concorde is in the interests of the country and the travelling public and that BOAC should lead the world in the introduction of civil supersonic flight. This being so, I am glad to tell the House that BOAC is today announcing that it will shortly place an order for five Concordes for use on a major part of its route network. The investment in these aircraft with spares and associated equipment will be £115m at estimated 1974 prices. The purchase has been endorsed by the British Airways Board, and I have given my approval. The whole House is aware that in the pioneering of any radical departure in types of aircraft there are bound to be financial risks. With supersonic flying many of these uncertainties are at present quite unquantifiable, and, indeed, some are outside airlines' control. The Government will, of course, do all they can at the appropriate time to ensure the successful operation of Concorde on BOAC's chosen routes. The Air Corporations will be faced, like other major world airlines, with a number of challenging uncertainties over the next few years, of which Concorde is only one. It is the Government's task to provide the British Airways Board with a capital structure which will enable it to meet this challenge. We have in particular to give the Board a reasonable debt/equity ratio having regard to that existing in foreign airlines.*

"The Board's present debt-to-equity ratio, 66:34, is considerably higher than is normal in world airlines. I consider a ratio of around 35:65 to 50:50 would be appropriate for the Board. To achieve this range the Government have decided that the requirements of the Board for new external capital will be met in the next two or three years by means of issues of public dividend capital, under section 43 of the Act, to a total of approximately £200 million that this facility will be available for financing existing debt as it matures as well as for financing new capital projects; and that the ratio will be maintained within the range thereafter by means of future issues of public dividend capital to finance a part of the Board's capital expenditure. "I have also agreed that the Government will be prepared to review the financial position periodically with the Board in the light of the outcome of the Board's operations and of the conditions prevailing in international civil aviation at the time; and, if necessary, to take steps to ensure that the Board maintains a sound financial performance. It would, of course, be perfectly open to the BAB to refer in its published reports and accounts to the way in which various assumptions made at the time of its equipment purchases had worked out in practice."

On the same day, 25 May 1972, a Press Conference was held by BOAC at the Headquarters of BAB in the presence of the Minister for Aerospace, David Nicolson of BAB, Sir George Edwards, Chairman of BAC, Geoffrey Knight of BAC's Commercial Aircraft Division, Ross Stainton, Managing Director of BOAC and Charles Abell, Engineering Director and others from BOAC and BAB. Granville still in his capacity as Chairman of BOAC gave a short speech to the Press telling them that BOAC had now entered into a formal commitment to buy five Concordes and congratulating the teams of BAC and Aerospatiale for a brilliant technological achievement; he said that the decision in his view was *'in the best interest of the travelling public, of the country and of BOAC'*, that BOAC was preparing to fly Concorde to the USA, Japan, Australia and South Africa, but was relying on Government help in the negotiation for supersonic corridors over land; he added that Concorde was likely to be configured in one class with about 100 to 104 seats.

The formal contract between BOAC and BAC was to have been signed on 26 June 1972 but was postponed at BAC's request, and by July the cost of the order had increased to about £121m. Eventually, the contract was signed on 28 July after confirmation by Government that the price increase did not alter their intention that BOAC should proceed with the order. But by September 1972 the price had further increased to £123m, primarily due to escalation.

Within BOAC a Concorde coordinating committee was set up to progress the application of the aircraft, as had been so successfully done with the 747. The G12 BOAC plan for 1975/76 to 1979/80 issued in November 1972 provided for the introduction of Concorde in summer 1975 at a twice-daily frequency London-New York, and by 1977/78 plans included in addition low frequencies to Australia, the Far East and South Africa. But for all these operations the assumption was taken that Government would have cleared the way by then with foreign Governments and that delivery dates for the aircraft would be one each in March, May, June, October and November 1975.

It was not long before a new situation had arisen affecting estimates of Concorde results. Early in 1973 two major US carriers, Pan Am and TWA, decided not to take up their options so the Corporation had to rethink its evaluations in the likelihood that there would be no competing Concorde service on the North Atlantic; in consequence it believed a small operating surplus could be achieved on the London - New York route. But BOAC was becoming increasingly concerned about lack of progress by Government in securing en route clearance both as to noise level at airports and as to supersonic corridors over land and these anxieties were repeatedly emphasised to Government officials. There was no such pressure on other Governments except the French for no other airlines except BOAC, and Air France showed any real signs of acquiring Concorde.

The delivery situation began to present problems by the autumn of 1973 when BAC told Stainton that there could be a delay of four months in the delivery of the first three aircraft. A new hurdle then appeared shortage of aviation fuel, and BOAC was being asked by DTI to consider the effects of a reduction in supplies of from 10% to 35%. That was serious enough but for Concorde, it was likely to be compounded as the price of aviation fuel would rise steeply, and Concorde costs with its high fuel consumption would suffer. By November 1973 BOAC was estimating the effect of Concorde as a worsening of about £4 million per aircraft per annum. At about the same time the Minister told members of BAB of the desirability of proceeding slowly in obtaining clearances for Concorde, which increased the Board's anxiety about the aircraft's future.

In February 1974 BOAC's Planning Director, now Alec Finlay, produced a review of progress on Concorde. This showed a deterioration in financial prospects due to escalating fuel prices and other costs, with the capital investment, excluding the cost of a simulator, likely to be 523m at least, up to £127m for the five aircraft. Delivery had gone back to November 1975 through March 1976 and the first service, not before January 1976. By this time environmental objections were increasing, and it was evident that supersonic flight overland was unlikely in many areas. In February 1974 a Labour Government came to power in an atmosphere of economic crisis. The Prime Minister divided DTI into two separate Departments, the Department of Industry under Anthony Wedgwood Benn and the Department of Trade with Peter Shore as Minister. The Concorde was Wedgwood.Benn's responsibility with Shore as the sponsoring Minister for British Airways. On 18 March 1974 Benn made a statement in the House of Commons on Concorde which referred to British Airways' forecasts and ended with the words *'in view of the size of the sums of public money involved and the importance of the decisions that must now be made, I thought it right to place all these facts before the House and the country before any decisions are reached'*. This gave the clear impression that government was having second thoughts about going on with Concorde and caused considerable alarm in BAC, as Geoffrey Knight explains in his book. But the project continued, and so Concorde's operations became the task of the new British Airways to plan and implement, and in December 1974 Gordon Davidson was appointed Director, Concorde to ensure departmental coordination towards a smooth introduction into service.

The first fare-paying passenger service took place on 21 January 1976 from London to Bahrain, but Concorde remained restricted in use to Air France and BA. With operations to New York banned initially, BA with restricted payload flew to Dulles Airport, Washington in open country, but eventually after much argument and hearings held by the US Department of Transportation the aircraft was permitted to land at JFK Airport, New York and regular passenger flights began on 22 November 1977. An attempt to extend operations to Singapore succeeded after a false start following the banning by Malaysia of supersonic flight. This did not survive very long for in the unfavourable economic climate of 1980/81 BA decided it could not continue to bear the losses, and the route to Bahrain and Singapore was withdrawn in November 1980.

Concorde went on to have an excellent operational record for reliability, in British Airways service but its restricted use has meant a very low annual utilisation. Despite its high operating costs and low utilisation, it has made an increasingly strong impression on the transatlantic premium markets as well as on specific charter activities.

Chapter Twelve

From BOAC to BA

The British Airways Board took over the responsibility for the two Corporations and their subsidiaries on 1 April 1972, but prior to that the Secretary of State for Trade and Industry had appointed as BAB's Chairman, David Nicolson (later Sir David), an engineering consultant, and BAB held its inaugural meeting on 10 January 1972.

Inevitably there were changes to the BOAC Board following the formation of the BAB. Sir Arthur Norman, BOAC's Deputy Chairman, resigned on his appointment to the main Board (BAB) as a part-time Member, and the same applied to Alan Fisher, who had joined BOAC on the retirement of Ron Smith from the Board in January 1970. BOAC's Financial Director, Derek Glover, who had joined the BOAC Board in February 1969, was appointed Group Financial Director and left the BOAC Board to move on to the BAB.

Keith Granville as Chairman and Ross Stainton as Managing Director of BOAC were both appointed to the BAB but also remained on the BOAC Board. These appointments left gaps on the BOAC Board which were filled by Richard Hilary as Commercial Director, by Captain Frank Walton as Flight Operations Director in December 1971, by Charles Abell as Engineering Director and by Peter Hermon as Managing Services Director in April 1972: in September 1972, however, Peter Hermon became Group Management Services Director, relinquishing his BOAC Board membership to join BAB and Bob Forrest, Secretary and Solicitor, then moved onto the BOAC Board.

In February 1972 the DTI issued a White Paper on civil aviation policy guidance to the new CAA. This was a very different document from the Labour Government's policy White Paper issued as recently as November 1969. The CAA were now being encouraged to licence more than one British carrier to serve the same scheduled service route, and the White Paper went so far as to instruct the CAA to give preference to the newly formed British Caledonian Airways (BCAL) *'over other airlines in allocating licences for new scheduled service routes'*. This blatant advantage to the new second force airline formed by the merger of BUA and Caledonian Airways was to be protested by BA subsequently: on the face of it, it was an odd requirement bearing in mind CAA's responsibility to accept evidence from air carriers seeking licences and to base its decisions on the evidence submitted.

BOAC did not have long to wait to see the way the wind was blowing for in February 1972 the ATLB (still operating until the CAA took over in April 1972) granted applications from BCAL for scheduled services from Gatwick with optional stops at Birmingham, Manchester and Prestwick to New York, and from Gatwick with an optional stop at Chicago to Los Angeles.

The BAB began to impinge on the activities of the two Corporations from its offices in Air Terminal. On February 11 1972 the new Chairman, David Nicolson, issued a message to staff in both Corporations telling them he had formed a joint Managing Directors' Committee composed of Ross Stainton (BOAC), Kenneth Wilkinson (BEA), plus Derek Glover in his new position as Group Financial Director, together with Stephen Wheatcroft, newly appointed as Group Planning Director. The purpose of this Committee was to coordinate a series of studies by senior managers of BOAC/BEA into a range of airline activities such as aircraft procurement, commercial activities, cargo terminals, catering and motor transport, marketing charters and so on. All this was background work prior to the review of the Group's affairs called for by the 1971 Act. It was already becoming increasingly clear that total merger would be the outcome at the end of the day and this was formally recognised by the BOAC Board at its April 1972 meeting.

The Secretary of State for Trade and Industry was presented with the BAB's first report on its organisation on 7 April 1972, and it was placed in the House of Commons a week later. So BAB assumed responsibility for BOAC and BEA on 1 April 1972 including their subsidiaries, with BOAC-AC made into the Group Company for all Group investments in subsidiaries and associates, but with its name altered to British Airways Associated Companies Limited (BAAC) in due course. Appendix VII sets out the BAB initial Group arrangements which comprised a series of separate operating divisions reporting to the Group Managing Director. The constitution of the new Group Board provided for David Nicolson as full-time Chairman, Keith Granville as Deputy Chairman and Henry Marking as the new Group Managing Director. Full-time members of the Board with one exception came from BOAC and BEA and included Derek Glover, Ross Stainton, Kenneth Wilkinson plus Stephen Wheatcroft from outside. There were additionally at that time five non-executive Board members, Alan Fisher and Sir Arthur Norman (both formerly BOAC Board members), Peter Parker and Sir Kenneth Keith, together with Sir Ronald Edwards who was seeing some of his recommendations put into practice. So for the time being BOAC and BEA, designated as Divisions of the Group, kept their identities and separate Boards but the intention was to limit Board membership solely to full-time executives. In the light of past history it is instructive to note that in its first report on its organization the Group Board emphasized its obligations to conduct its affairs *'on strictly commercial lines'* and goes on to state *'the national interest is, of course, of overriding importance and may necessitate some adjustment to the commercial policies of the Board. The Board does not, however, set itself up as a judge of the national interest. There are others whose function that is — the competence of the Board lies elsewhere':* and so the oft-quoted letter of 1 January 1964 from Julian Amery to Sir Giles Guthrie was still influencing policy. Having achieved the first major step of setting up separate operating Divisions, the Group Board appointed a special committee made up once more of senior management from the two Corporations to consider the question of a common trading identity for the Group. At the close of its work this committee unanimously recommended the adoption of a common brand name, 'British Airways', for marketing purposes. It was then obvious to the Board that only by setting up a common trading identity with only one legal entity trading would the full benefits be obtained. In this way, a common flight designation could be used and a common ticket type adopted. The new brand name of British Airways was to be introduced immediately wherever that was practicable. BOAC estimated the cost merely of changing fascia and other identity signs to the new British Airways logo at £316,000 and as from 1 April 1974 the two Corporations would be dissolved and all staff then directly employed by the Group Board. All these proposals were included in the BAB's second report on the organisation and put to Michael Heseltine the new Minister for Aerospace at theDepartment of Trade and Industry, on 10 January 1973; he endorsed the entire report which was laid in the House of Commons on 22 January 1973. This led to the Air Corporations (Dissolution) Order 1973 which took effect on 1 April 1974 and under which *'all property, rights and liabilities of either of the Corporations shall by virtue of this Order become the property, rights and liabilities of the British Airways Board'.*

When the British Airways Board assumed control of BEA and BOAC the two airlines had a combined staff of 58,177, of which 27,680 were in the BEA group and the remainder in BOAC and including some 6,500 in non-airline activities. BOAC had just completed a challenging year in which an overall deficit of £1.4m had been shown on the profit and loss account, while BEA had shown a profit of £1.5m.

BOAC was offering some 2,499 million capacity ton-miles (CTMs) per year and BEA's output was 932 million CTMs. The contrast between BOAC's long-haul and BEA's short haul operations is made more evident if these CTM's are considered in relation to total passengers carried - over 10 million by BEA (on scheduled services only) and only 2 million by BOAC.

BOAC was operating a fleet of Boeing 747s, 707-320s, 707-420s and VC10/Super VC10s. BEA was operating a mixture of Trident Ones, Twos and Threes, BAC One-Elevens, Vickers Vanguard/Merchantmen and Vickers Viscounts, plus a variety of helicopters used by BEA Helicopters and Comets and Boeing 707-420s used by BEA Airtours. The last-mentioned type was, therefore, the only aircraft common to BEA and BOAC; not only were all the other aircraft types different but so also were their engines. At the time of the British Airways take-over, BEA had several subsidiary companies: BEA Helicopters, which was operating one scheduled service (between Penzance and the Isles of Scilly); BEA Airtours, specialising in non-scheduled operations - principally holiday charter flights - and British Air Services, which controlled the primarily regional operations of Cambrian Airways, Northeast Airlines and the BEA Scottish Airways and Channel Islands Airways Divisions. BOAC had no subsidiaries operating their aircraft, although its non-scheduled operations were handled, on paper at least, by British Overseas Air Charter Ltd; it operated its Treforest engine overhaul plant through BOAC Engine Overhaul Ltd and it had interests in a number of overseas airlines, hotels and other companies grouped in BOAC Associated Companies. BOAC and BEA jointly controlled International Aeradio and this also now came under the aegis of British Airways.

The conjoined airline had the most comprehensive route network in the world, extending half a million miles with two hundred destinations served in eighty-four countries - and with 220 aircraft would be the world's largest passenger fleet. It soon became evident that rationalising BOAC and BEA into British Airways would take longer than the Government expected.

An exasperating problem was the necessity of cautious co-operation with trade unions fearful of redundancies, yet it was essential to increase productivity closer to that of the USA. For the time being the individual names of BOAC and BEA would still be used, though the combined aircraft fleet would adopt a new-style livery and colour scheme.

The difference in the type of operations conducted by BOAC and BEA had resulted in a distinctive character developing in each airline between 1946 and 1972, and although any rivalry was no more than friendly, both the management and the staff of each airline had in general resisted earlier suggestions that a merger could be beneficial, many arguments being produced to refute the idea. Although the history of air transport has seen many, many mergers between airlines, these have almost always been between companies having something - equipment, market, operating style - in common or have been in effect the taking over of the weak by the strong. None of these conditions prevailed in the case of British Airways, whose board had the task of fusing two large and inherently strong airlines with no common equipment, virtually no overlapping of routes, different advance booking systems (both computer-based) and different staff outlook; almost the only thing shared by BOAC and BEA on 1 April 1972 seemed to be the English language!

The task confronting the BAB was, therefore, no easy one, but given the co-operation of staff at all levels in both airlines, it appears to have been proceeding with relatively little disruption. The first step was the setting up of a Group Planning Committee, and one of the earliest and most important decisions of this committee was to retain as much as possible of the original character of the constituent companies through the creation of mostly autonomous Divisions, operating as profit centres and within guidelines and targets set by the British Airways Group Board. A small top management team was set up at Group level, drawing upon personnel from both BOAC and BEA; this team has overall responsibilities in finance, planning, staff and public relations, all of which functions remain to be performed at Divisional level; only in management services have all the previously separate functions been brought together to provide a centralised service.

By September 1972, the BAB was ready to put its first organisational changes into effect, and seven Divisions were set up.

British Airways Overseas Division - the original BOAC scheduled operations, plus British Overseas Air Charter.

British Airways European Division - the original BEA scheduled operations, plus BEA Airtours (now known as British Airways Airtours) but excluding the regional services that BEA had grouped in British Air Services.

British Airways Regional Division—the former BEA operations handled by British Air Services that still included four separate companies within the division - Northeast, Cambrian, Scottish Airways and Channel Islands Airways.

British Airways Helicopters - the original BEA Helicopters, responsible for all helicopter operations, scheduled or charter.

British Airways Engine Overhaul - the original BOAC company at Treforest.

British Airways Associated Companies - controlling all investments in subsidiary and associated companies, with specified exceptions.

International Aeradio Ltd - the original company of the same name.

While these divisions were left to continue operating mostly undisturbed and under their original management the Group turned its urgent attention to the creation of a what was termed a 'common trading identity'. Up to this time, there had been no specific decision to adopt the title British Airways for trading purposes, although to many this seemed to be an inevitable consequence of implementing the merger. The BAB recommended that the step should be taken in a report to the Secretary of State in January 1973 and that approximately one year should be allowed to establish the new name and image before BOAC and BEA ceased to trade as such.

The change of image involved much more than just substituting the name British Airways for BOAC or BEA. A complete new livery had to be designed, and this not simply for the aircraft in the fleet, but for all the many items of ground equipment, for office fronts, stationery, cutlery and the many other items through which an airline presents its image to the public. Details of this new livery were published in July 1973. The colours, almost inevitably, were red, white and blue - white for the fuselage top and sides, to below the window line and blue for the underside. The top half of the fin was red, and the lower half incorporated a highly stylised representation of the Union Jack, continuing a theme started in the final BEA livery. Wing and tailplane surfaces, top and bottom, were finished in protective grey paint. The name British airways (with a lower case 'a') appeared on each side of the fuselage, near the cockpit, in blue, together with the Speedbird symbol originated by Imperial Airways and widely identified with BOAC.

Though now effectively a single unit, the past year's financial reports for BEA and BOAC were issued individually in September for the last time. BEA had made an operating profit of £114 million which after paying interest became a net profit of £477,000 - but that excluded the now independent Scottish Airways and Channel Islands Airways whose accounts turned the profit into a net loss of £12 million. Over 180,000 flights had been made, carrying 10·5 million passengers, and a further 12,500 flights carried 140,000 tons of freight.

By comparison, BOAC had an operating surplus of £227 million compared with its disconcerting low of £2 million in the previous year, so there was now a net profit of £82 million. Entry into the low-fare market resulted in 2.8 million passengers and 88,000 tons freight and mail carried in the course of over 3,000 million miles. Nevertheless, BEA and BOAC were running at far under capacity, and so was every other airline. On the North Atlantic alone there were 46 competing carriers, and the result was a confusion of surplus seats and lost revenue.

And so BOAC and BEA lost their identities and disappeared for good just five years after Edwards, although both Edwards and the then Labour Government had been against this happening. It was ironic that a Report instituted to provide a solution to the troubles of the private sector should in the final process bring about the dissolution of both the Corporations. The long BOAC story had come to an end, but although the two years before this great change were difficult for BOAC staff and management with a cloud of uncertainty hanging over all their heads, yet the Corporation finished its life on a high note of success.

Appendices

Appendix 1 - BOAC Chairmen

Sir John Reith	24 November 1939 to 5 March 1940	Resigned before formal creation of BOAC
Hon. Clive Pearson	6 March 1940 to 24 March 1943	Resigned
Sir Harold Howitt	24 March 1943 to 25 May 1943	Temporary Appointment
Viscount Knollys	26 May 1943 to 30 June 1947	Retired
Sir Harold Hartley	1 July 1947 to 30 June 1949	Retired
Sir Miles Thomas	1 July 1949 to 30 April 1956	Resigned
Gerard d'Erlanger *	1 May 1956 to 28 July 1960	Resigned
Sir Matthew Slattery	29 July 1960 to 31 December 1963	Resigned
Sir Giles Guthrie	1 January 1964 to 31 December 1968	Retired
Charles Hardie*	1 January 1969 to 31 December 1970	Retired
Keith Granville*	1 January 1971 to 31 August 1972	To British Airways Board
Ross Stainton (later Sir Ross)	1 September 1972 to 31 March 1974	Dissolution of BOAC Board

* = knighted while in service

Appendix 2 - List of BOAC Board Members

Name	Appointment from	Member until
Sir John Reith	24 November 1939	29 February 1940
Clive Pearson	24 November 1939	24 March 1943
Leslie Runciman (later Lord Runciman)	24 November 1939	24 March 1943
Harold George Brown	24 November 1939	24 March 1943
Gerard d'Erlanger	6 March 1940	31 July 1946 (to BEA)
Irving Geddes	6 March 1940	24 March 1943
Sir Harold Howitt	24 March 1943	31 March 1948
Simon Marks (later Sir Simon)	24 March 1943	22 January 1946
John Marchbank	24 March 1943	23 March 1946 (died)
Viscount Knollys	26 May 1943	30 June 1947
B'Gen A Critchley	26 May 1943	22 January 1946
Miss Pauline Gower (Mrs Fahie)	26 May 1943	22 January 1946
Sir Harold Hartley	1 January 1946	31 July 1946 (to BEA)
Lord Burghley (later Marquess of Exeter)	10 January 1946	31 March 1946
G M Garro-Jones (later Lord Trefgarne)	10 January 1946	16 December 1947
Major R. H. Thornton	10 January 1946	31 March 1955
Major J. R. McCrindle	10 January 1946	30 June 1958
Sir Clement Jones	22 January 1946	31 March 1954
Lord Rothschild	24 April 1946	30 September 1960
H. L. Newlands	18 May 1946	31 March 1960
Sir Harold Hartley	1 July 1947	30 June 1949
Whitney Straight	1 July 1949	31 October 1955
Sir Miles Thomas (later Lord Thomas of Remenham)	1 April 1948	30 April 1956
Lord Douglas of Kirtleside	1 October 1948	14 February 1949 (to BEA)
John Booth	21 March 1949	30 April 1965
Sir Francis Brake	1 January 1950	31 December 1958
Sir John Stephenson	1 January 1950	31 December 1959
Basil Smallpeice (KCMG when in office)	30 September 1953	31 December 1963
Lord Rennell of Rodd	1 April 1954	20 November 1964
Lord Tweedsmuir	6 June 1955	6 June 1964
Gerald d'Erlanger (Knighted when Chairman)	1 July 1956	30 July 1960
Sir George Cribbett	1 July 1956	30 April 1960
Sir Wilfred Neden	11 December 1958	31 December 1963
Francis Taylor (later Sir Francis)	11 December 1958	31 October 1960
Keith Granville (Knighted when Chairman)	1 January 1959	31 August 1972 (to BAB)
John A. Connell	20 June 1960	23 July 1963 (died)
Lionel Poole	20 June 1960	19 June 1964
Sir Walter Warboys	20 June 1960	31 March 1964

Sir Matthew Slattery	29 August 1960	31 December 1963
Kenneth Staple	18 January 1961	7 January 1964
Gilbert Lee	6 November 1961	31 March 1974
Sir Giles Guthrie	1 January 1964	31 December 1968
A. H. Milward (later Sir Anthony)	1 January 1964	31 December 1970
Ron Smith	1 January 1964	1 January 1970
Charles Hardie (Knighted when Chairman)	18 January 1964	1 January 1970
Sir Duncan Anderson	1 April 1964	31 December 1970
Sir Arthur Norman	20 June 1964	21 December 1971 (to BAB)
Lord Normanbrook	6 June 1964	14 June 1967 (died)
Beverley Shenstone	1 December 1964	31 December 1966
G. R. Chetwynd	1 January 1966	31 December 1973
Ross Stainton (later Sir Ross)	16 September 1968	31 March 1974
Sir Richard Way	1 November 1967	31 December 1973
Derek Glover	6 February 1969	21 December 1971 (to BAB)
Alan Fisher	2 January 1970	10 February 1972 (to BAB)
Henry Marking (later Sir Henry)	1 January 1971	1 September 1972 (to BAB)
Winston Bray	22 January 1971	31 December 1973
Richard Hilary	21 December 1971	31 March 1974
Frank Walton	21 December 1971	31 March 1974
Charles Abell	1 April 1972	31 March 1974
Peter Hermon	11 April 1972	1 September 1972 (to BAB)
Kenneth Reddish	1 September 1972	31 March 1974
Robert Forrest	1 September 1972	31 July 1973 (to BAB)
KennethWilkinson	1 September 1972	30 November 1972 (to BAB)27
Philip Lawton	December 1972	31 December 1973
Roy Watts	1 January 1974	31 March 1974
Basil Bampfylde	1 January 1974	31 March 1974

Appndix 3 - British Prime Ministers during the time of BOAC's existance

Neville Chamberlain (Conservative)	28 May 1937 to 11 May 1940
Winston Churchill (National)	11 May 1940 to 23 May 1945
Winston Churchill (Conservative)	23 May 1945 to 26 July 1945
Clement Attlee (Labour)	26 July 1945 to 26 October 1951
Winston Churchill (Conservative)	26 October 1951 to 7 April 1955
Anthony Eden (Conservative)	7 April 1955 to 13 January 1957
Harold Macmillan (Conservative)	13 January 1957 to 20 October 1963
Alexander Douglas-Home (Conservative)	20 October 1963 to 16 October 1964
Harold Wilson (Labour)	16 October 1964 to 19 June 1970
Edward Heath (Conservative)	19 June 1970 to 3 March 1974
Harold Wilson (Labour)	4 March 1974 -

Appendix 4 - Ministers and their Deputies responsible for Civil Aviation.

Secretary of State for Air - 1939 to September 1944   *Parlimentary Secretary*

Until May 1940	Sir Kingsley Wood	Harold Balfour
May 1940	Sir Archibald Sinclair	Harold Balfour

Minister of Civil Aviation - September 1944 to September 1951

September 1944	Viscount Swinton	
July 1945	Lord Winster	George Lindgren
October 1946	Lord Nathan	George Lindgren
June 1948	Lord Pakenham	George Lindgren
		George Beswick (1950)
May 1951	Lord Ogmore	George Beswick

Minister of Transport and Civil Aviation - October 1951 to 1959

October 1951	John Maclay	Reginald Maudling
May 1952	Alan T Lennox-Boyd	Reginald Maudling
		John Profumo (1953)
October 1954	John A Boyd-Carpenter	John Profumo
December 1955	Harold A Watkinson	John Profumo
		Airey Neave (1956)

Minister of Aviation October - 1959 to July 1966

October 1959	Duncan Sandys	Geoffrey Rippon
July 1960	Peter Thorneycroft	J Woodhouse
July 1962	Julian Amery	Basil de Ferrari
		Neil Marten (1963)
October 1964	Roy Jenkins	John Stonehouse
January 1966	Fred Mulley	John Stonehouse

President of the Board of Trade - July 1966 to October 1970

July 1966	Douglas Jay	J Mallalieu
August 1967	Anthony Crossland	Edmund Dell
1969	Roy Mason	Goronwy Roberts
June 1970	John Davies	Michael Noble

Department of Trade and Industry - October 1970 onwards

October 1970	John Davies	Michael Noble (Trade)
		Fred Corfield (Aerospace)
April 1972	Peter Walker	Michael Heseltine
		(Aerospace & Shipping)
March 6 1974	Peter Shore	S C Davis

Appendix 5 - Fleet Names

In the early days, the aircraft of BOAC followed the accepted practice already used by Imperial Airways and were divided into 'Fleets', known as Classes, and many were given an individual name that utilised the first letter of the class name. This tradition continued after the war with other aircraft types up to, but not including the Britannia and Comet, and many of the old Imperial Airways names were re-cycled by BOAC. Some aircraft fleets marked * had some aircraft not transferred from Imperial to BOAC.

A Class - Armstrong Whitworth Atalanta*
Atalanta G-ABTI   *Arthemis* G-ABTJ   *Astraea* G-ABTL

A Class - Canadair C-4 (DC-4M)
Ariadne G-ALHC   *Ajax* G-ALHD   *Argo* G-ALHE   *Atlax* G-ALHF   *Aurora* G-ALHG
Attica G-ALHH   *Antares* G-ALHI   *Arcturus* G-ALHJ   *Atalanta* G-ALHK   *Altair* G-ALHM
Argosy G-ALHN   *Amazon* G-ALHO   *Aethra* G-ALHP   *Antilope* G-ALHR   *Astra* G-ALHS
Athena G-ALHT   *Artemis* G-ALHU   *Adonis* G-ALHV   *Aeolus* G-ALHW   *Astraea* G-ALHX
Arion G-ALHY

B Class - Boeing 314
Bristol G-AGBZ   *Berwick* G-AGCA   *Bangor* G-AGCB

B Class - Lockheed Constellation
Bristol II G-AHEJ   *Berwick II* G-AHEK   *Bangor II* G-AHEL   *Balmoral II* G-AHEM   *Baltimore* G-AHEN
Bedford G-AKCE   *Boston* G-AMUP   *Barnstable* G-AMUR   *Brentford* G-ALAK   *Banbury* G-ALAL
Belfast G-ALAM   *Beaufort* G-ALAN   *Braemar* G-ALAO   *Buckingham* G-ANNT   *Berkeley* G-ANTF
Bournemouth G-ANTG   *Branksome* G-ANUP   *Basildon* G-ANUR   *Blantyre* G-ANUV   *Bala* G-ANUX
Belvedere G-ANUZ   *Blakeney* G-ANVA   *Blackrod* G-ANVB   *Beverley* G-ANVD

C-Class - Short S.23*
Canopus G-ADHL   *Caledonia* G-ADHM   *Centaurus* G-ADUT   *Cavalier* G-ADUU   *Cambria* G-ADUV
Castor G-ADUW   *Cassiopea* G-ADUX   *Capella* G-ADUY   *Cygnus* G-ADUZ   *Capricornus* G-ADVA
Corsair G-ADVB   *Courtier* G-ADVC   *Challenger* G-ADVD   *Centurion* G-ADVE   *Coriolanus* G-AETV
Calpurnia G-AETW   *Ceres* G-AETX   *Clio* G-AETY   *Circe* G-AETZ   *Calypso* G-AEUA
Camilla G-AEUB   *Corinna* G-AEUC   *Cordelia* G-AEUD   *Cameronian* G-AEUE   *Corinthian* G-AEUF
Coogee G-AEUG   *Corio* G-AEUH   *Coorong* G-AEUI   *Carpentaria* G-AFBJ   *Coolangatta* G-AFBK
Cooee G-AFBL

C-Class - Short S.30*
Champion G-AFCT   *Cabot* G-AFCU   *Caribou* G-AFCV   *Connemara* G-AFCW   *Clyde* G-AFCX
Clare G-AFCZ   *Cathay* G-AFKZ

C Class - S.33*
Clifton G-AFPZ   *Cleopatra* G-AFRA

C Class - Boeing 377 Stratocruiser
Caledonia G-AKGH   *Caribou* G-AKGI   *Cambria* G-AKGJ   *Canopus* G-AKGK   *Cabot* G-AKGL
Castor G-AKGM   *Cathay* G-ALSA   *Champion* G-ALSB   *Centaurus* G-ALSC   *Cassiopea* G-ALSD
Cleopatra G-ANTX   *Coriolanus* G-ANTY   *Cordelia* G-ANTZ   *Cameronian* G-ANUA   *Calypso* G-ANUB
Clio G-ANUC   *Clyde* G-ANUM

D Class - DH.86 Express Air Liner
Delphinius G-ACPL   *Delia* G-ACWC   *Dorado* G-ACWD   *Draco* G-ADCM   *Daedalus* G-ADCN
Dione G-ADFF   *Dardanus* G-ADUE   *Dido* G-ADUF   *Danae* G-ADUG   *Dryad* G-ADUH
Denebola G-ADUI   *Demeter* G-AEAP

E Class - Armstrong Whitworth AW.27 Ensign
Ensign G-ADSR   *Egeria* G-ADSS   *Elsinore* G-ADST   *Euterpe* G-ADSU   *Explorer* G-ADSV
Eddystone G-ADSW   *Ettrick* G-ADSX   *Empyrean* G-ADSY   *Elysian* G-ADSZ   *Euryalus* G-ADTA
Echo G-ADTB   *Endymion* G-ADTC

F Class - DH.91 Albatross
Faraday G-AEVV   *Franklin* G-AEVW   *Frobisher* G-AFDI   *Falcon* G-AFDJ   *Fortuna* G-AFDK
Fingal G-AFDL   *Fiona* G-AFDM

G Class - Short S.26
Golden Hind G-AFCI   *Golden Fleece* G-AFCJ   *Golden Horn* G-AFCK

H Class - Handley Page HP.42
Horsa G-AAUC   *Hanno* G-AAUD   *Hadrian* G-AAUE   *Heracles* G-AAXC   *Horatius* G-AAXD
Helena G-AAXF

H Class - Short S.25 Sunderland 3
Hadfield G-AGER   unnamed G-AGES   unnamed G-AGET   *Hampshire* G-AGEU   *Hailsham* G-AGEV
Hanwell G-AGEW   *Hamble* G-AGHV   *Hamilton* G-AGHW   *Hawkesbury* G-AGHZ   *Haslemere* G-AGIA
unnamed G-AGIB   *Henley* G-AGJJ   *Howard* G-AGJK   *Hobart* G-AGJL   *Hythe* G-AGJM
Hudson G-AGJN   *Honduras* G-AGJO   *Huntington* G-AGKV   *Hostpur* G-AGKW   *Himalaya* G-AGKK
Hungerford G-AGKY   *Harwich* G-AGKZ   *Hunter* G-AGLA

J Class - Junkers Ju.52
Juno G-AERU   *Jupiter* G-AERX   *Jason* G-AFAP

K Class - DH.95 Flamingo
King Arthur G-AFYE   *King Alfred* G-AFYF   *King Harold* G-AFYG   *King Henry* G-AFYI   *King Richard* G-AFYJ
King James G-AFYK   *King Charles* G-AFYL   *King William* G-AGBY

L Class - Lockheed 10 Electra, Lockheed 12 , Lockheed 14 and Lockheed 18 Lodestar
Leith G-AEPR   *Lea* G-AFCS   *Livingstone* G-AFGP   *Lafayette* G-AFGR
Loch Invar G-AFKD   *Lothair* G-AFKE   *Leander* G-AFMR   *Loch Lomond* G-AGDC
Loch Leven G-AGDF   *Loch Lyon* G-AGDK   *Loch Loyal* AE581   *Lanark* G-AGBO
Leicester G-AGBP   *Lewes* G-AGBP   *Lake George* G-AGBP   *Lichfield* G-AGBS
Lincoln G-AGBT   *Lake Victoria* G-AGBT   *Lowestoft* G-AGBU   *Ludlow* G-AGBV
Lake Albert G-AGBV   *Lyndhurst* G-AGBW   *Llandaff* G-AGBX   *Lake Edward* G-AGBX
Lake Baringo G-AGEH   *Lake Karoun* G-AGIG   *Lake Nyasa* G-AGIL   *Lake Tanganykia* G-AGIM
Lake Timsah G-AGIN   *Lake Tana* G-AGJH

M Class - Avro 685 York
Mildenhall G-AGJA   *Marathon* G-AGJB   *Malmesbury* G-AGJC   *Mansfield* G-AGJD   *Middlesex* G-AGJE
Mersey G-AGNL   *Murcheson* G-AGNM   *Madras* G-AGNN   *Manton* G-AGNO   *Manchester* G-AGNP
Moira G-AGNR   *Melville* G-AGNS   *Mandalay* G-AGNT   *Montgomery* G-AGNU   *Morville* G-AGNV
Morecambe G-AGNW   *Moray* G-AGNX   *Melrose* G-AGNY   *Monmouth* G-AGNZ   *Montrose* G-AGOA
Milford G-AGOB   *Malta* G-AGOC   *Midlothian* G-AGOD   *Medway* G-AGOE   *Macduff* G-AGOF
Morley G-AGSL   *Malvern* G-AGSM   *Marlow* G-AGSN   *Marston* G-AGSO   *Marlborough* G-AGSP

N Class - Avro 691 Lancastrian 1 & 2
Nelson G-AGLS   *Newcastle* G-AGLT   *Northampton* G-AGLW   *Norfolk* G-ALGY   *Nottingham* G-ALGZ
Newport G-AGMA   *Norwich* G-AGMB   *Nairn* G-AGMD   *Newhaven* G-AGME   *Nicocia* G-AGMG
Naseby G-AGMJ   *Newbury* G-AGMK   *Nicobar* G-AGML   *Nepal* G-AGMM   *Natal* G-AKPY
Nile G-AKPZ   *Nyanza* G-AKRB

P Class - Short S.25 Sandringham 5
Portsmouth G-AHYY *unnamed* G-AHYZ *Penzance* G-AHZA *Portland* G-AHZB *Pembroke* G-AHZC
Portmarnock G-AHZD *Portsea* G-AHZE *Poole* G-AHZF *Pevensey* G-AHZG *Perth* G-AJMZ

S Class - Short S.25 Sandringham 7
St George G-AKCO *St David* G-AKCP *St Andrew* G-AKCR

S Class - Short S.45 Solent 2
Salisbury G-AHIL *Scarborough* G-AHIM *Southampton* G-AHIN *Somerset* G-AHIO *Sark* G-AHIR
Scapa G-AHIS *Severn* G-AHIT *Solway* G-AHIU *Salcombe* G-AHIV *Stornoway* G-AHIW
Sussex G-AHIX *Southsea* G-AHIY

City Class - Short S.45 Solent2/3
City of London G-AKNO *City of Cardiff* G-AKNP *City of Belfast* G-AKNR *City of Liverpool* G-AKNS
City of Edinburgh G-AHIX

Appendix 6 - Abbreviations Used

ABAMEL	Associated British Airlines (Middle East) Ltd
ACC	Airlines' Chairman's Committee.
ADG	Assistant Director-General
AIOC	Anglo-Iranian Oil Company.
ANA	Australian National Airways
AOA	American Overseas Airlines
AOC-in-C	Air Officer Comander-in-Chief
ARB	Air Registration Board
ATA	Air Transport Auxiliary
ATAC	Air Transport Advisory Council
ATC	Air Training Corps
ATC	Air Traffic Control
ATLB	Air Transport Licensing Board
BAAC	British Airways Associated Companies Ltd
BAB	British Airways Board
BAL	Bahamas Airways Ltd
BALPA	British Airline Pilots Association
BAS	British Air Services
BCL	BOAC-Cunard Limited
BCPA	British Commonwealth Pacific Airlines
BEA	British European Airways
BIAL	British International Airlines Ltd
BOAC-AC	BOAC Associated Companies Ltd
BSAA	British South American Airways
BWIA	British West Indian Airways
CAA	Central African Airways
CAB	Civil Aeronautics Board
CPA	Canadian Pacific Airways
CPA	Cathay Pacific Airways
DDG	Deputy Director General.
DG	Director General
DGCA	Director General Civil Aviation
DoT	Department of Trade
EAAC	East African Airways Corporation.
EAMS	Empire Air Mail Scheme
ECAC	European Civil Aviation Conference
FAA	Federal Aviation Administration
HMG	His Majesties Government
IAL	Imperial Airways Ltd
IASTA	International Air Services Transit Agreement
IATA	International Air Transport Association
ICAO	International Civil Aviation Organization
IFALPA	International Federation of Airline Pilots Associations
KLM	Koninklijke Luchtvaart Maatschappij
MAP	Ministry of Aircraft Production
MASCO	Mideast Airways Service Company
MCA	Ministry of Civil Aviation
MEA	Middle East Airlines
MoS	Ministry of Supply
NAC	National Air Communication
PAA	Pan American Airways (Pan Am)
PERA	Propeller and Engine Repair Auxiliary
QANTAS	Queensland and Northern Territory Aerial Services
RAE	Royal Aircraft Establishment
SAA	South African Airways
SABENA	Societé Anonyme Belge d'Exploitation de la Navigation Aérienne
SAS	Scandanavian Airline System
SBAC	Society of British Aircraft Constructors
SILA	Svensk Interkontinental Lufttrafic
TAA	Trans Australian Airlines
TCA	Trans Canada Airlines
TEAL	Tasman Empire Airways
THY	Turk Hava Yollari
TWA	Trans World Airlines
WAAC	West African Airways Corporation

Bibliography

Official Publications
Hansard: House of Lords

Volume	Columns	Date
130	437-70	19.1.42
136	16-7	24.4.45
137	651-737	6.11.45
157	1101-71	21.7.48
167	373-81	15.3.49
164	274-301	20.7.50
168	313-334	13.7.50
174	820	5.12.51
195	721-6	1.2.56
197	454-99	16.12.56

Hansard: House of Commons

Volume	Columns	Date
378	746-69	4.3.42
378	1485	18.3.42
383	803-4	30.9.42
385	701-2	25.11.42
385	1304-5	3.12.42
385	218-56	17.12.42
388	155-6	31.3.43
388	1346	14.4.43
388	1513-4	20.4.43
390	100	1.6.43
397	1321-82	22.2.44
398	161-2	14.3.44
424	1326-1447	26.6.46
425	41-194	8.7.46
425	593-713	11.7.46
430	813-24	19.11.46
439	1344	2.7.47
439	2183	9.7.47
444	358	12.11.47
446	25	21.1.48
447	2139/51/53	26.2.48
451	995	2.6.48
466	179	22.6.49
466	1260	29.6.49
467	1103-20	18.7.49
473	1405-15	6.4.50
477	728-88	6.7.50
489	54	20.6.51
501	1152-57	27.5.52
503	2180/84/85	16.7.52
523	1147-56	10.2.54
524	1741-8	8.3.54
524	1769-70	8.3.54
527	367	5.5.54
547	1927	20.12.55
546	1921-8	20.11.55
552	552	9.5.56
563	174-5	30.1.57
568	172-94	1.5.57
570	82	22.5.57
571	397-400	29.5.57
572	38-40	26.6.57
574	46-8	24.7.57
574	110-156	24.5.57
574	194	24.7.57
579	869-73	9.12.57
588	226-232	13.5.58
617	957-64	15.2.60
626	956-7	11.7.60
663	548-50	18.7.62
685	786-913	2.12.63
702	420-1	18.11.64
703	32	10.12.64
707	34	25.2.65
797	911-19	9.3.70
798	417-530	18.3.70
808	1254-75	15.12.70
809	195	18.1.71
814	55	29.3.71
814	1172-1182	29.3.71
817	20-1	10.5.71
834	839	10.4.72

BOAC Annual Reports 1940/41 to 1973/74
British Airways Annual Reports 1972/73 to 1980/81
7th Report of the Select Committee on Estimates
Reports of the ATAC and ATLB
Report from the Select Committee on Nationalised Industries - The Air Corporations - May 1959
Select Committee on Nationalised Industries- Outcome of Recommendations and Conclusions - February 1962
Report from the Select Committee on Nationalised Industries - BOAC Volumes I and II - June 1964
Report from Select Committee on Nationalised Industries- British Airways. The merger of BEA and BOAC- December 1975
CAA Report on the Committee on Flight Time Limitations - June 1973
BAB - First Report on Organisation - July 1972
    Second Report on Organisation - January 1973
    Third Report on Organisation - November 1973
    Fourth Report on Organisation - November 1976

Legislation
British Overseas Airways Act 1939
Ministry of Civil Aviation Act 1945
Civil Aviation Act 1946
Civil Aviation Act 1949
Air Corporations Act 1949
Air Corporations Act 1953
Civil Aviation (Licensing) Act 1960
Air Corporations Act 1966
Air Corporations Act 1967
Civil Aviation Act 1968
Civil Aviation Act 1971

Command Papers

5685-1938	The Cadman Report.
6561-1944	International Air Transport.
6605-1945	British Air Transport.
6614-1945	Interim Agreement on International Civil Aviation Chicago, 7 December 1944.
6712-1945	British Air Services.
6747-1946	The Bermuda Agreement.
7307 & 7478	Interim and final Reports of the Committee of Inquiry into Tudor Aircraft
105-1957	Report of a Court of Inquiry into the Causes and Circumstances of a Dispute between BOAC and MNAOA.
608-1958	Report of a Court of Inquiry into a Dispute at London Airport - The Jack Report.
1337-1962	The Financial and Economic Obligations of Nationalised Industries.
1916-1962	Agreement between the Government of the UK and the Government of the French Republic regarding the Development of a civil Supersonic Transport Aircraft.
2853-1965	Report of the Committee of Inquiry into the Aircraft Industry - the Plowden Report.
3428-1967	Report of a Court of Inquiry into the Dispute between BALPA and the NIC - the Scamp Report.
3551-1968	Report of a Court of Inquiry into the Dispute between BOAC and BALPA - the Pearson Report.
3789-1969	Prices and Incomes Board Inquiry into Pay of Pilots employed by BOAC.
4197-1969	Prices and Incomes Board Inquiry into the Pay of Pilots employed by BOAC
4213-1969	Civil Aviation Policy.
4018-1969	British Air Transport in the Seventies - Report of the Committee of Inquiry into Civil Air Transport- the Edwards Report.
4449-1970	Final Report of the Committee of Inquiry into the Disruption of Operations and Industrial Relations at Heathrow Airport
4899-1972	Civil Aviation Policy Guidance
5316-1973	Exchange of Notes US/USA Concerning Procedures for Implementation and Enforcement of Rules for Advance Booking Charters
6400-1976	Future Civil Aviation Policy

Books

707 Group Graham M Simons: Pen & Sword Aviation 2017
Air Transport Policy Stephen Wheatcroft: Joseph 1964
Beaverbrook A. P. Taylor: Hamish Hamilton 1972
Boeing 747 Martin W Bowman Pen & Sword Aviation 2014
Bristol Aircraft since 1910 C. H. Barnes: Putnam 1970
Bristol Brabazon Graham M Simons: History Press 2012
Britain's Imperial Air Routes 1918-1939 Robin Higham: Shoe String Press 1960
British Aircraft Industry - What went Right and What went Wrong Arthur Reed - Dent 1973
British Civil Aircraft 1919-1972 Volume 1 A J Jackson: Putnam 1973
British Civil Aircraft 1919-1972 Volume 2 A J Jackson: Putnam 1973
British Civil Aircraft 1919-1972 Volume 3 A J Jackson: Putnam 1973.
Carrying British Mails Overseas H. Robinson: Allen & Unwin 1964.
Comet! The World's First Jet Airliner. Graham M Simons: Pen & Sword Aviation 2013.
Concorde Conspiracy Graham M Simons: History Press 2012.
Concorde, The Inside Story Geoffrey Knight: Weidenfeld & Nicolson 1976.
De Havilland Enterprise Graham M Simons: Pen & Sword Aviation 2017.
Diamonds in the Sky K. Hudson and J Pettifer: Bodley Head 1979.
Famous Airliners - 2 Bristol Britannia Derek Harvey: Cassell & Co 1958.
Famous Airliners - 3 The Seven Seas. Derek Harvey: Cassell & Co 1959.
Famous Airliners - 4 The Comet Derek Harvey: Cassell & Co 1959.
Fly Me, I'm Freddie! Roger Eglin and Berry Ritchie; Weidenfeld and Nicholson 1980.
High Flight The politics of the air. Adam Thomson: Sidgewick & Jackson 1990.
History of the World's Airlines R E G Davies: Oxford 1964.
Into the Wind Lord Reith 1949.
Merchant Airmen John Pudney: HMSO 1946.
Ministry of Transport and Civil Aviation Sir Gilmour Jenkins: Allen & Unwin 1959.
Mosquito The original Multi-Role Combat Aircraft Graham M Simons Pen & Sword Aviation 2011.
Nationalisation in Practice John Longhurst: Temple Press 1950.
New Air Transport Policy for the North Atlantic J. Friedman: Athenaeum 1976.
Of Comets and Queens Sir Basil Smallpeice: Airlife 1981.

FUTURE

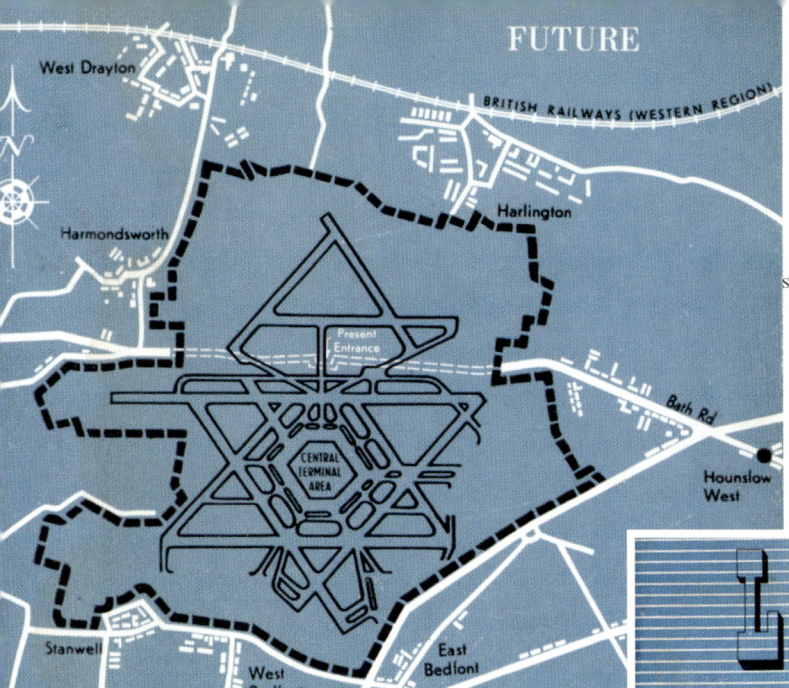

313

What might have been - and what could still be! A guide to London Airport as prepared by the Ministry of Civil Aviation and its plans for the eventual size and shape of BOAC's base. It could still be that of BOACs successor British Airways. The date of the future plan? 1948!

Out on a Wing Sir Miles Thomas: Michael Joseph 1964.
Pictorial History of BOAC and Imperial Airways K. Munson: Ian Allan 1970.
Railway Air Services John Stroud; Ian Allan Ltd 1987.
Reith Diaries Ed. Charles Stuart: Collins 1975.
Short's Aircraft since 1900 C. H. Barnes: Putnam 1967.
The Airline Encyclopedia 1909-2000 Volume 1 Myron J Smith Jr.: Scarecrow Press 2002.
The Airline Encyclopedia 1909-2000 Volume 2 Myron J Smith Jr.: Scarecrow Press 2002.
The Airline Encyclopedia 1909-2000 Volume 3 Myron J Smith Jr.: Scarecrow Press 2002.
The Concorde Fiasco Andrew Wilson: Penguin 1973
The Seven Skies John Pudney: Putnam 1959
The Speedbird Book Barbara Cooper: Golden Pleasure Books, 1962
The Waterjump David Beaty: Secker & Warburg 1976
Who's Who in British Aviation 1928: Airways Publications Ltd.
Who's Who in British Aviation 1933: Airways Publications Ltd.
Who's Who in British Aviation 1935: Bunhill Publications 1936.
Who's Who in British Aviation 1955: Temple Press 1956.
Who's Who in British Aviation 1963: Temple Press 1964.
Who's Who in British Aviation 1965: Temple Press 1966.
Wings Across the World Harald Penrose: Cassell 1980.

Miscellaneous Publications, Journals and Records
Air Transport World
Airline in Action - BOAC account - 1945
Annual Reports by IATA on the State of the Industry
Atlantic Bridge - Ministry of Information 1945
BOAC Board Papers 1940/1974
BOAC Contemporary files
BOAC Handbooks Vols. I and II 1945/1946
BOAC News 1959-1973
BOAC Newsletter 1940-1946
BOAC Review 1951-1967
BOAC Staff News Letter 1946-1950
Corbett Report, May 1963
De Havilland Gazette - Staff News 1945-1962
Flight International and the *Aeroplane* Press Cuttings
Gateway to the World Maurice Housego: PRO London Airport
Highways in the Air — British Airways 1979
Imperial Airways Staff News 1931-1939
Imperial Airways Weekly News Bulletin 1931-1939
Institute of Transport papers
London Airport - MCA 1948
London Airport - PRO London Airport: PRO London Airport
London's Airports in Colour.
Northolt Airport - MCA 1948
Speedbird 1946

Index

A

A-Class: *see* Armstrong Whitworth AW.15 Atalanta
Abell, Charles: 79, 133, 155, 192, 226, 239, 257, 268, 269, 298, 299, 300, 301, 303
Adams, Clive: 28
Aden Airways: 168, 170, 172, 173, 178, 189
Aer Lingus: 58, 94, 148, 168
Aga Khan, The: 41
Air Canada: 270, 295
Air France: 147, 263, 273, 293, 294, 295, 298, 299, 302
Air Freight Ltd: 206
Air India: 295
Air Jamaica: 178, 180
Air Mauritius: 179
Air Pacific: 180
Air Registration Board (ARB): 108, 193, 196
Air Transport Advisory Council (ATAC): 61, 104, 138, 187, 188, 190
Air Transport Licensing Board (ATLB): 207, 215, 220, 244, 283, 290, 291, 303
Airwork Ltd: 188, 189, 190, 286, 287, 289
Alabaster, Captain Robert Charles: 119
Alamuddin, Sheik Najib: 172
Alderson, Captain M J R 'Roly': 113, 115, 125
Alitalia: 147, 180
Allen, William 'Bill': 280
Allied British Airways: 9
American Airlines: 194
American Export Airlines: 28, 37, 41, 44
American Overseas Airlines: 71, 86, 87
American Overseas Airlines: 71, 89
Amery, Harold Julian: 5, 226, 240, 242, 245, 246, 261, 262, 300, 304
Anderson, Sir Duncan: 262
Andrew, Captain Jimmy: 298
Anglo-Iranian Oil Company: 91
Appleby, Sir Humphrey: 252
Arab Airways: 170
Argonaut airliner: *see* Canadair C.4 Argonaut (DC-4M)
Argüello, Patrick: 237
Arterton, First Officer B A: 119
Associated British Airlines (Middle East) Ltd: 166, 173
Atalanta, Armstrong Whitworth: 16, 191
Attlee, Clement, British Prime Minister: 100
Australian National Airways: 150, 151
Aviation Traders: 49
Avro Lancaster: 37, 38, 40, 43, 62, 70, 86, 90, 93, 94, 102
Avro Lancastrian: 42-49, 51, 57, 63, 72, 85, 86, 90, 93, 94, 102
Avro Tudor: 35, 37, 62, 63, 64, 65, 66, 74, 82, 83, 84, 85, 86, 89, 93
Avro York: 29, 33, 35, 37, 39, 42, 43, 50, 51, 53, 62, 63, 69, 72, 85, 87-93, 102, 172, 173, 206

B

B-Class: *see* Boeing 314
Bahamas Airways Ltd: 166, 175, 178, 179, 180, 209
Bahrain Gulf Aviation: 172
Bailey, Captain: 263
Bailey, E C: 119
Baldwin, George: 71
Balfour, Capt. Harold H; Lord Balfour Inchrye: 16-20, 26, 30, 45, 46, 54
Ballard, Florence: 210
Bamberg, Harold: 206, 208, 209, 288
Bampfylde, Basil: 239
Barker. Viva: 41
Barnby, James W 'Jim': 189
Battye, Eng Off G A: 68
BEA Airtours: 212, 304
BEA Helicopters: 304
BEA Scottish Airways: 304
Beatles, The: 210
Beatty, Sir Edward: 21
Beaverbrook, Lord; William Maxwell Aitken: 19, 20, 47, 51, 52, 53
Beharrell, Sir George: 13, 14
Bembridge, John: 179
Bennett, Air Vice-Marshal Donald 'Don': 21, 69
Bennett, Engineer Officer W I: 118
Bergin, Dr Ken: 239, 268
Bergman, Ingrid: 31
Berle, Adolf A: 45
Bertlin, P D: 118
Bevan, Ken: 239, 257
Bird, Kenneth: 10
Bishop, Ronald: 120
BKS: 189
Blackett, E: 114
Blue Star Shipping Line: 51, 288
BOAC Associated Companies Ltd (BOAC-AC): 165, 166, 169, 171, 173-180, 242, 244, 268, 303
BOAC Engine Overhaul Ltd: 180
BOAC Restaurants Ltd: 180
BOAC-Cunard Ltd: 179
BOAC-Cunard: 208, 209, 210
Board of Trade (BoT): 47, 49, 212, 219, 259, 260, 269, 276, 278, 283, 285, 288, 289, 290, 291, 296, 297
Boeing 314: 19-21, 25, 27, 28, 35, 37, 40, 44, 63, 64
Boeing 707: 135, 140, 157, 161, 162, 163, 191-224, 227, 229, 230, 232, 237, 238, 244, 250, 252, 264, 265, 266, 268, 269, 271, 272, 282, 279, 291, 304
Boeing 727: 280
Boeing 747: 219, 236, 258, 261-282, 295
Boeing Aircraft: 113
Boeing B-17: 40
Boeng Stratocruiser: 5, 59, 61, 64-66, 70, 71, 81-83, 86, 87, 98, 99, 102, 103, 108, 114, 131, 135, 136, 138, 143, 161, 184, 193, 245
Bone, Captain B E P: 141
Booth Lines: 51
Booth, John: 73, 262, 268
Borneo Airways: 178
Bowyer, Ray: 238
Boyd-Carpenter, John: 133
Brabazon of Tara, Lord: 12, 33, 34, 45, 62
Brabazon, Bristol: 62, 65, 66, 80, 82, 83, 84, 85, 95
Bracker, John: 35, 73
Brackley, Air Commodore H G: 91, 92
Brackstone Brown, First Officer E: 114, 156
Branker, John: 263
Bray, Winston: 230, 244, 268, 285, 298, 299, 300
Brealey, S C: 119
Brenet, Albert: 10
Brigish, C H: 119
Bristol Aircraft Co.: 221
Bristol B.175 Britannia: 5, 96, 97, 99, 102-110, 130, 131, 133, 135, 136, 138-144, 147, 148, 154, 155, 161, 192, 193, 197, 198, 202, 203, 221, 240-248, 252, 254, 288
Bristow Helicopters: 287, 288
Britavia: 289
British & Commonwealth Shipping: 288
British Air Services (BAS: 285
British Air Services: 304
British Airline Pilots Association (BALPA): 11, 13, 275, 276, 277, 278
British Airways (later) 181, 271
British Airways (original): 9, 12, 15, 17
British Airways Associated Companies Ltd: 303
British Airways Associated Companies: 305
British Airways Board (BAB): 180, 278, 282, 293, 301, 302, 303, 304, 305
British Airways Engine Overhaul: 305
British Airways European Division: 305
British Airways Helicopters: 305
British Airways Overseas Division: 305
British Airways Regional Division: 305
British Airways: 303-305

British Caledonian Airways (BCAL): 5, 220, 289, 290, 291, 292, 303
British Commonweath Pacific Airlines: 166, 167
British Continental Airways: 9
British European Airways (BEA): 67, 70, 73, 74, 79, 91, 167, 168, 169, 170, 176, 202, 212, 236, 244, 247, 248, 254, 258, 262, 268, 269, 284, 285, 286, 289, 290, 292, 293, 303, 304, 305
British International Airlines Ltd (BIAL): 170, 172
British International Airways (Kuwait) Ltd: 173
British Latin American Air Lines: 51
British Overseas Air Charter Ltd: 304
British South American Airways (BSAA): 69, 73, 74, 83, 86, 97, 98, 99, 102, 166, 175, 178, 179, 231, 268
British United airways (BUA): 232, 288, 289, 290, 291, 292, 293
British West Indian Airways BWIA): 166, 175, 176, 177, 178, 179, 203, 209, 241, 254
British West Indian Airways: 166, 175, 176, 177, 178, 203
Brocklebank, Sir John: 209
Brook, R: 119
Brown, Dr. J M: 119
Brown, Harold G: 16
Brown, Sir Max: 291, 298, 299
Bryce, Jack: 227
Buchanan, 'Buck': 99
Buchanan, Captain W G: 68
Buggé, Peter: 114, 121, 150, 156
Burchall, Colonel Harold: 18, 21, 262
Burghley, Lord: 73
Burke, Aubrey: 150, 157
Burns Robert: 219
Burns, Robert: 291
Butterfield & Swire: 174

C

C-Class: 9, 15, 17, 20, 21, 23, 25, 27, 28, 35, 36, 39, 43, 44, 63, 78, 89, 90, 91, 93

Cadman, Lord John: 11, 13, 14
Caledonian/BUA: 216, 289
Cambrian Airways: 188
Cambrian Airways: 304
Campbell-Orde, Alan: 30, 33, 42, 71, 72, 73, 105, 126, 191
Canadair Argonaut C.4 (DC-4M): 62, 79, 84, 85, 87, 90, 100, 102, 103, 108, 114, 128, 135, 136, 138, 174, 184
Canadair CL.44: 203, 212, 263, 264, 265
Cane, Captain Peter: 150, 151
Capital Airlines: 136
Caravelle: 157
Carlyon, Bill: 192
Carter, D: 118
Cartmell, A: 119
Casey, Miss Olive: 256
Cass, Annie: 13
Catalina *see* Consolodated PBY Catalina
Cathay Pacific Airways: 174, 175, 179
Cayzer, Lord William: 287, 288
Central African Airways: 168, 190
Chamberlain, Neville, British Prime Minister: 13, 16
Chandler, Radio Officer R W: 118
Chandler, Ted: 133
Channel Island Airways Division: 304
Channel Island Airways: 305
Charlwood, E W: 118
Chetwynd, George: 268
Churchill, Winston Spencer, British Prime Minister: 20, 21, 32, 59, 100
Clan Line: 288
Clark, Radio Officer I M: 125
Cobham, Sir Alan: 69
Cokely, Ralph: 280
Coleridge-Taylor, Miss T: 118, 119
College of Air TrainingLtd: 180
Colonial Airways Ltd: 24, 87, 168
Consolodated PBY: 20, 25, 26, 27, 28, 35, 36, 44
Concorde: 5, 157, 269, 293-302
Constellation airliner *see* Lockheed Constellation
Convair 880: 237
Cookman, O A: 118
Corbett Report: 138, 240, 248, 250, 256

Corbett, John: 138, 240, 248, 250, 256
Corfield, Frederick: 290, 291
Coryron, Air Marshal Sir Alec: 113
Courtney, E P: 119
Coutts, Radio Officer G L: 119
Canadian Pacific Airways (CPA): 130, 147, 150, 151,
Craig, David: 261, 268, 274, 294
Cribbett, Sir George: 73, 98, 99, 104, 138, 150, 166, 257
Cripps, Sir Stafford: 59
Critchley, Brig. Gen A C: 30, 31, 38, 42, 51, 60, 71
Crossland, Anthony: 285, 286
Crossley, J S: 118
Crudge, Vernon: 71, 73
Cumming, Sir Duncan: 166, 173
Cunard Eagle (Bahamas) Ltd: 209
Cunard Eagle (Bermunda) Ltd: 206
Cunard Eagle: 207, 208, 211, 244
Cunard Steamship Co.: 206
Cunningham, John: 114, 116, 120, 121, 122, 150, 156
Curtiss-Wright CW.20/C.46: 29, 43
Cyprus Airways: 169, 170, 172

D

D-Class (Diana): *see* DH.86 Express Air Liner.
D'Erlanger, Gerard: 17, 30, 137, 138, 144, 150, 158, 161, 197, 198, 199, 202, 225, 226, 240, 250, 256, 257
Davidson, Gordon: 302
Davies, John: 292
Davison, Kensington: 288
De Gaulle, Charles: 24
De Havilland Aircraft Co: 112-120, 221
De Havilland, Captain Sir Geoffrey: 33, 116, 120, 121, 150
De Havilland, Lady: 121
De la Haye, John: 289
Dean, Sir Patrick: 260
Department of Trade and Industry (DTI): 216, 220, 266, 278, 298, 299, 300, 301, 302, 303
DH.86 Express Air Liner: 9, 24
DH.89 Dragon Rapide: 44, 88

G-AGLA Mildenhall was the first of BOAC's Yorks. It later passed to BSAA as Star Fortune. (BOAC via author's collection)

DH.91 Albatross: 8, 15, 18, 22, 24, 27, 28, 34, 37
DH.98 Mosquito: 35, 36, 37, 40
DH.104 Dove: 170, 288
DH.106 Comet: 5, 85, 95, 96, 97, 101, 102, 103, 105, 106, 112-164, 182, 191, 192, 193, 196, 198, 202, 212, 221, 222, 231, 241, 245, 246, 250, 252, 254, 255, 304
DH.114 Heron: 178
DH.121 Trident: 304
Dietrich, Marie Magdalene 'Marlene': 231
Dismore, R A V: 125
Dismore, S A: 262
Dolman, Radio Officer R J: 119
Douglas DC-2: 22
Douglas DC-3/C-47 Dakota: 18, 21-23, 29, 32,35, 37-44, 50, 53, 63, 78, 87, 91-93, 168, 170-173, 176-179, 286
Douglas DC-4: 34, 45, 62, 71, 85, 89, 92, 167
Douglas DC-5: 15
Douglas DC-6: 62, 147, 148, 166, 167, 241, 287
Douglas DC-7: 5, 64, 75, 105-108, 131, 137, 139-144, 147, 148, 161, 183, 184, 185, 198, 203, 241, 243, 245, 247, 250, 255, 264, 289, 292
Douglas DC-8: 139, 162, 191, 192, 213, 237, 238, 271, 272, 273
Douglas DC-10: 269, 273, 280
Douglas, Lord: 209, 258
Dunphie, Sir Charles: 236

E

E-Class (Armstrong Whitworth AW.27 Ensign): 15, 18, 25, 28
Eagle Airways (Bermuda) Ltd: 205, 206
Eagle Airways: 169, 170, 206, 287, 288, 289
East African Airways Corporation (EAAC): 88, 89, 165, 168, 178, 244
East African Airways: 44
Eastern Airlines: 58
Edwards, Captain W A: 151
Edwards, Sir George: 221, 222, 226, 235, 301
Edwards, Sir Ronald: 283, 284, 285, 300, 304
Egyptian Airways: 51
Egyptian Engineering Company: 170
Eisenhower, Gen. Dwight D: 32
El Al: 147, 237
El-nur-Osman, Lt Col Babakr: 238
Elder Dempster Lines: 54, 56, 168, 169
Emmott, First Officer K: 119
Ethiopian Airlines: 170

F

F-Class (Frobisher) see DH91 Albatross
Fairey FC.1: 10, 15
Farndell, Captain C: 153
Farnsworth, Tom: 239
Field, Captain H J 'Dexter': 192
Fielden, Commodore Edward J H 'Mouse': 121
Fiji Airways: 180
Fillingham, Pat: 156
Finlay, Alec: 302
Fisher, Alan: 283, 303, 304
Fleming, Ian: 161
Flying Tigers: 264
Focke-Wulf 200 Condor: 21
Fokker F.12: 10
Foote, Captain: 129
Forrest, Bob: 284, 303
Forrest, Robert: 268
Freeman, Gerry: 287, 288, 289
Furness Withy: 56
Fysh, Hudson: 94

G

G-Class: 25, 35, 40, 91
Gardner, Dr: 139
Garlick, J: 118
Garlick, O: 119
Geddes, Irvine: 54
Geddes, Sir Eric Campbell: 7, 12, 13, 17, 262

Ghana Airways Ltd: 169
Gillan, Tom: 192
Gillman, Freddie: 102, 142, 155
Glover, Derek: 239, 267, 268, 273, 283, 293, 294, 303
Goebbels, Josef: 32
Gorman, John: 99, 268
Gorney, Serge: 192
Goulborn, Captain Cyril: 238
Gower, Pauline: 30, 31
Granville, Keith: 71, 73, 178, 209, 219, 222, 236, 239, 262, 265, 268, 276, 289, 290, 291, 294, 298, 299, 300, 301, 303
Groves, Freddie: 262
Grumman Goose: 178
Guiness, Loel: 287, 288
Gulf Aviation Ltd: 173, 179, 180, 237
Guthrie, Mary: 69
Guthrie, Sir Giles: 5, 179, 230, 232, 233, 234, 235, 254, 256, 258, 261, 262, 264, 266-268, 270, 273 274, 275, 284, 285, 293, 295, 296, 300, 304
Gwyther, R D: 118

H

H-Class: see Handley Page HP.42
Haddon, Captain Maurice: 41
Haig, Captain D D: 68
Halaby, Najeeb: 281
Hales, A C: 119
Hall, Sir Arnold: 133
Hamadallah, Majour Farouk: 238
Hanbury William Committee: 84
Handley Page Aircraft Co.: 221
Handley Page Halifax: 33, 44, 64, 206
Handley Page Hastings: 64
Handley Page Hermes: 63, 64, 84, 85, 88, 89, 93, 102, 118, 121, 188, 189
Handley Page HP.42: 7, 8, 25
Handley Page HP.70 Halton: 35, 49, 64, 68, 78, 87, 88, 93
Handley Page HP.97: 221
Handley Page, Sir Frederick: 68
Handover, Dennis: 30, 262
Hannaford, Miss D: 118, 119
Hansen, Captain H: 21
Hardie, Sir Charles: 262, 267, 268, 275, 285, 291, 292, 297, 298
Hardman, Sir Henry: 242
Hardy, B: 119
Harrison, Barbara John: 212
Hartley, Sir Harold: 30, 73, 85, 89, 90, 97, 102
Hawtin, G G 'Gerry': 78
Henshaw, Alex: 118
Hermon, Peter: 259, 260, 269, 303
Heseltine, Michael: 266, 282, 299, 300, 304
Hessey, Ernest: 41
Hilary, Richard: 216, 303
Hildred, Sir William: 33, 42, 47, 52, 53, 59, 60, 65
Hoare, Sir Samuel: 7
Hoffa, James 'Jimmy': 270
Hong Kong Airways (HKA): 173, 174
Horizon Holidays: 287
Horsey, Captain: 263
Howard, Leslie: 31, 32
Howitt, Sir Harold: 30, 73
Hughes, Howard: 136
Humphery, George Woods: 14, 21, 262
Hunting Air Transport: 288
Hunting Clan: 172, 173

I

Iberia: 147, 148
Imperial Airways: 1 9-2, 15, 17, 168
Indian Airlines: 286
International Aeradio Limited: 75, 114, 305
International Aeradio Ltd: 305
International Air Transport Association (IATA): 6, 46, 69, 70, 72, 87, 104, 145, 146, 147, 182, 189, 206, 213, 214, 215, 258, 284, 286, 295
International Federation of Airline Pilots Associations (IFALPA): 130
Iranair: 169
Iraqi Airways Ltd: 169, 172
Iraqi State Railways: 169

Irwin, T D: 119
Isreal, Wildrid B: 31

J

J-Class: see Junkers Ju.52
Jackson, B I D: 119
James, Captain Jimmy: 41
Japan Air Lines: 273
Jardine Matheson: 174
Jay, Douglas: 283
Jenkins, Roy: 234, 255, 256, 267, 283, 294, 295
Johnson, Amy: 288
Johnson, E A: 125
Johnson, Engineering Officer J A: 119, 156
Johnstone, Captain Andrew Colin Paul: 29
Jones, Captain Oscar P: 66, 263
Jordan Arab Airways: 172
Jubb, Barbara: 153
Junkers Ju.52: 10, 24, 25, 28

K

K-Class: see DH95 Flamingo
Kauffer, Edward McKnight: 180
Keegan, T D 'Mike': 189
Kendall, Dr M G: 283
Kennedy, A S: 150
Kennedy, President John F: 255
Kennet, Lord: 52
Khaled, Leila: 237 , 238
Kieth, Sir Kenneth: 300, 304
Kingston, Engineering Officer J H: 125
KLM: 23, 28, 31, 36, 40, 90, 142, 148, 199, 244, 263
Knatchbull-Hugessen, The Hon Alica Mary Dorothea: 13
Knight, Geoffrey: 295, 300, 301, 302
Knight, P F: 118
Knollys, Viscount; Edward George William Tyrwhitt: 30, 31, 34, 46, 50-54, 57, 59, 60, 62, 65-67, 71, 72, 93
Kuwait Airways: 171, 246, 254
Kuwait National Airways: 170, 171, 272

L

L-Class: see Lockheed 10A Electra, Lockheed 14.
Laker Airways: 220
Laker, Sir Freddie: 220, 287, 288, 289
Lamport & Holt Shipping Line: 51
Lancaster Finishing School: 42
Lawson, W E: 119
Le Goy, Ray: 292
Lee-Elliott Theyre: 180, 181
Lee, Gilbert: 71, 166, 180, 209, 239, 262, 268
Liberator, Consolodated B-24: 21, 22, 27, 28, 29, 35, 37, 39, 40, 44, 63, 64, 79, 87
Linstead, John: 180
Lloyd, Frank: 150
Lockheed 10A Electra: 10, 15, 25
Lockheed 14: 10, 13, 15, 16, 22, 23, 24, 28
Lockheed 18 Lodestar : 18, 21, 28, 29, 33,

35, 36, 39, 63, 90, 175
Lockheed C-5A: 271
Lockheed Constellation: 5, 34, 45, 56, 58-60, 62, 64, 65, 66, 71, 72, 79, 81, 82, 86, 87, 93, 94, 108, 121, 136, 138, 142, 145, 147, 148, 165, 167, 184, 245
Lockheed Hudson: 28, 29, 35, 40
Lockheed L.1011 TriStar: 269, 280
Londonderry, Lord: 47, 52
Londonderry, Lord: 7
Lorriane, Captain Anthony C. P.: 23, 24
Lothian, Lord: 19
Lubbock, Mr: 255, 256
Lufthansa: 148, 180, 263, 273
Lunn Poly: 206
Lusty, Ivor: 192
Lyle, Robert: 68

M

Macmillan, Maurice Harold: 252, 288
Majendie, Captain Alastair M A: 115, 116, 119, 120, 125, 126
Major, E R: 270
Malabott, Captain: 263
Malaya-Singapore Airlines: 178, 180
Malayan Airways: 165, 173, 174, 178
Manzie, Gordon: 283, 284
Marchbanks, John: 30
Marking, Henry: 289, 303
Marks, Sir Simon: 30, 45
Marsden, Captain J T A: 119
Marshall, J: 156
Marten, Neil: 254
Mason, Roy: 290, 291
Maude, Mr: 256
Maudling, Reginald: 135
Maxwell, Robert: 25, 26
May, Captain W S: 23
Mayo, Major Robert H: 11, 45
McCormack, A C J: 119, 153
McCrindle, Maj. Ronald: 14, 18, 29, 30, 33, 45, 47, 59, 69, 72, 73, 88, 142, 165, 170, 174, 176, 182
McMeeking, Captain: 263
Meagher, Captain: 99, 100
Mellor, G T: 71
Middle East Airlines (MEA): 166, 171, 172, 173, 176, 241, 254, 258
Mideast Airways Service Company (MASCO): 172, 173, 239, 254
Miller, J: 153
Millichap, Captain Roy: 154, 156, 161
Mills, Lord: 143
Milward, Sir Anthony: 254, 258, 262, 285
Ministry of Civil Aviation (MCA): 42, 75, 77, 80, 81, 82, 83, 85, 86, 170
Ministry of Supply (MoS): 64, 80, 82, 96, 97, 98, 103, 141, 142, 192, 193, 221, 222
Moore-Brabazon, John: see Brabazon of Tara.
Moore, Doris: 262
Morrell, Captain: 99, 100
Morris, L: 119
Morris, Lord, Labour Peer: 59

Morton Air Services: 287, 288
Morton, Captain Sammy: 288
Moss, Captain: 212
Mostert, Captain M C P: 170
Movshon, G: 119
Mulley, Fred: 236
Murrant, Geoffrey: 287
Murrant, Sir Earnest H: 54

N
Nathan, Lord; Major Harry Louis Nathan, 1st Baron Nathan: 85, 102
National Airlines: 220
Naude, S: 119
Neden, Sir Wilfred: 197, 248, 250, 252, 254
Nehru, Pandit: 122
Nicholson, Sir David: 269, 299, 300, 301, 303
Nigeria Airways: 241
Nixon, First Lady Pat: 281
Nixon, W E: 150
Noble, Michael: 291, 292, 297
Norman, Sir Arthur: 262, 303, 304
Normanbrook, Lord: 262, 269
Northeast Airlines: 304
Nucholl, Captain Jack: 228

O
Olley, Captain George: 263
One-eleven, BAC: 235, 289, 304
Orrell, James 'Jimmy': 64
Orton, Captain B W B: 36
Orton, L: 119
Overseas Air Travel Ltd: 216

P
Pacific Steam Navigation Co.: 51
Page, Captain R. H.: 21, 27
Pakenham, Lord: 73, 187
Pan American World Airways (Pan Am): 19, 28, 37, 40, 41, 44, 47, 71, 86, 87, 148, 157, 162, 166, 171, 191, 218, 219, 220, 241, 271, 273, 278, 279, 280, 281, 282, 295, 296, 302
Parker, John Lancaster: 11
Parker, Peter: 304
Partridge, Derek: 31, 32
Peacock, Captain Denis: 268
Pearson, The Hon Bernard Clive: 13, 14, 16, 17, 21, 30, 45, 71, 287
Peason, G W: 119
Peason, Weetman Dickinson; 1st Viscount Cowdray: 13

Pentland, Captain Charles R: 130
Percy, Captain James Tate: 29
Perkins, Robert: 11, 13
Pern, Harry: 133
Perry, Captain: 263
Peters, W K: 118
Pinkney, E T: 118
Piper, Harold: 11
Plowden, Lord: 234
Polytechnic Touring Association: 206
Poole, Lionel: 248, 254
Poole, Lord: 287
Portal, Lord: 228
Pound, Admiral Sir Dudley: 20
Powell, Air Commodore: 176
Powell, Captain: 263
Prince Phillip, The Duke of Edinburgh: 99, 100, 122, 133, 219, 270
Princess Alexandra: 270
Princess flying boat: see Saunders Roe SR.45 Princess
Princess Margaret: 121, 122
Pritchard, Sir Fred: 207
Provost James Strachan of Falkirk: 68

Q
QANTAS Empire Airways see QANTAS
QANTAS: 5, 8, 9, 10, 17, 26, 27, 36, 39, 42, 44, 50, 57, 60, 86, 93, 94, 136, 147, 150, 151, 165, 167, 266, 274, 275, 295
Queen Elizabeth I: 99
Queen Elizabeth II: 99, 100, 101, 121, 219
Queen Elizabeth the Queen Mother: 121, 122
Quinn Harkin, A J: 18, 71, 263

R
Rae, Captain Gilbert: 36
Raitz, Vladimir: 287
Ralfe, Captain I D V: 151
Rankin, Captain A: 151
Rawson, Captain B: 151
Reith, Lord John Charles Walsham, 1st Baron Reith 6, 12, 13, 14, 15, 16, 17, 30, 45, 60, 71
Rendall, Captain Athristan Sigrid Mellersh 'Flaps': 155, 228, 230
Rennall, Lord: 138, 254, 262
Rhodesia and Nyasaland Airways Ltd: 168
Richardson, Bill: 289
Roberts, D T: 36
Roberts, Geronwy: 290
Rodley, Captain Ernest E: 125, 153
Rodney, Captain: 115, 124
Rogers, Captain Kelly: 20

The year was 1953 and both BOAC and QANTAS were pushing The Kangaroo Route between the UK and Australia. The journey by Lockheed Constellation incorporated nine intermediate stops in each direction

319

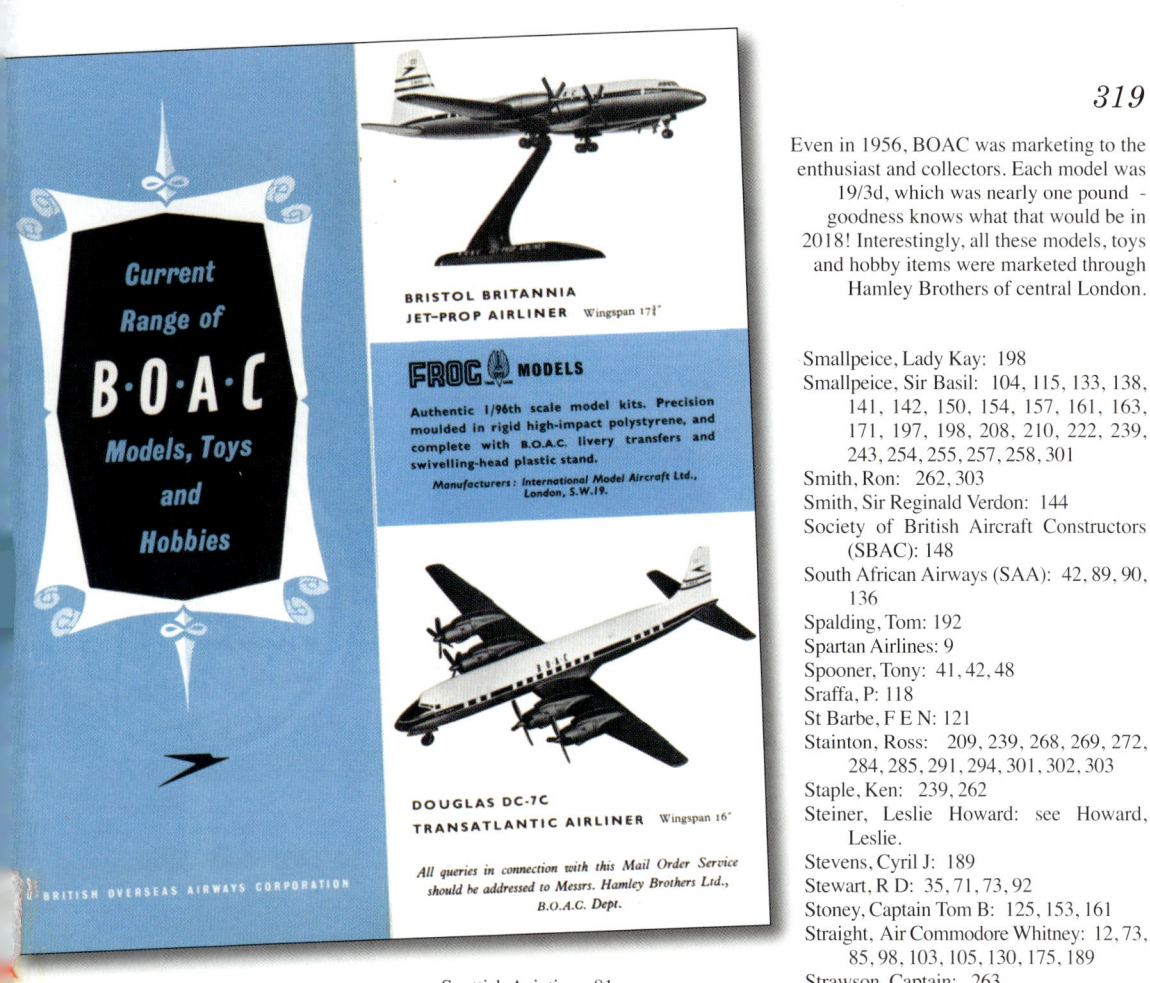

Even in 1956, BOAC was marketing to the enthusiast and collectors. Each model was 19/3d, which was nearly one pound - goodness knows what would be in 2018! Interestingly, all these models, toys and hobby items were marketed through Hamley Brothers of central London.

Smallpeice, Lady Kay: 198
Smallpeice, Sir Basil: 104, 115, 133, 138, 141, 142, 150, 154, 157, 161, 163, 171, 197, 198, 208, 210, 222, 239, 243, 254, 255, 257, 258, 301
Smith, Ron: 262, 303
Smith, Sir Reginald Verdon: 144
Society of British Aircraft Constructors (SBAC): 148
South African Airways (SAA): 42, 89, 90, 136
Spalding, Tom: 192
Spartan Airlines: 9
Spooner, Tony: 41, 42, 48
Sraffa, P: 118
St Barbe, F E N: 121
Stainton, Ross: 209, 239, 268, 269, 272, 284, 285, 291, 294, 301, 302, 303
Staple, Ken: 239, 262
Steiner, Leslie Howard: see Howard, Leslie.
Stevens, Cyril J: 189
Stewart, R D: 35, 71, 73, 92
Stoney, Captain Tom B: 125, 153, 161
Straight, Air Commodore Whitney: 12, 73, 85, 98, 103, 105, 130, 175, 189
Strawson, Captain: 263
Street, Miss Mavis: 256
Street, Sir Arthur: 16
Sud Aviation: 293, 295, 296, 298
Supremes, The: 210
Svensk Interkontinental Lufttrafic (SILA): 86
Swash, S V: 239
Swinton, Viscount; Philip Cunliffe-Lister,: 11, 38, 42, 57, 67
Swire, Sir Adrian: 179
Swissair: 148, 180, 237

T

Tait, Air Vice-Marshal Sir Victor: 68, 71, 73, 142
Tapp, Captain Russell B: 36
Tasman Empire Airways: 150, 165, 167
Taylor, Captain F A: 283
Taylor, Captain: 212
Taylor, Engineering Offer T W: 119
Tedder, Air Marshal Arthur William; 1st Baron Tedder: 25, 29
Tett, Sir Hugh: 283
Thomas Cook & Son: 206, 263, 283
Thomas, Sir Miles: 73, 98, 99, 101, 102, 104, 107, 112, 119, 121-124, 126, 127, 131, 133-135, 137, 138, 142, 154, 161, 168, 171, 176, 178, 188, 250, 257, 288
Thompson, Marshall: 14
Thomson, Adam: 289, 292

Rogers, William: 260
Romeril, John: 269
Ross, Brg.G: 118
Ross, Diana: 210
Rothchild, Lord: 298
Rothchild, Lord: 95
Roxborough, Captain W S: 151
Royal Aircraft Establishment (RAE): 117, 130, 139, 143, 157, 223
Royal Mail: 51
Runchiman, Walter Leslie: 12, 14, 16, 17, 25, 31, 45
Russell, Sir Archibald: 113
Ryan, Col. E P J: 119

S

SABENA: 28, 39, 44, 148, 244, 289
Salisbury, Lady: 121
Salisbury, Lord: 121
Sandys, Duncan: 197, 224, 288
SAS: 30, 70, 86, 148, 244
Saturn Airways: 265
Saunders Roe SR.45 Princess: 73, 82-86, 102, 103
Sawle, Captain North: 130
Scott, Captain Charles William Anderson: 47
Scott, J B: 239
Scott, J B: 45
Scott, Jim: 244
Scottish Airways: 305
Scottish Aviation: 81
Seaboard & Western Airlines: 137, 148
Seaboard World Airlines: 264
Self, Sir Henry: 19, 97
Severne, Sqn Ldr John de M: 133
Shakespere, Captain, 20
Shelborn, Phillip: 283
Shelmerdine, Sir Francis: 16
Shenstone, Beverley: 268
Shorts Hythe: 86, 90, 93, 94
Shorts Plymouth: 89, 102, 174
Shorts S-30: 25, 27
Shorts S-33: 25
Shorts S.23: 25, 28
Shorts S.25 Sunderland: 23, 24, 35, 36, 37, 39, 40, 43, 72, 90
Shorts S.25: 27
Shorts Sandringham: 54, 89
Shorts Seaford: 83
Shorts Shetland: 33, 35
Shorts Solent: 52, 55, 88, 95, 102
Silver City Airways: 287
Sims, Charles: 117
Sinclair, Sir Archibald: 20, 21, 30, 45, 51, 52
Sir Henry Lunn Ltd: 206
Skytrain: 220
Skyways: 92, 170, 172, 173, 190
Slattery, Sir Mathew: 144, 196-199, 202, 209, 226, 227, 229, 242, 243, 245-248, 250, 252, 254, 256-259, 261, 264, 268

Thorn, Bill: 64
Thorne, Peggy: 153
Thorneycroft, George Edward Peter: 197, 242, 243
Thornton and McRindle: 174
Thornton, Major: 33,
Todd, Miss J: 125
Townsend, Grp Capt. Peter: 121
Trans Australian Airlines: 150, 151
Trans Canada Airlines (TCA): 87, 148, 150, 151
Trans World Airlines (TWA): 142, 148, 170, 218, 219, 220, 237, 278, 281, 282, 295, 302
Transair Ltd: 287, 288, 289
Tripp, Terry Juan: 19, 157, 162, 282
Trubshaw, Brian: 227
Tu.144: 294
Turk Hava Yollari (THY): 178, 179
Tweedsmuir, Lord: 254, 262
Tymms, Sir Frederick: 177
Tyrrell, Sir William: 74

U
Uncles, J L: 192
United Airways: 9
United Arab Airlines: 286

V
Vanguard, Vickers: 195, 222, 304
VC10, Vickers: 5, 202, 212, 221—238, 244, 246, 250, 252, 254, 256, 258, 264, 267, 268, 271, 282, 289, 304

Vickers Armstrongs Co.: 221
Viking, Vickers: 175, 176, 206, 286
Viscount, Vickers: 139, 168, 169, 171, 172, 173, 176, 177, 178, 179, 188, 189, 221, 287, 304

W
Waddell, Jack: 280
Walker, W A: 119
Wallick, Jess: 280
Walton, Captain Frank: 228, 303
Warwick, Vickers: 41
Watkins, Harold: 206
Watkinson, Harold: 138, 156, 221, 224, 247, 256, 288
Way, Sir Richard: 233, 267, 269
Webb, Everette L: 280
Wedgewood-Benn, Anthony: 302
Weeks, Captain Robert M 'Bob': 281
Weir, Captain James N: 139, 141, 157
Welch, Air Marshal Sir William: 38
Wellington, Vickers: 28, 36
Welsh, Sir William: 68
Wemsley, Alderman J H: 119
West African Airways Corporation (Nigeria): 169, 237
West African Airways Corporation (WAAC): 87, 165, 168
West, T: 119
Wheatcroft, Stephen: 228, 283, 303
Whitehall Securities: 288
Whitley, Armstrong Whitworth: 27, 28, 29, 33, 37

Whitteham, First Officer D T: 119
Wighton, Frazer, : 119
Wigmore, Helen: 41
Wilcockson, Captain Arthur S: 21, 263
Wilkins, C T: 150
Wilkinson, David: 183
Wilkinson, Kenneth: 303
Willis, D: 119
Wilson Airways: 168
Wilson, Andrew: 298
Wilson, First Officer P A: 125
Wilson, John: 156
Wilson, Mary: 210
Wilson, Sir Horace: 14
Wilson, Sir Reginald: 283
Wilson, Thornton A: 280
Winster, Lord, Reginald Thomas Herbert Fletcher, 1st Baron Winster, Labour peer: 42, 59
Wolfson, Captain V: 166
Wood, Sir Kingsley: 7, 13, 14
Woodhill, First Officer J G: 118
Worboys, Sir Walter: 248, 262
Wright, G N: 119
Wright, Wilbur: 47
Wyatt, Miles: 189, 287, 288, 289
Wygle, Bryen: 280

Y
Young, Bertie: 262

Z
Ziegler, Henry: 298